The University of Wisconsin

PUBLICATIONS IN MEDIEVAL SCIENCE

MARSHALL CLAGETT, *General Editor*

NICOLE ORESME

and the Kinematics of Circular Motion

NICOLE ORESME
and the Kinematics of
Circular Motion

Tractatus de commensurabilitate
vel incommensurabilitate motuum celi

EDITED WITH AN INTRODUCTION,

ENGLISH TRANSLATION, AND COMMENTARY BY

EDWARD GRANT

THE UNIVERSITY OF WISCONSIN PRESS

MADISON, MILWAUKEE, AND LONDON

1971

Published 1971
The University of Wisconsin Press
Box 1379, Madison, Wisconsin 53701

The University of Wisconsin Press, Ltd.
27–29 Whitfield Street, London, W.1

First printing

Printed in the Netherlands
Koninklijke Drukkerij G. J. Thieme N.V., Nijmegen

ISBN 0-299-05830-1; LC 79-133238

To the memory of

Norwood Russell Hanson

Contents

Preface

That Nicole Oresme was one of the most significant authors of mathematical and physical works during the Middle Ages is now beyond dispute. Recent editions of some of his major treatises bear incontestable witness to his intellectual powers. In light of this, it seems highly desirable to make available in modern editions as many of his works as possible. The critical text established here from seven manuscripts represents not only the first printed Latin edition of Oresme's *Tractatus de commensurabilitate vel incommensurabilitate motuum celi*, but also the first English translation (an earlier Russian translation was made in 1960 by V. P. Zoubov).

In the three-parted *De commensurabilitate*, Oresme is largely concerned with the conditions that determine whether or not two or more bodies moving with uniform circular motion can meet or conjunct in one or more specific points of a circle (or concentric circles). If they can so meet, will this occur only once? or more than once but a finite number of times? or an infinite number of times? What are the consequences of these possibilities for celestial motions? Following a general introduction, Oresme presents answers to these questions by assuming in Part I that the uniform circular motions are mutually commensurable and in Part II that they are incommensurable. In Part III he seeks to answer the fundamental question posed in the very title of the treatise, Are the celestial motions commensurable or incommensurable? This sole problem of Part III is argued in the framework of a great debate presided over by Apollo, the Judge, and involving as direct protagonists Arithmetic (Commensurability) and Geometry (Incommensurability). In translating this treatise, every effort has been made to present Oresme's thought as faithfully and intelligibly as possible.

As the present volume is intimately related to my earlier *Nicole Oresme: "De proportionibus proportionum" and "Ad pauca respicientes"* (Madison, Wis., 1966), I have not repeated certain relevant discussions and summaries already given in the earlier volume. Thus, the biographical sketch of Oresme, on pp. 3–10, has been reduced here to a single paragraph containing only the most basic data. Furthermore, since I have had no cause to alter my earlier interpretation on pp. 74–80, that Oresme's *De commensurabilitate* is very likely an expanded and revised version of his earlier *Ad pauca respicientes*, I thought it

xi

best not to repeat these arguments, but to provide citations to my earlier treatment. However, the differences and similarities in the propositions, ideas, and concepts of the *Ad pauca respicientes* and *De commensurabilitate*, which had been noted only briefly in the earlier volume, are thoroughly discussed in the present volume wherever appropriate.

The reader will observe, in Chapters 2 and 3 of the Introduction, that rather than commence with a historical account of the application and interpretation of the concept of celestial incommensurability, I have chosen to summarize and analyze the *De commensurabilitate* prior to presenting the history of the problem before and after Oresme. By first acquiring some familiarity with Oresme's treatise, the reader will undoubtedly appreciate more fully its significance when finally seen in historical context. Material that seemed inappropriate for inclusion in the "Summary and Analysis of the *De commensurabilitate*" (Introduction, chap. 2), but which is important to a fuller appreciation of the Latin text, has been systematically discussed in the Commentary, which is keyed to the part and line numbers of the text. So that the reader may know precisely what lines are considered in the Commentary, he need only note the superscript reference numbers accompanying the translation.

That this book has reached completion now rather than some years in the future is largely a consequence of the generous support I have received from the Division of Social Sciences of the National Science Foundation. My debt to this great American institution is a large one. By virtue of its support I was able to engage in uninterrupted research for a number of summers and obtain essential microfilms, photographic reproductions, and secretarial assistance, and through its subsidy for publication my labors have been brought to final fruition in the present volume.

Various libraries, individuals, and other institutions have been of very real assistance to me. For permission to reproduce pages from manuscripts in their possession, I would like to express my thanks to the following libraries: Magdalene College, Cambridge; Bibliothèque Nationale and Bibliothèque de l'Arsenal, Paris; Biblioteca Medicea Laurenziana, Florence; Biblioteca Vaticana, Vatican City; and Universiteits Bibliotheek, Utrecht. As one who also helped expedite completion of this volume, I must gratefully acknowledge the role of the late Vassily P. Zoubov (Institute of the History of Science and Technology, Academy of Sciences, U.S.S.R.), who, as mentioned above, translated Oresme's *De commensurabilitate* into Russian in 1960. Although ignorance of Russian prevented me from utilizing his translation, I was able to use fully the valuable footnotes to the translation which correctly identify most of the sources of Oresme's numerous quotations and allusions—especially in Part III—to classical, Christian, and Arabic authors. Zoubov's profound knowledge of this vast literature saved me much labor and frustration. It is a distinct pleasure to express my indebtedness to the John Simon Guggenheim Memorial Foundation for financial aid awarded me as a Guggenheim Fellow during 1965–66, which enabled me to take leave and actually complete this volume during the early months of the fellowship. To the School of Historical Studies of The Institute for Advanced Study, Princeton, New Jersey, which was kind enough to offer me a Visiting Membership and grant-in-aid during 1965–66, I must acknowledge a special sense of gratitude. The final stages of research and writing were completed there in an atmosphere of hospitality, warmth, and unparalleled intellectual stimulation. And, finally, to Marshall Clagett, Professor in the School of Historical Studies of The Institute for Advanced Study, I am under obligation, as so often in the past, not only for many fruitful discussions on particular problems, but also because his kindness and helpfulness made my sojourn at the institute eminently pleasant, memorable, and productive.

Edward Grant

Bloomington, Indiana
August, 1969

Editorial Procedures

All manuscripts of the *De commensurabilitate vel incommensurabilitate motuum celi* that were known to me have been used throughout their extent and fully collated in this first Latin edition of that work. Of the six complete manuscripts employed, it quickly became evident that MS *B* (see the list of manuscripts and their sigla, pp. 162–69, below, for full identification of this and all other manuscript sigla) stood apart from the other five, namely, MSS *A, F, L, P,* and *U*. In literally hundreds of instances, the five stood in agreement but differed from *B*'s reading. In many of these differences, *B* has additions of a non-substantive nature that seem intended, for the most part, to clarify or make more explicit what is already perfectly intelligible in the other five manuscripts.* Often *B* has a quite different reading, but one that conveys the same sense as *A, F, L, P,* and *U*.† Although it has been fully collated, MS *R* contains only Part III, and its impact on the final text is not such as to require further discussion.

From such an alignment of manuscripts, it was first necessary to decide which of these two groups should be taken as representative of the most reasonable approximation to some hypothetical original (hypothetical because no autograph manuscript is among them). This decision was made difficult by the fact that a sound and intelligible text could have been generated from either grouping. After much vacillation, I chose to base the text on the five-manuscript group and to utilize *B* only where its reading seemed correct or more appropriate, or where it was useful to include some of its expanded readings in square brackets. In the absence of conclusive criteria, my choice was largely determined by the reasonable probability that five manuscripts in agreement so many times constituted a safer guide than a single manuscript that differed repeatedly from

---

* Only a few examples, of many, can be listed here; all can be found in the variant readings under the appropriate part and line number (for an explanation of the following abbreviations, see below, p. xix): Prol. 33; I.51, 138, 154, 220, 223, 224, 233, 243, 252, 276, 298, 337, 353, 357, 359, 387, 404, 492, 553, 572, 623, 635, 644, 691, 698, 742–43; II.42–43, 74, 164, 171, 177, 197, 383; III.23, 40, 102, 142, 211, 228, 262, 269, 271, 275, 277, 305, 345, 349, 354, 397, 401, 423, 449, 453, 479, 480.

† As examples, see I.284, 315, 316, 367, 485, 490, 734; II.8, 44–45, 68, 72–73, 77, 167, 169, 174, 278–79, 395–96; III.92–93, 101, 180, 290, 293, 358, 369, 380. 456.

them in only trivial ways. Obviously, such a decision is not founded upon firm and seemingly undeniable evidence.

The character of B's text, and the nature of its differences from $A, F, L, P,$ and U, indicated that its scribe—either the scribe who copied B, or the scribe who copied the version from which B is derived—had carefully read and absorbed his copy, or copies, of Oresme's *De commensurabilitate* and decided to improve his text wherever possible by additions, transpositions, omissions, and slight rearrangements. A significant instance of an addition that seems to reveal an intervention by someone other than Oresme appears near the end of Part I, Proposition 16, where Oresme is concerned with determining the time in which three mobiles will successively conjunct. After the word *coniungentur* in the variants for I.645 only MS B includes a brief section giving the *place, or location,* of the next point of conjunction. This is superfluous, and merely adds a bit of information that is not required by the proposition and is lacking—rightly, it seems—in the other manuscripts. This addition, though longer than most, is typical of many other elaborations and extensions found in the text of B. It seems more plausible to explain MS B as the work of an intelligent, but over-zealous, scribe, than to explain the relationships between A, F, L, P, and U as deriving ultimately from extensive scribal surgery in which Oresme's pristine version was consistently abbreviated and made less explicit, albeit still intelligible.

But perhaps it is also plausible to argue that MS B represents a revision made by Oresme himself from the basic version found in MSS A, F, L, P, and U. Could it be that after releasing a copy of the *De commensurabilitate* that served as the ultimate basis of manuscripts A, F, L, P, and U, Oresme subsequently revised the treatise and that B represents this later reworking? This is possible, of course, but it seems highly improbable that Oresme would have troubled to make hundreds of slight revisions—many are mere transpositions, while others indicate alterations from correct to incorrect, or less desirable readings in quotations from classical or later authors*—without adding anything substantive.

A more telling argument against this position, however, can be based on a postscript in B immediately after the *De commensurabilitate* (see below, pp. 164–65) and which seems to have been added by a scribe. Admittedly, the words "Cum scriberem hec…" ("When I was writing these things…") are ambiguous and could refer either to a scribe or to the author, Oresme. That this postscript was composed by Oresme is a hypothesis that is difficult to entertain seriously

---

* See, for example, the variants for B in III.76, 180, 184, 188, 189, 200, 211, 290, 294, 306, 314. These instances—there are not many—are perhaps only ordinary scribal errors and may not signify conscious alteration. Only when they are taken in conjunction with the hundreds of other instances where B's readings stand in isolation does the argument for deliberate alteration become somewhat more compelling and reasonably convincing.

in the face of the very structure of Part III of the *De commensurabilitate*, the part which immediately precedes the postscript. After carefully and studiously constructing the arguments of Part III (see below, pp. 69–77) and consciously excluding himself from the debate—even where he refers to his own *De proportionibus proportionum* (see below, III,457–66 and p. 73, n. 112—is it reasonable to suppose that Oresme would then append, as an afterthought, a postscript in which he interjects himself in the first person singular in order to present to the reader two more quotations (one from Abraham ibn Ezra, the other from Henry Bate) that seem related to the *De commensurabilitate*, and then make reference to a possibly more extensive discussion in another work by Henry Bate? If he were desirous of bringing these quotations before his readers, would he not have incorporated them skillfully into the structure of Part III in order to preserve the character of the treatise, rather than place them out of context after the treatise itself? This idea seems so inappropriate, that we may properly conclude that it was a scribe who added this postscript, perhaps the same scribe, if our conjecture is reasonable, who made the extensive additions and revisions found in MS *B*. For if he was author of the postscript, it reveals his genuine interest in, and understanding of, the problems of the *De commensurabilitate*, which renders plausible the conjecture that he might also have added to, and altered, the version before him as he attentively followed its arguments in the process of copying. Directly or ultimately, we may conclude that MS *B* represents the outcome of this process.

Of the five manuscripts that served as the basis for the text, MS *U* is quite defective, with omissions in some places that extend to quite a number of lines (see below, p. 169). Among the four that remain, *F* and *P* seem most closely related, but have not contributed more to the final text than *A* or *L*. Throughout, the primary objective has been to establish an intelligible text reflecting, as accurately as possible, Oresme's substantive thoughts and, hopefully, his actual words. To achieve this, it was essential to utilize all manuscripts, including *B* on a number of occasions.

Since orthographic practices vary between the manuscripts, and indeed frequently within one and the same manuscript, arbitrary decisions were made in order to arrive at uniform spellings for quite a number of terms. A list of the most significant of these follows, in which the first spelling is the one adopted and the second, in parentheses, the major variant version: pulchrum (pulcrum); verisimile (verissimille); perhemne (perenne); nihil (nichil); sidera (sydera); dyameter (diameter); peryodus (periodus); spatium (spacium); umquam (unquam); symphonia (simphonia); arismetica (arismetrica); giratio (gyratio); almanach (almanac); chorda (corda).

In many places in the text of the *De commensurabilitate* further elucidation is required that would be inappropriate to the Introduction, as for example, discussions of the numerous quotations given by Oresme at the beginning of the treatise and, more especially, in Part III. Explanatory material of this kind is relegated and confined to the Commentary. Where further discussion of the Latin text is required or warranted, a superior reference number in the parallel translation refers the reader to the Latin tag and part and line numbers in the Commentary where it is considered.

Oresme's method of quotation from classical and later authors requires a brief explanation. In selecting words or passages for citation, Oresme was primarily interested in material that was relevant to his arguments and did not hesitate to omit words or lines from any quotation when he deemed them inappropriate to the desired effect. Nowhere, however, did he indicate these omissions—it was not customary to do so—and all his quotations are presented consecutively, even where parts of them are separated by a number of omitted lines. All such omissions have been indicated in the Latin text and English translation by use of the conventional points of ellipsis. Where omissions might be of interest to the reader, they are cited or discussed in the Commentary. Furthermore, wherever possible, every quotation has been compared with at least one authoritative printed edition. When manuscripts of the *De commensurabilitate* contain readings at variance with modern editions, I have preferred the manuscript readings where these are intelligible and reasonable, since they reflect editions and versions actually used by Oresme.

Editorial Apparatus

In order to indicate unambiguously textual corrections and additions, the following procedures have been adopted. Where all the manuscripts have an erroneous reading, the correction is entered into the text without resorting to brackets of any kind. The erroneous manuscript readings are then cited in the variants at the bottom of the page preceded always by the correct reading and the abbreviation *corr. ex.* Additions to the text which are thought to have been in the original version are enclosed within angle brackets ⟨ ⟩. The translation of such passages or words is free of brackets. Square brackets [ ] are employed in all instances where a word or phrase has been included that was almost certainly lacking in the pristine text. In the translation such additions are also contained within square brackets, as are the many editorial expansions and elaborations that have been incorporated into the translation for the sake of clarity and intelligibility.

Many references to the Latin text of the *De commensurabilitate*, both in the Introduction and the Commentary, have necessitated the use of an abbreviated form of citation. Since in the Prologue and in each of the three parts of the *De commensurabilitate* the successive lines are numbered consecutively, all references cite first the part number in roman numerals—Prol., I, II, or III—followed by a period and the line numbers to which reference is made. For example, III.21–28 signifies that reference is being made to lines numbered 21 to 28 in Part III of the *De commensurabilitate*. Formal propositions, which are found only in Parts I and II, are referred to by the part number, followed by the proposition number expressed as an arabic cardinal number. Although propositions are numbered in the manuscripts by ordinal numerals, these have been systematically rendered in the translation as arabic cardinal numbers.

Titles of books and articles are cited fully in the bibliography. In the notes and body of the work shortened, although readily identifiable, titles are used. A few books, however, have been cited with sufficient frequency to warrant use of the following abbreviated forms:

Euc.-Campanus	*Euclidis Megarensis mathematici clarissimi Elementorum geometricorum libri XV cum expositione Theonis in priores XIII a Bartholomaeo Veneto Latinitate donata; Campani in omnes, et Hypsicles Alexandrini in duos postremos.* Basel, 1546. (Unless otherwise stated, all references to this work are to Campanus of Novara's Latin version of Euclid's *Elements*.)
Grant, *Oresme PPAP*	*Nicole Oresme: "De proportionibus proportionum" and "Ad pauca respicientes."* Edited with introductions, English translations, and critical notes by Edward Grant. Madison, Wis.: University of Wisconsin Press, 1966.
Menut, *Oresme du ciel*	*Nicole Oresme: "Le Livre du ciel et du monde."* Edited by Albert D. Menut and Alexander J. Denomy, C.S.B.; translated with an introduction by Albert D. Menut. Madison, Wis.: University of Wisconsin Press, 1967.
Zoubov, *Orem O soizmerimosti*	Vassily P. Zoubov, *Nikolaĭ Orem i ego matematiko-astronomicheskiĭ traktat "O soizmerimosti ili nesoizmerimosti dvizhenii neba." Istoriko-astronomicheskie issledovaniĭa*, no. 6. Moscow, 1960, pp. 301–400.

Although all references to Oresme's *Le Livre du ciel et du monde* are to the new edition and translation listed above, they are given in a form that will also allow their easy location in the first edition of the French text published in *Mediaeval Studies*, vols. 3–5 (1941–43). This is possible because both editions have identical book and chapter numbers for the text of *Le Livre du ciel*, as well as folio numbers added by Menut and Denomy. Thus a reference in the form "bk. 1, chap. 29, 44c" can be readily located in either edition. Page references, however, are exclusively to the new edition and translation.

References to Euclid, whether to the Greek text or the Latin version of Campanus of Novara, have the book number in arabic numerals followed by a period and the proposition number also in arabic numerals (for example, "Euclid 9. 11" signifies Book 9, Proposition 11 of the *Elements*).

In the variant readings, manuscripts *A*, *B*, *F*, *L*, *P*, and *U* (and *R* for Part III) have been collated in their entirety. A negative apparatus has been used whereby the notation provides readings only for those manuscripts that diverge from the preferred (textual) reading. Thus "152 postea: deinde *BU*" indicates that, in line 152 of the given page, manuscripts *B* and *U* substitute "deinde" for the preferred reading, "postea"; it is implied that manuscripts *A*, *F*, *L*, and *P* agree with the textual reading. Whenever the preferred reading and the variant, or variants, are sufficiently close in form that there can be no doubt or hesitation concerning the word in a particular line for which the variant is given, the lemma has been omitted. Hence, instead of "84 numerat: numeret *B* numerantur *L*," the form is "84 numeret *B* numerantur *L*" if line 84 contains only one word of which either "numeret" or "numerantur" could be a variant. When a given word appears more than once within a line, the word for which there is divergence is signaled by an exponent after the lemma, e.g., "14 sunt^2" indicates that the variant occurs with the second appearance of "sunt" in line 14. An italicized question mark precedes the siglum or variant of a manuscript with a doubtful reading; for example, "6 commendandi: *?F*" indicates that in line 6 the preferred reading, "commendandi," is in question in manuscript *F* (and, by implication, is in agreement in all other manuscripts). A somewhat more complicated notation serves most other cases; for example, "284 ad unitatem *om. A* unitate *F et post* unitate *scr. et del. F* necessitate" indicates that in line 284 the preferred reading, "ad unitatem," is omitted by *A*, while *F* substitutes the word "unitate" and after it the scribe of *F* has written and deleted the word "necessitate," and *B*, *L*, *P*, and *U* are implied to follow the textual reading.

All variants have been included except those involving *ergo* and *igitur*. Legends to the figures included in the text cite the manuscripts from which the figures are derived and indicate where changes have been made. Figures in the manuscripts that are not an integral part of the text, or that do not properly elucidate the text, have been omitted but their locations are given in the variants. The following abbreviations have been used in the variant readings:

add.	=	addidit	*om.*	=	omisit
corr. ex	=	correxi ex	*rep.*	=	repetivit
hab.	=	habet	*scr. et del.*	=	scripsit et delevit
mg. hab.	=	in margine habet	*tr.*	=	transposuit
obs.	=	obscuravit			

Introduction

1

Oresme and the Composition of the
De commensurabilitate

Since there are now at least two recent biographical summaries of Oresme's life, only the salient facts will be presented here.[1] Oresme was probably born sometime between 1320 and 1325 (but see chap. 3, n. 56) in Normandy, perhaps near Caen. Of his early life, nothing is known until he entered the College of Navarre in the University of Paris as a theological student in 1348, presumably having already obtained a Master of Arts degree. In 1356 he became grand master of the College of Navarre, an office for which the doctorate in theology was prerequisite. It is presumed that Oresme probably received his doctorate in 1355 or 1356, after pursuing theological studies for seven or eight years. In 1360, during his tenure as grand master, he was sent to Rouen by the dauphin, the future Charles V, to seek a loan from the municipal authorities. On November 23, 1362, Oresme was made canon of the Cathedral of Rouen, at which time he probably resigned as grand master of Navarre. Then, in succession, he became a canon of La Sainte Chapelle in Paris on February 10, 1363, dean of the Cathedral of Rouen on March 18, 1364, and, finally, at the command of Charles V, was named Bishop of Lisieux[2] on August 3, 1377. On July 11, 1382, he died in Lisieux, where he was buried in the cathedral church.

None of the manuscripts used in establishing the text of this edition have precisely the title *De commensurabilitate vel incommensurabilitate motuum celi*, the one that I have officially adopted for the treatise. There is, however, reason-

---

1. See Grant, *Oresme PPAP*, pp. 3–10; see also Marshall Clagett, ed. and trans., *Nicole Oresme and the Medieval Geometry of Qualities and Motions* (Madison, Wis., 1968), p. 4.

2. In his *Le Livre du ciel et du monde*, bk.4, chap.12, 203c, Oresme informs his readers that while completing his French translation of that treatise, Charles V appointed him Bishop of Lisieux. "Thus, with God's help, I have finished the book on *The Heavens and the World* at the command of the very excellent Prince Charles, the fifth of this name, by the grace of God, King of France, who, while I was doing this, has made me Bishop of Lisieux [1377]."—Menut, *Oresme du ciel*, p. 731.

able justification for its selection, since Oresme himself cites it in his *Le Livre du ciel et du monde*,[3] a work completed in 1377[4] and probably representing Oresme's last extant scientific treatise prior to his death in 1382. Inclusion of both terms, *commensurabilitate* and *incommensurabilitate*, lends even further support to the authenticity of this title,[5] for they properly reflect the threefold division of the treatise, where the first part assumes that the celestial motions are exclusively commensurable, the second part that they are exclusively incommensurable, and the final part considers whether they are commensurable or incommensurable. Any title that omitted one or the other of these terms would only partially reflect the contents, and to that extent be misleading.[6] It seems reasonable, then, to accept as genuine the title quoted a number of times in *Le Livre du ciel et du monde* and adopted in this edition.

The date of composition of the *De commensurabilitate* can only be estimated as lying within rather widely separated terminal years. Although it was probably written earlier, the year 1377 must stand as the ultimate *terminus ante quem*, for in that year Oresme completed his *Le Livre du ciel et du monde* (a translation from Latin into French of, and commentary on, Aristotle's *De caelo et mundo*) in which, as we have already seen, he cites the *De commensurabilitate* explicitly by title.[7] That it may have been written considerably earlier, while Oresme was an active master at the University of Paris, is perhaps indicated by his own statement that he "did not release this little book without [first] submitting it for correction to the Fellows and Masters of the most sacred University of Paris..." (Prol.45–47). Vague as it is, this remark suggests that Oresme sought the advice

---

3. In Book 1, Chapter 29, it is mentioned three times in this very form (at 44d, 45c, and 46c). See Menut, *Oresme du ciel*, pp. 196, 200, and 202. All three of these instances are also quoted in this volume, below in chap. 2, nn. 59, 97, and 113. For the variety of titles found in the manuscripts of the *De commensurabilitate*, see the variant readings on pp. 172 and 322. The closest approximations to the title I have adopted appear in MS *B* (for its identification, see below, p. 162), which has *Tractatus de commensurabilitate motuum celi*, and MS *U* (see p. 169 for full description), which has in its colophon *Tractatus de commensurabilitate et incommensurabilitate motuum celestium*.

4. Menut, *Oresme du ciel*, p. 6.

5. Of the manuscripts used in this edition, only *U* has both terms (see above, n. 3) in the title that is found in its colophon. However,

only the word *commensurabilitate* was included in the title that appears at the beginning of it (see variant readings for title on p. 172).

6. It is possible that where Oresme says "hunc libellum edidi de commensurabilitate motuum celi in quo premisi...." (Prol.41–42), he may have intended that "de commensurabilitate motuum celi" be taken parenthetically as a title; indeed it is the very title found in MS *B*. But it is equally plausible to interpret this as a shortened and more convenient way of referring to the treatise, rather than to its precise title. However, this interpretation does not accurately reflect the contents and nature of the treatise, and we cannot easily set aside the numerous references to it by title in the later *Le Livre du ciel et du monde*.

7. See above, pp. 3–4 and n. 5.

and judgment of present, rather than former, *colleagues*. If so, then it may have been written in or before 1362, the year Oresme probably relinquished the grand mastership of Navarre[8] and presumably withdrew from full participation in university affairs. There is also reason to believe that Oresme wrote the bulk, if not all, of his technical Latin treatises before and during his tenure at the College of Navarre[9]—that is, before he partially or fully severed his association with the university to take up his affairs at the Cathedral of Rouen and sometime before the period 1369–77, when he was occupied with a series of French translations undertaken at the specific request of King Charles V of France.[10]

Conditionally, a *terminus post quem* of 1351 may also be suggested. This would truly obtain if it could be established that the *De proportionibus proportionum*, which is cited by this title in the *De commensurabilitate* (II.201), was written after the *Algorismus proportionum*, a treatise that Oresme composed sometime between 1351 and 1361.[11] On admittedly inconclusive evidence, I have previously argued the plausibility of assuming the priority of the *Algorismus* over the *De proportionibus*,[12] and so tentatively offer 1351 as a *terminus post quem*. Indeed, the 1350's probably represent the height of Oresme's productive and intellectual powers and many of his most provocative treatises may derive from that period. His association with Rouen in 1362 may signify a shift into other activities and the termination of his truly creative period. Moreover, if the *De proportionibus* was not written after the *Algorismus proportionum*, the *De commensurabilitate* might even have been written in the 1340's. It is reasonable to assume that Oresme was already active at the University of Paris prior to his entry into the College of Navarre as a theological student in 1348, since we may justly presuppose a period of study toward a Master of Arts degree and perhaps even previous teaching experience as a regent master prior to 1348. In the broadest sense, the *De commensurabilitate* may, therefore, have been written as early as the 1340's and as late as the 1370's, but no later than 1377.

8. See above, p. 3.
9. Grant, *Oresme PPAP*, p. 5.
10. Ibid., p. 9 and above, n. 4.

11. Grant, *Oresme PPAP*, pp. 12–13.
12. Ibid., pp. 13–14.

2

Summary and Analysis of the
De commensurabilitate

The *De commensurabilitate vel incommensurabilitate motuum celi* consists of a prologue and three parts. After the presentation of a few preliminary definitions and assumptions, the remainder of Part I is devoted to 25 propositions in which bodies, or mobiles, are assumed to move with commensurable speeds on concentric circles (see I.60–62; in Proposition 20 the circles are eccentric). Part II contains 12 propositions in each of which at least two of the motions are incommensurable. In these two parts, Oresme's objective is to derive various consequences from the motions of two or more bodies whose speeds are first assumed to be commensurable and then incommensurable. In most propositions, these kinematic consequences are applied arbitrarily to astronomical aspects—conjunction is used as the paradigm case—to determine the times in which, and places where, these can occur in terms of the specific data and assumptions. Oresme concludes the treatise by presenting Part III in the form of a debate presided over by Apollo and involving, as antagonists, personifications of Arithmetic and Geometry, who argue whether the celestial motions are commensurable (Arithmetic) or incommensurable (Geometry).

Prologue

In something akin to an ode to the heavens, Oresme extolls the diverse but uniform nature of the celestial pageant. Quotations from Seneca and Cicero help convey the wonders of the celestial region that have uplifted and attracted the attention of man, who, despite his inability to grasp the underlying and hidden rationale of the celestial order (Prol.31–33), has ceaselessly inquired into its behavior. A desire to place the study of the heavens in proper perspective prompted Oresme to compose the *De commensurabilitate*, so that its study would prevent despair and distrust, and enable men to guard themselves against those who would seduce them into believing that they know, or can know, things that are *de facto* unknowable (Prol.39–42).[1]

---

1. Sentiments quite similar to Prol.39–42 are found in the opening lines of Oresme's

Part I

The first part commences with a series of definitions (I.2–23) of terms and concepts such as prime number, composite number, mutually non-prime numbers, proportionality, commensurability and incommensurability between proportionalities, and commensurability and incommensurability between magnitudes in general (see discussion in the Commentary, pp. 328–29). These strictly mathematical definitions are followed by others specifically relevant to the subject matter of the treatise, i.e., relevant to relationships between circular motions (I.24–55).

Commensurability and incommensurability between motions are to be expressed either in terms of (1) angles described around the centers of circles, which are, in effect, comparisons between parts of circles, or (2) the relationships between the number of total circulations traversed by each moving body, where a circulation (*circulatio*) is defined as "the return of *one mobile* along a circular path from any point to that same point" (I.42–43).[2] Both modes of comparison are used by Oresme. To illustrate each mode, let us assume that bodies, or mobiles, *A* and *B* are in motion on concentric circles with unequal, but respectively uniform, velocities;[3] let the motions be measured from some point *d*, which

---

Ad pauca respicientes (see Grant, *Oresme PPAP*, p. 382, lines 1–6, 11–13). That some men, primarily astrologers, rashly deceive themselves and others in claiming knowledge about celestial events and configurations that is essentially unknowable was a popular theme in Oresme's writings. Its ultimate justification lay in his own mathematical demonstration in the *De proportionibus proportionum* that the celestial motions are probably incommensurable and precise knowledge about their motions unattainable. See below, *De commensurabilitate*, III.10–31, 435–66 and p. 67, n. 105 and p. 76, n. 113 of this chapter.

2. A revolution (*revolutio*) is said to occur when two or more bodies, or mobiles, have departed from and returned to some definite aspect (e.g., conjunction, opposition, etc.) or point (I.43–44). Thus if two mobiles are in conjunction in some point *d*, they will have completed one revolution upon their next simultaneous arrival in *d*. A *circulatio*, which is restricted to the motion of one body only, is analogous to a sidereal period in

astronomy, but a *revolutio*, which concerns a relationship between two or more bodies, has no direct counterpart in astronomy since it requires that the mobiles return to the very same point or configuration—taken as absolutely fixed in space—from whence they commenced their motion. Actually, a *revolutio* is more like a Great, or Perfect, Year in ancient or medieval astrology (see below, p. 103 and n. 64).

In the *Ad pauca respicientes*, Oresme did not make, and probably had not yet formulated, the distinction between *circulatio* and *revolutio*, for the term *circulatio* does not even appear, and *revolutio* is used in both senses (for examples, see Grant, *Oresme PPAP*, p. 396, line 144; p. 408, line 65; p. 410, lines 66 and 72).

3. Each body in motion will have a different angular velocity from every other body but every motion is assumed to move with a uniform speed, although combinations of uniform motions can produce an irregular motion (I.56–59). In the earlier *Ad pauca respicientes*, Oresme explained that if two

lies on a radius drawn from the common center and intersects both circles; and, finally, let T_A and T_B represent the times of the motions of A and B, respectively. The speeds of A and B will be commensurable (I.26–29) when $T_A = T_B$ and $\angle A : \angle B = m : n$, where $m : n$ is a ratio of integers, or when the number of circulations of A equals the circulations of B, and $T_A : T_B = m : n$. The motions of A and B will be incommensurable (I.29–31) when $T_A = T_B$ and $\angle A : \angle B \neq m : n$, or when the number of circulations of A and B is equal and $T_A : T_B \neq m : n$.

After defining the terms *circulatio* and *revolutio* (I.42–44; see n. 2 of this chapter), Oresme tells us that a conjunction—in almost all propositions conjunctions serve as the paradigm for all other astronomical aspects (see I.738–40 and II.290–92)—occurs when a line drawn from the center of the world passes through the exact centers of two or more bodies. This definition reflects Oresme's exclusive concern with the motion of idealized bodies, which are but mathematical points moving in a single direction only. He states emphatically that "my purpose is to consider *exact* and *punctual aspects* of mobiles that are moved circularly" (I.45–46). The primary concern of astronomers, however, is to seek approximate results and avoid sensible error (I.45–49; see also I.204–14). This difference of purpose explains why Oresme's definition of conjunction, when extended to actual physical planetary conjunctions, would eliminate many conjunctions normally accepted by astronomers on the basis of observation. To qualify as a conjunction in terms of Oresme's definition, it must be theoretically possible to draw a line from the center of the world directly to the centers of two or more planets. This could be achieved only when their centers are exactly on the same celestial meridian. Now it might happen that two planets are in the first point of Aries and yet would not conjunct since their centers would not lie on exactly the same meridian (I.53–55). To observational astronomers this would constitute a physical conjunction, but not for Oresme in the *De commensurabilitate*, where, in Part II, Proposition 10, he demonstrates that on the assumption of incommensurable speeds, two or more planets "would, at some time, arrive in the same minute, and at some time in the same second, and at some time in the same third, or fourth, and so on, approaching closer and closer into infinity, and yet they will never conjunct exactly" (II.283–86). It is obvious, then, that when Oresme applies the results of his mathematical and kinematic propositions to planetary motions, the planets are conceived as mathematical points or idealized geometrical bodies and not as actual celestial bodies observed by astronomers.[4]

(Note 3 continued)
bodies moved with equal angular velocities they would forever maintain their relative positions (see Grant, *Oresme PPAP*, p. 388,

lines 60–64).

4. For another contrast between the more approximate approach of astronomers and the more mathematical approach taken by

In Proposition 1, Oresme is concerned with the criteria for determining the common prime factors in geometric progressions beginning with unity. The proof depends on Euclid 9. 11,[5] where it is shown that if a prime number measures the last number in a geometric progression, it must also measure the number immediately following unity. Oresme's proof shows the converse of this, namely, that if a prime number measures the number immediately following unity, it will also measure the last number in the series, and, of course, all the intervening numbers as well. The form of the proof is by denial of the consequent ("a destructione consequentis"; I.77). If 1, A, B, C, D,..., are the terms of the series, and G is some prime number, then, by Euclid 9. 11, if G measures D, it will also measure A. But if the consequent is denied, i.e., if G does not measure A, then G will not measure D or any other number in the series. Therefore any prime number that measures A must also measure D (see also below, the Commentary, n. 4 for I.68–79, p. 330).

Relations between two geometric progressions are considered in Proposition 2. Since Oresme will deal almost exclusively with circular continua, he first asserts (I.91–96) that statements true of straight-line continua are also true of circular continua. The sole difference between the two types of continua lies in the fact that where only one point is required to divide a rectilinear continuum into two parts, it takes two points to achieve this in a circle.[6]

The proposition itself demonstrates that if a given continuum were divided in accordance with two different geometric proportionalities, there would be no common points of division unless the numbers following immediately after unity in both proportionalities are not mutually prime or communicant. Let A be a continuum that is divided into D^n number of successive equal parts, where $n = 0, 1, 2, 3, 4,...$; and let E^n represent another proportionality that divides A into successive equal parts; finally let numbers D^n and E^n be mutually prime or incommunicant.[7] Now should p/D^n and q/E^n parts of A be taken, where p and q are integers and $p < D^n$ and $q < E^n$, it is clear that if they are to share at least

---

Oresme, see Oresme's remarks in I.210–17.

5. Oresme's citation is correct in Campanus of Novara's thirteenth-century edition of Euclid (*Euc.-Campanus*, pp. 114–15), the version that Oresme consulted; but in the modern edition of the Greek text, as translated by Sir Thomas L. Heath, it appears as 9. 12 (*The Thirteen Books of Euclid's Elements*, vol. 2, pp. 397–98).

6. Radii, or chords, drawn to two separate points on the circumference of a circle will divide it into two sectors or parts.

7. Although in the proof itself, Oresme speaks of the division of A solely in terms of parts equal to numbers D and E, it is clear from I.97–98 that any part of D can be subdivided further into D parts, and so on (the same is said of E in I.99–100). Therefore, D^n and E^n are more general and also more appropriate representations of the two proportionalities. Although n can equal zero in this representation, Oresme had no concept of a zero exponent.

one common point of division, then p/D^n must equal q/E^n for at least one value of p and one value of q. But this is impossible since D^n and E^n are mutually prime numbers. The test for determining the existence of common points of division when continuum A is divided by two different proportionalities is furnished by Proposition 1 and the definitions of communicant and incommunicant proportionalities (see I.122–31, and I.10–17 for definitions of these proportionalities).

In an example (I.132–40), Oresme divides a continuum into 1, 2, 4, 8, 16,... (i.e., 2^n), successive equal parts, and then into 1, 3, 9, 27,...(i.e., 3^n), successive equal parts. Since the numbers immediately following unity, namely, 2 and 3, are prime to each other, division of A by these two proportionalities cannot produce common points of division (except for 1 which, however, represents the whole of A before any divisions have been effected in either proportionality). But if the continuum is divided by proportionalities, 1, 3, 9, 27,...(i.e., 3^n), and 1, 6, 36, 216,...(i.e., 6^n), there will be common points, since 3 and 6 are not mutually prime numbers (I.144–48). Thus, points of division corresponding to $\frac{1}{3}$ and $\frac{2}{6}$ are common to both.

The second proposition contains an important concept that recurs in later propositions. Oresme remarks (I.141–43) that a continuum divided by any given geometric proportionality—as, for example, 2^n or 3^n, where $n = 1, 2, 3,...\infty$— ought to be exhausted as the successive divisions into smaller and smaller equal parts continue to infinity. And yet this does not happen, for although numerous different proportionalities can be employed to divide the continuum and seemingly completely exhaust it, one can imagine that infinite points yet remain on which no point of division has yet fallen.[8]

The objective of Proposition 3 is to justify Oresme's decision to use vulgar, or common, fractions rather than sexagesimal fractions to express the parts of circles traversed by mobiles. The limitations of the sexagesimal system are responsible for this. Let us assume that a continuum is divided by a sexagesimal proportionality. First it is divided into 60 equal parts, then into 60^2 equal parts, then into 60^3 equal parts, and so on to 60^n equal parts. In this proportionality 60, the number following immediately after 1, has three prime factors, namely, 2, 3, and 5. Then, by the first proposition, no number in a sexagesimal proportionality can be measured by any prime number except 2, 3, and 5. Thus, fractional parts of 60^n can represent integral numbers of equal parts only if the denominators of those fractions are 2, 3, or 5, or some multiple of these prime numbers. For example, $\frac{1}{5}$, $\frac{1}{3}$, and $\frac{1}{2}$ parts of 60 will yield 12, 20, and 30 equal parts, re-

8. The same paradox appears with respect to circular continua in Part II, Proposition 4 (II.116–21; see also below, pp. 46–47) and Part II, Proposition 12 (II.339–47 and II. 356–63; see also below, pp. 59 and 60).

spectively; $\frac{1}{10}$, $\frac{1}{6}$, and $\frac{1}{4}$ parts will likewise yield 6, 10, and 15 parts, respectively. But $\frac{1}{9}$, whose denominator is a multiple of 3, cannot produce an integral number of equal unit parts of 60. In this case we could take $\frac{1}{9}$ of 60^2 to obtain 400 parts of 3,600. However, should 60^n equal parts be divided by fractions that are reciprocals of prime numbers other than 2, 3, and 5, no exact number of parts can be obtained. Thus $\frac{1}{7}$, $\frac{1}{11}$, $\frac{1}{13}$, $\frac{1}{14}$, and so on, fractional parts of 60^n cannot produce an exact number of parts since these denominators are not exactly divisible into any term in the series 60^n. Therefore, during the course of a day, if a body traveling on a circle divided sexagesimally should traverse an integral number of degrees plus $\frac{1}{7}$ of a degree, its motion could not be represented precisely by any integral number of equal unit parts in the sexagesimal system (I.181–87). However, if circles were divided into 17 signs—instead of 12 as in the sexagesimal system—and tables were compiled from a geometric proportionality based on 17, such tables would disagree with the tables derived from the sexagesimal system (I.198–202). The same problem would arise for any particular geometric proportionality that might be selected as a basis for constructing tables. Certain fractional parts will remain inexpressible by any exact number of equal parts into which the continuum may be divided.

Since Oresme seeks to express exact punctual velocities and distances, he rejects the sexagesimal system and, indeed, the use of any particular proportionality where the range of employable fractions is restricted to fractions whose denominators are exactly divisible by the term immediately following unity or by its prime factors. Instead, he employs vulgar, or common, fractions, for these embrace the whole range of rational fractions and will permit him to relate any rational velocities (I.214–17), since vulgar fractions used in any particular example can be reduced to some common denominator and expressed as an integral number of parts of that denominator.

With Proposition 4 the actual sequence of kinematic propositions begins. Assuming commensurable motions, Oresme shows that two mobiles, or bodies, presently in conjunction will conjunct repeatedly in that same point,[9] and will have conjuncted there repeatedly in the past. Let V_A and V_B represent the veloci-

---

9. In I.221–22, Oresme observes that, for the purposes of this treatise, conjunction in a point is equivalent to conjunction on a line or surface. Since his discussion is concerned for the most part with points in motion on concentric circles (I.60–62), it was more convenient to speak of conjunctions as if they occurred in single fixed points that served to terminate radii drawn from the common center through the mobiles or moving points (cf. I.50–51). When, as frequently happens, Oresme applied his kinematic results to planetary motions, these conjunctions could be imagined as occurring in a single point on the celestial sphere (they would then have the same right ascension and declination) or on the same celestial meridian (the planets in question would have the same right ascension) but at different celestial latitudes (they would have different declinations).

ties of mobiles A and B, respectively. Since the velocities are commensurable, it follows by Euclid 10. 5^{10} that $V_A : V_B = C : D$, where $C : D$ is a ratio of integers (I.222–24). After some time, A will have completed C circulations and B will have moved through D circulations, and they will conjunct again in their present point of conjunction. Thus if $V_A : V_B = 5 : 3$, A will have completed 5 circulations when B completes 3 and they will again conjunct. This must have occurred regularly in the past and will occur regularly in the future.[11]

After demonstrating that two mobiles with commensurable velocities will conjunct repeatedly in their present point of conjunction, Oresme, in Proposition 5, shows how to determine the time interval between two successive conjunctions in the same point. Since he was actually seeking the period of revolution for the two mobiles (I.268–70), Oresme required a temporal unit and chose to express his results in terms of the day (I.248–54).

Assuming that $V_A > V_B$ and that $V_A : V_B = C : D$, where $C : D$ is a ratio of integers in its lowest terms (I.238–39),[12] Oresme says that if A completes one circulation in D days, B will require C days. But by Proposition 4, A must make C circulations and B make D circulations before they conjunct again in the same point. Therefore the period of revolution must be $C \cdot D$ days. As in Proposition 4, Oresme sets $C : D = 5 : 3$, so that $C \cdot D = 5 \cdot 3 = 15$ is the period of revolution.[13] At the end of each period of revolution, every conjunction and aspect in which these mobiles participate will be repeated (I.271–72).

10. "Any two communicant quantities are related as a ratio of a number to a number." ("Omnium duarum quantitatum communicantium est proportio tanquam numeri ad numerum."—*Euc.-Campanus*, p. 247.)

11. The substance of this proposition appears in the first part of Part 1, Proposition I of the *Ad pauca respicientes* (see Grant, *Oresme PPAP*, p. 388, lines 65–71). But where in the *De commensurabilitate*, Oresme assumes that the circles are concentric (I.60–62) and, presumably, unequal, and where the velocities are expressed in terms of total numbers of circulations (or ultimately in terms of angular velocities), in the *Ad pauca* he distinguishes two conditions under which the proposition may be applied (ibid., pp. 386–88, lines 55–64). The circles may be unequal but commensurable and the curvilinear velocities equal, or unequal, and commensurable; or the circles may be assumed to be equal and the velocities unequal but commensurable.

12. In Proposition 4 Oresme did not specify that ratio $C : D$ was in its lowest terms.

13. Proposition 5 of the *De commensurabilitate* is also found in Part 1, Proposition I of the *Ad pauca respicientes* (Grant, *Oresme PPAP*, pp. 388–90, lines 72–85). A comparison reveals that in the *De commensurabilitate* Oresme has again simplified the procedures of the *Ad pauca* (see above, n. 11 of this chapter). In the latter treatise he determines the time interval between successive conjunctions in the same point but distinguishes two basic parts: (1) where the circles are unequal and the curvilinear velocities are equal or unequal, and (2) where the circles are equal and the curvilinear velocities are unequal. Under (1) he treats two cases: first the case (ibid., p. 390, lines 77–80) where one circle is to the other as $m : n$ (m is multiple to n and the circles are related as 2 to 1), and the curvilinear velocities are equal; in the second case (ibid., p. 390, lines 81–83)

In Proposition 6 Oresme assumes first that the two mobiles are in conjunction and then explains how to determine the time that will elapse before their very next conjunction, at whatever point on the circle this might occur. Once again $V_A > V_B$ and $V_A:V_B = C:D$. Therefore, $V_A:V_B = 1/D:1/C$, signifying that A traverses $1/D$ part of its circle when B traverses $1/C$ part of its circle. In order to represent these fractional parts of the circle by a common denominator, Oresme multiplies C by D and divides the circle into $C \cdot D$ equal parts. He then uses the specific data of Proposition 5 and divides the circle into 15 parts, since A traverses $\frac{1}{3}$ of its circle in 1 day (it completes one circulation in 3 days) and B traverses $\frac{1}{5}$ of its circle in the same time (it completes one circulation in 5 days). Since A moves with a greater velocity than B, there will be a conjunction when A, the quicker mobile, completes one more circulation than B and overtakes it. Now in the course of any given day, the distance which A gains over B is $1/D - 1/C = (C-D)/CD$, or $\frac{1}{3} - \frac{1}{5} = (5-3)/(5 \cdot 3) = \frac{2}{15}$ of a circle. The time of the very next conjunction is then obtained by dividing the numerator into the denominator. Thus $CD/(C-D) = \frac{15}{2} = 7\frac{1}{2}$ days, at the end of which time A will have gained a full circle over B and they will conjunct for the first time since departing from their initial conjunction. All this is summarized concisely in a rule given at the conclusion of the proposition: "The motion of one [mobile] is subtracted from the motion of another and the remainder has a numerator and denominator. The time sought is produced by dividing the denominator by the numerator" (I.307–9).[14]

Once we have utilized Proposition 6 to determine the time of the first conjunction following some given conjunction, we know *ipso facto* the time between any two successive conjunctions for, on the assumption that the veloci-

---

the circles remain related as 2 to 1, but now the curvilinear velocities are unequal. Only this latter case corresponds to Proposition 5 of the *De commensurabilitate*. In his *De configurationibus qualitatum* (chap. II.iv), but not in the *De commensurabilitate* or *Ad pauca*, Oresme does explicitly make the distinction between curvilinear and angular velocity (see Clagett, *Nicole Oresme and Medieval Geometry*, pp. 276–78, lines 7–18).

14. The same rule is found in Part 1, Proposition II of the *Ad pauca respicientes* where it is given as follows: "Divide one circle by the difference of the speeds, and the resulting number is the time of their first conjunction" (see Grant, *Oresme PPAP*, p. 393; Latin text, p. 392, lines 97–98; an example is

given on p. 392, lines 104–6). In the same proposition of the *Ad pauca*, Oresme utilizes the rule to determine the total number of conjunctions and their locations in a period of revolution. In the *De commensurabilitate*, however, the rule itself constitutes a separate proposition, the sixth, which is then employed in Part I, Proposition 7 to determine the total number of conjunctions in a period of revolution and in Part I, Proposition 8 to find the locations of all these conjunctions. The material embodied and compressed into one proposition of the *Ad pauca* is thus expanded and distributed over three consecutive propositions of the *De commensurabilitate*.

ties of the mobiles are respectively uniform, these time intervals must be equal. Therefore, in order to find the total number of conjunctions in a period of revolution, we are told in Proposition 7 that it is only necessary to divide the period of revolution (found by Proposition 5) by the time interval between any two successive conjunctions (found by Proposition 6). In the example cited above from Proposition 6, there would be two conjunctions in every period of revolution, since $15/7\frac{1}{2} = 2$.

After finding the *time* of the very next conjunction in Proposition 6, the manner of determining the *place* of that same conjunction is shown in Proposition 8. This depends upon a knowledge of the distance that either of the mobiles travels in a given day and the time between successive conjunctions. Building upon the same example employed in Propositions 4–7, Oresme assumes that A traverses $\frac{1}{3}$ and $B\frac{1}{5}$ of a circle per day and that $7\frac{1}{2}$ days will elapse before the occurrence of the very next conjunction. At this point, the daily distance of either A or B may be used. Choosing the distance of B, we multiply $7\frac{1}{2} \cdot \frac{1}{5}$ and obtain $\frac{3}{2}$, which signifies that B has moved around the circle $1\frac{1}{2}$ times in the time that has elapsed between the initial conjunction and the immediately succeeding conjunction. Finally, we must subtract the whole circle from $1\frac{1}{2}$ circles, and this leaves $\frac{1}{2}$. Consequently, A and B will conjunct in a place halfway around the circle from the place of the immediately preceding conjunction, i.e., directly opposite to it. The same result is obtainable when A is used (I.338–39). Although Oresme omits the calculations, we see that $7\frac{1}{2} \cdot \frac{1}{3} = \frac{5}{2}$, which means that A has moved around the circle $2\frac{1}{2}$ times in the interval between the two successive conjunctions. We must now eliminate the integer by subtracting 2 from $2\frac{1}{2}$, which leaves the fraction indicating that A and B will conjunct in a point directly opposite their previous place of conjunction.

Expressed in terms of the letters C and D (see above, p. 12, for the description of Proposition 5), for mobile A we have $CD/(C-D) \cdot 1/D = C/(C-D)$, and for B we have $CD/(C-D) \cdot 1/C = D/(C-D)$. Where the division of $(C-D)$ into either C or D produces a quotient consisting of an integer plus a fraction, the integer must be eliminated. The fraction alone reveals the degree of separation between the two successive points of conjunction, and thereby indicates the location of the next point of conjunction.[15] The denominator of the fraction also

15. Oresme describes the elimination of the integer when he says (I.330–31) that the whole circle should be subtracted as many times as possible from the total distance traversed by either A or B during the interval between the two successive conjunctions. This procedure and the substance of Part I, Proposition 8 of the *De commensurabilitate* constitute a portion of the second proposition of the first part of the *Ad pauca respicientes* (see above, n. 14) where Oresme says (Grant, *Oresme PPAP*, p. 393; Latin text, p. 392, lines 101–3): "Now you will be able to find the places [of conjunction] by multiplying the speed of one mobile by the time of one conjunction, and then subtract the

signifies the number of equal parts into which the circle may be divided, where each point of division represents a place of conjunction. Thus, the places of all conjunctions in one revolution are located—not merely that of the first conjunction.[16] For example, if a fraction of $\frac{1}{3}$ remains, this tells us not only that the first conjunction will be separated by $\frac{1}{3}$ of the circle from the initial point of conjunction, but also that there are 3 equidistant points of conjunction, the locations of which can easily be found with reference to the initial point of conjunction.

Prior to Proposition 9 Oresme had shown how to determine the time and place of some first conjunction between two mobiles initially in conjunction at a given point. But in the ninth proposition, both the time and place of some first conjunction are found when, initially, a given distance separates the mobiles. The distance between any two mobiles is to be measured from the slower mobile counterclockwise to the quicker mobile, and, except when in conjunction, the faster moving mobile is *always* conceived as being behind the slower moving mobile. Let us assume that a circle is divided into 12 signs and that A is quicker than B. Then, if A is one sign ahead of B, measured clockwise, the distance separating them in a counterclockwise direction would be 11 signs. Oresme would simply say that B precedes A by 11 signs (I.342–45).[17]

---

circle as many times as possible from this product, and the remainder will indicate the place [of conjunction]." For an example, see ibid., p. 392, lines 109–11, and, for further discussion, pp. 91–92.

16. This may be implied in I.331–32 where Oresme equates the *subtractive* process described in I.330–31 (see above, n. 15) with *dividing* the whole circle by the distance traversed by either mobile. If by distance, Oresme means here the final angular separation of the two successive points of conjunction, rather than the total distance traversed by either mobile, then all the places of conjunction can be found by using the initial point of conjunction as reference. Cf. I.750–74 and the discussion of these lines below, pp. 29–30; see also chap. 26 of App.1 and the discussion of it below, p. 96, n. 44.

17. Oresme's decision to measure distance relationships from the slowest moving mobile and to treat the quicker mobile as if it were always situated behind the slower mobile was probably motivated by a desire to conceive of the quicker moving body as closing a gap between itself and a slower

mobile. Thus, immediately after a conjunction with a slower mobile, the quicker mobile passes it and is then 11^+ signs behind and thereafter is imagined to continually close the gap, moving constantly "forward" and diminishing the distance between them until the next conjunction. Apparently this seemed conceptually "more natural" than supposing that as the swifter passes the slower it is 0^+ signs distant and would thereafter constantly increase the gap to 11^+ signs prior to the next conjunction. From this standpoint the swiftest would overtake the slowest when it is farthest removed from it, an obviously unsatisfactory situation. But (see below, chap. 2, n. 34), circumstances compelled Oresme to measure distances clockwise, as well as counterclockwise, from slower to faster mobiles; or, as we may also express it, he was compelled to measure counterclockwise from quicker to slower mobiles as well as from slower to quicker. The problem of proper directional measurement on a circle led Johannes de Muris into serious error (see below, pp. 92–93).

As in many of the preceding propositions, let us suppose that $V_A:V_B = C:D$, where $C:D$ is a ratio of mutually prime numbers and $C > D$.[18] Oresme distinguishes two cases. In the first case (I.356–58), the difference between the numbers representing the ratio of velocities is set equal to the distance separating the mobiles so that $C-D = S_{B \to A}$, where $S_{B \to A}$ is the distance separating A and B when measured counterclockwise from B to A and expressed positively in degrees, or signs, or any other suitable unit. Should these conditions obtain, A and B will conjunct after A moves C degrees and B moves D degrees. In his example, Oresme sets $V_A:V_B - C:D - 8:3$ and $S_{B \to A} - 5$ degrees. Therefore, when A moves 8 degrees B will have moved 3 degrees, and they will conjunct.

In the second case (I.358–63), $C-D \neq S_{B \to A}$ and Oresme proposes the following proportional relationship to determine the distances that must be traversed before conjunction can occur: $(C-D):S_{B \to A} = C:Z$ or $D:Z$, where Z is the unknown distance which either A or B must traverse in order to conjunct. If, again, the ratio of velocities is $C:D = 8:3$, but now $S_{B \to A}$ is 2 degrees, we find the distance that A must travel to conjunct with B by using C/Z in the proportional relationship given above. Thus $(8-3):2 = 8:Z$ so that Z equals $3\frac{1}{5}$ degrees. Substituting $D:Z$ for $C:Z$, we find that B must travel $1\frac{1}{5}$ degrees to conjunct with A.

Oresme has not actually calculated or determined the specific time and place of conjunction of A and B, although this was the stated objective of the proposition. But the time and place could easily be found by Propositions 6 and 8, respectively.[19]

In Proposition 10 Oresme shows how to find the *number and sequence* of the points of conjunction of two mobiles during the course of one or more complete revolutions.[20] This proposition depends on Propositions 7 and 8 and the fact that the times between any two successive conjunctions must be equal—a condition that follows from the assumption that the velocities, though unequal, are respectively uniform or constant (see I.314–15 and 367–68). By Proposition 7 the total number of conjunctions can be determined for one revolution. Because the time interval between any two successive conjunctions is necessarily equal, it follows that the points of conjunction distributed around the circle must be equidistant and divide the circle into an equal number of parts.[21]

---

18. In order to avoid a "lengthy demonstration," Oresme provides only a specific numerical example to illustrate the proposition (I.354–55). By using letter designations from earlier propositions, I have given a somewhat more general representation of the types of cases exemplified by Oresme.

19. In Oresme's examples the place is not determinable because the initial positions of mobiles A and B are not furnished.

20. In a preliminary way, we find an order of conjunctions given at the conclusion of Proposition 7 (I.322–25).

21. Oresme's remark in the *De commensurabilitate* that there are just "as many conjunctions in one revolution as there are

Thus, if there are five conjunctions in a revolution there must be five different points of conjunction that divide the circle into five equal parts. Conjunctions can occur only in these five points. Now by Proposition 8 we can determine the place of the first conjunction after some given initial conjunction. This enables us to determine the distance separating any two successive conjunctions, and therefore to locate the position, or place, of any point of conjunction. Should the number of points of conjunction (Proposition 7) and the distance separating any two successive conjunctions and, therefore, all the places of the points of conjunction, be known (Proposition 8), we can determine the sequence in which these conjunctions will occur.

Let us assume that the ratio of velocities is $V_A : V_B = 12:5$. Oresme now determines the number of conjunctions that can occur during one revolution, and he does this not by Proposition 7, but rather by the next proposition, Proposition 11, where it is shown that the difference of the velocities equals the number of points of conjunction.[22] Therefore, $12-5 = 7$ and 7 is the number of distinct points of conjunction. By Proposition 8 we are shown that any conjunction occurs $\frac{5}{7}$ of a circle away from the immediately preceding conjunction.[23]

---

places, or points, of conjunction equidistant from one another" (I.368–70) was asserted earlier in Part 1, Proposition II of the *Ad pauca respicientes* (Grant, *Oresme PPAP*, p. 329, lines 107–9) where, however, the equidistance of the points of conjunction is only implied but not expressed.

22. To utilize Proposition 7, Oresme would have had to supply the period of revolution and the time between any two successive conjunctions and then divide the latter into the former to obtain the number of points of conjunction. Instead, he gives the ratio of velocities of the mobiles and then appeals directly to Proposition 11 where the number of points of conjunction is found by subtracting velocities. It is not clear why Oresme chose to invoke Proposition 7 at the beginning of Proposition 10 (I.366) and then proceeded to ignore it by citing Proposition 11 (I.387–89), which shows the same thing in another way but had not yet been formally demonstrated. His move may have been dictated by the fact that in his example (I.387–95) he required straightaway the total number of points of conjunction to determine the order in which conjunctions would occur in

these points. Hence he wished to avoid an elaborate consideration involving both the period of revolution and the time between any two successive conjunctions—and both of these would have been required if Proposition 7 were utilized in determining the total number of points of conjunction—and thereby deliberately chose to ignore Proposition 7 in favor of Proposition 11 where the same result was obtainable by a simple subtraction. Although this move required the citation of a proposition that had not yet been demonstrated, it did avoid an extended discussion.

23. By Proposition 5, the period of revolution is found by multiplying the lowest terms representing the ratio of velocities. Thus $12 \cdot 5 = 60$, so that 60 days is the period of revolution. Since there are 7 points of conjunction, the time between successive conjunctions is $\frac{60}{7}$, or $8\frac{4}{7}$ days (by Proposition 6). Now by Proposition 8, mobile A travels $\frac{1}{5}$ of its circle every day so that in $8\frac{4}{7}$ days it will have completed $1\frac{5}{7}$ circulations; and mobile B, which travels $\frac{1}{12}$ of its circle every day, will have completed $\frac{5}{7}$ of a circulation during the same $8\frac{4}{7}$ days. Therefore, after $8\frac{4}{7}$

The conjunctions can now be arranged sequentially. When a conjunction occurs in any point, say, C, the next conjunction must occur $\frac{5}{7}$ of a circle away from C. Let us number the seven equidistant points of conjunction clockwise from C_1 to C_7. If the first conjunction of a period of revolution is assumed to occur in C_1, the second conjunction must occur in point C_6, $\frac{5}{7}$ of the circle away from C_1. The third conjunction will occur in C_4 with the remaining four conjunctions taking place, in order of occurrence, in C_2, C_7, C_5, C_3. The cycle is then repeated beginning with C_1. Thus, the order of conjunctions does not necessarily follow the order of the points (I.380–86). One, or two, or more points may be omitted with regularity as successive conjunctions occur, say, in the first, fourth, seventh, etc., points.[24]

In Proposition 11 Oresme presents another way of determining the number of points of conjunction in a complete period of revolution. Instead of dividing the time between any two successive conjunctions into the period of revolution of two mobiles, as in Proposition 7, Oresme now shows that the number of points of conjunction equals the difference between the prime numbers that represent the ratio of velocities. If $V_A : V_B = C : D$, where C and D are mutually prime and $C > D$, then $C - D = n$, where n represents the total number of points of conjunction in every revolution of mobiles A and B.[25] This is evident by an inference from Proposition 5 (I.268–70)[26] where it was shown that a revolution will have been completed when A and B make C and D circulations, respectively. But when this occurs A will have completed $C - D = n$ more circu-

---

(Note 23 continued)
days, A will have made one more circulation than B and conjunct with it $\frac{5}{7}$ of a circle away from their last point of conjunction. Similarly, their very next conjunction will also occur $\frac{5}{7}$ of a circle away from the present conjunction, and so on with all successive conjunctions.

24. The omission of points in sequences of conjunctions is discussed further in Part I, Proposition 11 (see I.449–62 and the Commentary, n. 7 for I.423–62). Proposition 10 of the *De commensurabilitate* has no counterpart in the *Ad pauca respicientes*. Indeed, in the latter treatise, Oresme goes no further than merely finding the place of the first conjunction following some initial conjunction (Grant, *Oresme PPAP*, p. 392, lines 101–3 and 109–11). However, Proposition 10 of the *De commensurabilitate* is but an extension of this procedure, since the method

of finding the second conjunction is the same as that for finding the first, and so on. By carrying the process far enough, the order and places of all the possible conjunctions can be determined for one revolution, and then for as many as desirable. It is probable, however, that Oresme had not yet formulated all this until he revised the *Ad pauca respicientes* (see *Oresme PPAP*, pp. 79–80).

25. No anticipation of Proposition 11 appears in the *Ad pauca respicientes*.

26. In my article "Nicole Oresme and the Commensurability or Incommensurability of the Celestial Motions" (*Archive for History of Exact Sciences*, vol. 1 [1961], p. 429, n. 22), I maintained that although the manuscripts cited Proposition 5, it seemed more appropriate to invoke Proposition 4. It is clear to me now that the textual reference is appropriate.

lations than did B (I.405–7). And in Proposition 6 we were told (I.287–88) that every time A gains one circulation over B there will be a conjunction, from which it follows that A and B must conjunct n times, since A has gained n circulations over B. For example (I.414–22), if $V_A : V_B = 5:3$, then in one revolution A and B will conjunct twice, since their difference is 2. The first conjunction will occur in the point opposite their present point of conjunction—i.e., when A makes $2\frac{1}{2}$ and B, $1\frac{1}{2}$ circulations—and the second will occur in their present point of conjunction when A completes 5 and B, 3 circulations. Since these are the only points where conjunctions can occur, this sequence of conjunctions will recur as long as the same conditions obtain.

After drawing upon earlier propositions to formulate a concise five-step procedure[27] that will permit the proper ordering of conjunctions in any period of revolution, Oresme applies Proposition 11 to the widely held belief that the velocities of the planets are related in harmonic ratios and so produce the celestial harmonies or music of the spheres.[28] He singles out the diapason, or octave (i.e., 2:1), the diapente, or fifth (i.e., 3:2), the diatessaron, or fourth (i.e., 4:3),

---

27. See below, the Commentary, n. 7 for I.423–62 for a summary and discussion of this five-step procedure.

28. This is the Pythagorean theory of the harmony of the spheres. In reporting and rejecting this view, Aristotle says: "the theory that the movement of the stars produces a harmony, i.e. that the sounds they make are concordant, in spite of the grace and originality with which it has been stated, is nevertheless untrue. Some thinkers suppose that the motion of bodies of that size must produce a noise, since on our earth the motion of bodies far inferior in size and in speed of movement has that effect. Also, when the sun and the moon, they say, and all the stars, so great in number and in size, are moving with so rapid a motion, how should they not produce a sound immensely great? Starting from this argument and from the observation that their speeds, as measured by their distances, are in the same ratios as musical concordances, they assert that the sound given forth by the circular movement of the stars is a harmony. Since, however, it appears unaccountable that we should not hear this music, they explain this by saying that the sound is in our ears from

the very moment of birth and is thus indistinguishable from its contrary silence, since sound and silence are discriminated by mutual contrast."—*De caelo* 2. 9. 290b12–28 (trans. J. L. Stocks, *The Works of Aristotle* [Oxford, 1930]).

Oresme was no doubt aware of this passage since he alludes to it in Part III of the *De commensurabilitate*, III.232–34. This passage and one from Macrobius on the music of the spheres (*Commentary on the Dream of Scipio* [trans. W. H. Stahl] 2. 1. 2–3), quoted in Part III of *De commensurabilitate* (III.162–67), probably constitute his primary sources for this Pythagorean belief. The passage in Macrobius is a quote from Cicero's *Dream of Scipio*, which formed part of Book 6 of Cicero's *Republic*. The doctrine of the harmony of the celestial spheres appears in Plato's *Republic* 10. 617B and in *Timaeus* 35B. The ratios assigned to the planets may have had musical significance for Plato, although this suggestion, and any connection with the doctrine of celestial harmony, is denied by Francis M. Cornford, trans., *Plato's Cosmology* (London, 1937), p. 72 (the discussion in 35B would have been known to Oresme in Chalcidius's translation and

and the tone (i.e., $9:8$).[29] Since the difference between the velocities of any two planets related by any one of these four ratios is 1, it would follow by Proposition 11 that any two planets whose velocities are related as one of these principal harmonic ratios can conjunct in only one point. For example, if the mean motions of Mars and the sun were as $2:1$, they could conjunct in only one point of the heavens—and nowhere else. But this is contrary to experience, since no celestial configuration is found to occur in only one point in the sky. We must conclude that no two celestial motions are related as any one of the principal harmonic ratios (1.475–78).[30] Therefore, if the celestial bodies produce harmonies, it may be in virtue of something other than their ratios of velocities (I.478–80). Oresme will suggest later that such harmonies may arise from the ratios of the volumes of the planetary spheres or, perhaps, the magnitudes (weights?) of the orbs (III.394–97; see below, p. 69, n. 107 of this chapter).

By granting the basic assumptions of the celestial harmony theory, Oresme has drawn from it an empirically testable consequence that is contrary to observation. Thus any celestial harmony theory would be untenable if it supposed that the motions of any two planets are related commensurably in harmonic ratios, and that it is this relationship that produces the harmony of the spheres.

Beginning with Proposition 12, and extending through Proposition 19, Oresme considers three or more mobiles in motion simultaneously. The object of Proposition 12 is to show that under certain initial conditions no more than two of the three or more mobiles may ever conjunct in any point. Let us assume that A, B, and C are three mobiles, or bodies. From previous propositions[31]

(Note 28 continued)

commentary which, in another connection, he cites elsewhere in the *De commensurabilitate*). For a discussion of this influential doctrine see Sir Thomas L. Heath, *Aristarchus of Samos* (Oxford, 1913), pp. 105–15; see also below, p. 68, n. 107, for Oresme's repudiation of the view that celestial music is produced by planetary velocities that are related as musical concordances.

29. Various consonances are discussed by Macrobius in his *Commentary on the Dream of Scipio* 2. 1. 14–25 (trans. Stahl, pp. 188–89), where in a final list (p. 189) he says that the "consonant chords are five in number, the fourth, the fifth, the octave, the octave and fifth, and the double octave." Oresme's list of principal consonances includes the fourth, the fifth, and the octave, but departs from Macrobius's by including the tone, or

$9:8$, as the fourth and final consonance, thereby ignoring Macrobius's octave and fifth (i.e., $3:1$) and double octave (i.e., $4:1$). In his later *Le Livre du ciel et du monde*, bk. 2, chap. 18, 125d–126a (Menut, *Oresme du ciel*, p. 480), Oresme expressly limits the principal consonances to four, the same four mentioned in the *De commensurabilitate*, I.465–66. For a table of harmonic intervals see Boethius, *De institutione musica*, ed. Godofredus Friedlein (Leipzig, 1867), p. 201; for additional bibliography see Stahl's translation of Macrobius, p. 188, n. 8.

30. Nor could they be related as $256:243$, a lesser semitone or diesis, for then there could be only 13 places of conjunction (I.470–74). See also III.384–97.

31. Especially relevant are Propositions 7, 8, 10, and 11.

concerned with only two mobiles, we know that A and B can conjunct in a fixed number of different points and in no others. Let d represent any one of these points. Similarly, B and C can conjunct in a limited number of points only, any one of which may be represented by e. If it can be shown that no d is an e, it would follow that A, B, and C could never conjunct simultaneously in any point. This reasoning could be extended to any number of mobiles, for if they are taken two at a time it might happen that the points of conjunction of any two mobiles differ from those of every other pair of mobiles. Should this occur, no more than two mobiles could conjunct simultaneously in the same point. However, changes in the initial conditions might completely alter this situation. Thus if we had six mobiles (I.491–93) it is possible that three, four, five, or even all six might conjunct simultaneously in the same point if, when taken two at a time, they shared in common one or more points of conjunction.

As an example (I.494–511), combine mobiles A, B, and C into three pairs: A with B, A with C, and B with C. Let the ratios of velocities for these three pairs of mobiles be: $V_A : V_B = 4:2$, $V_A : V_C = 4:1$, and $V_B : V_C = 2:1$. In Fig. 20 (see below, p. 215) point e is separated from d by 2 signs, or $\frac{1}{6}$ of a circle,[32] and points f, g, and e divide the circle into three equal parts. Finally, point h is separated from d by 3 signs, or $\frac{1}{4}$ of the angular measure of the circle. Initially, A and B are in conjunction in d while C precedes them by $1\frac{1}{2}$ signs, or by $\frac{1}{8}$ of the angular measure of the circle.

By Proposition 9 A and C will conjunct in e, since C will traverse $\frac{1}{2}$ a sign while A moves 2 signs from d to e. In fact, Proposition 10 shows that they can conjunct only in points e, f, and g (I.507–8), for after departure from conjunction in e, A will have traversed 16 signs ($1\frac{1}{3}$ circles) when C has traversed 4 signs

---

32. A concise earlier version of Proposition 12 was given by Oresme in Part 1, Proposition VII of his *Ad pauca respicientes* (see Grant, *Oresme PPAP*, pp. 398–400 for text and 95–96 for discussion), where he also assumed that two mobiles are initially in conjunction with a third mobile separated from them by $\frac{1}{6}$ of a circle. In the *Ad pauca*, Oresme considers mobiles A, B, and C as moving with equal curvilinear speeds on their respective concentric circles, A, B, and C, which are related as follows: $A:C = 6:3$, $A:B = 3:2$, and $B:C = 4:3$. Since the curvilinear velocities are equal, the inverse ratios of the circumferences of the circles give us, in effect, the ratios of the angular velocities

between mobiles A, B, and C (in the *De commensurabilitate* angular velocities are assumed directly eliminating the need to introduce ratios between the circles). In his proof Oresme considers only the ratios between mobiles $A:C$ and $A:B$, quite properly ignoring the relationship between mobiles B and C. For if the pairs of mobiles (A, C) and (A, B) shared no common points of conjunction, then whenever B and C conjuncted, A could not be there at the same time. In the *De commensurabilitate* Oresme expanded his demonstration unnecessarily by including a discussion of all three pairs of possible relationships. Cf. App. 1, Johannes de Muris, *Quadripartitum*, chap. 25.

($\frac{1}{3}$ of a circle), and the two will conjunct in f; by the same calculations, they will conjunct next in g, then in e, then in f, and so on ad infinitum.

By Proposition 9 mobiles B and C can conjunct only in h. This is apparent from the fact that their ratio of velocities is 2:1 and that C precedes B, which is initially in conjunction with A in d, by $1\frac{1}{2}$ signs. Obviously they will conjunct in h which is separated from d by 3 signs. Thereafter B and C can conjunct only in h.

Finally, mobiles A and B can conjunct only in d, their initial point of conjunction. This is evident from Proposition 11, for when their ratio of velocities is reduced to lowest terms the difference in their velocities is 1 (i.e., $4:2 - 2:1$ and $2-1 = 1$), signifying that only one conjunction can occur in every period of revolution. Since A and B are already in conjunction in d this must be their one and only point of conjunction.

Thus we see that the only points of conjunction for the three mobiles taken two at a time are d, e, h, f, and g; but in none of these can A, B, and C conjunct simultaneously.

If, as we have seen, mobiles A, B, and C will never conjunct, how close together can they come; or, to put it another way, what is the minimum space that will encompass them? Proposition 13 seeks to determine this.[33] Assuming, as in the preceding proposition, that $V_A > V_B > V_C$, Oresme asserts (I.517–19) that only when the fastest and slowest of the mobiles— namely, A and C—are in conjunction can all three be "squeezed" within a minimum possible space. This can occur in two ways: (a) when A and C are in conjunction and B precedes them (I.519–22); and (b) when A and C are in conjunction and B follows, or trails (I.522–23).

This can be verified by reference to Fig. 21, on p. 219, below. Excluding conjunctions between A and C, we see that only six arrangements of the mobiles are possible (I.529–32). The motions of the mobiles are assumed to occur on concentric circles and their direction, represented by the arrows, is clockwise. Thus the mobiles represented by the letters in the extreme right-hand column are to be taken as the "preceding," or lead, mobiles; the letters in the extreme left column represent the rear mobiles; and the letters in the center column represent mobiles located somewhere between the leading and trailing mobiles in any particular case.

Let us now see why, in situation (a) above, mobiles A, B, and C are encompassed within a minimum space. Now *immediately before* the conjunction of A and C in any point the order of the mobiles is represented by (2) in Fig. 21. Obviously, as A, the fastest mobile, overtakes C, the slowest, the distance between A and B will continuously diminish until it reaches a minimum with the conjunction of A and C. But this produces situation (a). In a similar manner,

---

33. Nothing like Proposition 13 is found   in the earlier *Ad pauca respicientes*.

immediately after conjunction, when *B* leads or precedes, the order of mobiles is represented by (5). But in this arrangement the distance between *B*, the lead mobile, and *C*, the rear mobile, is greater than the distance separating *A* and *C* from *B* in situation (a).[34] We may therefore conclude that when *B* precedes and *A* and *C* are in conjunction, i.e., when (a) obtains, the three mobiles are as close as possible. Thus a minimum distance results only when (2) terminates in a conjunction between *A* and *C* with *B* preceding. But under the conditions represented by (2) and (5), where *B* precedes but no conjunction is involved, the minimum distance separating the leading mobile from the rear mobile exceeds that which separates the three mobiles represented in configuration (a), where *A* and *C* are actually in conjunction and preceded by *B*.

When situation (b) obtains, i.e., when *A* and *C* are in conjunction and *B* follows, a minimum distance must separate them. Otherwise, either arrangements (3) or (4) must, at some point, represent a minimum distance. But, as a consequence of the fact that $V_B > V_C$, (3) cannot produce a minimum separation since *immediately before* conjunction of *A* and *C*, *B* will be farther from *C*, the lead mobile, than during conjunction. And because $V_A > V_B$, (4) cannot effect a minimum separation *immediately after* conjunction since the distance separating *A* and *B* will then be greater than it was when *A* and *C* were in conjunction.

---

34. In these comparisons of distance, the previously mentioned method of measuring distances counterclockwise from slower to quicker mobiles (see above, p. 15 and n. 17) had to be abandoned in cases (4), (5), and (6). In all six cases the distance of separation must be measured between the extreme mobiles—that is, between the first, or preceding, mobile and the last, or following, mobile. Now in cases (4), (5), and (6), the quicker mobile precedes the slower, so that if Oresme followed the procedure described earlier in I.342–45 and measured counterclockwise from slower to quicker mobile, he would have obtained greater distances of separation than if he had measured clockwise from slower to quicker or counterclockwise from quicker to slower. This is immediately evident, for example, in case (5) of Fig. 21 when Oresme says that *B* will be more distant from *C* immediately after conjunction than when *C* was in conjunction with *A*. This would be false, however, if Oresme faithfully adhered to the procedure in I.342–45, for by using that method—i.e., measuring counterclockwise from *C*, the slower mobile, to *B*, the quicker mobile—the distance between *C* and *B* immediately after *C*'s conjunction with *A* would be less than the distance separating *C* from *B* when *C* was in conjunction. This must be so because that method requires that the distance between a quicker and slower body continuously diminish once the quicker has passed the slower. In all cases where the slower mobile is assumed to precede the quicker—i.e., in (1), (2), and (3)—Oresme's procedure will yield the desired results.

It is clear that Oresme has abandoned his rule—perhaps unknowingly—and taken for his measurements the shortest possible absolute distance separating any two extreme mobiles, regardless of whether this entailed measuring clockwise or counterclockwise, or whether it involved measuring from the slower or quicker mobile. The very nature of the proposition itself seems to have required absolute measurements since its objective is to determine minimum possible distances between mobiles.

Up to this point Oresme has eliminated (2), (3), (4), and (5). But we might still deny that (a) and (b) represent the minimum separation (I.526–29), and opt for (6). This is rejected because (6) arises directly from (5) when A, the quickest mobile, conjuncts with and then passes B. But it has already been shown (I. 520–21) that the mobiles as arranged in (5) are separated by a greater distance than when A and C are in conjunction and B precedes. Therefore, *a fortiori*, their separation in (6) must be even greater. Although (1) is not specifically mentioned, similar reasoning would eliminate it. In this case, however, (1) is transformed to (3) when A, the fastest mobile, passes B. But the mobiles in (3) are spread over a greater distance than when they enter disposition (b) (i.e., when A and C conjunct and B trails). *A fortiori*, then, the mobiles are separated by an even greater distance when in disposition (1).

Using the data of Proposition 12, Oresme proceeds to illustrate Proposition 13 (I.542–51). Recalling that initially A and B were in conjunction in d and that C preceded them by $1\frac{1}{2}$ signs, we saw that A and C (where $V_A : V_C = 4:1$) conjuncted later in e, separated from d by 2 signs. But in that very same time B has moved only 1 sign from d because its velocity is only half that of A. Thus only 1 sign separates B from A and C in conjunction at e. This, Oresme shows, is the minimum space that can separate these three mobiles; or, to put it another way, this is the closest that they can approximate to conjunction. But this minimum space does not occur for every conjunction of A and C. Indeed, the very next conjunction in f will occur when A has traversed 18 signs after its initial departure from d, and B, whose velocity is half that of A, will have traversed 9, so that 3 signs now separate them (i.e., A and C are in conjunction in f, which is halfway around the circle, or 6 signs removed, from d, while B is separated from d by $\frac{3}{4}$ of the circle, or by 9 signs). If we make similar calculations for every conjunction of A and C during one period of revolution (there is only one other—namely, in g—where they are separated by 5 signs; I.549–51), and know that the conjunctions of each revolution are identical, we can clearly see that one sign is the minimum distance of separation which will be realized in only one of the points of conjunction of A and C, namely, in e.

The results of Proposition 13 are applied to hypothetical planetary motions when Oresme conjectures (I.558–62) that three or four planets, moving commensurably as the mobiles, or points, in Proposition 13, might never conjunct, falling short of conjunction by perhaps two or three degrees.

In Propositions 14 through 19, Oresme extends to three or more mobiles results from propositions demonstrated earlier for two mobiles only.[35] This is easily achieved by taking the mobiles two at a time.

---

35. The correspondence between the later   14 and 4   16 and 6   18 and 8
and earlier propositions is as follows: 15 and 5   17 and 7   19 and 10

Throughout Propositions 1–19 the circular motion of every mobile was assumed to be concentric with the motions of all other mobiles to which it may have been compared or related (I.60–62). In Proposition 20, Oresme investigates the consequences deriving from the motions of two mobiles moving on non-concentric circles where one mobile has the center of the earth, or world, as its center and the other does not.[36] He shows that if the velocities are commensurable, the number of places, or points, of conjunction will be the same in this situation as it is when both circles are concentric, but the eccentricity will affect the intervals of time and the distances between successive conjunctions.

In Fig. 22 (p. 233), A and B are mobiles moving on two non-concentric circles. The center of the world, or the center of the earth itself, is point c, the center of A's motion; and point d is the center of B's motion. Although their velocities are not made explicit, it is clear from the distances traversed that $V_A : V_B = 5 : 1$. Initially A and B are in conjunction on line cdg, where g represents the aux point, or point of apogee.[37] On B's circle let us assume that line dh is separated from line cdg by $\frac{1}{4}$ of the circle, and that on A's circle the same distance separates line ck from cdg. Let us assume further that when B traverses $\frac{1}{4}$ of its circle, A will have moved around its circle $1\frac{1}{4}$ times. Therefore, if the circles were concentric, B would reach line dh when A reached line ck and they would be in conjunction (I.716–20). But since one of the circles is eccentric with reference to point c, the center of the world, they will conjunct *before A* moves $1\frac{1}{4}$ times around its circle and B moves $\frac{1}{4}$ of the way around its circle. Indeed, because of the eccentricity and A's greater velocity, A cannot reach point k until it has conjuncted with and passed B prior to B's arrival in h. This

---

Propositions 14 through 19 are summarized below in the Commentary, nn. 8–13 on pp. 332–35.

36. No proposition of this kind appears in the *Ad pauca respicientes*.

37. The line of aux (from the Latin *aux, augis*) connects the points of apogee (or aux point) and perigee (or "oppositum augis"; see I.712–13) of a planet. For Sacrobosco's definitions of these terms in his *Tractatus de spera* see Lynn Thorndike, ed. and trans., *The Sphere of Sacrobosco and Its Commentators* (Chicago, 1949), pp. 113 (Latin) and 140 (English); for definitions of the same terms in the anonymous *Theorica planetarum* (in some manuscripts and the edition cited

here, it is falsely ascribed to Gerard of Cremona), see below, p. 61, n. 95; and see also the glossary of terms in Derek J. Price, ed. and trans., *The Equatorie of the Planetis* (Cambridge, 1955), pp. 168 and 174–75.

In the particular case described here in Proposition 20, the line of aux with its points of apogee and perigee applies only to mobile B whose motion is eccentric to the center of the earth. It would be vacuous to speak of points of apogee and perigee for body A which moves on a circle whose center is the center of the earth, or world, so that its distance from that center is always constant.

conjunction will occur regularly at some point in the first quadrant of each circle, but nearer to g than to points h and k.[38]

Oresme does not describe any other conjunctions. But it is obvious that the very next conjunction must occur in point f (see Fig. 22), since A and B will depart simultaneously from their respective quarter points in k and h, so that when B traverses $\frac{1}{4}$ of its circle, A will have traveled $1\frac{1}{4}$ times around its circle and they will meet in f, the point of perigee. Only in the points of apogee and perigee (g and f, respectively) will A and B conjunct after traversing exactly $1\frac{1}{4}$ and $\frac{1}{4}$ of their respective circles (I.732–34). Thus the time and place of the conjunction in f is identical for concentric and eccentric circles. However, the very next conjunction will not occur exactly $\frac{1}{4}$ of the circle from f, but somewhere beyond in the last quadrant. This is evident because calculations from f reveal that the situation in the fourth quadrant will be the reverse of what happened in the first quadrant (see Fig. 22). For when B and A traverse $\frac{1}{4}$ and $1\frac{1}{4}$ of their respective circles from f, B, the slower mobile, precedes A, the faster, and they will not yet have conjuncted. Conjunction will occur later at some point beyond n in the fourth quadrant of both circles. Thus the time and place of the third conjunction is quite different when the circles are eccentric—or to put it more precisely, when one of them is eccentric—than when they are both concentric. The fourth and final conjunction will occur once again in g after A moves $1\frac{1}{4}$ times around its circle, and B, $\frac{1}{4}$ of its circle.

In this example, Oresme has shown that the number of conjunctions would be four, whether the circles are concentric or eccentric. But when the circles are eccentric the points of conjunction are not equidistant nor is the time between any two successive conjunctions equal. Two conjunctions occur at opposite ends of the line of aux, gf, whereas the other two take place in the first and fourth quadrants, with none occurring in the first and third.

Thus Oresme has demonstrated his proposition that mobiles with unequal but respectively uniform velocities will conjunct differently on concentric circles than on eccentric circles. He implies that this is a consequence of the fact that on concentric—but not eccentric—circles the mean and true motions are identical.[39] In Fig. 22, if A and B moved on concentric circles, their mean (and true) motions as measured from c, their common center, would produce a conjunction in k. But since A and B move on eccentric circles, their true motions are measured only from c,[40] and after departure from g their next conjunction will occur some-

38. In the absence of any data or explanation, it is not clear why Oresme should hold that this particular conjunction should occur nearer to g than h (I.722–23).

39. When A and B conjunct on the line of aux, cdg, Oresme says that they are "related

as if the motions and circles were concentric, for the true and mean motions are the same at that place" (I.712–15).

40. Oresme's use of the terms *motus verus* and *motus medius* (e.g., I.714 and 727) must not be understood or interpreted in a

where between *g* and *h*. In this case, the true motions produce a conjunction on eccentric circles sooner than their mean motions on concentric circles (also sooner than their true motions on concentric circles since mean and true motions are identical on concentric circles for the conditions given in the proposition). Oresme expresses this by saying (I.726–28) that, with respect to *c*, the mean motions on concentric circles "add" to—i.e., are relatively slower than—the true motions on eccentric circles on one side of the diameter *gdcf*. However, if *A* and *B* were moving toward their next conjunction from *f* (see Fig. 22) on the opposite side of diameter *gdcf*, the true motions on the eccentric circles would effect a conjunction later than one which would be produced by the mean (or true) motions on concentric circles. Here Oresme says that the true motions on eccentric circles "add" to—i.e., are relatively slower than—the mean motions on concentric circles.

Thus far all propositions and demonstrations have been expressed in terms of conjunctions only. Proposition 21 formally extends the results of the previous propositions to oppositions, and to all other astronomical aspects or relationships. One proviso must be kept in mind. Oresme notes that conjunctions and oppositions can occur in only one way, but that all other aspects have a twofold character, since they can occur either before or after any given conjunction or opposition.[41] Although Oresme applies this distinction to all aspects, he men-

---

strictly astronomical sense. In astronomy these terms signified that angular measurements were to be made with the first point of Aries, or vernal equinox, as the basic point of reference (see *Theorica planetarum*, fol. 169r [mistakenly paginated as 161r] for the sun, fol. 170r [paginated as 162r] for the moon, and fol. 171v for the planets; for a recent discussion see Price, *Equatorie of the Planetis*, pp. 97–98, 100–101). Oresme's angular measurements, however, are made with respect to the line of aux. Since Oresme does not include epicycles in his proposition, the only celestial body with respect to which his discussion could have any astronomical relationship is the sun. But even here there is little direct resemblance, for in addition to the different basic reference points used, Oresme is not representing the motion of one body on a single eccentric circle, but rather comparing the motions and positions of two distinct bodies moving on two separate non-concentric circles, one of which is

eccentric to the earth and the other having the earth as its center. In Oresme's scheme, the true and mean motions of *A* are always identical because its motion is uniform and the center of the earth, or world, is its center of motion. In the case of *B*, however, Oresme measures its mean motion with reference to *d*, the center of its motion, but measures its true motion from *c*, the center of the earth.

Actually, the mean motion of *B* on its eccentric circle plays no role in Proposition 20 since the only comparisons made by Oresme are between the mean motions of *A* and *B*, assumed to occur on concentric circles (where the mean and true motions would be identical for each mobile as measured from a common center) and their true motions on non-concentric circles as measured from *c*, the center of the earth and center of *A*'s motion. But the center of *B*'s motion, namely, *d*, is not used as the basis for comparisons.

41. Oresme seems here to be drawing upon his *Ad pauca respicientes* where great em-

tions specifically only the trinal aspect ("trinus aspectus"; I.741–42) where two or more planets are separated by a third part of the zodiac, or 4 signs.[42]

In this very brief proposition, Oresme elaborates no further, but on the basis of statements made in the *Ad pauca respicientes*[43] he seems to mean that

---

(Note 41 continued)

phasis was placed upon this distinction: "It can happen that several mobiles... have been related at some time or other in three ways: either in conjunction or opposition, and then they form no angle or angles; or in another disposition, and then they describe an angle or angles. A conjunction can occur in only one way and likewise for an opposition. Any other disposition is twofold: in one way before conjunction or opposition; in the other way after [conjunction or opposition]."— Grant, *Oresme PPAP*, p. 405 (p. 404, lines 2–7 for Latin text). The category of angular dispositions is then further subdivided into those that are "properly similar" and those that are "improperly similar." Properly similar dispositions obtain when several mobiles, or bodies, occupy or assume identical relative positions to one another. For example, whenever three bodies enter into conjunction anywhere, it is an essential preliminary that they occupy the same positions relative to each other. The same may be said about post-conjunctive situations; or about their relationships before every opposition; or after every opposition. Now just before conjunction they will occupy different relative positions than they will after that same conjunction, and Oresme designates this an improperly similar disposition (ibid., p. 404, lines 8–12 and pp. 98–100).

Oresme applied and utilized this distinction in most of the propositions of Part II of the *Ad pauca respicientes*, but abandons it completely in the *De commensurabilitate*. Rather than use the vague and confusing comparisons of angular dispositions before and after conjunctions (ibid., p. 99, n. 23 and p. 102), Oresme chose to couch his propositions in the *De commensurabilitate* almost exclusively in terms of precisely locatable

conjunctions. Then in order to extend to all other astronomical aspects the results derived from an exclusive consideration of conjunctions, he devoted for this purpose one proposition in each of the first two parts of the treatise (pt. I, prop. 21 and pt. II, prop. 11).

Although conjunctions and oppositions can occur in only one way—i.e., they are always joined by a straight line through the center of the world—Oresme explains in the *Ad pauca respicientes*, but not in the *De commensurabilitate*, that the possible number of ways that mobiles can be in opposition to each other at any particular time depends upon the total number of mobiles in motion simultaneously. Thus two mobiles can oppose each other in only one way—i.e., one mobile opposes the other; three mobiles can enter into opposition in three ways; and generally when the mobiles are taken two at a time they can oppose in $p(p-1)/2$ ways, where p is the total number of mobiles involved (ibid., pp. 96–97 and 406, lines 23–27; for an analysis of a more sophisticated consideration of combinations by Oresme, see Clagett, *Nicole Oresme and Medieval Geometry*, pp. 444–47).

42. Four such aspects were distinguished and applied to distances between celestial bodies: (1) opposition, (2) trinal, (3) quartile (*quartilis*), where two or more planets are separated by a fourth part of the zodiac, or three signs, and (4) sextile (*sextilis*), where a sixth part of the zodiac, or two signs, separate them. Definitions of these aspects are given by Robert Anglicus in Thorndike, *Sphere of Sacrobosco*, pp. 176 (Latin) and 226 (English); see also Grant, *Oresme PPAP*, pp. 92–93.

43. See above, n. 41.

the relative positions or sequence of the mobiles before conjunction will be wholly reversed after conjunction. In a trinal aspect, to use Oresme's illustration, two bodies are separated by 4 zodiacal signs, with the swifter body located to the left of the slower, so that a conjunction will occur when the swifter overtakes the slower as the former moves to the right (I.744–45). After conjunction, the swifter body will gradually pull away from the slower and when they are separated once again by 4 signs they will enter into a second trinal aspect. This time, however, it will occur from the left, or opposite, side (I.745), since the swifter body is on the other side of the slower as it moves steadily away. Thus the two-fold nature of a trinal aspect, or any other aspect, signifies that a reversal of relative positions has occurred between the quicker and slower mobiles before and after conjunction—or opposition.[44] The same reversals would occur if more than two mobiles entered into and departed from conjunction. Thus in Proposition 21 Oresme alerts his readers to bear in mind this twofold distinction when applying any of the earlier propositions to aspects other than conjunction or opposition.

The extension of previous propositions to include all aspects must be understood to apply to both concentric and eccentric circles. In eccentric circles, however, the same qualifications made in Proposition 20 for conjunctions must also be made for all other aspects. That is, just as the points of conjunction are not equally spaced, so the other aspects will not be equally spaced; and the same must be said of the time intervals.

In all previous propositions, every mobile was assumed to have a single motion, but in Proposition 22, Oresme applies some of the results of earlier propositions to the case of a single mobile moving with several simultaneous motions, where each motion is itself uniform and regular. The double motion of the sun, diurnal and annual, is used to illustrate the proposition.[45]

In Fig. 23 (p. 237) circle a is the tropic of Cancer (summer tropic) and A_n, where n can be any integer, represents any point on circle a; circle b is the ecliptic on which the center of the sun, B, moves with a uniform motion and completes one circulation in a year; and, finally, d is imagined as fixed absolutely in space and serves as the only point of contact between circles a and b. Therefore,

44. As we have already seen (n. 41, above) in the *Ad pauca respicientes*, when a group of mobiles reverse their relative order and positions, the occurrence of one and the same aspect before and after conjunction was called by Oresme an "improperly similar" disposition. Proposition 21 of the *De commensurabilitate* is basically all that remains of Oresme's preoccupation with such relationships in the *Ad pauca respicientes*. Even the earlier terminology is abandoned.

45. Although he confines his discussion to two simultaneous motions, Oresme says that three or more motions could be treated if they were taken two at a time, as was done earlier for the simultaneous motions of three bodies (I.784–87).

any point A_n will become the first point of Cancer (i.e., the summer solstice) when it is in d simultaneously with B, the center of the sun.[46]

At the outset, assume that B and some particular point on the tropic of Cancer—say, A_1—are in d simultaneously.[47] Since the motions of circles a and b are assumed to be commensurable, their ratio of velocities is rational. If V_a and V_b represent the velocities of circles a and b, respectively, let $V_a : V_b = n : 1$, where n is an integer so that $n : 1$ is a multiple ratio (I.756). Since A_1 is carried around by circle a and B by circle b, it follows that when a makes n circulations and b one circulation they will again meet in d. Indeed, they can only meet in point d. For example (I.757–60), if circle a should complete 100 circulations in the time that b makes one circulation, points A_1 and B will meet in d. Thus, of all the points on the tropic of Cancer, B will meet only A_1.

However, if the ratio of velocities of the circles is not exactly a multiple ratio but includes also some fractional part of a circulation, the denominator of the fraction will indicate the exact number of points of circle a that B could meet in d (I.761–63). Thus if $V_a : V_b = P\frac{m}{n} : 1$, n will represent the total number of points on circle a that will meet B in d. In the mixed fraction, P is the integral number of circulations that circle a will complete in some given time, and m/n represents an additional fractional part of a circulation. Since the n points on circle a will divide it into equal parts, we need only determine the distance separating two successive meetings in d in order to arrange the sequence of points on circle a through which the sun can enter the first point of Cancer.[48] For example (I.767–70), should circle a complete $100\frac{1}{4}$ circulations to 1 for b, there would be four equidistant points on a that would meet B in d at equal intervals. If A_1 and

---

46. Despite the fact that Oresme speaks as if A, the first point of Cancer, and B, the center of the sun, are in motion on stationary circles, the context makes it obvious that it is the circle of the tropic of Cancer that must be assumed to be in motion; otherwise, only the single point of contact between the circles could serve as the first point of Cancer. But Oresme's purpose is to determine the number of different equidistant points on the tropic of Cancer that will function as the first point of Cancer for different ratios of rational velocities. For this reason, I have distinguished the circle of the tropic of Cancer, circle a, from the points on that tropic —namely, A_n—that can meet B in d. For consistency, B is also treated as if it were a stationary point carried around by circle b, the ecliptic.

47. Oresme does not call this a conjunction. The use of that term is confined to a meeting of two separate mobiles, or bodies, in a single point, or line, but is not applied to the meeting of two mathematical points each of which represents a component of a single motion. Except for a minor lapse in Part II, Proposition 12 (II.332–33), which is concerned with a single motion analyzed into simultaneous incommensurable components and is the counterpart of Part I, Proposition 22, Oresme seems to have consistently and quite consciously avoided extending the term *conjunction* to this special context.

48. In this proposition, Oresme does not actually determine the order or sequence of these points, but says only that it can be done (I.775–76). Presumably he would have used for this purpose Part I, Proposition 10.

B are initially in d, then after $100\frac{1}{4}$ circulations A_2 will meet B in d; after another $100\frac{1}{4}$ circulations of a, A_3 will meet B in d; and, finally, A_4 will arrive in d simultaneously with B after the next $100\frac{1}{4}$ circulations of circle a. This will be followed, once again, by the meeting of A_1 and B in d, after which the cycle will repeat; and so on ad infinitum. If, however, circle a completes $100\frac{2}{5}$ circulations to every 1 for circle b (I.770–74), there will be five equidistant points on a that will meet B in d, but the order of the points meeting B will not be A_1, A_2, A_3, A_4, and A_5, where each point is separated from its immediate neighbors by $\frac{1}{5}$ of the circle. For if, at the beginning, A_1 and B are together in d, after the next $100\frac{2}{5}$ circulations of circle a, point A_3—not A_2—will meet B in d; and after the next series of circulations A_5 will meet B; then A_2 and finally A_4. The order of points is, therefore, A_1, A_3, A_5, A_2, and A_4, after which the cycle is repeated ad infinitum.

On the basis of Proposition 22, Oresme concludes (I.775–83) that if the solar year were measured by an integral number of days, the sun could enter the first point of Cancer on one meridian only. If, however, the year contained exactly $365\frac{1}{4}$ days, or $365\frac{3}{4}$ days, only four equidistant points on the tropic of Cancer could serve as the first point of Cancer. In the course of four years, the sun would have entered the first point of Cancer in each of the four points and the cycle would then repeat ad infinitum. Similar applications could be made to the moon and planets with respect to their entry into any point of the zodiac.

Another interesting consequence of the sun's two simultaneous motions is the nature of the path it would trace as it moved between the two tropics, or solstices (I.793–813). Since the diurnal and annual motions of the sun are oppositely directed, its path on any given day will not be a circle but will form part of a long, finite spiral line that stretches on the celestial sphere from one tropic to the other.[49] For each day of the year there is a corresponding spiral line and the

---

49. That the oppositely directed motions of the planets produce a spiral line was mentioned by a number of authors before Oresme. Perhaps the earliest statement of this appeared in the *Timaeus* 39A–B, where Plato says: "For the movement of the Same, which gives all their circles a spiral twist because they have two distinct forward motions in opposite senses, made the body which departs most slowly from itself—the swiftest of all movements—appear as keeping pace with it most slowly."—Cornford, trans., *Plato's Cosmology*, p. 112; see also p. 114 for discussion.

The "two distinct forward motions in opposite senses" are the motion of the circle of the Same (i.e., the celestial equator which undergoes a daily rotation from east to west, carrying all the planets with it) and the proper motion of each planet on the circle of the Different or Other (i.e., the ecliptic or, more properly, the zodiacal band, on which each planet moves in its period of revolution from west to east; see also *Timaeus* 36C–D).

Plato's meaning and the generation of the resultant spiral are admirably described by Heath as follows: "Suppose a planet to be at a certain moment at the point F. It is carried by the motion of the Same about the axis GH, round the circle $FAEB$. At the same time it has its own motion along the circle $FDEC$ [i.e., the ecliptic]. After 24 hours ac-

pattern which emerges at the end of every year is similar to a fabric woven in a pattern of crisscrossing lines. Thus in Fig. 1 (see n. 49), a long spiral will be generated as the sun moves from point *D*, a point on the tropic of Cancer, to *C*, a point on the tropic of Capricorn. But during the second half of any year, when the sun returns from point *C* to *D*, each daily spiral will intersect a spiral made previously as the sun moves from *D* to *C*. The result is a crisscross pattern that will be exactly retraced ad infinitum (I.811–12), since the sun's two motions are commensurable. Now if we take the diameters of the circles of the tropic of Cancer (line *KD* in Fig. 1), the tropic of Capricorn (line *CL*), the celestial equator (line *AB*), and the ecliptic or zodiac (line *CD*), and then join points *KC* and *DL* by straight lines, we have a right-angled quadrilateral figure. This figure, says Oresme, can be squeezed between the two tropics (I.808–9) and serves as a frame for the crisscrossing spiral patterns.[50] But because this quadrilateral

(Note 49 continued)

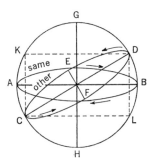

Fig. 1

cordingly it is not at the point *F* on the latter circle, but at a point some way from *F* on the arc *FD*. Similarly after the next 24 hours, it is at a point on *FD* further from *F*; and so on. Hence its complete motion is not in a circle on the sphere about *GH* as diameter but in a spiral described on it. After the planet has reached the point on the zodiac (at *D*) furthest from the equator it begins to approach the equator again, then crosses it, and then gets further away from it on the other side, until it reaches the point on the zodiac furthest from the equator on that side (at *C*). Consequently the spiral is included between the two small circles of the sphere which have *KD*, *CL* as diameters."—*Aristarchus of Samos*, p. 169; the figure is Heath's and

appears on p. 160. The brackets are mine.

Plato's text was available to Oresme in Chalcidius's translation of, and commentary on, the *Timaeus* (see *Timaeus a Calcidio translatus commentarioque instructus* (ed. J. H. Waszink, p. 31.16–20 for text and pp. 160.28–161.21 for the commentary). Others who mention this celestial spiral in one way or another are Theon of Alexandria, Averroes, al-Bitruji, Albertus Magnus, and Sacrobosco. For references to all but Sacrobosco see Francis J. Carmody, ed., *al-Bitruji: "De motibus celorum"* (Berkeley and Los Angeles, 1952), pp. 52–54. Sacrobosco mentions it briefly in Chapter 3 of his *Tractatus de spera* (see Thorndike, ed. and trans., *Sphere of Sacrobosco*, pp. 101 [Latin] and 133 [English]). Since a specific citation by Oresme is lacking, the source, or sources, of his knowledge of celestial spiral motion must be left unidentified, although Chalcidius's version of the *Timaeus* would be the most plausible candidate.

50. In *Timaeus* 36C, Plato also seems to have thought of the oppositely directed motions as occurring within a rectangular frame fitted between the two tropics. See Cornford's translation and interpretation in *Plato's Cosmology*, p. 73. Chalcidius translated this passage (*Timaeus a Calcidio translatus* [ed. Waszink], p. 28.14–20) but provided no commentary.

frame lies in the interior of the sphere on whose surface the sun moves, Oresme observes (I.810–11) that the sun will never actually move anywhere within the area of the quadrilateral figure itself.

In concluding Proposition 22 (I.814–24), Oresme designates as a Great Year the period of the sun's double motion—i.e., the interval between successive meetings in d of B, the center of the sun, and some point A_n on the tropic of Cancer. Thus, just as two or more distinct mobiles with commensurable motions have a fixed period of revolution, or Great Year, in which all motions are repeated identically, so also can a single body have a Great Year if it has two or more simultaneous and commensurable motions (I.814–18). However, if the world were assumed to be eternal and to have no beginning in time, and if the motions of all celestial bodies were commensurable, there would be an infinite sequence of Great Years (I.821–24), since every point, or position, on the celestial sphere could arbitrarily serve as a starting and terminal point. Oresme's thought here seems based on the assumption that if the world had no beginning in time, we would, consequently, be ignorant of the unique point, or configuration, from which all the motions commenced initially and to which a single unique Great Year could be referred. Without this knowledge, all points, or configurations, are equally privileged to serve as reference points.[51]

In the next two propositions, 23 and 24, Oresme focuses on the relationships between the angular distances that separate mobiles from a given point of conjunction (and from each other). He shows that these angular separations are always commensurable when the velocities are commensurable.[52] In Proposition 23, the three mobiles, A, B, and C, depart from a simultaneous conjunction in point d, with $V_A > V_B > V_C$. At any instant thereafter, says Oresme, the arcal

51. Oresme may have been influenced here by a statement of Macrobius, who, in his *Commentary on the Dream of Scipio* 2. 11. 13, observes: "Then, just as we assume a solar year to be not only the period from the calends of January to the calends of January but from the second day of January to the second day of January or from any day of any month to the same day in the following year, so the world-year begins when anyone chooses to have it begin...."—Stahl, trans., p. 221.

Remarking on this very passage in Macrobius, Cornford (*Plato's Cosmology*, p. 117) writes: "Since the celestial clock was never set going at any moment of time, there was never any original position to serve as start-ing-point." The World Year, Great Year, or Perfect Year, was a widely discussed and accepted doctrine throughout antiquity and the Middle Ages; for references to Plato, Cicero, Chalcidius, and Honorius of Autun, see Stahl, trans., *Macrobius*. The Great Year is discussed more fully below, p. 103 and especially n. 64.

52. In the *Ad pauca respicientes*, pt. 2, props. IX–XII (Grant, *Oresme PPAP*, pp. 412–18; discussion on pp. 105–7), the same type of problem is considered but the initial conditions differ, since the three mobiles do not have mutually commensurable velocities. There the velocity of A is incommensurable to the velocities of B and C while the latter two are commensurable.

distances dA, dB, and dC (where $dA > dB > dC$) are commensurable (see Fig. 2). Although Oresme offers no explanation, this is obvious, since the velocities are assumed to be commensurable and distance traversed is proportional to velocity.

Expressed as angles, since $doA > doB > doC$, it follows that $\angle doA - \angle doB = \angle AoB$; $\angle doB - \angle doC = \angle BoC$; and $\angle doA - \angle doC = \angle AoC$. In the second part of the proposition (I.832–35), Oresme declares, without offering details, that

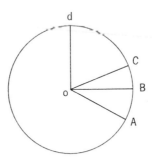

Fig. 2
This figure does not appear in the manuscripts, but has been added for convenience.

by Euclid 10.9 it can be demonstrated that the angles separating the mobiles themselves—namely, angles AoB, BoC, and AoC—are also mutually commensurable. By Fig. 2, we see that $\angle doA = \angle doC + \angle AoC$. It has already been shown that angles doA and doC are commensurable; therefore, by Campanus's commentary on Euclid 10.9, doA, the whole, must also be commensurable to $\angle AoC$, the remainder,[53] and by 10.9 itself angles doC and AoC are commensurable. Similarly, $\angle doB = \angle doC + \angle CoB$, and since we know that angles doB and doC are commensurable, it follows from Campanus's commentary that doB, the whole, must also be commensurable to $\angle CoB$, the remaining angle; then, by 10.9, angles doC and CoB are also commensurable. Since angles AoC

53. The enunciation of Euclid 10.9 reads: "If there are commensurable quantities, the whole composed of them will be commensurable to each of them. Also, if the whole is commensurable to each of them, both of them will be commensurable." ("Si fuerint duae quantitates communicantes, totum quoque ex eis confectum utrique earum erit communicans. Si vero fuerit totum utrique commensurabile, erunt ambae commensurabiles."—*Euc.-Campanus*, p. 251.) After repeating the substance of the enunciation in his

commentary, Campanus adds the following statement on which Oresme's claim is based: "Furthermore, if the whole composed of these [two magnitudes] should be commensurable to one of them, I say that it will also be commensurable to the other magnitude; and these two magnitudes will also be mutually commensurable." ("Adhuc quoque si totum ex eis compositum uni earum communicet, dico quod communicabit alteri; et ipsae similiter inter se etiam communicabunt."—Ibid., p. 252.)

and *CoB* are both commensurable to *doC*, it follows by Euclid 10.8[54] that they are commensurable to each other. Now $\angle AoC = \angle CoB + \angle BoA$, and since angles *AoC* and *CoB* are commensurable, it follows, once again by Campanus's commentary on 10.9, that $\angle AoC$, the whole, is also commensurable to $\angle BoA$, the remainder. Therefore, angles *AoC*, *CoB*, and *BoA* are commensurable.

In Proposition 24, the three mobiles are not separated from each other by independent angles, but rather two of them are assumed to be in conjunction,

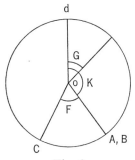

Fig. 3

This figure does not appear in the manuscripts, but has been added for convenience.

while the third is separated from them by an angle. This angle of separation, says Oresme (I.837–39), is always commensurable to a right angle or to the whole circle—i.e., to four right angles.

Initially, mobiles *A*, *B*, and *C* are in conjunction in some point—say, *d*—and at some other time only *A* and *B* are in conjunction while *C* is elsewhere on the circle (see Fig. 3). On the assumption that the velocities are mutually commensurable, Oresme now demonstrates that the angle separating *C* from the point of conjunction of *A* and *B* must be commensurable to a right angle, or to the whole circle.

By Proposition 10, the places where *A* and *B* conjunct during every period of revolution divide the circle, or circles, into equal segments with equal central angles, and each of these segments, or central angles, is commensurable to the whole circle (see I.366–74). In Fig. 3, let *G* be any one of these angles which, when multiplied a certain number of times, equals the whole circle—i.e., four right angles. Obviously, if angle *G* is commensurable to the whole circle, or four right angles, it is also commensurable to one right angle. Consequently, when-

---

54. This proposition is not actually cited by Oresme but reads as follows: "If there are two magnitudes commensurable to another magnitude, it is necessary that they also be commensurable to each other." ("Si fuerint duae quantitates uni quantitati communicantes, ipsas quoque invicem commensurabiles esse necesse est."—*Euc.-Campanus*, p. 251.)

ever *A* and *B* conjunct elsewhere than in *d*, the place where all three mobiles conjuncted, they will be separated from *d* by an angle that is commensurable to a right angle. Let this angle be *K*, where $K = n \cdot G$ and *n* is an integer such that $n \cdot G$ is less than, but commensurable to, four right angles.[55]

Angles *doC* and *do(A,B)*, where the latter equals *K*, represent the angles separating the three mobiles from *d*. Let $\angle doC - \angle K = \angle F$, where *F* is the angle separating *C* from the conjunction of *A* and *B*. From Proposition 23, we know that angles *doC* and *do(A,B)*, or *K*, are commensurable; and by Campanus's commentary on Euclid 10.9 (see above, n. 53), we see that *doC* must also be commensurable to angle *F*, which with *K* comprises *doC*. Since the whole angle, *doC*, is commensurable to each of its parts—namely, angles *K* and *F*—it follows by Euclid 10.9 that *K* and *F* are commensurable to each other. But since *F* is commensurable to *K*, and, as we have already seen, a right angle is also commensurable to *K*, it follows by Euclid 10.8[56] that *F* is also commensurable to a right angle, and, of course, to the whole circle. And this is what was to be demonstrated.

The final proposition of Part I, Proposition 25, contains an application of Propositions 2 and 3 to the results obtained in some of the later propositions of this first part. On the basis of Propositions 10, 11, 17, 19, and 21, we know that two or more bodies, or mobiles, moving with commensurable velocities will have a finite and fixed number of dispositions and points of conjunction. Indeed, the points of conjunction can be represented on the circle, or circles, on which the motions occur by dividing the circles into as many equal parts as there are points of conjunction. If we subsequently divide the circle into equal parts according to some second mode of division, or proportionality, whereby the second group of points is so divided that none of them coincides with any point in the first division (except the first point, since the divisions are started from a common point), the conjunctions of the first division cannot be represented exactly in the second division, or proportionality. If, for example (I.871–76), the sun could enter the first point of Cancer on only seven equidistant points on the tropic of Cancer (see Proposition 22), these points could not be located exactly by the common astronomical tables which use the sexagesimal system. Indeed, if the length of the year were calculated from two successive entries into the first point of Cancer, the common tables could not precisely determine

---

55. If $n \cdot G$ were to equal four right angles, the conjunction of *A* and *B* would be located in *d*, the place where all three mobiles conjuncted. Indeed, even if *C* were not in *d*, it is obvious that Oresme did not wish angle *K* to be such that it would locate *A* and *B* in *d*, since he explicitly assumes (I.845–47) that the function of angle *K* is to separate the point of conjunction of *A* and *B* from point *d*.

56. Although Oresme omits mention of Euclid and fails to identify the proposition of the tenth book (I.851), it is obvious that 10.8 is intended. For the enunciation of this proposition see above, p. 35, n. 54.

this. The same may be said for conjunctions of the sun and moon, or of any planets (I.876–79), for if they conjuncted in only 7, or 19, or 23 equidistant fixed points, these could not be expressed exactly in the sexagesimal system. This is explained by the fact that if these circles were divided into 7^n, 19^n, or 23^n successive equal parts, where $n = 1, 2, 3, \ldots$, no points in any of these divisions would be common to any of the points resulting from a sexagesimal division by 60^n, since the terms following immediately after 1 in the proportionalities based on 7, 19, and 23, are all prime to 60 (see Proposition 2).

These differences, however, pose no problem for the astronomer who does not expect to obtain exact punctual knowledge of conjunctions, or any other dispositions. He is satisfied if the margin of error is not visually detectable through the use of an instrument (I.879–82).[57]

Part II

In each of the 12 propositions of Part II, some or all of the motions are assumed to be incommensurable, and in the first five propositions two bodies only are in motion simultaneously.

It is assumed in Proposition 1 that two bodies, A and B, have moved with incommensurable velocities and are presently in conjunction in point d. Oresme demonstrates that (1) not only will A and B never again conjunct in d, but also that (2) they had never before conjuncted there.[58] In order to conjunct again in d each body must complete an integral number of circulations during the same time interval. Now if A makes e and B, g circulations, respectively, it follows that $S_A : S_B = e : g$, where S is distance. Furthermore, when $T_A = T_B$, where T is time, the velocities are related as the distances so that $V_A : V_B = S_A : S_B$ and consequently $V_A : V_B = e : g$. Since e and g represent integral numbers of circulations, the velocities must be commensurable, which is contrary to the assumption that they are incommensurable. Thus by an indirect proof Oresme

---

57. Cf. I.45–49. The approximate nature of astronomical observations is also emphasized in the concluding lines of the *Ad pauca respicientes*, pt. 2, prop. XX (Grant, *Oresme PPAP*, p. 428, lines 263–70). For a similar statement by Johannes de Muris, see App. 1, chap. 24, lines 45–47.

58. In Part I, Proposition 4, the motions of bodies A and B were also calculated from conjunction in a particular point. There, on the assumption of mutually commensurable

speeds, it was demonstrated that A and B have conjuncted, and will conjunct in the future, repeatedly and regularly in that same point. In Part II, Proposition 1, Oresme proposes the same initial conditions, but now the speeds of A and B are assumed to be mutually incommensurable and an opposite conclusion is derived, namely, that A and B have never before conjuncted, and will never again conjunct, in that same point.

shows that *A* and *B* cannot again conjunct in *d*. Although Oresme merely states without comment (II.18–19) the second case distinguished above, it is clear that only if their velocities were commensurable could *A* and *B* have conjuncted previously in *d*.[59] The same type of argument could be extended to show the

59. A similar indirect proof appears in Oresme's *Ad pauca respicientes*, Part 1, Proposition IV, where, however, *A* and *B* move with equal curvilinear speeds on circles that are incommensurable (Grant, *Oresme PPAP*, p. 394, line 136), thereby producing incommensurable central angles in equal times (ibid., p. 394, lines 125–27 and pp. 93–94 and 432). In Part II, Proposition 1 of the *De commensurabilitate*, nothing is said about the manner in which the incommensurability is produced, but from an earlier remark (I.30–31) it seems apparent that it must arise from the fact that in equal times the bodies have traversed incommensurable central angles; nothing is said about the circles themselves, although we know that Oresme assumed them to be concentric (I.60–62), unless stated otherwise. See below, the Commentary, n. 2 for I.24–31.

In *Le Livre du ciel et du monde*, bk. 1, chap. 29, 46a–46d (Menut, *Oresme du ciel*, pp. 200–203), Oresme specifically cites Part II, Proposition 1 of the *De commensurabilitate* as the basis of a counterargument to Aristotle's declaration in the *De caelo* that whatever had a beginning must eventually cease to be (actually, *De caelo* 1. 11. 281a26–27 is but the initial point of departure, for Oresme soon argues against the whole of Aristotle's position in Book 1, Chapter 12, that [1] whatever had a beginning must cease to be, and [2] whatever will ultimately perish must have had a beginning; in what is quoted below, Oresme is arguing only against the first of these assertions).

"(46a) Now, I should like to show next that it is possible both in fact and in uncontradictable theory that some motion has a beginning and lasts forever. First, with regard to circular motion: I assume that a wheel of any kind of material is like the wheel of a clock [see Fig. 4 reproduced here

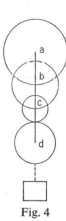

Fig. 4

from Menut, *Oresme du ciel*, p. 203]; this we will designate as *a* and will call its center *a*. *b* (46b) Let us next set in *a* another wheel / smaller than the first, in the manner of an epicycle, with its center labeled *b*. Let us add a third wheel set on the second, having its center on the circumference of the second like the moon on its epicycle, and let us call this third wheel *c*. Now let us place a fourth wheel outside these three so fixed that *c* can touch it, and let its center be marked *d*. Then, I posit that *a* be moved around its center, that *b* be moved with the motion of *a* on which it is set, and, with this, that *b* be moved also with its proper motion around its center, while *c* has no proper motion but is moved with the motion of *a* and *b*. This is the way the moon moves in its epicycle. Now I posit that *d* be so adjusted or controlled by counterweights and otherwise that it is inclined to move but is not moved until touched by *c*; this contact removes the hindrance and *d* begins to move regularly. Something of this sort or similar to it could be performed artificially or by skilled craftsmanship. Next, I posit that the two motions of *a* and *b* // (46c) should be incommensurable, regular, and perpetual. Such perpetual motion could not

impossibility of *A* and *B* entering into opposition twice in the same two points; indeed, no aspect will repeat in exactly the same points (II.19–21).

Oresme concludes (II.21–22) that the period of revolution (for Oresme's definition, see I.43–44) of *A* and *B* must be infinite; or perhaps it is more preferable to say that they have no period at all.

---

be created by any skill if the idea of certain persons, who at one time have conceived and attempted to make something called the *rota viva*, were not true. At this present moment, I posit that the wheel *c* touches wheel *d*, and I say that it is impossible that it should have touched it previously or should touch it again, for it can touch it only when the centers *b*, *c*, and *d* are in conjunction exactly in one line; and this cannot have happened before nor again in future time, as appears from geometrical demonstration in the first conclusion of the second part of the treatise *On the Commensurability and Incommensurability of the Motions of the Heavens* [*De commensurabilitate vel incommensurabilitate motuum celi*; note that Menut's translation of this title differs from mine]. From this it follows necessarily that wheel *d* would now begin to move and will never stop; and, although such a series of events cannot occur in nature, nor be shown by material art or in destructible matter, nor endure so long [as forever], nevertheless, it contains or implies no contradiction whatever, nor is it within its own frame of reference incongruous to reason, but it is possible if we grant the nature of the motions. All / (46d) the incongruity arises by reason of the material or from something outside the frame of reference."

While the centers of wheels *a*, *b*, and *d* would conjunct every time that *a* completed a rotation (this occurs because *a* carries *b*), wheel *c* could never again join them in conjunction, since wheel *b*, to which it is fixed, will never complete its own rotation simultaneously with that of *a*, since their motions are assumed to be incommensurable.

A proposition analogous to the one just described appears in Oresme's *Ad pauca respicientes*, pt. 1, prop. V (Grant, *Oresme*

PPAP, pp. 396–98 and 433–34). Here, instead of wheels we have two mobiles moving with incommensurable velocities on two intersecting circles. Should the mobiles meet in one of these points of intersection, or nodal points, they will never conjunct there again.

A similar proposition—actually a corollary—is found in Question 7 of Oresme's *Quaestiones super geometriam Euclidis* (ed. H. L. L. Busard). Here again two mobiles move on separate but intersecting circles, but this proposition differs from the version in the *Ad pauca respicientes* by assuming that the mobiles move with equal, not incommensurable, speeds, and the intersecting circles are now incommensurable—they are related as the diagonal of a square to its side. As in the *Ad pauca*, Oresme demonstrates that if the mobiles conjunct in one of the points of intersection, they will never conjunct there again. See my quotation and discussion of the Latin text in Grant, *Oresme PPAP*, p. 435 and n. 9 (my quotation is an emended and corrected version incorporating John E. Murdoch's corrections to H. L. L. Busard's edition of Oresme's *Quaestiones super geometriam Euclidis*, fasc. 1, p. 18. Murdoch's corrections and emendations are given in his review of Busard's edition in *Scripta Mathematica*, vol. 27 (1964), pp. 67–91; the corrections to Question 7 appear on p. 81).

Another proposition similar to those just cited from the *Ad pauca respicientes* and *Quaestiones super geometriam Euclidis* is found in an anonymous *Questio* called *De proportione dyametri quadrati ad costam ejusdem*. Although the editor of this brief tract ascribed it to Albert of Saxony (see Heinrich Suter, "Die Quaestio *De proportione dyametri quadrati ad costam ejusdem* des Alber-

Should bodies A and B conjunct in some point other than d—say, e (I have specified this point for convenience)—Oresme demonstrates in Proposition 2 that on the assumption of incommensurable speeds the angular distance separating d and e must be incommensurable to the whole circle (i.e., to 360°).[60]

(*Note 59 continued*)
tus de Saxonia," *Zeitschrift für Mathematik und Physik*, vol. 32 [1887], pp. 41–56), it may have been written by Oresme himself (see Grant, *Oresme PPAP*, pp. 77–78, n. 101 and below, p. 63, n. 97 of this chapter), or the substance of it taken from one or more of his works. The relevant proposition in the *Questio* is more closely related to the similar proposition in the *Quaestiones super geometriam Euclidis* than to the related proposition of the *Ad pauca respicientes*. In the *Quaestiones super geometriam Euclidis*, the circumferences of the intersecting circles are not only incommensurable but also related as the diagonal of a square to its side. Because the circumferences are incommensurable, it is shown that two mobiles, E and F, initially in conjunction in d, one of the two points of intersection, would never again meet there, although they could meet in the other point of intersection. ("Item dico secundo, quod si sint duo circuli A et B intersecantes se in puncto d, et habeat se circumferentia unius ad circumferentiam alterius sicut dyameter quadrati et costa ejusdem, et essent jam duo mobilia E et F conjuncta in puncto intersectionis d et moveatur E super circumferentiam A, et F super B, et equaliter moveantur; dico quod si ista in eternum moverentur, nunquam amplius in puncto d conjungerentur, quamvis bene in alio;..." Suter, "Die Quaestio," p. 49.) The author says later that it does not matter whether circles A and B move equally or unequally (the circles are assumed to move and carry their otherwise stationary mobiles), for in equal times they would describe incommensurable angles around the center and E and F would never again meet in d. This would, of course, follow solely from the incommensurability of the two circumferences. ("3° dico, quod qualitercunque moveantur A et B

sive equaliter sive inequaliter, dum tamen in temporibus equalibus describerent angulos incommensurabiles circa eorum centra, nunquam in eternum se invenirent in isto puncto, si jam essent in ipso, nec unquam fuerunt in ipso, si ab eterno fuissent mota isto modo...."—Ibid.)

The propositions cited above from the *Ad pauca respicientes*, *Quaestiones super geometriam Euclidis*, and the anonymous *Questio* share the same conclusion as Part II, Proposition 1 of the *De commensurabilitate*—i.e., all assert that two mobiles moving incommensurably will never meet twice in the same point. In the first three mentioned treatises, however, the two mobiles move on intersecting circumferences (in two of them the circumferences are assumed to be mutually incommensurable; in the *Ad pauca* incommensurability is asserted only for the motions of the mobiles), whereas in the related proposition of the *De commensurabilitate* two mobiles move with incommensurable angular velocities on concentric—not intersecting—circles. Nowhere in the *De commensurabilitate* does Oresme consider a case involving two intersecting circles, but he does include two propositions (pt. I, prop. 22 and pt. II, prop. 12) where two distinct circular components of a single motion are in contact at only one point. Cf. bk. IV, tract 1, chap. 14, of the *Quadripartitum numerorum* of Johannes de Muris, below in App. 1 and see the discussion of it on pp. 89–90.

60. Although there is no direct counterpart to this proposition in the *Ad pauca respicientes*, Oresme did consider, in Part 2, Proposition IX, the case of three mobiles that are assumed to depart from some initial unique conjunction but could never thereafter traverse angles that are mutually commensurable and also commensurable to a right angle. Of the three mobiles, one moves

If the angle of separation were commensurable to the whole circle, it follows that, after departure from conjunction in *d*, *A* and *B* would each have traversed a certain integral number of circulations with respect to *d* plus some part of a circulation. Let us call this part angle *doe*, where *o* is the center of the concentric circles.[61] Now if angle *doe* is commensurable to the whole circle, or 360°, and is added to the number of circulations made by *A* and *B*, respectively (say, $h \cdot 360°$ for *A* and $k \cdot 360°$ for *B*, where *h* and *k* are integers), the total distances traversed by *A* and *B* must be commensurable, i.e., $(h \cdot 360° + \angle doe) : (k \cdot 360° + \angle doe) = p:q$, where *p:q* is a rational ratio; Oresme justifies all this by appeal to Euclid 10. 8 and 9.[62] But if their total distances are commensurable so must their velocities be commensurable (II.33–35), which is contrary to the initial assumption that their speeds are incommensurable. Thus, when the velocities of two bodies are incommensurable, it follows that the distance or angular separation between any two successive conjunctions is incommensurable to the whole circle, or 360° (II.45–46).

A similar argument may be formulated for any other aspect (II.40). For example (II.40–44), if *A* were in *d* and *B* in opposition at point *e* (this and the following points are specified for convenience only and do not appear in the text), then, should another opposition occur at points *f* and *g*, the equal vertical angles resulting from the intersection of the diameters (see Fig. 5) would be incommensurable to the whole circle, or 360°.

Where the velocities are assumed to be incommensurable, it also follows—and this is Oresme's final point in the proposition—that the times between two such successive conjunctions (or between successive occurrences of any aspect) of *A* and *B* must also be incommensurable (II.47–49). This is obvious because the times are inversely proportional to the velocities.

Extending his consideration of the relationship of parts of circles, representing

---

commensurably with respect to the second and incommensurably with respect to the third. See Grant, *Oresme PPAP*, pp. 412–14 and p. 105.

61. Angle *doe* is introduced here for convenience only. Oresme does not designate any angles by letters in this proposition but speaks only of a "part of a circle" or "part of a circulation" (see II.37, 37–38, and 43–44).

62. The general appeal to Euclid 10. 8 and 9 (for their enunciation, see pp. 35, 34, nn. 54 and 53, respectively) is made in II.26–28, where Oresme says that from these two prop-

ositions it follows "that if something is added to each of two commensurable quantities, and is commensurable to each of them, the wholes will be commensurable." As Oresme indicates, this is not made explicit by Euclid (or Campanus) but follows directly from the two propositions cited. By hypothesis $\angle doe$ is assumed to be commensurable to angles $h \cdot 360°$ and $k \cdot 360°$, and the latter two angles are assumed to be commensurable to each other, so that their respective sums must be commensurable— i.e., $h \cdot 360° + \angle doe$ is commensurable to $k \cdot 360° + \angle doe$.

distances of separation, to whole circles, Oresme considers next, in Proposition 3, the distances separating the two mobiles as measured from a given point. Once again he assumes that bodies *A* and *B* have departed from an initial and unique conjunction in *d* (II.52) and then demonstrates that when *A* or *B* returns to *d* it will be separated from the other body by an angular distance that is incommensurable to a whole circle. The argument is quite similar to that of the preceding proposition.

For example, whenever *A* returns to *d* it will have traversed an integral number of circulations. However, *B*, since it can never again conjunct with *A* in *d* (pt. II, prop. 1), will have traversed *n* circulations (where *n* may be zero or any

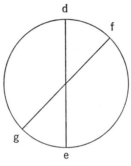

Fig. 5

This figure does not appear in the manuscripts, but has been added for convenience.

integer) plus some part of the circle. As Oresme shows, this part must be incommensurable to the whole circle, or 360°. If not, it will be commensurable to a whole circle and, when added to the *n* total circulations of *B*, that body will have traversed a total distance that is commensurable to the total distance traversed by *A*, since *A* completed an integral number of circulations. But if, in a given time, their total distances are commensurable, so must their velocities be commensurable, which is contrary to the initial assumption of their incommensurability. The same result is obtained when *B* returns to *d* and *A* is elsewhere. Therefore, whenever *A* or *B* is in *d* they are separated by a central angle that is incommensurable to a whole circle and to a right angle.[63] Consequently (II.

63. *The Ad pauca respicientes* does not include this proposition, but in considering the motion of three mobiles in Part 2, Proposition X, Oresme shows that "Whenever two such mobiles are in conjunction, the third mobile would produce with them an angle in the center which is incommensurable to a right angle" (Grant, *Oresme PPAP*, p. 415; Latin text, p. 414; for the confusion in this proposition, see ibid., p. 106). Other than the fact that the angle separating the third mobile from a point of conjunction is shown to be incommensurable to a right angle—and therefore also incommensurable to the whole

60–63), they could never be related in the aspects called sextile (separated by an angle of 60°), quartile (separated by 90°), trinal (separated by 120°), or opposition (separated by 180°); nor can they be related in any aspect whose degree of separation is commensurable to any of these.

On the assumption of incommensurable velocities, Oresme shows in Proposition 4 that there is no part, or sector, of a circle so small that two mobiles, *A* and *B*, could not have conjuncted there in the past, and in which they cannot conjunct in the future.

Let mobiles *A* and *B* depart from conjunction in *d* and let point *e* represent their very next point of conjunction (see Fig. 6, below). Although the motions of

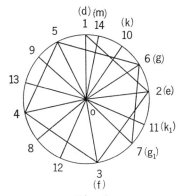

Fig. 6

This figure is placed here as an aid to the reader and does not appear in the manuscripts.

A and *B* are incommensurable, each moves with a uniform speed so that between any two successive conjunctions equal time intervals elapse and equal distances are traversed (II.71–76). Consequently, the arcal distances separating the points of any two consecutive conjunctions are equal. In Fig. 6, let arcs *de* and *ef* represent the distances that separate the first and second and second and third points, respectively. Since the distance that separates any two successive conjunctions will be equal (II.72–76), arc *de* = arc *ef*. Let the conjunctions be numbered in order of occurrence beginning with 1 at *d*, 2 at *e*, 3 at *f*, etc., so that the arcal distance separating any two successive numbers—i.e., conjunctions—is always equal to arc *de* (e.g., arc 5 → 6 equals arc *de*).

Since the velocities of *A* and *B* are incommensurable, it follows by Part II, Proposition 2, that arc *de* will be incommensurable to the whole circle. As *A* and *B* pass through successive conjunctions, a sequence of points will be laid off

circle, or four right angles (although this is not stated)—there is almost no substantive relationship between the two propositions.

around the circle beginning with d (in Fig. 6, these have been numbered in order of occurrence). Since arc de is incommensurable to the whole circle and because the distance between any two successive conjunctions (i.e., between any two successively numbered points in the figure) is equal to arc de, none of these arcs can terminate in d,[64] where a unique conjunction has already occurred. Consequently, when as many arcs have been marked off as can be accommodated by the circle in the first sequence of conjunctions, some particular arc—say, $5 \rightarrow 6$ (see Fig. 6)—will overlap the circle and cross beyond d into sector de where it will terminate in some point g (II.77–79), which is implicitly assumed to lie more than halfway through arc de, since arc $dg >$ arc ge (II.80–82).

Proceeding now from g, the next, or seventh conjunction, must occur in arc ef, or $2 \rightarrow 3$, at some point g_1 such that $gg_1 =$ arc de, arc $eg_1 =$ arc dg, and arc $g_1f =$ arc ge. Thus arcs de and ef (or $1 \rightarrow 2$ and $2 \rightarrow 3$) are subdivided identically by these successive conjunctions. The remaining arcs ($3 \rightarrow 4$ and $4 \rightarrow 5$) will subsequently be divided in the same manner by conjunctions 8 and 9, respectively, so that at this juncture Oresme could say that no arc greater than dg has been left undivided on the circle (II.80–82). But the next, or tenth point of conjunction, occurs once again within arc de, and, more precisely, must fall within arc dg, say, at point k.[65] The same process is repeated again. Conjunction 11 must fall at some point, say, k_1, such that arc $kk_1 =$ arc de, $ek_1 = dk$, and $k_1f = ke$; the remaining arcs will be divided in the same manner by conjunctions 12 and 13 so that arc dk is the largest arc on the circle left undivided by any point of conjunction. With conjunction 14—say, at m—the process will repeat so that, after the next time around, arc dm will represent the largest arc undivided by any point of conjunction.

Since this process can be continued ad infinitum, the circle will be divided into smaller and smaller arcs as the number of conjunctions becomes greater and greater. However small the arcs become, we can imagine them being made even smaller,[66] since A and B can always conjunct between any two given points

---

64. In II.76–77, Oresme chose to appeal to Part II, Proposition 3, but this consequence could have been justified equally well by Part II, Proposition 1, where it was demonstrated that with incommensurable velocities no more than one conjunction is possible in one and the same point.

65. This is obvious, for in terms of Fig. 6, arc $9 \rightarrow 10$ must equal arc de, whereas arc $9 \rightarrow 10 +$ arc $10 \rightarrow 6 >$ arc de; and since arc $9 \rightarrow 10 >$ arc $5 \rightarrow 1$, it follows that k must fall within arc dg, or arc $1 \rightarrow 6$.

66. To convey the sense in which the circle

should be conceived as divisible into smaller and smaller arcs that are always capable of further subdivision, Oresme offers an analogy in terms of the superposition of straight lines (II.88–95). He assumes that the side of a square is superposed, or laid off, on its diagonal as many times as possible. Since the side of a square is incommensurable to, and shorter than, its diagonal, it cannot be laid off an integral number of times and eventually a segment of the side will extend beyond the diagonal. If this excess is cut off and then superposed lengthwise as many

of conjunction (II.96–97). Indeed, despite the impossibility of conjuncting twice in the same point, they will, nevertheless, conjunct an infinite number of times in the future where the distance between any two successive conjunctions will always be equal (II.98–101). The same may be said of conjunctions through an infinite past time (II.97–98). And if these points of conjunction were linked in order of occurrence by drawing chords from the first to the second point, and the second to the third, and so on, a figure would result consisting of an infinite number of mutually intersecting equal angles each of which is incommensurable to a right angle (II.101–4; see Fig. 6).[67] Indeed, even between any two given points, however close together they may lie, an infinite number of such angles could be drawn so that no part of the circle, however small, would lack such angles (II.104–9).

That a combination of incommensurable but respectively uniform and regular

---

times as possible, it too will ultimately extend beyond the diagonal. This excess can then be cut and subsequently superposed on the diagonal as many times as possible until another excess results. Obviously, after each superposition of a given length, the excess extending beyond the diagonal diminishes in length so that each successive diminished excess length can be superposed, or laid off, more times than its immediate predecessor. Thus with each successive superposition, the diagonal can be divided into smaller and smaller parts. Since theoretically this process can be extended ad infinitum, the diagonal is potentially divisible into infinitely small lengths as the successive excesses become infinitely small.

67. At the conclusion of this proposition, Oresme demonstrates that each angle formed as a consequence of joining successive points of conjunction is incommensurable to a right angle (II.124–36). In Fig. 6, o is the center of the circle, d the initial point of conjunction of A and B, e is their first point of conjunction after departure from d, and f (Oresme says g, but since he had previously used g in another connection, I have substituted f to correspond with the letters employed in the figure used to represent this proposition), the second point of conjunction. After connecting these points with chords, the angle def

will have been described. This angle, says Oresme, is incommensurable to a right angle.

First, draw lines do and eo which, with line de, form triangle doe. Now $\angle doe + \angle ode + \angle oed = 180°$ and since $\angle doe$ is incommensurable to a right angle (demonstrated in pt. II, prop. 2) and therefore to 180°, it follows that $\angle ode + \angle oed$ must also be incommensurable to 180° and to a right angle. (This is based on Campanus's commentary on Euclid 10. 9, where it is stated that if c, the sum of quantities a and b, were incommensurable to either a or b, it would also be incommensurable to the other, and a and b would be mutually incommensurable. "Idem quoque in contrario, si enim a et b sint incommunicantes, dico quod c utrique earum erit incommunicans; et econverso, si c alteri earum sit incommunicans, erit quoque incommunicans et alteri, et ipsae etiam inter se."—*Euc.-Campanus*, p. 252.) But $\angle ode = \angle oed$ (the triangle is isosceles), so each of them must be incommensurable to a right angle. And because $\angle oed = \angle oef$ (line oe bisects $\angle def$), $\angle oef$ must also be incommensurable to a right angle, as must the sum of $\angle oed + \angle oef$, which is $\angle def$. Therefore, as any angle formed by linking three successive points of conjunction is equal to $\angle def$, it will be incommensurable to a right angle.

motions could produce such remarkable consequences deeply impressed Oresme. He saw that out of this combination there would arise not only regularity, in the form of successive conjunctions occurring at equal time intervals and separated by equal arcal distances, but also total disorder, since conjunctions cannot possibly occur twice in the same point and all the angles described must be incommensurable to the whole circle. These extraordinary consequences, arising from seemingly antithetical concepts, incommensurability and regularity, justify the use of such paradoxical descriptive expressions as "rational irrationality," "regular non-uniformity," "uniform disparity," and "harmonious discord" (II.112–14). Universal order seems to be preserved by the generation of inequality from equality (II.114–15).

Other paradoxical consequences also follow from this combination of motions (II.116–23). For instance, if we assume an eternity of past time, no part of the circle should remain undivided after an infinite number of such successive conjunctions. But if the future alone is considered, it is equally true that there must be an infinite number of points in which no conjunction has yet occurred, since through an infinite future time there will also be an infinite number of conjunctions.[68] How is this possible, though, if conjunctions occurred in every point in the infinite past and no conjunction can occur twice in the same point? And yet, on the assumption of an eternal future in which the same unceasing uniform but incommensurable motions continue, it is clear by Proposition 4 that an infinite number of different points can become loci of future conjunctions. This set of infinite points in which there have been no conjunctions must also have existed in the past. Paradoxically, although each point of the infinite number distributed around the circle should have had conjunctions through an eternal past, thereby exhausting all possible points of conjunction, it is yet the case that at this very moment there exists, and has existed, an infinite number of points in which conjunctions have never occurred but which can occur through an eternal future. Thus Oresme distinguishes two infinite sets of points that have

68. This follows from Part II, Proposition 4, where it was demonstrated that "no sector of the circle will be so small but that at some time in the future the mobiles could not conjunct in some point of it..." (II.96–97). On the paradox of an apparent exhaustion of the points of division of a continuum which at the same time leaves an infinite number of points in which no division has occurred, see below, I.141–43 and above, p. 10. In Part 1, Proposition V of the *Ad pauca respicientes*, Oresme shows that for two bodies moving with incommensurable velocities "There are an infinite number of points" in which they "have conjuncted, and the points in which they will conjunct are [also] infinite [in number]" (see Grant, *Oresme PPAP*, p. 397 [Latin on p. 396, lines 146–47] and discussion on p. 94). The proof depends on the impossibility of any conjunction occurring twice in the same point. Apart from this similarity, there is nothing else in the *Ad pauca* that bears even a remote resemblance to this lengthy, subtle, and intricate proposition of the *De commensurabilitate*.

coexisted through an infinite past and will coexist through an eternal future.[69] The first embraces all points of conjunction that occurred during the eternal past; the second includes all points in which conjunctions did not occur in the infinite past and in which conjunctions could occur in the infinite future. This second group contains all points that are separated from any point in the first group by an arcal distance that is incommensurable to the whole circle (II.121–23). The existence of a second category of points constitutes the paradox, since through an eternal past every point on the circle should have served as the locus for one conjunction with none remaining to accommodate future conjunctions.

In Proposition 5[70] Oresme makes explicit what was implicit in Part II, Proposition 4. If points of conjunction c and d are assumed to be close to one another, it is obvious from Part II, Proposition 4 that bodies A and B will conjunct somewhere between these points. Furthermore, should f be a point of conjunction halfway between c and d, it would follow by Proposition 4 that A and B would eventually conjunct between d and f. As time passed, conjunctions would fall ever closer to d, and so on ad infinitum. Hence, through an eternal past conjunctions have been approaching infinitely close to d; and through an infinite future time they will continue to approach infinitely closer to it.

Moving on to a consideration of three mobiles, or bodies, moving with mutu-

---

69. Although infinite past time can be conceived as terminating at this moment, Oresme insists that it is the equal of infinite time through an eternal future. In his *Le Livre du ciel et du monde*, bk. 1, chap. 33, Oresme argues against Aristotle's view "that infinity in one direction is neither infinite nor finite" (*De caelo* 1. 12. 283a10–11, trans. Stocks) and more specifically attacks Averroes (bk. 1, chap. 33, 55a–56b), who insisted that an infinite time (or any quantity) having neither beginning nor end must be greater than an infinite past time (or any quantity) that has an end but no beginning, and is also greater than an infinite future time that has no end but did have a beginning (see Averroes, *De caelo* 1, Text 133, B–C, *Aristotelis opera cum Averrois commentariis*, Junctas ed., vol. 5, fol. 91r, col. 1). According to Oresme, Averroes denied that one can even divide time into past and future, since time is a continuum that is never wholly actualized. But Oresme insists that however many dissimilar infinite quantities there may be one is neither

greater nor smaller than another, from which he concludes (and this is relevant to his treatment of infinite time in the *De commensurabilitate*) that the totality of infinite time embracing both past and future is neither greater nor less than all the time that will come in the future only; nor, indeed, can it be said that they are equal. ("Or appert donques clerement que de plusseurs quantités infinies, combien que elles soient dessemblablement infinies, une n'est pas plus grande ne plus petite que l'autre; et est verité, pousé que les quantités soient permanentes et successives, quar une meisme rayson est des unes et ⟨des⟩ autres. Et donques tout le temps infiny passé et a venir n'est pas plus grant que le temps infini seullement a venir. Item encor, monstray je autrefoys que .ii. infinis ne sont pas egualz, mais il souffist a present a cest propos ce que dit est contre Averoÿs."—*Le Livre du ciel*, bk. 1, chap. 33, 56a–56b, Menut, *Oresme du ciel*, pp. 236–38).

70. Nothing like Proposition 5 appears in the *Ad pauca respicientes*.

ally incommensurable but respectively uniform velocities, Oresme, in Proposition 6, demonstrates that three such mobiles now in conjunction could conjunct again in some other point.[71] Let us assume that mobiles *A*, *B*, and *C* are presently in conjunction in point *d*. From all the previous propositions of this part we know that any two of these bodies will conjunct at points other than *d* (but never twice in the same point). Should *A* and *B* conjunct in some point *e*, it follows from Part II, Proposition 2 that arc *de* must be incommensurable to the whole circle. But prior to conjunction in *e*, both *A* and *B* must have completed a certain number of total circulations with respect to *d*. For example,[72] if, with respect to *d*, *A* completed 7 circulations and *B*, 5, it only remains to add arc *de* to each total number of circulations to produce a conjunction of the two bodies in *e*. But arc *de* is incommensurable to the whole circle so that when it is added to 7 and 5 circulations, respectively, the total distances traversed will be incommensurable (II.151–54); and yet, after traversing these incommensurable distances, the two mobiles will nevertheless conjunct in *e*. If the third mobile, *C*, should join them in *e*, this could be explained in the same way, since, in the same time, *C* might have completed 3 total circulations plus arc *de*. Therefore, it too would arrive in point *e* and the three mobiles would be in conjunction simultaneously at *e* after traversing mutually incommensurable total distances.[73]

In Proposition 7 Oresme assumes conditions the consequences of which negate Proposition 6. He demonstrates that three (or more) mobiles now in conjunction and moving with mutually incommensurable velocities will never conjunct again.[74] Thus "in one case of incommensurability it follows that the mobiles can be in conjunction an infinite number of times,[75] but in another case this does not follow" (II.168–69).

In establishing this proposition, Oresme refers at the outset to his earlier

71. Proposition 6 has no counterpart in the *Ad pauca respicientes*.

72. This example is mine; Oresme offers none in this proposition.

73. Not until Part II, Proposition 9 (II. 235–37) does Oresme make explicit the consequence, based on this sixth proposition, that these three mobiles would conjunct an infinite number of times (without mentioning Part II, Proposition 6, he makes the same assertion in Part II, Proposition 7 [II.168–69]). In effect, this is equivalent to saying that they will conjunct in an infinite number of different points.

74. Although in Part 1, Proposition IX of the *Ad pauca respicientes*, Oresme also considered conditions that would produce only one conjunction of three mobiles, the mutual relationships included commensurable, as well as incommensurable, velocities. See Grant, *Oresme PPAP*, p. 402. In a subsequent proposition (Part 2, Proposition XVIII), he drew the obvious inference from Part 1, Proposition IX that "It is possible that three, or four, or more planets could be in conjunction only once through all eternity."—Ibid., p. 423; Latin text, p. 422, lines 210–11.

75. This is implied in Part II, Proposition 6 (see above, n. 73).

discussion in Part I, Proposition 2 where the different possible modes of dividing a continuum were considered. There he showed the possibility of dividing a continuum according to different proportionalities such that no point—except the first—could serve as a common point of division in two or more proportionalities unless the numbers immediately following unity were not mutually prime. Oresme *then* confined his attention to rational proportionalities, but *now* concentrates on the division of continua by irrational proportionalities in order to apply the results to the present proposition.

As with rational proportionalities, there are some irrational proportionalities which share common points, and others which do not. As an example of the former, Oresme cites $[(2:1)^{1/2}]^n$ and $[(8:1)^{1/2}]^n$ (see II.180–81), where $n = 1, 2, 3, \ldots$. All points represented by $[(2:1)^{1/2}]^{3n}$ and $[(8:1)^{1/2}]^n$ are common to both proportionalities, since for any value of n both ratios would be equal and represent the very same point in their respective proportionalities.[76] Thus, when $n = 2$ we get $2^{6/2}$ and $8^{2/2}$, which represent identical points. However, two proportionalities that do not share any points, except the initial point represented by 1, are $[(2:1)^{1/2}]^n$ and $[(3:1)^{1/2}]^n$ (II.181–83). If a circle were divided by two such proportionalities beginning from point d, the divisions would be separated by distances incommensurable to the whole circle and never meet in any point other than d, their starting point. Translated into the conditions of this proposition, it could happen that A, B, and C starting from conjunction in d

---

76. Oresme uses the terms *medietas duple* and *medietas octuple* to designate the ratios $(2:1)^{1/2}$ and $(8:1)^{1/2}$, respectively (II.180–81). In medieval mathematical terminology and usage *medietas* could signify either the arithmetic half of a number, or quantity, or the exponent $\frac{1}{2}$. In the proposition described above, it is employed in an exponential sense. Both usages are illustrated in Oresme's *De proportionibus proportionum* (see under "medietas" and "medietas duple" in "A Selective Index of Mathematical Terms and Expressions" in Grant, *Oresme PPAP*, pp. 454–55).

By analogy with the division of continua by rational proportionalities in Part I, Proposition 2, the division of the same continua by irrational proportionalities would be as follows. For example, to divide a continuum in accordance with the proportionality $[(2:1)^{1/2}]^n$, divide it first into $(2:1)^{1/2}$ equal parts (when $n = 1$), and then successively into 2 equal parts (when $n = 2$), $(2:1)^{3/2}$

equal parts (when $n = 3$), 4 equal parts, i.e., $(2:1)^{4/2}$ (when $n = 4$), and so on. Thus, while the parts in any particular division—i.e., for any particular value of n—are all equal, the basic part continually diminishes as n increases. Obviously, when the whole expression is irrational [as it would be, for example, in $(2:1)^{1/2}$], it is not possible to divide a finite continuum into an exact number of equal parts. Here the analogy with rational proportionalities is inapplicable. That Oresme did not raise, or even allude to, this point is perhaps a consequence of his overriding interest in determining whether or not two proportionalities could divide a continuum and share common points of division. In fact, since this could be detected by mere inspection of the proportionalities, it was not necessary that Oresme concern himself with the exact or inexact subdivision of any particular continuum by one or another rational or irrational proportionality.

might never conjunct elsewhere because the infinite number of points in which *A* and *B* would conjunct are completely independent of the infinite set of points in which *B* and *C* would conjunct (II.189–93).

To illustrate this, Oresme furnishes the following example (II.194–204). Initially, bodies *A*, *B*, and *C* are in conjunction in point *d* (see Fig. 25, p. 265). Subsequently, *A* and *B* will conjunct in points *e* and *f* which are equidistant because the speeds of the mobiles are respectively uniform though unequal. Hence arc *de* equals arc *ef* and these will equal the arcal distance separating any two successive conjunctions of *A* and *B*. Finally, the ratio of the whole circle to arc *de* is as $(3:2)^{1/2}/1$.[77] However, mobiles *B* and *C*, after departure from *d*, will conjunct in *g*, and then *h*, and so on, where arc *dg* equals arc *gh*; indeed, all subsequent arcs formed by successive conjunctions of *B* and *C* will equal arc *dg*. The ratio of the whole circle to arc *dg* is now assumed as $(4:3)^{1/2}/1$.

Now ratios $(3:2)^{1/2}$ and $(4:3)^{1/2}$, which represent the ratios of the whole circle to arcs *de* and *dg*, respectively, are irrational ratios that are incommensurable to each other, as Oresme showed in his *De proportionibus proportionum*.[78] Con-

---

77. Since the velocities of *A* and *B* are respectively uniform but incommensurable, the angular distances separating successive conjunctions will be equal (by Part II, Proposition 4), and every such angular distance will be incommensurable to the whole circle (by Part II, Proposition 2). Here, for the first time in the treatise, Oresme specifies such incommensurable relationships in terms of particular ratios.

78. Among a number of distinctions made in the *De proportionibus*, Oresme differentiates between irrational ratios that are mutually commensurable and those that are mutually incommensurable. Any two irrational ratios that can be related by a rational exponent are considered by Oresme to be commensurable; and if not so relatable are classified as incommensurable. For example, $(4:1)^{1/3}$ and $(2:1)^{1/2}$ are commensurable because they are relatable by the rational exponent $\frac{4}{3}$ when expressed as $(4:1)^{1/3} = [(2:1)^{1/2}]^{4/3}$. But in the example cited above in the *De commensurabilitate*, the irrational ratios $(3:2)^{1/2}$ and $(4:3)^{1/2}$ are incommensurable because no rational exponent can relate them in the sense of making them equal. In Grant, *Oresme PPAP*, see pp. 31–33, 39, and chap. 2, pt. 1, prop. IX, of the *De propor-*

tionibus, pp. 200–204, lines 360–404; for a further discussion in this volume, see below, p. 74, n. 113 of this chapter.

If we should represent these two proportionalities as $[(3:2)^{1/2}]^n$ and $[(4:3)^{1/2}]^n$, it becomes obvious that Oresme has restricted *n* to 1, whereas in the division of continua made earlier in Part I, Proposition 2, *n* was permitted to range ad infinitum. In this example in the *De commensurabilitate*, the conditions imposed preclude any variation in *n*, since the distance separating any two successive conjunctions must remain equal as long as the velocities remain respectively the same. Any change of *n* would alter the angular distance that separates successive conjunctions.

In connection with the division of the circle, Oresme explains (II.205–10) that the circle should not be divided in accordance with the ratios of velocities of *A* and *B* or *B* and *C*. Accordingly, if $V_A:V_B = 4:1$, the circle will not be divided into four equal parts, but only into three (i.e., $4 - 1 = 3$; in this connection, he cites Part I, Proposition 11, discussed above, pp. 18–19), so that the ratio of the whole circle to any one of these parts would be as $3:1$.

sequently, no point except *d* can ever have served, or will ever serve, as a common point of conjunction for *A*, *B*, and *C* since all points of conjunction for *A* and *B* will differ from those of *B* and *C*. Without calling attention to it, Oresme has applied the conditions of Part II, Proposition 4, involving two mobiles, to Proposition 7, involving three mobiles sorted into two pairs. As long as the successive conjunctions for each pair are separated by unequal arcs, the three bodies can never conjunct again, nor could they have conjuncted in the past (II.202–4).

Varying his initial conditions by assuming that *A* and *B* now move with commensurable velocities while *B* and *C* continue to move with incommensurable speeds, Oresme shows in Proposition 8 exactly what he had demonstrated in Proposition 7—namely, that *A*, *B*, and *C* never conjuncted simultaneously prior to conjunction in *d* and will never conjunct again after departure from *d*.[79] Since *A* and *B* move commensurably, it follows from a corollary of Part I, Proposition 10 that any point in which *A* and *B* conjunct must be separated from *d* by an angular distance that is commensurable to the whole circle.[80] In contrast, *B* and *C*, which move with mutually incommensurable velocities, will of necessity be separated from *d* by an angular distance that is incommensurable to the whole circle. Thus, *A* and *B* will share no points of conjunction with *B* and *C* and the three can never conjunct simultaneously; nor in the past did they ever conjunct simultaneously prior to their initially assumed conjunction in *d*.

For example (II.221–26), if $V_A : V_B = 2:1$ and *A* and *B* depart from conjunction in *d*, through all eternity they will conjunct only in *d* (Part I, Proposition 11). But after departure from *d*, *B* and *C* will never again conjunct there, nor could they have conjuncted there prior to their unique conjunction in *d* (this is

79. Substantially the same proposition appears in Oresme's *Ad pauca respicientes*, pt. 1, prop. IX (Grant, *Oresme PPAP*, p. 402). In the *Ad pauca*, however, Oresme has mobiles *A*, *B*, and *C* move with equal curvilinear speeds on their respective circles—namely, *A*, *B*, and *C*—where circle *B* is double circle *C* so that mobiles *B* and *C* move with commensurable speeds, while circle *A* is to circle *B* as the diagonal to the side of its square so that *A* and *B* have incommensurable velocities (these conditions were laid down in Part 1, Proposition VIII [ibid., p. 400] and adopted in Proposition IX). In the *De commensurabilitate* the manner in which commensurability obtains between the speeds of the mobiles is expressed only as "*A* is moved twice as quickly as *B*" (II.221). The ratio

between the speeds of *B* and *C* is not given, but merely assumed to be incommensurable. Nothing is said about relationships between the circles on which the mobiles move, and presumably the mobiles are not moving with equal curvilinear velocities, but with direct and unequal angular velocities on concentric circles. However this may be, the consequences following from both propositions are identical.

80. In Part 1, Proposition 10, it was shown that a circle would be divided into as many equal parts as there are points of conjunction, and since each part is commensurable to the whole circle, it follows that the distance from *d* of any point of conjunction for *A* and *B* must also be commensurable to the whole circle.

shown in Part I, Proposition 1). Obviously, through an infinite past and an infinite future, A, B, and C conjunct only once.

Drawing on a number of earlier propositions, Oresme announces in Proposition 9 that in every case involving three or more mobiles there exist only three possibilities for the number of conjunctions that may occur through an eternal past and future time. Three mobiles will (1) never conjunct, (2) conjunct only once, or (3) conjunct an infinite number of times. Never could they conjunct twice only, or three times, or any exact finite number of times.

Each of the three possibilities just mentioned can arise from incommensurable velocities, from commensurable velocities, or from a combination of such motions.[81] That three or more mobiles might never conjunct when moving with mutually commensurable speeds was shown in Part I, Proposition 12. By a simple alteration of the conditions of Part II, Proposition 7, it could be shown that three mobiles moving with mutually incommensurable speeds might never conjunct at all (II.229–33). Here we merely assume that the three mobiles are not initially in conjunction.[82] In situations where only one conjunction is possible, Oresme cites Part II, Propositions 7 and 8. In the first of these propositions, the effect is achieved from motions that are exclusively incommensurable; in Proposition 8 this is attained by a combination of commensurable and incommensurable motions. Should all the motions be mutually commensurable, however, it could never happen that only one conjunction occurs through all eternity. For even if they conjuncted in only one point, it would repeat ad infinitum (see below). Thus Oresme could cite no proposition from Part I. At this juncture, then, the symmetry and correspondence between exclusively com-

---

81. That three mobiles moving with mutually commensurable speeds may, on the assumption of certain initial conditions, conjunct an infinite number of times or never at all is shown in Part 1, Propositions VI and VII, respectively, of the *Ad pauca respicientes*. The possible occurrence of only one conjunction through all eternity is not considered. Both commensurable and incommensurable motions are involved in Part 1, Propositions VIII and IX of the same treatise, where Oresme shows that conjunction may never occur or occur only once; but he fails to include the case involving an infinite number of conjunctions. Omitted entirely are situations involving three mobiles whose speeds are all mutually incommensurable (see below, n. 82). Furthermore, nowhere in the *Ad pauca respicientes* does Oresme present a summary of these three possibilities, or demonstrate explicitly that a finite number of conjunctions—two or more—is impossible.

82. At first glance Oresme seems to have demonstrated this very proposition in Part 1, Proposition VIII of the *Ad pauca respicientes*, where he shows that "It is possible there could be three mobiles moved incommensurably with respect to the center which will never conjunct" (Grant, *Oresme PPAP*, p. 401). However, in his proof, the motions of two bodies, B and C, are made commensurable, while the motions of A and B are incommensurable. In Part II, Proposition 7 of the *De commensurabilitate* and the version that Oresme says could be derived from it (II.229–33), all the mobiles are assumed to move with mutually incommensurable velocities.

mensurable and exclusively incommensurable motions collapse. But these conditions are re-established for the case involving an infinite number of conjunctions. Here, Part I, Proposition 14 supports the claim for commensurable speeds, and Part II, Proposition 6, for incommensurable velocities.

Oresme now explains (II.238–49) why an exact and finite number of conjunctions (two or more) is impossible. Without invoking or appealing to commensurability or incommensurability, Oresme builds his demonstration on the fact that all motions have been assumed to be uniform and regular, though, of course, the speeds differ. Let mobiles A, B, and C be in conjunction at d and sometime later conjunct in e, and then g, and so on, where angular distance de equals eg, and in general the angles separating any two successive conjunctions are equal. That these angles must necessarily be equal follows from the fact that the motions are respectively uniform. Now if K represents the time interval between conjunctions in d and e, it is a consequence of the uniformity of the motions that during the next interval K each mobile will traverse the same distance as when moving from conjunction in d to conjunction in e. But for each mobile, g is as far from e as e from d, so that in time K they will arrive in conjunction at g. At the end of another interval K, they will conjunct a third time in a point as far from g as g is from e, and so on ad infinitum. Thus no finite upper limit can be set for such conjunctions and through an infinite time there would occur an infinite number of them.

The manner in which this infinite sequence is produced, however, is significantly different depending upon whether the motions are commensurable or incommensurable. If the velocities are commensurable (II.250–51), the possible points of conjunction will be finite in number (see Part I, Proposition 17), but the three mobiles will conjunct ad infinitum in each one of them, producing an infinite number of conjunctions. If the velocities are incommensurable, however, no conjunction can occur twice in the same point, so that an infinite number of conjunctions results only because mobiles A, B, and C will conjunct in an infinite number of different points (see Part II, Propositions 6 and 9).

In Proposition 10 Oresme raises the same question considered earlier with respect to commensurable motions in Part I, Proposition 13. If the velocities of three or more mobiles are mutually incommensurable, how close can they approximate to each other short of actual conjunction? He shows that however small a space may encompass them at any given time, it will be possible at other times for them to approximate even closer, and yet fall short of conjunction.[83]

With respect to some given point d, mobiles A and B will return to it an infinite number of times and, since their velocities are incommensurable, the times in which they complete their respective circulations will be incommensu-

---

83. No form or version of this proposition   appears in the *Ad pauca respicientes*.

rable (II.263–64). Furthermore, since their velocities are unequal (and incommensurable), it must happen that some time when A is in d, a short time later B will arrive in d so that A and B are rather close together. But at some later time, B will arrive in d within an even shorter interval of time after A was in d. The time elapsing between the entry of A, and then B, into d will become less and less so that A and B approximate nearer and nearer to one another. Regardless of their proximity at any given time, however, the intervening distance can become even less at another time.[84] The same may be said of C when it is compared to A and B, respectively, so that what has been said about two mobiles can apply as well to three or more mobiles.

It is now apparent that, short of actual conjunction, the proximity possible between three or more mobiles depends on whether the motions are commensurable or incommensurable. If the speeds are commensurable *and the mobiles could never conjunct*,[85] then, by Part I, Proposition 13 there must be a minimum distance of approximation beyond which the mobiles cannot approach. But when the speeds are incommensurable no such minimum limits exist, for however close these bodies approach they can and will come even closer without entering into conjunction.

Oresme now investigates how these results might apply to planetary motions (II.280–89). We might assume that all the planets are moving incommensurably, but as a minimum condition we must presuppose that no three planets have

---

84. From Part II, Proposition 1, we know that A and B can conjunct no more than once in point d. And by inference from Part II, Proposition 6, as expressed in Part II, Proposition 9 (II.234–37, 251–53), we also know that three bodies moving with incommensurable velocities can conjunct no more than once in a given point. Because Oresme fails to specify whether A and B (or, for that matter, A, B, and C) have as yet had their unique conjunction in point d, we might ask whether a minimum distance of separation was perhaps reached just prior to such a conjunction. But it is unnecessary to take into account whether or not a conjunction had occurred, since Oresme divides time into two infinites (above, pp. 46–47 and n. 69)—an infinite past prior to conjunction in any particular point, and an infinite future subsequent to conjunction in any particular point. The events described in Part II, Proposition 10 could have occurred either through an in-

finite past prior to conjunction in d, or through an infinite future following such a conjunction, for at some time or other any previous minimal distance short of conjunction would become even smaller.

Oresme notes (II.272–74) the similarity between the ever diminishing approximation of these mobiles and the demonstration in Part II, Proposition 4, showing that through an eternal time no part of an arc could remain undivided by points of conjunction.

85. This is a necessary condition, as Oresme recognizes (II.274–78), for if three or more mobiles moving commensurably could conjunct in a particular point, the distance separating them must approach zero and become zero as they enter conjunction. And in virtue of the commensurability of the motions, this process would repeat ad infinitum and preclude any fixed minimal distance of separation.

mutually commensurable motions.[86] Through an infinite time it would happen that at some time or other all the planets would arrive in the same degree, and then in the same minute, and the same second, and so on, approaching ever closer to one another. And yet they will never conjunct exactly. This could be said of any position in the sky, so that at some time or other they might be in the first minute of Cancer and at another time in the first second of Capricorn. But however small a unit encompasses them, they can approach ever closer and squeeze within an even smaller unit.

Up to this point all the demonstrations in Part II are concerned exclusively with conjunctions. In Proposition 11 Oresme extends the range of applicability of most of the previous propositions to cover all other astronomical aspects. In this way, Proposition 11 corresponds to Part I, Proposition 21, in which earlier propositions involving commensurable motions and applying only to conjunctions were shown to have application to all other astronomical aspects.

In extending earlier propositions of Part II, Orseme does little more than specify individual propositions and assert that they also apply to other aspects. Just as Part II, Proposition 1 demonstrated that no conjunction of two mobiles ever occurs twice in the same point, so it is also true to say that no other aspect can occur twice in exactly the same way. For example (II.296–98), if these two mobiles were in opposition, they would never again oppose one another in the very same two points.[87] Similarly, Part II, Propositions 4–8, which were originally formulated for conjunctions only, can easily be extended to any other astronomical relationship or aspect (II.298–300).[88] A general consequence is that in any given past or future instant three or more mobiles, or bodies, can never be related in precisely the same way (II.300–302).

Extension to all aspects and relationships can also be made for Part II,

---

86. Thus if two of three planets—say, A and B—move with commensurable speeds, the third planet—say, D—cannot move commensurably with respect to *both* A and B, for then A, B, and D would have mutually commensurable velocities. However, if D were commensurable to A only, then A, B, and D "could be commensurable if taken two at a time" (II.282–83), in the sense that D and A might move with commensurable speeds but not A and B or B and D (cf. pt. 1, prop. VIII, of the *Ad pauca respicientes* in Grant, *Oresme PPAP*, pp. 400–402).

87. In Part 1, Proposition IV of the *Ad pauca respicientes*, which is the closest counterpart to Part II, Proposition 1 of the *De commensurabilitate* (see above, p. 38, n. 59),

Oresme did not explicitly extend the application of the proposition to any aspect other than conjunction.

88. In the *Ad pauca respicientes* a number of propositions are generalized to embrace all possible aspects or relationships. See pt. 1, prop. V (Grant, *Oresme PPAP*, p. 397; its closest counterpart in the *De commensurabilitate* is Part II, Proposition 4 [see above, p. 46, n. 68], Part 2, Proposition VII (ibid., p. 413; this is an extension of Part 1, Proposition IX of the *Ad pauca* and with it corresponds to Part II, Proposition 8 of the *De commensurabilitate*, when the last mentioned is extended to all aspects), and Part 2, Propositions XII, XIII, XV, and XVII (ibid., pp. 417–19, 421, and 423, respectively).

Proposition 9 (II.303–6). In that proposition we saw that three mobiles moving with mutually commensurable velocities would either never conjunct or conjunct an infinite number of times; should they move with incommensurable velocities they might never conjunct, conjunct only once, or conjunct an infinite number of times. If we apply the same initial conditions to cases of opposition or any other aspect or relationship, the same consequences would follow; that is, whatever the aspect it might never occur, occur but once, or take place an infinite number of times.

Motions that are incommensurable and eternal produce consequences which are truly amazing and inexplicable (II.307–23). Where the conditions of motion are such as to produce only one conjunction through all eternity (past and future), how can we explain its necessary occurrence at one particular time rather than another?[89] Perhaps this is due to the regularity of motions and motive

89. Oresme raises the same puzzling question in the *Ad pauca respicientes*, Part 2, Proposition XVIII, where he says (Grant, *Oresme PPAP*, p. 425): "And perhaps it will seem amazing to some how such a conjunction could happen necessarily in the [very] instant in which it does occur, so that it was true through all eternity that this would necessarily occur with mobiles coming to this [configuration] by means of their regular motions and being predisposed for this through all eternity. And one ought not to seek another cause [to explain] why more are conjuncted now than then, or why [it happens] in this instant rather than in another. But knowing this it will seem more credible that a free agent like God could arrange and ordain through eternity that something be done or produced for a certain [particular] instant. Now...one ought not to seek for an explanation of this."

In Book 1, Question 24 of his *Questiones super de celo*, Oresme was concerned with the same problem, this time in connection with an eclipse of the moon: "And when it is said further that there is no reason why it will begin in one instant rather than in another, I say that in many things there is no reason unless it is the sole will of God; but still, naturally speaking, one ought not to seek the reason why an eclipse of the moon took place today and not yesterday unless it

is because the celestial bodies were so disposed from eternity." ("Et cum dicitur ultra quod non est ratio quare magis incipet in uno instanti quam in alio, dico quod in multis nulla est ratio nisi sola dei voluntas sed adhuc loquendo naturaliter non est querenda causa quare eclipsus lune fuit hodie et non heri nisi quia corpora celestia sunt ita ab eterno ad hoc disponebantur."—Claudia Kren, ed. and trans., "The *Questiones super de celo* of Nicole Oresme" [Ph.D. diss., University of Wisconsin, 1965], p. 420 [English] and p. 419 [Latin].) It would seem that *eclipsus* should be spelled *eclipsis*. The Latin text may also be found in Erfurt, Wissenschaftliche Bibliothek, Amplonius Q. 299, fol. 26r.

Toward the end of his life, Oresme was to raise this perplexing problem once again and offer substantially the same response in his *Le Livre du ciel et du monde*, bk. 1, chap. 34, 57a–57c (Menut, *Oresme du ciel*, pp. 240–43). He utilizes the argument of necessity to repudiate Aristotle's contention that no cause could be assigned to explain why something that had always existed should be destroyed at this or that moment (Oresme appears here to be grappling with *De caelo* 1. 12. 283a 10–12). But if the celestial motions are eternal and there occurs a unique conjunction of three planets that has never happened before nor will ever occur again, Oresme argues that such a conjunction had, of necessity, to

forces. Furthermore, if celestial dispositions or configurations are the cause of effects in the sub-lunar or terrestrial regions of the world, a unique disposition might occur that could generate a wholly new individual constituting a new species. Such an event might occur after a great and unique conjunction of all the planets. Such a new species might itself be eternal or perhaps it would pass away with the advent of some new and powerful celestial configuration.[90]

---

occur at some particular moment rather than another. Perhaps somewhat more emphatically, Oresme offers substantially the same response in the *Ad pauca respicientes*, *Questiones super de celo*, and *De commensurabilitate*, when he explains that this necessity arises from the regularity, order, and interrelationships of the celestial motions, which, moving under the guidance of the celestial intelligences or movers, produce this unique conjunction necessarily at this time and not at any other time. In the final analysis, however, Oresme says (Menut, *Oresme du ciel*, 58b, p. 224)—as he did earlier in the *Ad pauca*, *De celo*, and *De commensurabilitate*—that if it pleased Him, God could arrange for the occurrence of an event at one time rather than another, and it is vain to inquire into His reasons.

90. The formation of new species by a unique conjunction is also mentioned in the earlier *Ad pauca respicientes*, pt. 2, prop. XVIII (Grant, *Oresme PPAP*, p. 423): "And now if it be assumed that such a [unique] conjunction could be a natural cause of the generation of some species by means of putrefaction or in some other way, perhaps some species can be newly created which will last through eternity. And the same thing may be considered about corruption."

In his *Questiones super de celo*, Oresme devotes all of Book 1, Question 24 to the problem "whether something created *de novo* could be perpetuated and also whether something that existed from eternity could be corrupted." ("Queritur utrum aliquid de novo genitum possit perpetuari et utrum etiam aliquid quod fuit ab eterno corrumpi." —Kren, ed. and trans., p. 409 [Latin] and p. 410 [English]; for a manuscript source,

see Erfurt, Wissenschaftliche Bibliothek, Amplonius Q. 299, fol. 25v.) Toward the end of Question 24, Oresme includes much of the substance of *De commensurabilitate*, II.315–22—even the reference to Pliny's *Natural History*: "Fifth, as to the beginning of species, one can say the same especially regarding species which do not contribute to the perfection of the universe so that it is likely there may be some conjunction of stars at some time which has not been similar [to one] in the past nor will be [to one] in the future. This is perhaps the cause of the beginning of some [species] which never ceases to be or the end of some [species] which never begins to be, naturally speaking. Again it is possible that some species either of substance or accident does not exist for a long time and this may exceed memory, and afterward [the species] may begin to exist at another time and thus many times, as Pliny says in regard to certain sicknesses." ("Quinto de inceptione secundum speciem potest dici idem maxime quantum ad speciem que non sunt de perfectione universi unde verisimile est quod sit quandoque aliqua coniunctio stellarum que non potuit esse simile in preteritum nec poterit in futurum que forte est causa inceptionis alicuius speciei que nunquam desinit esse vel finis alicuius que nunquam incipit esse naturaliter loquendo. Iterum possibile est quod aliqua species sive substantie sive accidentis per multa tempora non sit et quod transeat memoria et postea incipiat esse alio tempore et sic multotiens sicud dicit Plinius de quibusdam infirmitatibus."—Kren, ed. and trans., p. 423 [Latin] and p. 424 [English]; in Erfurt, Q. 299, see fol. 26r.) A possible reference to this section of the *Questiones super de celo* appears in

Part II, Proposition 12, the final proposition of Part II, is the direct counterpart of Part I, Proposition 22, but where in the latter proposition one body, the sun, was assigned two simultaneous commensurable and eternal motions, diurnal and annual, these same two solar motions are now assumed to be incommensurable. The initial conditions (II.328–33) are practically identical with those in Part I, Proposition 22. Let A_n, where $n = 1, 2, 3, 4, \ldots \infty$, represent the first point of Cancer which makes a daily rotation as it is carried around by circle a, the tropic of Cancer; assume that B, the center of the sun, traverses a stationary ecliptic, or, to achieve the same effect, assume that B is motionless on the ecliptic but carried around by it (see above, n. 46) once every year. Finally, imagine that A_1 and B are now simultaneously in a stationary point d, the only point of contact between the two circles (see above, pp. 29–30 and Fig. 23, p. 237).

Since the consequences that follow from these incommensurable components of a single motion are directly analogous to those derivable from two separate incommensurable motions, Oresme makes direct appeal to a number of earlier propositions in Part II. Thus, after departure from d, points A_1 and B will never

---

(*Note 90 continued*)

Oresme's *Quodlibeta*, where we find a brief mention of unique conjunctions and the generation of new species. Although only a finite number of individual qualities or events exists at any particular time in the heavens, Oresme insists that on the assumption of an eternal time there would have been and will be an infinite number of celestial events as well as an infinite number of species "because it is possible, as I have shown elsewhere, namely [in the *Questions*] on the *De celo*, that if some motions in the sky were incommensurable, there were and will be an infinite number of configurations and no one of these will be similar to another; and thus also, if you wish, understand [this] as species newly made there." ("quia possibile est sicud alibi, scilicet super *De celo*, est probatum quod si sint in celo aliqui motus incommensurabiles quod infinite fuerunt constellationes et erunt quarum una non erit similis alteri; et sic etiam nove species vide ibi si vis."—Oresme, *Quodlibeta*. MSS Paris, Bibliothèque Nationale, fonds latin 15126, fol. 103r; see also Florence, Laurentian Library, Ashburnham 210, fol. 48r, col. 2.)

Without reference to *De commensurabili-* tate, II.315–22, or any other of his earlier treatises mentioned in this note, Oresme repeated substantially the same speculation —including the quotation from the twenty-sixth book of Pliny's *Natural History* given in II.319–20 of the *De commensurabilitate*— in his much later *Le Livre du ciel et du monde*, bk. 1, chap. 34, 57c (Menut, *Oresme du ciel*, pp. 242–43).

In discussing the formation of unique new species in the *De commensurabilitate* and *Questiones super de celo*, Oresme was careful to explain that he was "speaking naturally" (*loquendo naturaliter*; see II.315–16 and above, this note), that is, in accordance with natural philosophy, or, perhaps more broadly, in conformity with rules of natural reason (see Anneliese Maier, *Studien zur Naturphilosophie der Spätscholastik*, vol. 4, *Metaphysische Hintergründe der spätscholastischen Naturphilosophie* [Rome, 1955], p. 9). This was in contrast to speaking according to faith, revelation, or dogma. As a Christian, Oresme could not accept the eternality of motion or created things. But when "speaking naturally," it was wholly proper to introduce hypothetical eternality, as Oresme frequently did.

meet there again (by Part II, Proposition 1) so that the sun could never enter the first point of Cancer twice on the very same meridian (II.334–37); nor could it ever enter the first point of Cancer on any meridian that is separated from the first meridian (at point A_1) by a distance commensurable to the whole circle (by Part II, Proposition 2; II.338–39). Furthermore, through an infinite past there will have been an infinite number of different points on which B entered Cancer; and yet through an infinite future there must exist infinite other points on which B will enter Cancer (II.339–41). It follows then (by Part II, Proposition 4) that no arc or sector of the circle is so small that it does not contain a meridian on which B entered the first point of Cancer in the past; and yet, however small it be, such an arc will contain a meridian that has not yet served as the first point of Cancer in the infinite past but will do so at some time in the infinite future (II.342–45). The same argument is applicable to any point of the zodiac.

Oresme now turns to a consideration of the "Platonic spiral" (see pp. 31–33). In direct contrast with the spiral traced by the sun when it was assumed to move with two commensurable motions in Part I, Proposition 22, the spiral resulting from the assumption of two incommensurable components of the sun's motion would be infinite; it would have neither beginning nor end, since it has described a new spiral every day through an infinite past and will describe a new daily spiral through an infinite future (II.348–54). Although Oresme does not elaborate or elucidate, it is clear that the spiral must be infinite because the sun always enters the tropic of Cancer at a different point (II.335–37) and then proceeds to spiral down toward the tropic of Capricorn (the formation of the spiral is explained above, p. 31, n. 49), which it always enters at a different point. It then spirals upward once again toward the tropic of Cancer, entering at a new point. Under these circumstances the spiral will never terminate, since the sun arrives and departs every time from a different point on each of the tropical circles. Furthermore, the sun will never retrace a spiral described previously within the fixed space in which Oresme imagines all spirals to be formed. However, intersections between spirals are possible, for as the sun moves down in its annual motion from Cancer to Capricorn it sweeps out one spiral at the completion of each daily motion and will intersect a point on each of these spirals as it moves upward from Capricorn to Cancer. It can then truly be said that through all eternity, B, the center of the sun, will have been in every one of these points of intersection twice, but no more.[91] In all other points, however, B may pass through only once or never, through all eternity.[92]

---

91. This phenomenon could not occur more than twice because each annual spiral —from Cancer to Capricorn and back again to Cancer—will differ from every previous and subsequent spiral, for in each a unique set of points has been intersected.

92. The class of points through which B passes only once includes all points lying on

Continuing to describe startling consequences from such simultaneous incommensurable motions, Oresme remarks (II.356–63) that B's eternal motion through the vast expanse between the tropical circles can be conceived as leaving behind a large crisscross pattern or netlike figure. Since, in the past, these spirals have been infinite in number, they must have formed an infinitely compressed or thickened figure which, nevertheless, will continue to thicken through an infinite future as additional daily spirals are described. But despite this infinite thickness and the fact that B must have passed between any two assignable points (i.e., between any two immediately proximate spirals), it follows from what has been said that between any two such points there are yet an infinite number of points in which B never was or will be.

Up to this point, Oresme imagines B to be in motion around the center of the world or the earth. Now he assumes that the sun moves either on an eccentric circle or an epicycle. An immediate consequence is that the spirals would now approach and recede as the sun approaches and recedes from the center of the world. Indeed, the distance of the center of the sun from the center of the world would be unique for every one of the infinite times that it appeared and will appear on one and the same meridian (II.370–74). From this general conclusion, it follows that, having once entered the point of perigee for any given meridian, the sun would forever after be farther from the center of the world when on that very same meridian (II.374–76).[93]

Similar consequences can be derived with respect to the nadir of the sun ("nadir solis";[94] II.378), which is the point of the earth's shadow lying at the terminus of a straight line extending from the center of the sun through the center of the eccentric earth to the apex of the cone of the earth's shadow. Thus just as the sun describes a new spiral daily, so also, but in an opposite direction, does the entire cone of the earth's shadow describe a new spiral (II.377–80). Consequently, parts of the sky on which the sun had previously never ceased to shine will now be plunged into darkness (II.380–81). For example (II.381–88),

---

(Note 92 continued)
the infinite spiral line, exclusive of points of intersection. Points through which B never passes, for example, are presumably those on either tropical circle that cannot serve as the first point of Cancer or Capricorn. Thus, the sun will never enter Cancer on any point that is separated from one on which it has entered by an angular distance that is commensurable to the whole circle, or four right angles (see II.335–39).

93. Perhaps Oresme is here taking into account the motion of the line of apsides, or aux line as it was called in the Middle Ages.

94. Oresme uses and defines this expression in his *Traitié de l'espere*, chap. 48 ("Maistre Nicole Oresme: *Traitié de l'espere*," ed. Lillian M. McCarthy [Ph.D. diss., University of Toronto, 1943], p. 266), and *Le Livre du ciel et du monde*, bk. 1, chap. 29, 45b (the relevant passage is quoted below, n. 97). The expression also appears in Sacrobosco's *Tractatus de sphera* (see Thorndike, *Sphere of Sacrobosco*, p. 115).

if the sun, which has been assumed to move on an eccentric circle, were as close as possible to the center of the earth ("centrum mundi";[95] II.382), the cone of the earth's shadow would extend into the sky farther than ever before or ever thereafter (II.384–86). Although Oresme has erred in supposing the earth's shadow to be longest when the earth-sun distance is minimum[96]—i.e., when the sun is at perigee—this would occur when the sun reaches apogee, or its maximum distance from the earth, thereby validating his point that "some part of the heavens would become darkened which was never at any other time wholly deprived of the sun's light" (II.380–81).[97]

95. That the expression *centrum mundi* signifies the center of the earth even when the earth is itself eccentric seems clear from the context. As confirmation, however, I cite this brief passage from the *Theorica planetarum*, sometimes ascribed to Gerard of Cremona (its author is simply unknown; see Olaf Pedersen, "The Theorica Planetarum-Literature in the Middle Ages," *Classica et Mediaevalia*, vol. 23 [1962], pp. 230–32): "Pars eccentrici quae maxime elongatur a centro mundi dicitur aux vel longitudo longior. Sed pars quae maxime appropinquatur centro terre vocatur oppositum augis vel longitudo propinquior."—*Gerardi Cremonensis Theoricae planetarum veteres* in *Spherae tractatus Ioannis de Sacrobusto Anglici viri clariss.* (n.p., 1531), fol. 161r. It is obvious that *a centro mundi* and *centro terre* are synonymous.

96. Although he makes no mention of his error in the *De commensurabilitate*, Oresme corrected himself in the *Le Livre du ciel et du monde*, bk. 1, chap. 29, 45a, where he asserts that when the sun is at its maximum distance from the earth, the latter produces its longest shadow. "Item, tant est le solleil plus loing de la terre, de tant est l'ombre de la terre plus grant et entre plus avant ou ciel."— Menut, *Oresme du ciel*, p. 198. A few lines below (ibid., 45c; see below, n. 97), Oresme remarks that the moon's shadow is longest when it is farthest from the sun.

97. As part of a lengthy argument against Aristotle's contention that whatever had a beginning must at some time cease to be and whatever will ultimately pass away must

have had a beginning, Oresme, in *Le Livre du ciel et du monde*, chap. 29, 44d–45c (see above, n. 59 and Maier, *Metaphysische Hintergründe*, p. 30), observes that if the sun had simultaneous incommensurable motions (he assumes three such motions of which one, at least, is assumed to be incommensurable to the other two) and reached its maximum distance from the earth on some particular celestial meridian, the earth's shadow would also reach its greatest possible length (see above, n. 96) when the sun is on that particular meridian. Because of the incommensurability of the motions this cannot have happened previously through an infinite past time, and cannot occur again through an infinite future. Thus, on the assumption of infinite past and future time, we can truly conclude that a light that had no temporal beginning has ceased to exist. Simultaneously with the sun's motion back toward the earth and away from its most distant point on that meridian, the earth's shadow ceases to extend as far as it did, so that the point darkened previously by its apex is now bathed in sunlight. Because the solar motions are assumed to be incommensurable, the earth's shadow will never extend that far again, and we may conclude, therefore, that a light begins that will never cease to exist through an eternal future. Indeed, a few lines before, Oresme makes the more general assertion that, under the assumption of incommensurable solar motions, the earth's shadow shifts incessantly to points which it will occupy only once through all eternity. Consequently, there is

The sun's diurnal and annual motions are used once again in a final consequence involving two simultaneous incommensurable motions of one and the same body (II.389–93). If these two motions were actually incommensurable, as is possible, the annual mean solar motion would contain a certain integral

---

(Note 97 continued)

an unending process whereby eternal light that had no beginning will pass away, and new light commences that will never cease to exist. Since the moon's motions are probably incommensurable (undoubtedly based on *De commensurabilitate*, III.457–66, and *De proportionibus proportionum*, IV.566–82 in Grant, *Oresme PPAP*, p. 305; see also pp. 61–63 and 88), Oreme suggests that the lunar shadow may actually exemplify this very situation. The entire argument can be demonstrated, says Oresme, from his *De commensurabilitate*.

In light of the summary presented above, only the French text is cited here:
"Et semblablement, se un meisme corps du ciel estoit meu de .iii. ou de plusseurs mouvemens, pousé que un seul de ⟨c⟩es mouvemens fust inconmensurable as autres, je di que, ce posé ce corps est touz- // (45a) jours en nouvelle disposicion et son centre en tel endroit ou en tel point inmobile ou onques mes ne fu ne jamais ne sera, et descript touzjours nouvelle ligne et fera perpetuelement. Et pour ce que le solleil est meu de .iii. ou de pluseurs mouvemens et est possible et vraysemblable, comme dit est, que aucun de ces mouvemens soit inconmensurable as autres, il s'ensuit par necessité que en chascun movement le centre du corps du solleil est en nouveau point ou onques ne fu, et la pointe de l'ombre de la terre continuelment en aucun endroit ou point ou onques ne fu et jamais ne sera. Et par consequant, aucune lumiere ou ciel cesse estre selonc soy toute ou cesse estre toute laquelle n'eust onques commencement, et aucune commence qui jamés ne cessera. Item, tant est le solleil plus loing de la terre, de tant est l'ombre de la terre plus grant et entre plus avant ou ciel. Et par l'inconmensurableté desus dicte, puet estre que le solleil en un merid⟨i⟩an est si

loing de la terre que onques ne puet ne ja ne porra estre autrefoys si loing en cest meridian. Et donques l'ombre de la terre est si longue et si en parfont ou ciel en celle partie que / (45b) onques ne fu tant ne jamais ne sera. Et par consequant, une lumiere cesse que onques ne commença, et une commence qui ja ne cessera. Et l'inconmensurableté qui pourroit estre es mouvemens du ciel de la lune la ou est ceste ombre fait encor ceste conclusion estre plus vraysemblable. Item, se telle chose estant, il avenoit que en ce meridian la lune estoit le plus pres de terre que elle puisse estre quant elle est eclipsee et qu'elle fust droitement opposite au solleil ou point appellé *Nadair solis*, si comme il est possible, il seroit adonques la plus tres grande eclipse de lune qui peust estre [see Fig. 7, below, which is taken from p. 201 of Menut's edition]. Et se aucuns des mouvemens de la lune ou du solleil ⟨ou de la lune et du soleil⟩ sont inconmensurables, si comme dit est, et c'est vraysemblable, ce seroit impossible que autrefoys eust esté ou // (45c) fust ou temps avenir si grande eclipse de lunne. Item, posons que la lune soit le plus loing du solleil que elle puisse estre sanz estre eclipsee, et pour l'inconmensurableté desus dicte qui est vraysemblable, il s'ensuit que le solleil et la lune luisant ne furent onques si loing l'un de l'autre ne jamais ne seront. Et donques convient il que l'ombre de la lune soit ou ciel plus grant et plus loing que onques ne fu et que jamais ne puet estre, quar la lune fait ou ciel plus grant umbre de tant comme elle est plus loing du solleil [see Fig. 7]. Et par consequant, aucune lumiere cesse ou ciel qui onques n'ost commencement et aucune commence qui ja ne finera. Or avons donques .iii. cas ou telle chose est possible d'aucune lumiere du ciel. Et tout ce que dit est qui touche l'inconmensurableté dessus dicte est evidenment declairé par demonstracions geometriques ou traytié dessus dit, *De commensurabilitate vel incommensurabilitate motuum celi*, ou s'ensuit de ce que est en ce traitié par telles demonstracions."—Menut, *Oresme du ciel*, pp. 198, 200.

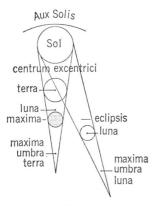

Fig. 7

In his much earlier extensive consideration of Aristotle's *De caelo*, Oresme treated the very same problem much more briefly. Thus, in his *Questiones super de celo*, bk. 1, question 24, his counterargument to Aristotle's position is exemplified by, and restricted to, unique lunar eclipses arising from the probable incommensurable and assumed eternal motions of the sun and moon. The essence of the argument is once again that, on the basis of their incommensurable motions, the earth's shadow will never return twice to the very same place on the moon's surface, so that some light on the lunar surface that had previously existed from eternity is extinguished. But when the shadow moves from that very location, a new light will appear that had never existed before and which will never cease to exist, since the earth's shadow will never extend there again.

"Second, if the motion of the sun and moon are incommensurable, as is likely, then the pyramid of the earth's shadow always crosses the sphere of the moon by some path through which it will never cross [again], nor would be able to cross naturally nor could be able [to cross] in the future, therefore continuously some light is categorically and wholly corrupted which has existed as an entity from eternity, and also by the same [reasoning], something is generated which will never be corrupted, and this could be mathematically demonstrated." ("Secundo si motus solis et lune sunt incommensurabiles sicud est verisimile, tunc semper piramis umbre terre in orbe lune per aliquam viam per quam nunquam transibit nec potuit transire naturaliter nec poterit in futurum, ergo continue aliquod totum lumen categorice corrumpitur quod fuerat ab eterno secundum aliquod sui et etiam per idem aliquod generatur quod nunquam corrumpetur quod possit mathematice demonstrari."— Kren, ed. and trans., pp. 421–23 [Latin] and pp. 422–24 [English]; in Erfurt manuscript Q. 299, the passage is on fol. 26r.)

If the *De commensurabilitate* had been written prior to the *Questiones super de celo*, it would be reasonable to construe Oresme's assertion that "this could be mathematically demonstrated" as a reference to Part II, Proposition 1 of the former treatise—especially II.19–21, which extends the proposition to include oppositions. What Oresme says in the *Questiones super de celo* is only true if the sun and moon are never twice in opposition in exactly the same way—and the impossibility of two identical oppositions is explicitly asserted in *De commensurabilitate*, II.19–21. But even if the *De commensurabilitate* had not yet been written, it is more than likely that when Oresme claimed that "this could be mathematically demonstrated," he had in mind a proposition akin to Part II, Proposition 1 of the *De commensurabilitate*.

The argument that something that had been bathed in light through an eternal past might be darkened momentarily and thereafter never darkened again is found in Oresme's *Quaestiones super geometriam Euclidis* and the anonymous *Questio de proportione dyametri quadrati ad costam ejusdem* (see above, p. 39, n. 59). The substance, and even the wording, of the discussion in these two treatises is almost identical. Where the

number of days plus some part of a day that is itself incommensurable to a whole day. Obviously, the exact length of the solar year would be inexpressible and the determination of an exact calendar or almanac would be impossible.[98]

In concluding Part II, Proposition 12, Oresme considers one body moving

----

(Note 97 continued)
De commensurabilitate and *Le Livre du ciel et du monde* consider this matter in terms of the maximum length of the earth's shadow, the *Quaestiones super geometriam Euclidis* and the anonymous *Questio* make the same point with respect to the deprivation of light on the lunar surface. On the assumption that the motions of the moon and sun are incommensurable, they will never be in opposition twice in precisely the same way (see below, p. 66, n. 103 of this chapter), so that at some time or other a portion of the lunar orb that had never previously been in shadow will be darkened. But thereafter, through an eternal future, that portion of the moon's surface will never again be without light. If the world had no beginning and will never end, it would follow that some light that had existed through eternity will cease to be, while a new light that had never before existed will be generated at this very point, or place, on the lunar orb (i.e., after the shadow cast over that particular portion of the moon has passed beyond it, never again will it be darkened). To reveal the great similarity of these two passages, I shall quote both, giving first the one from the *Quaestiones super geometriam Euclidis* (Busard, ed., p. 19): "Ex quo eciam sequitur, quod aliquando aliqua pars orbis lune erit obumbrata, que nunquam alias fuit obumbrata, nec umquam alias erit obumbrata, supposita eternitate mundi; et ideo si lumen sit aliqua res, aliquod lumen potest generari, quod durabit in eternum, et aliquod corrumpi, quod fuit ab eterno." In the anonymous *Questio* the text reads (Suter, ed., *Zeitschrift für Mathematik*, vol. 32, p. 50): "Ex quo ulterius sequitur, quod possibile est, aliquam partem orbis nunc esse obumbratam, quae nunquam alias fuit obumbrata, nec erit unquam, supposita etiam eternitate mundi. Ex quo ulterius

sequitur, quod aliqua res sive aliquod lumen(?) potest generari, quod durabit in eternum, et aliquod corrumpi quod fuit ab eterno."

A further consequence of the assumed incommensurability of the lunar and solar motions is found in the same two treatises expressed, once again, in similar, and sometimes identical, language. Thus if the lunar and solar day—i.e., the time in which each makes a complete daily revolution—were incommensurable and the moon rose at a particular instant of some hour, it would never again rise at the same instant of that very same hour. The judgments of astrology must, therefore, be very uncertain. For the texts of these two passages see Grant, *Oresme PPAP*, pp. 77–78, n. 101.

98. See also below, III.29–31. The conviction that a true and exact calendar is an impossibility if the annual and daily solar motions are incommensurable is also expressed by Oresme in the *De proportionibus proportionum*, chap. IV (see Grant, *Oresme PPAP*, p. 304, lines 566–72 and pp. 60–61), *Ad pauca respicientes*, pt. 2, prop. XVI (ibid., pp. 420–22, lines 192–97), *Quaestiones super geometriam Euclidis*, Question 9 (Busard, ed., p. 25, lines 5–10), and, if by Oresme, the anonymous *Questio de proportione dyametri quadrati ad costam ejusdem* (Suter, ed., *Zeitschrift für Mathematik*, vol. 32, p. 50).

That the solar year would be inexact if the day and the year were incommensurable was implied much earlier by Theodosius of Bithynia (born ca. 180 B.C.) in Proposition 19 of his *De diebus et noctibus* (Theodosius says only that under such conditions the sun would never return to the same place in equal intervals of time; see below, p. 84). Some later authors who mention or discuss this problem in terms of the possible incommensurability of the celestial motions are

with three simultaneous motions (II.394–418).[99] He assumes that the moon moves with (1) a daily motion, (2) a motion on its eccentric deferent, and (3) motion on its epicycle.[100] In Fig. 8, let *A* be the center of the moon, *B* the center of the

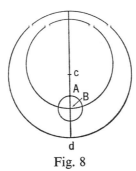

Fig. 8

epicycle, *d* a point superposed over the point of perigee ("longitudo propior";[101] II.401–2), and *c* the center of the world.[102] Draw line *cd* and let it describe a daily motion completing one circulation every day. If points *A* and *B* are initially on line *cd*, and the motions of *B* on the eccentric and of *A* on the epicycle are

---

Pierre d'Ailly, Nicholas of Cusa, and Jean Gerson. Indeed, d'Ailly, who plagiarized much of his *Tractatus contra astronomos* from Oresme's *De commensurabilitate*, includes II.29–31 of the *De commensurabilitate* almost verbatim. ("Primus est quorumdam praesumptuosorum qui jactant se almanach seu verum kalendarium perpetuum tradere, cum tale quid invenire forte sit impossibile, vel saltem impossibile cognoscere se aliquid huiusmodi reperisse."—*Petri de Alliaco Cardinalis Camaracensis Tractatus contra astronomos* in vol. 1, *Joannis Gersonii...opera omnia...* [Antwerp, 1706], col. 799.) Since Gerson relied heavily for his opinions on d'Ailly, it is very likely that Oresme influenced Gerson through d'Ailly. For d'Ailly and Gerson see below, pp. 130–32; and for Cusa, pp. 132–34, n. 123.

99. In Part I, Proposition 22 (I.784–87), Oresme considered one body moving with three or more simultaneous *commensurable* motions. In his later *Le Livre du ciel et du monde*, he ascribed a triple motion to the sun (see above, p. 61, n. 97).

100. In the *De commensurabilitate*, Oresme

is probably following the usual representation of lunar motion as found in the anonymous *Theorica planetarum* ascribed to Gerard of Cremona (fol. 161v) and in the *Sphere* of Sacrobosco. The latter speaks of the moon's motion on (1) an equant circle (ecliptic or zodiac?), (2) an eccentric deferent, and (3) an epicycle (see Thorndike, *Sphere of Sacrobosco*, pp. 114 [Latin] and 141 [English]). Robert Anglicus, in his commentary on Sacrobosco's *Sphere*, distinguished five lunar motions (ibid., pp. 194 [Latin] and 243 [English]).

101. In the *Theorica planetarum* the terms used for perigee are "oppositum augis" (fol. 161r), "longitudo propinquior" (fol. 161r), "longitudo proprior" (fol.171v), and "longitudo propior."

102. Although slightly altered, Fig. 8 is based upon figures in MSS *A* (Vat. lat. 4082, fol. 105r, in margin below col. 2) and *L* (Florence, Ashburnham 210, fol. 168v, in margin below col. 2). A somewhat different figure, but representing the same example, appears in MS *B* (BN 7281, fol. 268v, lower margin).

incommensurable, then it follows (by Part II, Proposition 8) that they will never again meet on line *cd*, which despite its daily motion is treated throughout as if it were stationary; in fact the moon's three motions will be related differently at every moment (by analogy with the motion of three mobiles in Part II, Proposition 11 [II.300–302]), so that the moon will bear a unique relationship to the earth at every moment (II.397–400). Furthermore, with only the two incommensurable motions of *A* and *B*, it can be shown by Part II, Proposition 1 that they will never meet simultaneously on line *cd* (II.403–9). If they did meet on line *cd* and *A* had initially left from the point of perigee (Oresme seems to imply this), *A* would have to move around its epicycle an exact number of times while, in the same time, *B*, the center of the epicycle, would have traversed the deferent an integral number of times. But if this happened and they coincided on line *cd*, their motions would be commensurable (as shown in Part II, Proposition 1 [II.6–12]), which is contrary to the assumption of their incommensurability.

Since it is impossible that *A* and *B* ever meet again on line *cd* on which lies the moon's point of perigee (the point of *A* in Fig. 8), and if *A* is at perigee when *A* and *B* depart initially from line *cd*, it follows that never in the past or future could the moon be closer to the center of the earth, for it can enter perigee only once through all eternity (II.409–12). If *A* is in its unique perigee when the sun on its eccentric circle is also at perigee but in direct opposition to the moon, there would occur a perfect total eclipse that could not have been exceeded in the infinite past nor ever surpassed through an eternal future (II.412–18).[103] Similar con-

---

103. Although the sun is not said to be in perigee, substantially the same conclusion appears in Oresme's *Le Livre du ciel et du monde*, bk. 1, chap. 29, 45b (the text is quoted above, n. 97). Immediately following this, Oresme introduces the case where the moon, without undergoing eclipse, is as far from the sun as possible. Under these conditions, the moon will produce its longest possible shadow. See the text of this discussion and Fig. 7 on p. 63, in n. 97 above. Statements concerning the uniqueness of a total lunar eclipse are also found in Oresme's *Ad pauca respicientes*, pt. 1, prop. V (Grant, *Oresme PPAP*, pp. 396–98, lines 153–56) and *Quaestiones super geometriam Euclidis*, Question 7 (for the text see ibid., p. 435, n. 11). In these two treatises, however, Oresme is concerned only with the overall incommensurability of the resultant motions of both sun and moon, and not with multiple simultane-

ous incommensurable motions of either sun or moon. But in all these treatises, an identical consequence is derived—a unique total lunar eclipse can occur only once through all eternity.

In the anonymous *Questio de proportione dyametri quadrati ad costam ejusdem* (see above, p. 39, n. 59), the same conclusion could be inferred from the general claim that if the motions of sun and moon were incommensurable and they once entered into conjunction or opposition, never through all eternity could this occur again in exactly the same way. Thus if a total lunar eclipse occurred once, it could never happen again. ("4°. dico, supposito quod motus solis respectu sui centri et motus lunae respectu sui sint incommensurabiles, sicut est verisimile, vel ad minus ejus oppositum nondum est demonstratum, sic quod centrum solis et centrum lunae eisdem temporibus describerent

sequences and conclusions about other planets would make it obvious that planetary tables involving conjunctions, oppositions, and other aspects could never be represented by numbers if the celestial motions are incommensurable (II.419–26).

Part III

With the conclusion of all formal propositions concerned with commensurable and incommensurable motions, Oresme now seeks to fulfill his earlier promise (I.66–67) to determine whether or not the celestial motions are actually commensurable. Instead of a formal discussion of this important issue in keeping with the mathematical and propositional character of the two preceding parts, Oresme resorts, in Part III,[104] to a debate form set within a personal dream in which Apollo and his Muses and Sciences appear before Oresme. The *dramatis personae* are four in number: Oresme, the mortal onlooker; Apollo, the judge; and two protagonists, personifications of two of Apollo's sciences, Arithmetic, representing the side of commensurability, and Geometry, defending incommensurability. Oresme, convinced by Apollo at the outset that mortal man is incapable of acquiring exact knowledge about sensible things (III.10–31) and therefore impotent to determine whether or not the celestial motions are commensurable,[105] begs Apollo to enlighten him. Turning to Arithmetic and Geometry, Apollo orders them to "teach him what he asks" (III.50).

---

angulos incommensurabiles circa centrum terrae, dico, si sol et luna nunc sunt conjuncti conjunctione rectissima vel oppositi oppositione rectissima, si mundus fuisset ab eterno, nunquam ita punctualiter fuerunt conjuncti vel oppositi, et si duraret in eternum, nunquam ita punctualiter conjungerentur vel opponerentur; patet sicut secundum. Ex quo etiam sequitur, quod per totum tempus eternum, si sol et luna sic hodie fuissent conjuncti conjunctione tali, consimilis eclipsis nunquam fiet, si etiam mundus duraret in eternum."—Suter, ed., *Zeitschrift für Mathematik*, vol. 32, pp. 49–50.)

104. A very brief description of this final part of Oresme's *De commensurabilitate* is given by A. Maier in her *Metaphysische Hintergründe*, pp. 28–30, and by Pierre Duhem, in *Le Système du monde*, 10 vols. (Paris, 1913–59), vol. 8, pp. 456–62. Duhem, however, conjectured that Pierre d'Ailly was the

author of the *De commensurabilitate*, not Oresme, whom he did not even suggest as a possibility (Duhem used only anonymous manuscript BN 7281, which, in this edition, has been assigned the siglum *B*). See below, App. 2.

105. See also Prol.39–42 and III.435–66. Statements concerning man's inability to acquire exact knowledge about sensible objects and his inability to determine whether or not the celestial motions are commensurable also appear in Oresme's *Ad pauca respicientes* (Grant, *Oresme PPAP*, pp. 62; 386, lines 49–50; 426, lines 239–46). Henry of Hesse and Pierre d'Ailly echoed this opinion directly from Oresme's works, while Jean Gerson repeated it from Pierre d'Ailly (ibid., p. 119, nn. 58, 59, and p. 120, n. 61; see also below, pp. 126–27 and 130–32 and the relevant footnotes. Cf. also p. 78, n. 1, above.

Arithmetic, who speaks first, offers a number of "basic arguments" to each of which Geometry will later respond. Their respective positions are summarized by bringing together Arithmetic's argument, followed immediately by Geometry's counterargument. At least four specific proposals and counterproposals may be distinguished:[106]

1. ARITHMETIC: Rational ratios are better than irrational ratios because they produce pleasure, whereas irrational ratios cause offensive effects (III.84–86, 100–102).
   GEOMETRY: True, rational ratios do produce a certain beauty in the world; but if they were united harmoniously with irrational ratios this would be far better, since from such a union there would result a rich variety of wonderful effects (III.340–42).

2. ARITHMETIC: Arithmetic is the firstborn of the mathematical sciences and her numbers are most worthy to represent the relationships between the celestial motions (III.135–47).
   GEOMETRY: Geometry represents magnitude in general, which includes numbers as a special category. Thus, Geometry embraces not only numerical or rational ratios but much more, so that it not only furnishes as much beauty to the heavens as does Arithmetic, but adds much more splendor (III.360–67).

3. ARITHMETIC: Only if the celestial motions are commensurable and represented by rational ratios can the music produced by the moving spheres be harmonious concords; if these ratios were irrational, the music would be discordant (III.148–53).
   GEOMETRY: If there really is celestial music, it would not vary as the velocities of the celestial motions but, rather, as volumes of the celestial spheres, or in some other way (III.384–88). But even if celestial music resulted from the celestial motions themselves, there is no evidence for assuming that the principal harmonic concordances would be produced. Furthermore, no one has yet determined whether celestial music is sensible or merely intelligible (III.392–99). But if it is sensible and created by fixed and rational ratios, it would be monotonous; only infinite variation is capable of producing interesting sounds (III.402–6).[107]

106. Oresme's quotations and sources for these arguments will be found in the footnotes to the translation. In most instances, there is further discussion in the Commentary.

107. In III.375–97, Oresme, whose genuine opinions are represented by Geometry, uses the widely accepted traditional account of the Pythagorean discovery of the principal musical concordances to offer a more consistent and plausible alternative to the equally widely held Pythagorean conviction that ratios of planetary velocities produce the principal musical concordances (see above, p. 19, n. 28). Thus Oresme repudiates the view expressed earlier by Arithmetic (III. 160–67), who was quoting from Cicero's *Dream of Scipio* in the Macrobian version and commentary. Oresme's counterargument is simple—if we assume that the principal musical concordances arise from the ratios of planetary velocities, consequences directly at variance with observable celestial phenomena would result (see I.463–80 and III.388–94). But even if musical concordances are produced in the sky, they would not arise from planetary velocities—since Oresme believes that these are probably incommensurable.

It is in seeking for an alternative explanation that Oresme utilizes and applies the traditional and false Pythagorean "experimen-

4. ARITHMETIC: Unless the celestial motions are commensurable, astronomical prediction and tables, as well as knowledge of future events, would be impossible (III.263–66, 275). Terrestrial effects would fail to repeat (III.304–11) and a Great Year could not occur (III.298–303).

GEOMETRY: But if all celestial motions are commensurable, conjunctions and other astronomical aspects could occur only in a certain finite number of fixed and different places in the sky (III.421–28).[108] Thus a peculiar consequence of commensurable celestial motions is that certain places in the sky would be preferred over other places. Would it not be better that such events be capable of occurring anywhere in the sky (III.420–34)? Man cannot attain to exact knowledge of astronomical phenomena and must rest content with approximations (III.435–40). Indeed, acquisition of exact knowledge would serve to discourage man from making continual observations (III.442–44); and if man had precise knowledge of future celestial positions he would become like the immortal gods themselves, a repugnant thought (III.451–54). Fortunately, this is not likely to pose problems since, as shown elsewhere by mathematical demonstration, it is probable that the celestial motions are incommensurable (III.457–66).[109]

Upon completion of the two orations, Oresme is bewildered and unable to determine the truth, which seems to agree with each side (III.469–74). Apollo, sensing Oresme's uneasiness and frustration, generously offers a final and true judgment on the issue, when the dream terminates and Oresme awakens (III.474–81).

It seems natural to inquire why Oresme chose a purely literary framework in

---

tal" account of the discovery of the musical concordances (considering the numerous quotations from Macrobius, it is not unlikely that Oresme drew his account from Macrobius's *Commentary on the Dream of Scipio* 2. 1. 9–15 [Stahl, trans., pp. 186–87; see below, the Commentary, n. 62 for III. 381–83 for the full text of this passage]; however, Stahl notes [ibid., p. 187, n. 6] that similar descriptions are offered by Boethius [*De institutione musica*], Chalcidius [commentary on Plato's *Timaeus*], and Adelard of Bath [*De eodem et diverso*]—authors whose works were available to Oresme). If, as Oresme believed, Pythagoras determined that a ratio of consonant tones was produced by a ratio of the weights of hammers striking an anvil, then by analogy, celestial musical harmony, if it exists, should arise from ratios between the weights or figures of the planets

(III.394–97; on the assumption of the homogeneity of planetary matter, their weights would be proportional to their spherical sizes or volumes). Thus, in one sense, Oresme's explanation (he offers the same one in *Le Livre du ciel et du monde*, bk. 2, chap. 18, 126a [Menut, *Oresme du ciel*, p. 480]; see below, the Commentary, n. 29 for III.162–67) is consistent with the weight relationships of the Pythagorean account; but in another sense, it is deficient, since it fails to describe how the planets, moving through a resistanceless ether, could produce musical sounds by analogy with the striking of hammers on an anvil.

108. See I.855–58.

109. The demonstration referred to here is most probably Chapter 3, Proposition X of Oresme's *De proportionibus proportionum*. See below, pp. 73–76, nn. 112 and 113.

which to argue a scientific question in a treatise which up to this point had been of a highly technical character. Any reasonable attempt to answer this question must turn on the fact that Oresme was perfectly aware that no precise solution or conclusive demonstration could decide whether or not the celestial motions are commensurable. In emphasizing man's inability to acquire *exact* knowledge about sensible things (III.10–31)—and this includes the movements of the celestial motions—Apollo convinces Oresme that no precise mathematical solution or demonstration could ever decide an issue that depended upon physical and sensible, rather than mathematical, criteria (III.32–37). The slightest alteration in the physical relations between any two or more physical things, however small and undetectable, would alter their true mathematical ratio or relationship but go unnoticed by those compelled to rely solely on sense perception.[110] Or conversely, if men attain exactness, they could not prove that they had done so. Thus, for example, even if men could and did produce an exact and absolutely true calendar, they would not know for certain that they had done so (III.29–31). Although we might resort to probable arguments in favor of one or the other hypothesis, such arguments would not answer satisfactorily the fundamental question raised here in the *De commensurabilitate*—namely, is it physically true that celestial motions are commensurable; or are some, or all, incommensurable? A categorical, not a probable, reply was required, since Oresme had himself stated that "it is necessary that all celestial motions be either mutually commensurable, or that some be mutually incommensurable" (I.62–64). Because no solution was possible, a formal scholastic discussion of the issue would have been fruitless. For this reason, it appears that Oresme wished to avoid a serious formal consideration of this insoluble problem.

Nevertheless, since physical, astronomical, and astrological arguments could be seriously influenced by the choice between the two alternatives—even if that choice were made in terms of probability and persuasiveness—Oresme realized how important it was to put before his readers the strongest arguments available for each of these two hypotheses. To achieve this, he hit upon the idea of utiliz-

---

110. For a brief consideration of *De commensurabilitate*, III.9–17 and 34–36, see Maier, *Metaphysische Hintergründe*, p. 402, n. 37. In deliberating the commensurability and incommensurability of celestial motions, Oresme expressed a similar idea in his later *Le Livre du ciel et du monde*, bk. 1, chap. 29, 44c (Menut, *Oresme du ciel*, p. 197): "Now, since it is impossible to know absolutely whether the number of all the stars is even or odd, in the same way the mortal men in the world, dead or alive or to be, could not discover nor know by their natural lights for certain whether all the movements of the heavens are commensurable or incommensurable, for by the part of a movement which would be imperceptible to the senses, even if it were a hundred thousand times larger, two such movements or similar motions could be incommensurable and yet could appear to be commensurable."

ing a literary form that did not commit him, to a definite position and which would yet allow him to propagandize effectively in behalf of his genuine conviction that the celestial motions were probably incommensurable. The debate, or oration, form of Part III served this purpose admirably, for it permitted him the appearance of objectivity in the guise of a seeker after truth listening with rapt attention as the arguments of Apollo's immortal muses were marshaled impressively on each side of this great issue.

But if the realization that a categorical solution was impossible, coupled with a desire to avoid obvious and direct identification with an opinion he favored and deemed probable, led Oresme to present Part III as a debate, why did he choose to employ two separate orations set within a dream rather than utilize a continuous but imaginary dialogue? Why is it that, after Apollo's rebuke to Oresme and all mortals on the futility of seeking exact knowledge through the senses, Arithmetic steps forward and delivers an uninterrupted oration in behalf of the commensurability of the celestial motions, and is then followed by Geometry's continuous plea in behalf of incommensurability? Would it not have been more stimulating and exciting to have had a continuous dialogue or conversation employing argument and immediate counterargument, subtle thrust and counterthrust, rather than long-winded and uninterrupted separate discourses?[111]

However more attractive and exciting the former procedure would have been it is one that Oresme might consciously have avoided since in a continuous conversational dialogue on this particular issue the arguments and counterarguments would have resulted in little more than a sequence of claims and immediate denials with little dramatic interplay of ideas. For example, Arithmetic claims that if the celestial motions were not represented by rational ratios "no one could ever foresee aspects, or predict conjunctions, or learn of effects beforehand. Indeed, astronomy would lie hidden [from us] in every age, unknown and even unknowable,..." (III.263–66). In response, Geometry would have emphasized that knowledge of future events would make man akin to the immortal gods, an unworthy thought (III.451–54); "it is enough to know beforehand that a future conjunction, or eclipse, of this mobile falls below a certain degree, minute, second, or third; nor is it necessary to predict the exact point or instant of time [in which these occur]..." (III.435–38). Arithmetic insists that without precise knowledge of the celestial motions, men would be in a state of perpetual ignorance (III.270); Geometry would counter by asserting that "it would be better that something should always be known about

111. Oresme was surely familiar with the continuous dialogue as exemplified by Cicero's *De natura deorum*, Chalcidius's trans- lation of Plato's *Timaeus*—both of which he quotes directly—and Adelard of Bath's *Questiones naturales*.

them, while, at the same time, something should always remain unknown, so that it may be investigated further" (III.444–46). Throughout Part III, Arithmetic lauds the beauty and harmony of the world founded on celestial motions that are regular and related rationally to produce a never-ending sequence of identical but magnificent celestial and terrestrial effects, while Geometry emphasizes the wonders of an unending sequence of unique and new effects stemming from the incommensurability of these same motions (see III.406–15).

Since a definite solution to this problem was beyond reach, a conversation or dialogue centering on such claims could not have been properly joined to yield any dramatic impact, but would have produced instead a series of tedious assertions and denials without substantive argument. To avoid so sterile a format, Oresme wisely chose to have each side represented by one uninterrupted oration. In this way, maximum effectiveness could be achieved as each side gathered all its arguments for consecutive presentation, so that not only was the greatest cumulative impact possible, but a considerable dramatic element was permitted to develop. Arithmetic could present, in rapid succession, one authoritative and rhetorical appeal after another. An impressive array of quotations from poets and philosophers could be paraded before the reader. In her turn, Geometry was free to reply in kind with arguments and appeals drawn from cosmic diversity.

Finally, we must consider why Oresme chose to have these orations occur within the framework of his own dream. The explanation seems obvious. It provided a convenient means of sidestepping the need for supplying a definite solution to an issue for which man could supply no categorical solution. None could be derived from the rhetorical appeals of Arithmetic and Geometry, for they served only to confuse Oresme, who remarks that "truth seemed consonant with each side" (III.469–70). Apollo explains that the great debate by these two "most illustrious mothers of evident truth" (III.475) had been but an amusing exercise and that now he, Apollo, would pronounce a final and true judgment. But "the dream vanishes, the conclusion is left in doubt" (III.480) and the reader remains in ignorance on this weighty question.

As a vehicle, the dream was a convenient literary device for avoiding a positive stand. Anyone could easily understand the abrupt and frustrating interruption of a dream in which some significant utterance or act was on the verge of occurrence. But the interrupted dream serves an ulterior motive. Not only has this rude awakening robbed us of a unique opportunity to acquire absolute truth, but it has made us dependent on the appeals, such as they are, of Arithmetic and Geometry for whatever insight we may acquire into this problem.

The reader now realizes that at best he must formulate his own judgment solely on the basis of these two orations, which incorporate the only kinds of arguments and appeals that men can know and understand. Compelled to return to the orations themselves, there is little doubt that the reader will settle for the probability—and this is all he can now hope for—that the celestial motions are incommensurable. And this is precisely what Oresme has planned. This is evident in a general way from the fact that Arithmetic was made to speak first and Geometry to have the last word. More important, and even conclusive, is the fact that Geometry alone invokes and appeals to a mathematical demonstration that emerges from all the rhetoric and appeals to authority as the one sound foundation on which to base a decision. As her final argument, Geometry tells us that it has been "demonstrated elsewhere" (III.459)—in Oresme's *De proportionibus proportionum*[112]—that "when any two unknown magnitudes have been designated, it is more probable that they are incommensurable than commensurable, just as it is more probable that any unknown [number] proposed from a multitude of numbers would be non-perfect rather than perfect" (III.459–63). Thus any two unknown celestial (as well as terrestrial) motions would probably be incommensurable and their ratio of velocities irrational.[113] This is beyond

112. In Part III of the *De commensurabilitate*, Oresme could not have appropriately cited his *De proportionibus proportionum*, or, for that matter, any other of his own works. It would have been strange had Oresme introduced quotations from, or titles of, his own works, into the discourses of Apollo, Arithmetic, or Geometry (however, Arithmetic and Geometry were permitted to quote freely and anachronistically from the likes of Cicero, Macrobius, Boethius, and many others). It is presumably for this reason that Oresme has Geometry say only that the proposition in question has been "demonstrated elsewhere." Earlier in the treatise, where such awkardness was not involved, Oresme did cite his *De proportionibus* by name (see II.201).

113. There is little doubt that Oresme intended to impress upon readers of the *De commensurabilitate* his conviction that celestial motions are probably incommensurable. This interpretation gains strong support from the fact that in the much later *Le Livre du ciel et du monde*, Oresme cited the *De commensurabilitate* as the place where he had offered several reasonable arguments to show the probable incommensurability of the celestial motions. ("Et que aucuns des mouvemens du ciel soient inconmensu-

rables, ce est plus vraysemblable que / (44d) n'est l'opposite, si comme je monstray jadys par plusseurs persuasions en un traitié intitulé *De commensurabilitate vel incommensurabilitate motuum celi*."—Menut, *Oresme du ciel*, p. 196; see also V. P. Zoubov, "Nicole Oresme et la musique," *Mediaeval and Renaissance Studies*, vol. 5 [1961], pp. 102–3.) It is, however, surprising that Oresme would cite the *De commensurabilitate*, which contained only "plusseurs persuasions," rather than the *De proportionibus proportionum*, which is the only treatise known thus far where he offered mathematical arguments to demonstrate not only the probable incommensurability of the celestial motions, but also the more general claim that any two ratios involving physical or mathematical magnitudes are probably incommensurable. Indeed, in the *De commensurabilitate* itself, Oresme cites another treatise (III.458–66; see above, n. 112)—almost certainly his *De proportionibus*—for demonstrative support of the general claim, and by implication the specific claim, about celestial incommensurability; it is again the *De proportionibus proportionum* to which Oresme appeals explicitly in support of the same two claims in his *Questiones de sphera* (for the text of this passage, see Grant, *Oresme PPAP*, p. 63,

(Note 113 continued)

n. 81). Because of their importance, Oresme's arguments in the *De proportionibus*, and their obvious relevance for both the *De commensurabilitate* and *Le Livre du ciel et du monde*, justify a brief description and summary at this point.

Proposition X of Chapter 3 of the *De proportionibus proportionum* constitutes Oresme's fundamental mathematical proposition, which was to underlie his many declarations in later treatises, that any two celestial motions would probably be incommensurable. In order to grasp the substance of that proposition, it is essential to understand that in the *De proportionibus*, Oresme was, to a large extent, concerned with demonstrations involving two ratios—say, $A:B$ and $C:D$—related by an exponent, which he calls a "ratio of ratios" (*proportio proportionum;* see Grant, *Oresme PPAP*, pp. 49–50). Thus, if $A:B = (C:D)^{p/q}$, Oresme seeks to determine whether p/q, the exponent, is rational or irrational. Should it be rational, he would then say that ratios $A:B$ and $C:D$ are commensurable and constitute a "rational ratio of ratios" (this obtains whether $A:B$ and $C:D$ are rational or irrational, or whether one is rational and the other irrational; ibid., pp. 38–39). If p/q is irrational, then $A:B$ and $C:D$ form an "irrational ratio of ratios" (here, too, it is immaterial whether the two ratios $A:B$ and $C:D$ are both rational, both irrational, or one rational and the other irrational; ibid., pp. 39–40). For example, $3:1$ and $27:1$ produce a rational ratio of ratios, since $27:1 = (3:1)^{3/1}$; but $3:1$ and $6:1$ form an irrational ratio of ratios, since $6:1 \neq (3:1)^{p/q}$, where p/q is rational—i.e., only if p/q is an irrational exponent can these two ratios be related.

In Chapter 3, Proposition X (ibid., pp. 247–55 for text and translation and pp. 40–42 for discussion), Oresme demonstrates (ibid., p. 247) that "it is probable that two proposed unknown ratios are incommensurable because if many unknown ratios are proposed it is most probable that any [one] would be incommensurable to any [other]."

The proof embodies an antecedent and consequent. The antecedent proclaims (ibid., p. 246, lines 343–47) that for any given sequence of ratios (for a qualification see ibid., p. 41, n. 54) more irrational than rational ratios of ratios can be formed. This is shown by a specific example (ibid., pp. 246–48, lines 349–55) involving 100 rational ratios from $2:1$ to $101:1$. When any two of them are related exponentially, the possible combinations among ratios of greater inequality only are 4,950, of which only 25 are rational ratios of ratios; all of the remainder are irrational. Consequently, the ratio of irrational to rational ratios of ratios is $197:1$, a disparity that would increase as more ratios are taken (ibid., p. 248, lines 355–58; the details of this example are given in a *practica conclusio* [chap., 3, prop. XI] on pp. 254–58, lines 440–98).

The consequent, which is actually the enunciation of Proposition X, asserts that any two unknown ratios are probably incommensurable. This follows immediately from the antecedent, for it is based on the probability that when any two unknown ratios are given, they would be incommensurable—i.e., relatable only by an irrational exponent—since there are more irrational than rational ratios of ratios. As Oresme expresses it (the very remark is made in the *De commensurabilitate* about perfect and non-perfect numbers [III.461–63]): "...we see that however many numbers are taken in series, the number of perfect or cube numbers is much less than other numbers and as more numbers are taken in the series the greater is the ratio of non-cube to cube numbers or non-perfect to perfect numbers. Thus if there were some number and such information as what it is or how great it is, and whether it is large or small, were wholly unknown... it will be likely that such an unknown number would not be a cube number....

"Now what has been said here about numbers may be applied to ratios of rational ratios, as was shown before, since there are many more irrationals than others, understood in the previous sense [of ratios of ratios].... Therefore, if any unknown ratio of ratios were sought, it is probable that it would be irrational and its ratios incommensurable"—(Ibid., pp. 249, 251).

Thus far, all this is purely mathematical and it is not until Chapter 4, Proposition VII, the final proposition of the treatise, that Oresme applies Chapter 3, Proposition X to celestial motion and, indeed, to all continuous magnitudes in general. On the assumption that ratios of velocities are relatable just as ratios of ratios (ibid., p. 286, lines 334–37 and p. 51), Oresme proceeds as if all ratios that represent changeable and continuous magnitudes truly reflect the mathematical behavior of ratios described in Chapter 3, Proposition X.

In order to represent a ratio of terrestrial velocities arising from ratios of force to resistance, Oresme adopted and used a widely accepted function formulated by Thomas Bradwardine in 1328 (ibid., p. 17) and which we may represent as $F_2 : R_2 = (F_1 : R_1)^{V_2/V_1}$, where F signifies motive force, R resistive power, and V velocity. By analogy with terrestrial motion, Oresme says that such ratios of force and resistance can also be applied to celestial motions. Kinematically, however, such celestial velocities are derived "from the ratio of the quantities of the motions or circles described, *and* from the ratio of the times in which they revolve" (ibid., p. 293). Thus if we know the ratio of distances traversed by any two planets in equal times (the distances are measured by angles swept out by radius vectors), we can determine their ratio of velocities, since $S_2 : S_1 = V_2 : V_1$, where S is distance; or, if the distances are equal but traversed in unequal times, the ratio of velocities will be inversely as the ratio of the times—i.e., $T_1 : T_2 = V_2 : V_1$ when $S_2 = S_1$ (ibid., pp. 52–53). Since ratios of distances, times, and velocities are interrelated in the manner just described, the ratio of velocities in the formulation $F_2 : R_2 = (F_1 : R_1)^{V_2/V_1}$ could be replaced by $S_2 : S_1$ when $T_2 = T_1$, or replaced by $T_1 : T_2$ when $S_2 = S_1$. But whichever of these possible ratios is used as the exponent, we

have in each instance a ratio of ratios. And since Chapter 3, Proposition X had demonstrated in a purely mathematical context the probability that any two unknown ratios selected at random from a properly chosen sequence (ibid., p. 41, n. 54) would probably be incommensurable—i.e., only relatable by an irrational exponent and constituting an irrational ratio of ratios—Oresme now believes it justifiable to extend the application of this demonstration to ratios of ratios representing both terrestrial and celestial physical magnitudes. Hence we have a straightforward extension to physics and cosmology of results obtained in pure mathematics.

As a first move in this direction, Oresme generalizes the range of application of the mathematical probability argument to embrace all continuous magnitudes: "When there have been proposed any two things whatever acquirable [or traversable] by a continuous motion and whose ratio is unknown, it is probable that they are incommensurable. And if more are proposed, it is more probable that any [one of them] is incommensurable to any [other]. The same thing can be said of two times and of any continuous quantities whatever" (ibid., p. 303). It follows from this that any proposed, but unknown, ratio of terrestrial or celestial velocities would probably be incommensurable, so that Oresme could conclude: "When two motions of celestial bodies have been proposed, it is probable that they would be incommensurable, and most probable that any celestial motion would be incommensurable to the motion of any other [celestial] sphere;..." (ibid., p. 305). And in what may be a reference to the *Ad pauca respicientes*, which in its corrected and expanded form probably became the *De commensurabilitate*, Oresme, a few lines below, declares that "...many very beautiful propositions that I arranged at another time follow, and I intend to demonstrate them more perfectly later, in the last chapter,..." (ibid.; despite Oresme's reference to the last chapter of the *De proportionibus*, I have argued [ibid., pp. 76–80] that this refers to the *Ad pauca* and, there-

question the strongest argument found in Part III and, since Oresme chose to conclude the orations with it, he leaves little doubt that he intended to guide the reader to his viewpoint[114] as skillfully and unobtrusively as possible.

If the interpretation given above is a reasonable reflection of the considerations that induced Oresme to choose the particular format of Part III, it is obvious that it represents a carefully constructed climax to a sophisticated and unusual treatise. But quite apart from the structure of Part III, we are impressed by the parade of quotations and the evident learning that is displayed here. Although on occasion Oresme composed a poem or became lyrical and enthusiastic on some particular idea or concept, nowhere else in his known works does there appear anything comparable to the literary effort of Part III. A striking, but upon reflection not really unexpected, feature of the concluding part of the *De commensurabilitate* is the fact that nowhere does Oresme inject any Christian

---

(Note 113 continued)
fore, indirectly to the *De commensurabilitate*).

The passage in *De commensurabilitate*, III. 457–66, and the arguments of Geometry in general, may have been in Oresme's mind when, in his *Livre de divinacions*, he says that astronomy (actually the word *astrologie* is used, but the terms are synonymous in the context of the discussion) "can be adequately known but it cannot be known precisely and with punctual exactness, as I have shown in my treatise on the Measurement of the Movements of the Heavens [*La Mesure des mouvemens du ciel*] and have proved by reason founded on mathematical demonstration."—G. W. Coopland, ed. and trans., *Nicole Oresme and the Astrologers* (Cambridge, Mass., 1952), p. 55. Since the title that Oresme cites here has only a general relationship to the title of the *De commensurabilitate*, it is possible that the *Livre de divinacions* was written prior to the *De commensurabilitate* and we have here a reference to Oresme's *Ad pauca respicientes*, which in one manuscript is called *Tractatus brevis et utilis de proportionalitate motuum celestium* (see Grant, *Oresme PPAP*, p. 380; this title does not precisely translate as *La Mesure des mouvemens du ciel* and I have argued that there is no concrete evidence for believing that it was the proper title of the *Ad pauca respicientes*, which is a title that I formulated

from the opening words of the treatise). In my view the reference is to one of these two treatises, but the title cited by Oresme seems definitely to preclude the *De proportionibus proportionum* as the intended work, even though it would have been the most appropriate reference—provided, of course, that it antedates the *Livre de divinacions*—for it is only in the *De proportionibus* that Oresme "proved" that we cannot obtain precise knowledge of celestial motions and positions.

114. From note 113 and the frequent use made of this key concept in many of his other treatises (e.g., *De proportionibus proportionum*, *Ad pauca respicientes*, *Questiones de sphera*, *Questiones super de celo*, *Quaestiones super geometriam Euclidis*, the *Quodlibeta*, and *Le Livre du ciel et du monde*), there is no doubt whatever of Oresme's conviction that the celestial motions were probably incommensurable. But in the *De commensurabilitate*, where he had to give the appearance of impartiality, this conviction could not be made explicit. I have tried to show, however, that the arguments and appeals are so formulated as to convince the reader that the celestial motions are probably incommensurable. Indeed, in his *Le Livre du ciel et du monde*, he tells us that he had shown this in the *De commensurabilitate* (the statement appears near the beginning of the immediately preceding note).

theme or special appeal to his Christian faith. Among the numerous quotations, offered for the most part by Arithmetic, only three are from Holy Scripture but are not different in kind from the mass of quotations drawn from such pagans as Plato, Aristotle, Cicero, and others. One biblical citation (Ecclesiastes 1:5–6 in III.294–97) is said to have been drawn from "divine pronouncements" ("divinis oraculis"; III.294), a description sufficiently vague to embrace the utterance of any pagan oracle or god. In choosing a pagan god and two of his attendants as the instruments for his debate, Oresme could not interject quotations that were patently Jewish or Christian in character. And it would have been grossly anachronistic and absurd to have Apollo and his muses cite by name specific biblical books. Although Oresme did not wholly refrain from the use of biblical quotes, these are masked and blended into the pagan setting. Many Christians who employed pagan themes had followed a quite similar procedure.[115]

---

115. See Stahl, trans., *Macrobius*, p. 8.

3

The Concept of Celestial Commensurability and Incommensurability from Antiquity to the Sixteenth Century

Before Oresme

In his prologue to the *De commensurabilitate*, Oresme declares that "if another has set out the more fundamental principles [or elements found in this book], I have yet to see them" (Prol. 48). Thus, apart from some commonly known and used mathematical principles and definitions that he admittedly borrowed from other treatises (Prol. 45–46; exemplified in I.2–23), and perhaps even a few unacknowledged propositions on celestial commensurability (discussed below in the section on Johannes de Muris), Oresme wanted it known that most of the propositions, and indeed the fundamental character of the book itself, were original. As will be seen in what follows, all the evidence of which I have become aware tends to uphold the reasonableness of this claim.[1]

I now know of only two authors who might properly qualify as Oresme's precursors in the mathematical treatment of celestial commensurability and incommensurability. One is Theodosius of Bithynia (born ca. 180 B.C.), often called Theodosius of Tripoli,[2] and the other is Johannes de Muris

---

1. A preliminary discussion in support of this position appears in Grant, *Oresme PPAP*, pp. 111–21. However, new evidence that has since come to light, especially in the *Quadripartitum numerorum* of Johannes de Muris, as well as a thorough reexamination of earlier materials in my previous summary, has necessitated a more comprehensive exposition and reevaluation of the history of this problem. Although important qualifications are in order, the available evidence does not warrant a denial of Oresme's claim to origin-

ality; it only requires proper characterization and definition.

2. The surname "of Bithynia" rather than "of Tripoli" is argued for in the most recent authoritative discussions by R. Fecht (see p. 3 of his edition cited in the next note) and Konrat Ziegler (*Paulys Realencyclopädie der classischen Altertumswissenschaft*, s.v. "Theodosios, Mathematiker und Astronom aus Bithynien," rev., 2d ser., vol. 10 [Stuttgart, 1934], cols. 1930–32).

(d. ca. 1350), a well-known mathematician and astronomer and an older contemporary of Oresme.

Of the extant works of Theodosius, only the five concluding propositions of his *De diebus et noctibus*[3] are relevant. Here, in Book 2, Propositions 15–19,[4] Theodosius was concerned exclusively with the relationships that obtain between the daily and annual motions of the sun. If these motions are commensurable, all solar events will repeat after certain intervals; if not commensurable, they will never repeat. In assuming commensurable solar motions and different lengths of the solar year, Theodosius seeks to determine the periods of time that must elapse between successive repetitions of solar occurrences for the different assumed lengths of the solar year.

In Proposition 15 Theodosius says:

If the total number of revolutions in a solar year consists of a rational number of days and nights, then the [corresponding] days and nights of each year will be equal in length and number; and the solstices, risings, and settings will occur at the same points of the horizon and solar circle [i.e., ecliptic]; and, moreover, the sun will reach the tropics and equator in the [very] same hour.[5]

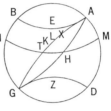

Fig. 9

3. The Greek text is in two books and has been edited and translated into Latin by Rudolf Fecht, *Theodosii De habitationibus; De diebus et noctibus* in *Abhandlungen der Gesellschaft der Wissenschaften zu Göttingen*, phil.-hist. Klasse, n.s. (Berlin, 1927), vol. 19, 4. The last five propositions, 15–19 of bk. 2, are on pp. 144–55.

4. These propositions were briefly summarized, but not translated, in Grant, *Oresme PPAP*, pp. 111–13. I here include complete English translations of these five propositions, citing the appropriate page numbers of the Greek text and Latin translation in Fecht's edition.

5. Fecht, *De diebus*, pp. 144–45. I now translate the remainder of this proposition (ibid., 144–47), which demonstrates what has been enunciated above. In this and the other remaining propositions of the *De diebus et noctibus*, I have replaced the Greek letters in the text and figures with roman letters.

"Let *ABGD* be the horizon circle [Fig. 9], *BEA* the summer tropic, *GZD* the winter tropic, and let the solar circle [i.e., ecliptic] be located at *ATGH*; and let the sun rise on a certain day at point *T* and when it has com-

pleted all of its revolutions on circumference *TAHGT* it rises again at point *T*. I say that the days and nights of each year would be equal in length and number.

"For in the first year, after it rises at *T*, the sun sets at point *K* and then rises at point *L*. Now since in the first year, arc *TK* becomes the visible hemisphere during the time in which it is traversed by the sun, then arc *TK* always becomes the visible hemisphere during the same time [of year]; and, moreover, the sun always traverses arc *TK* in an equal time, so that it also traverses it in the same time during the second year and arc *TK* becomes the visible hemisphere rising with point *T* and setting with *K*. Therefore, the time of day in which the sun traverses arc

Next, in Proposition 16, Theodosius complicates the problem somewhat by adding fractional parts to the length of the solar year:

But if the solar year does not consist of an integral number of revolutions, but a certain part [of a revolution] must be added to the whole number of revolutions, then the days and nights of the first year will not be equal in length to the days and nights of the following year; and neither the solstices nor risings and settings will occur at the same points of the horizon and solar circle; and the sun will not reach the tropics or equator at the [very] same hour.[6]

(Note 5 continued)

TK in the second year is the same as in the first year. Again, since in the first year, arc *KL* becomes the darkened hemisphere during the time in which it is traversed by the sun, then arc *KL* always becomes the darkened hemisphere during the same time [of year]; and, moreover, the sun always traverses arc *KL* in an equal time, so that it also traverses it in the same time during the second year and arc *KL* becomes the darkened hemisphere, setting with point *K* and rising with point *L*. Therefore, the time of night in which the sun traverses arc *KL* in the second year is the same as in the first year.

"In a like manner, one could demonstrate that the days and nights of each year would be numerically equal.

"I say also that the risings and settings would occur at the same points of the horizon and solar circle.

"It appears, then, that the sun always rises and sets at the same points on the solar circle. I say that this also occurs on the horizon. For let *NM* be the circle on which point *T* is carried. And because point *T* always rises at point *M*, the sun also will always rise at point *M*. Therefore, the risings and settings will occur at the same points of the horizon and solar circle.

"I say that the sun would also be in the tropics at the same hour. If the sun will have made a rising at point *A* after having made an integral number of revolutions in advancing from point *T* along arc *TA*, it would appear that the sun would be in the [summer] tropic at the [very] same hour of every year. For, indeed, having completed a whole num-

ber of revolutions after having advanced along arc *TX* from point *T*, the sun would traverse the remaining arc, *XA*, in a certain part of a revolution. But since the sun traverses arc *TX* in a whole number of revolutions, it also traverses arc *XA* in an equal time, so that in every year the sun will reach the tropic at the [very] same hour. We can also demonstrate that the sun will arrive at the equator and the winter tropic in the [very] same hour."

The opposite of this, involving a celestial body (the moon) that will never twice rise in the same place at the same corresponding time of any two years, is demonstrated by Oresme in his *Quaestiones super geometriam Euclidis* (Busard, ed., p. 24) and by the author of an anonymous *Questio*. See above, chap. 2, n. 97, and Grant, *Oresme PPAP*, pp. 77–78, n. 101.

6. Fecht, *De diebus*, pp. 148–49. The demonstration of this short proposition follows (ibid.):

"Let the solar circle be *GED* [Fig. 10], and [assume that] on a certain day the sun rises at point *Z*, and, upon completion of an integral number of revolutions on circum-

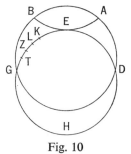

Fig. 10

But if solar events will not recur exactly in successive years when fractional parts are added to the length of the year, they will recur after the passage of some time, the time depending upon the magnitude of the rational fractional part, or parts, that is added to the integral number of solar revolutions, or days, in the year. In Propositions 17 and 18, Theodosius assumes that lengths of the year are equal to $365\frac{1}{4}$ and $365\frac{3}{19}$ days, respectively, and proceeds to determine the intervening period of time before exact recurrence. Proposition 17, which begins with a brief recapitulation of what has immediately preceded, reads as follows:[7]

If we assume that the revolutions of the sun are made in equal times—and this appears true to our senses—and the whole year consists of an integral number of solar revolutions, all [solar] events would recur in [exactly] the same way every year, as we showed above. But if the year does not consist of an integral number of revolutions, but adds a part of a revolution that is commensurable to a whole revolution, [then], as already stated, the same [solar] events would not recur in [exactly] the same way in immediately successive years, but this would happen only after [the passage of] some years. Precisely how many years must pass can be shown in this way: having taken two mutually prime numbers in the same ratio as one whole revolution to the part added, everything will recur in the same way after the passage of a number of years equal in magnitude to the greater number.[8]

---

ference *ZEHT*, it rises at point *T*. I say that the days and nights are unequal in length.

"Now since the two risings are at points *T* and *Z*, the setting after the point of rising at *T* will be nearer than the setting after the point of rise at *Z*. Let the setting after the rise at point *T* be at point *L*; and let the setting after the point of rise at *Z* be at point *K*. Now since an earlier day is shorter than a later day as the sun advances from the winter solstice, the day in which the sun traverses arc *TL* is shorter than the day in which it traverses arc *ZK*. Therefore the days are unequal.

"In a similar way, the nights can be shown to be unequal. And it appears that neither the solstices nor the risings and settings could occur at the same points of the horizon and solar circle; nor would the sun be on the tropics and equator at the [very] same hour."

7. Ibid., pp. 148–51.

8. Oresme's rule for finding the number of points in which the sun can enter Cancer (pt. I, prop. 22 [I.750–74]) is essentially the same as Theodosius's rule for determining

the number of years that must intervene between exact recurrences of solar phenomena. This is not surprising, since both are concerned with two simultaneous motions of a single celestial body, the sun (for Oresme the sun is exemplary, since he extends his results to the entry of any celestial body into any degree [see I.798–99]; but Theodosius, who is primarily concerned with days and nights, as the title emphasizes, formulates his rule solely for the two motions of the sun with which he is exclusively concerned).

What differences there are arise from somewhat divergent objectives. Oresme seeks to determine the number of fixed points on which the sun can enter Cancer and gives, accordingly, a ratio between the motions of the tropic of Cancer—this represents the sun's daily motion—and the sun's motion along the ecliptic, i.e., the sun's annual motion. Thedosius seeks the interval of time between the exact recurrence of solar phenomena and therefore gives a ratio of times between the length of the year and a day. The number of circulations made by the tropic of

But if the part added is incommensurable to a whole revolution, all [solar] events will not recur in the same way.

Now since according to Callippus, the year consists of 365 days plus a fourth part of a [solar] revolution, everything should recur after four years, because it is a fourth part that is added.

Let us assume these things [from Callippus], and [also assume] that the sun rises at point *E* on circumference *EDZH* [see Fig. 11] and, after completing 365 whole revolu-

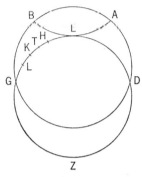

Fig. 11

tions, rises at point *H*; it would then traverse the remaining arc, *HE*, in a fourth part of a revolution. I say that after four years all [solar] events would repeat in the same way.

Let arcs *HT*, *TK*, and *KL* be assumed equal to arc *HE*. Now in the first year, since the sun has advanced from point *E*, arc *HE* would remain after 365 revolutions had been completed, so that the sun rises at point *H*. In the second year, the sun advances from point *H* and an arc equal to *HE*—namely, *HT*—would remain after 365 revolutions had been completed, so that the sun rises at point *T*. In the third year, the sun advances from point *T* and an arc equal to *HT*—namely, *TK*—would remain after 365

(Note 8 continued)
Cancer for every revolution by the sun on the ecliptic is, in Oresme's scheme, analogous to the number of days in the solar year in the propositions of Theodosius. In both approaches, however, when the number of days, or circulations, involves rational fractions, the greater number of the fraction—namely, the denominator—determines the fixed and regular relationship. For Oresme it indicates the sun's fixed points of entry into Cancer, while for Theodosius it signifies the number of years that must pass before solar phenomena will repeat exactly. If only multiple ratios, and no fractions, are involved, then in Oresme's system only one point can serve as

the first point of Cancer to receive the sun at regular intervals; for Theodosius it signifies that solar events will repeat exactly in successive years. To arrive at the sorts of results that were of interest to Theodosius, Oresme had only to supply the period of time that elapsed between successive entries of the sun into the tropic of Cancer. He could then easily calculate the number of years that would necessarily elapse before the sun entered the very same point of the tropic of Cancer (indeed, he does just this in I.793–99).

Also cf. Johannes de Muris, *Quadripartitum*, bk. 4, chap. 26, below in App. 1 and the summary of it on p. 96 (especially n. 44).

revolutions had been completed, so that the sun rises at point K. In the fourth year, the sun advances from point K and an arc equal to TK—namely, arc KL—would remain after 365 revolutions had been completed, so that the sun rises at point L. Now since the sun traverses each of the arcs LK, KT, TH, and HE, in a fourth part of a revolution [or day], it would traverse arc LE in one whole revolution. Therefore, when arc LE has been traversed by a whole revolution, the sun rises again at point E, where it rose in the first year. Thus after four years, all [solar] events would recur in the same way.

The interval of recurrence having now been calculated for a length of year involving a unit fraction, or single part of a day, Theodosius, in Proposition 18,[9] shows the procedure for determining such an interval when the length of the year adopted might involve an improper fraction, i.e., a fraction whose numerator, while less than the denominator, is greater than 1.

Again, since according to Meton and Euctemon, the year consists of 365 days plus $\frac{5}{19}$ parts of a [solar] revolution, all [solar] events should repeat after 19 years.

For let the solar circle be $EDZG$ [see Fig. 12], and [assume that] the sun rises at point E on circumference $EDZH$ and after completing all its integral revolutions [i.e., 365], it would traverse the remaining arc, HE, in $\frac{5}{19}$ of a revolution. I say that after 19 years everything would repeat in the same way.

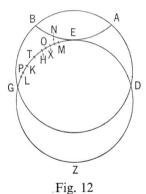

.Fig. 12

Let arcs HT, TK, and KL be assumed equal to arc EH; let arc LP be $\frac{1}{19}$ part [of a revolution]; and let arc EH be divided into $\frac{5}{19}$ parts at M, N, X, and O. Now in the first year, since the sun has advanced from point E, arc EH would remain after 365 revolutions had been completed and the sun rises at point H; in the second year, an arc would remain equal to HE—namely, HT— and the sun would rise at point T; in the third year arc TK would remain and the sun would rise at K; in the fourth year, the sun would rise at point L and when it had gone beyond and traversed arc LM, which is [equal to] one revolution,[10] it would rise again in point M and arc ME would remain

9. Fecht, *De diebus*, pp. 152–55.

10. That is, $LM = LK+KT+TH+HM = \frac{5}{19}+\frac{5}{19}+\frac{5}{19}+\frac{4}{19} = \frac{19}{19}$.

—namely, $\frac{1}{19}$ of a revolution. Therefore in four years, the sun adds one revolution, and $\frac{1}{19}$ of a revolution is left, so that in 16 years the sun adds $\frac{4}{19}$ of a revolution and will be at point O. Now in the three years remaining, $\frac{5}{19}$ of a revolution in each of these years would leave $\frac{15}{19}$ of a revolution and the sun will be at P. But after completing one revolution it will be at point E and the starting point will have been reached, so that in 19 years $\frac{19}{19}$ of a revolution would remain—namely, a whole revolution. Therefore, when this revolution has been made the sun would once again rise at point E and after 19 years everything would recur in the same way.

Exact recurrences would be impossible, however, if the part added were incommensurable to the whole. This is the substance of Proposition 19,[11] the concluding proposition of the treatise.

But if the part that is added is not commensurable to a whole revolution, [solar] events could never repeat, i.e., we shall demonstrate that the sun could never return to the same place [in any equal time intervals whatever].[12]

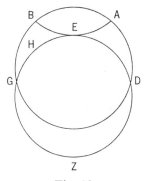

Fig. 13

Let the solar circle be $EGDZ$ [Fig. 13] and [assume that] the sun rises at E; after 365 revolutions had been completed it would rise again at point H and arc HE, which is incommensurable to a whole revolution, would remain. I say that all the [solar] events could never recur in the same way.

For if it could happen, let these events repeat. Then, since in every year an arc is left that equals arc HE, there would remain in a later year an arc equal to arcs composed of arc EH; and this [whole] arc should be measured by a whole revolution. Moreover, arc HE measures the arc composed of arcs equal to EH, and therefore arc EH is commensurable to a whole revolution; but it is also incommensurable. Therefore, all [solar] events could never repeat in the same way.

11. Ibid., pp. 154–55.   its path over the ecliptic—i.e., all solar events
12. The qualification in square brackets   will repeat, but not in equal time intervals.
must be added, since the sun always retraces

The fundamental assumption that links these five propositions in the *De diebus et noctibus* is that the sun's daily and annual motions must be either commensurable or incommensurable. Without hinting at a "true" opinion or judgment in this matter, Theodosius assumes their commensurability in Propositions 15–18 and derives the consequence of regular periodic recurrence of solar phenomena; and in Proposition 19 he concludes his treatise by assuming these motions to be incommensurable, demonstrating the impossibility of exact recurrence of solar phenomena over any periods whatever. The aims and essential characteristics of these five propositions are incorporated into parts of only two (of 37) propositions in the *De commensurabilitate*, where Oresme considers the simultaneous motions of a single body instead of the motions of separate bodies, as he does in all other propositions. Thus in Part I, Proposition 22, Oresme assumes that the daily and annual solar motions are commensurable; and in Part II, Proposition 12, he assumes that they are incommensurable.

Theodosius and Oresme had somewhat different immediate objectives in these propositions: where Oresme sought to determine the number of fixed points on which the sun could enter the tropic of Cancer, Theodosius wished to determine the times that must elapse before solar events could repeat exactly. They offer much the same rule, however, since the problem that underlies their divergent objectives is basically the same (see above, n. 8 of this chapter). But even in these two propositions, it is obvious that the consequences drawn by Oresme are more interesting, subtle, and detailed than those found in the five relevant propositions of Theodosius. In comparison with the entire *De commensurabilitate*, however, it is immediately apparent that Theodosius's *De diebus et noctibus* could not have served as its source, for other than the parts of the two propositions discussed above, concerning the sun's simultaneous motions, the remainder of the *De commensurabilitate* has no counterpart in the *De diebus et noctibus*.

And yet, is it not possible that Oresme inherited from this treatise of Theodosius, either directly or indirectly, the basic theme that is the leitmotif of the *De commensurabilitate*—namely, that celestial phenomena produced by commensurable celestial motions would be regular and periodic, while those produced by incommensurable motions would be non-repetitive? Discovering this initially in Theodosius's discussion of solar motions, perhaps Oresme was motivated to pursue the matter until he had extended this basic concept to the motions of separate bodies and derived consequences whose implications were applied to cosmological, philosophical, and astrological problems. Oresme's treatise would thus be original in its scope and in the richness of detail and consequence, but would have its roots, and perhaps its inspiration, ultimately in Theodosius's *De diebus et noctibus*. This is possible, but unlikely, for although Costa ben Luca translated the *De diebus et noctibus* into Arabic in the ninth century,[13] there is as yet no evidence of a medieval Latin translation from either Greek or Arabic.[14] It is also possible that the substance of these propositions was incorporated into some other work that was subsequently translated into Latin and read by Oresme. Until there is reasonable evidence for this suggestion, or until a hitherto unknown Latin translation of the *De diebus et noctibus* is discovered, the hypothesis that Book 2, Propositions 15–19 of the *De diebus et noctibus* exerted any influence on Oresme must be considered improbable, but nevertheless possible.

Although the essential content of Theodosius's five relevant propositions can be found in Oresme's *De commensurabilitate*, a more plausible potential source of direct influence may have been an assortment of chapters in Book 4, Tract 1, of the *Quadripartitum numerorum* completed in 1343[15] by Johannes de Muris. In the summary of these chapters below, it will be evident that Johannes de Muris did not consider, as did Oresme, propositions of the kind found in the *De diebus et noctibus*. Thus, even if the latter treatise were available in some Latin version, it seems highly improbable that it could have exerted any influence on the method and content of the chapters formulated by de Muris.

In the fourth book of the *Quadripartitum*, Johannes de Muris is concerned with the application of numbers and fractions to physical quantities, and in

13. See Moritz Steinschneider, *Die arabischen Übersetzungen aus dem Griechischen* (Graz, 1960), p. 221.

14. In *Die europäischen Übersetzungen aus dem Arabischen bis mitte des 17. Jahrhunderts* (Graz, 1956), pp. 19–20, Moritz Steinschneider cites translations from Arabic into Latin of Theodosius's *De sphaera* and *De habitationibus*, but makes no mention of the *De diebus et noctibus*. Furthermore, Francis J. Carmody, in an article on Autolycus of Pitane published in *Catalogus translationum et commentariorum* (ed. P. Kristeller), vol. 1 (1960), pp. 170, col. 2–171, col. 1, reports that certain Greek codices in the Vatican Library contain works of the minor Greek mathematicians and astronomers, including the *De diebus et noctibus*, *De sphaera*, and *De habitationibus* of Theodosius. He believes that a similar codex served as the basis for translations of these treatises into Arabic. He also observes that Latin manuscript codex Bibliothèque Nationale 9335 includes four treatises from among the works of these minor Greek scientific writers, making it a reasonable inference that these were translated from an Arabic codex similar in content to one of the earlier Greek codices. But whereas the *De sphaera* and *De habitationibus* are two of the four treatises, the *De diebus et noctibus* is missing. While it would be improper to infer from evidence of this kind that no such translation had been made, it seems that none has yet been identified or become known. Konrat Ziegler declares (*Paulys Realencyclopädie...*, s.v. "Theodosios, Mathematiker und Astronom aus Bithynien," rev., 2d ser., vol. 10, col. 1933) that prior to the time of Fecht's edition of *De diebus et noctibus*, the latter work was known only in a rather superficial and arbitrary Latin translation by Auria, published in Rome in 1591. Hence the earliest Latin translation may not antedate 1591.

15. So we are told on fol. 100v of manuscript codex Bibliothèque Nationale, fonds latin, 7190 (see Marshall Clagett, *The Science of Mechanics in the Middle Ages* [Madison, Wis., 1959], p. 120).

Tract 1 he devotes himself solely to problems involving rectilinear and circular motion.[16] Of the 32 chapters in Tract 1, only 10 are genuinely relevant to the subject matter of this volume (these are 12–14, 21–26, and 28).[17] I shall now summarize briefly the content of these chapters and indicate by way of footnotes whether or not there are corresponding or similar methods, ideas, and examples in Oresme's *Ad pauca respicientes*, *De proportionibus proportionum*, and *De commensurabilitate*. The Latin text and translation of these chapters from the *Quadripartitum* are given in Appendix 1 of this volume.

In Chapter 12, the first of the relevant chapters, Johannes assigns velocities to two mobiles, or bodies, and then, assuming that they move in opposite directions[18] after starting simultaneously from a common point on a circle, seeks to determine where, when, and how many times they will conjunct before meeting again in their starting point. If A makes one circulation in 3 days and B[19] in 4 days, their ratio of velocities is $V_A : V_B = 4 : 3$. Should the circle be divided into 7 equal parts, then when A will have traversed 4 parts in one direction, B will have traversed 3 parts in the opposite direction and they will conjunct. The time of conjunction is calculated by multiplying the parts traversed by one or the other of the bodies by the time required for that same body to complete one circulation. Thus for A we get $\frac{4}{7} \cdot 3 = 1\frac{5}{7}$ days; and for B, $\frac{3}{7} \cdot 4 = 1\frac{5}{7}$ days, so that they must conjunct in $1\frac{5}{7}$ days. Without providing the computation, Johannes tells us that A and B must conjunct 7 times before they complete a revolution and arrive simultaneously at their starting point.[20]

16. This first of five tracts in Book 4 is titled "De moventibus et motis secundum motum rectum et circularem una cum motu gravium et levium" ("On movers and things moved with rectilinear and circular motion [and having] only the motion of heavy and light bodies"; see BN 7190, fol. 72r).

17. I am indebted to Professor Marshall Clagett for discovering these chapters and bringing them to my attention.

18. In *De commensurabilitate*, I.59–60, Oresme says that he will only consider motions in the same direction, since determinations about other kinds—including, presumably, oppositely directed motions—can be derived from what will be said about similarly directed motions.

19. In this chapter, as in most, Johannes does not use letters to represent velocities, distances, or bodies. Only in Chapter 13 are letters used at all, and then only to represent

the moving bodies, which are called A and B. To facilitate the discussion, letters have been assigned throughout.

20. Johannes de Muris devotes all of Chapter 26 and part of Chapter 13 (see below) to the determination of the number of times that two or more mobiles will conjunct before completing a revolution (Oresme's methods are contrasted below in n. 24 of this chapter). Although Johannes does not formally define *revolutio*, his use and understanding of that term in this chapter and elsewhere (for example, chap. 13, line 12 and chap. 26, line 6) are the same as the definition in *De commensurabilitate*, I.43–44. However, just as Oresme was to do in the *Ad pauca respicientes* (but not in the *De commensurabilitate*), we find Johannes also using *revolutio* synonymously with *circulatio* (for example, chap. 14, lines 5–6 and chap. 23, line 5). See above, chap. 2, n. 2.

This method, we are told (chap. 12, lines 19–20), is applicable to all things moving (uniformly) in contrary directions. As specific illustrations, Johannes turns to celestial motions, offering the motion of the moon and the oppositely directed motion of its nodes, called the head and tail of the dragon.[21] Indeed he says this method can be applied to the motion of any planet and its nodes. Although the initial conditions and data involved contrary motions of two bodies, we see that the illustrations do not concern the motions of planets, since these do not move in contrary directions. Instead, Johannes cites as examples of such motions the conjunction, or meeting, of a single celestial body with a point moving in the opposite direction.

Chapter 13 may be conveniently subdivided into three parts which have direct counterparts in Oresme's *Ad pauca respicientes*, Part 1, Proposition 2 and *De commensurabilitate*, Part I, Propositions 6, 7, and 8. Here Johannes shows how to determine: (1) the time required for two mobiles to conjunct for the first time after simultaneous departure from a common point; (2) the place where they will conjunct; and (3) the total number of conjunctions that will occur in the course of a single revolution.

Proceeding as usual by means of concrete examples, Johannes divides the circle into 24 equal parts and assumes that A moves 5 parts per day and B 9 parts. If S is the distance or number of parts traversed by a mobile in a day, then, if we take $S_B - S_A = 9 - 5 = 4$ and divide this difference into the whole circle, or 24 parts, we obtain $\frac{24}{4} = 6$ signifying that B, the quicker mobile, will overtake A in 6 days and produce the first conjunction.[22] To find the place on the circle where the first conjunction occurs, multiply 5, the daily distance traversed by A, by 6, the time elapsed before the first conjunction. From this product, 30, subtract the total number of parts of the circle, or 24, which leaves 6, and signifies that A and B will conjunct in the 6th part of the circle located clockwise from their common starting point.[23] Superfluously, but perhaps as a check,

---

21. For Sacrobosco's definition of the "head and tail of the dragon," see his *Tractatus de spera*, chap. 4, in Thorndike, *Sphere of Sacrobosco*, p. 141; the Latin text appears on p. 114. See also Grant, *Oresme PPAP*, pp. 433–34, which includes a quotation of Sacrobosco's definition.

22. This is much the same method used by Oresme in *Ad pauca respicientes*, Part 1, Proposition II (in Grant, *Oresme PPAP*, see p. 392, lines 96–98 for a general rule, and lines 104–7 for a specific example) and in the more detailed account given in *De commensurabilitate*, Part I, Proposition 6 (see especially

the rule in I.293–96). The distances given above for A and B could equally well be represented as speeds, since velocity is proportional to distance when the times of A and B are equal. Johannes often speaks indifferently of the motions (speeds) of mobiles and the constant daily distances which they traverse.

23. In the *Ad pauca respicientes*, Part 1, Proposition II, Oresme employs the same method, furnishing both a general rule (Grant, *Oresme PPAP*, p. 392, lines 101–3) and an example (ibid., lines 109–11). A general rule and an example are repeated in

Johannes then calculates the same result for *B*. Finally, a total of 4 conjunctions will occur during the course of a single revolution. This is obtained by dividing the circle, or 24 parts, by the time required for the first conjunction, namely, 6 days.[24]

Once again, these purely kinematic procedures are applied to planetary motions[25] when Johannes insists that by such methods and with the appropriate data we could calculate, with respect to a fixed point, mutual conjunctions of planets as well as the daily and mean motions of the moon.

Incommensurable motions are considered in Chapter 14 where Johannes assumes two concentric but incommensurable circles or circumferences (they are related as the diagonal and side of a square whose sides are presumably taken as unity) on which two mobiles move with commensurable speeds.[26] When these conditions obtain and the two bodies start from the same point—i.e., from conjunction—they will never, through all eternity, conjunct again in the same point.[27] If they did, their motions would be necessarily commensurable,

---

greater detail in *De commensurabilitate*, Part I, Proposition 8.

24. A slightly different procedure is followed in the *Ad pauca respicientes*, Part 1, Proposition II, where Oresme derives the number of places of conjunction by dividing the total time of revolution (instead of the total number of parts into which the circle is divided, as with Johannes) by the time elapsed between any two successive conjunctions (see Grant, *Oresme PPAP*, p. 392, lines 105–9). In the *De commensurabilitate*, Part I, Proposition 7, the same method is repeated (see also I.366–76). But in Part I, Proposition 11 of the latter treatise, a quite different rule is given. Here, if the ratio of velocities of *A* and *B* is $V_A : V_B = C : D$, and $C : D$ is a ratio of prime numbers with $C > D$, then $C - D = n$, where *n* represents the total number of conjunctions.

In Chapter 26 (see below and App. 1), Johannes determines the number of conjunctions in a revolution when the conjunctions are assumed to occur in parts of the circle represented by both an integer and a fractional part. See above, n. 8 and below, n. 44 of this chapter.

25. Although Oresme applied many of his kinematic propositions to celestial motions,

he did not see fit to apply the propositions cited in notes 22–24 of this chapter. However, after offering an alternative method for determining the number of conjunctions in a period of revolution in Part I, Proposition 11 of the *De commensurabilitate*, he applies it to celestial motions in I.463–80, but in a manner wholly different from the very general application by Johannes de Muris in Chapter 13 of the *Quadripartitum*.

26. In the *Ad pauca respicientes*, Part 1, Proposition IV, the conditions assumed here —i.e., two mobiles moving commensurably on their respectively incommensurable but concentric circles—constitute the first of four ways in which Oresme says mobiles can move incommensurably with reference to a common center (see Grant, *Oresme PPAP*, p. 394, lines 127–29 and pp. 432–33). Also in the *Ad pauca* (pt. 1, prop. VIII; ibid., p. 400, lines 186–87), Oresme relates two circumferences incommensurably as diagonal and side with the mobiles moving with equal curvilinear velocities. Neither this nor the fourfold division of circular incommensurability is repeated in the *De commensurabilitate* (on the latter point see below, the Commentary, n. 2 for lines I.24–31).

27. Since the sense seemed to require it, I

contrary to the assumption.[28] Such incommensurable motions can have no common measure, so that if the celestial motions were incommensurable, their precise motions could not be known[29] and adjustments or corrections would have to be made, since continuous motion cannot be represented exactly by any fixed astronomical instrument.[30]

After devoting Chapters 19 and 20 to the conjunction of mobiles moving along the same straight line, Johannes seeks to determine in Chapter 21 the time that would elapse between two successive conjunctions of three mobiles moving with mutually commensurable speeds along the circumference of a circle. Dealing with three instead of two mobiles, Chapter 21 is but a straightforward extension of the first part of Chapter 13.

If the circle is divided into 60 equal parts and A moves through 9 parts daily, B through 7 parts, and C through 5, the time of conjunction for any pair of mobiles is determined by dividing their difference in velocities (or distances traversed in equal times) into the total number of parts into which the circle is divided. Hence $V_A - V_B = 2$ and $\frac{60}{2} = 30$, so that A and B will conjunct again in 30 days; $V_A - V_C = 4$ where $\frac{60}{4} = 15$, so that A and C will conjunct in 15 days; and, finally, $V_B - V_C = 2$ and $\frac{60}{2} = 30$, the number of days in which B

(Note 27 continued) have interpreted and translated "in parte numerali" (chap. 14, line 5, in App. 1, below) as in "the same part [or point]." In a strict sense, this translation seems wide of the mark, but it is made necessary by the fact that although the two mobiles will conjunct in an infinite number of points, they will never conjunct twice in the same point.

28. In more explicit form, the substance of Chapter 14, Book 4 of the *Quadripartitum* appears in the *Ad pauca respicientes*, Part 1, Proposition IV, where Oresme lets two mobiles move with equal curvilinear speeds on circles that are incommensurable. Later, in the *De commensurabilitate*, Part II, Proposition 1, Oresme arrives at the same conclusion but now assumes that the two mobiles move with incommensurable velocities and makes no mention of the relationship between the circles (see above, chap. 2, n. 59).

29. This opinion is also expressed by Oresme in much greater detail in the *Ad pauca respicientes*, Part 2, Proposition XVI (see Grant, *Oresme PPAP*, pp. 420–23) and

is repeated in a number of places in the *De commensurabilitate* (for example, II.389–93, 424–26, III.29–31, 435–40). For Oresme, who accepted the incommensurability of the celestial motions as probable, this necessarily inherent imprecision in astronomy and astrology played a significant, and even crucial, role in his treatment of this subject; in contrast, and as we shall see below (p. 95 and n. 40), for Johannes de Muris the question as to whether or not the celestial motions are commensurable was of no interest or importance.

30. At the conclusion of Chapter 14 (line 12; see App. 1 below), Johannes says that "a discussion of these matters will be made at another time." Was this promise fulfilled? If so, is the discussion in an extant treatise? And are we to assume that Johannes intended to consider more fully cases involving incommensurable celestial motions? Or did he mean to confine this statement only to the role of fixed astronomical instruments and their capacity to cope with continuous celestial motion?

and C will conjunct again.[31] Since 30 is the least common multiple of these three times, A, B, and C will conjunct again in 30 days.[32]

Their place of conjunction will be in the 30th part of the circle, or halfway around the circle from their initial point of conjunction. This is determined by multiplying the distance any one of the mobiles traverses daily by 30 days and then subtracting from this product the next lowest integer.[33] Thus, $30 \cdot 9 = 270$ parts traversed by A, which when divided by 60, the total number of parts of the circle, yields 4 plus 30 parts, or $4\frac{1}{2}$, signifying that A has moved around the circle $4\frac{1}{2}$ times. Subtraction of 4 from $4\frac{1}{2}$ leaves $\frac{1}{2}$, or 30 parts of the circle. At the end of 30 days, A will conjunct in the 30th part of the circle. By calculating in the same manner, we can see that B and C must also arrive at the same place and conjunct.

As a mere repetition of Chapter 21, Chapter 22[34] offers only another example, in which, however, the necessity of finding a least common multiple is stated quite explicitly.

In Chapter 23 Johannes presents a new set of conditions for the problem of determining the time of the first conjunction of three mobiles. He now assumes that after simultaneous departure from an initial point of conjunction, and prior to the next conjunction, the three mobiles have already been in motion a certain number of days and, because of their differing velocities, are in separate places. The objective is to find the time that must elapse before they conjunct from their present separate locations.[35]

Let V_A be 9 parts per day, V_B 7 parts, and V_C 5 parts, and assume further that all three mobiles have been in motion for 20 days since they departed from their initial point of conjunction. How many more days beyond 20 must pass before they conjunct again? In the 20 days since their simultaneous departure, A will have traversed $20 \cdot 9 = 180$ parts, and since the circle is implicitly assumed to be divided into 60 parts, A will have traversed $\frac{180}{60} = 3$ full circulations and be located in the point from which it started. But B will have

31. This is the same method used earlier in the first part of Chapter 13 to determine the time of conjunction for two mobiles (see above, p. 88 and n. 22).

32. Nowhere in the *Ad pauca respicientes* does Oresme find the time of conjunction of three mobiles, and that treatise contains no direct counterpart to Chapter 21 of the *Quadripartitum*. But Part I, Proposition 16 of the *De commensurabilitate* (see below, the Commentary, n. 10 for I.616–46), although much more detailed, employs substantially the same methods to determine the time of con-

junction for three mobiles that have commenced their motions from the same point.

33. The calculations that follow are not supplied by Johannes de Muris but have been included for convenience.

34. For Oresme's use of the methods and techniques of Chapter 22, see above, nn. 31 and 32 of this chapter.

35. In none of his major treatises on celestial commensurability does Oresme include any proposition comparable to Chapter 23 of the *Quadripartitum*.

moved through $20 \cdot 7 = 140$ parts and is now located 20 parts of the circle away from its starting point (i.e., $\frac{140}{60} = 2 + \frac{20}{60}$ parts); whereas C will have moved through $20 \cdot 5 = 100$ parts and has arrived in a point 40 parts away (in a clockwise direction) from its starting point (i.e., $\frac{100}{60} = 1 + \frac{40}{60}$ parts). Johannes now divides the difference in velocities of any two of these mobiles into the distance that separates them and obtains the time in which they will conjunct from their present locations. Thus $V_A - V_B = 9 - 7 = 2$, which is divided into 20, the number of parts separating B from A, and results in 10, signifying that A, the quicker mobile, will overtake B and conjunct with it in 10 days. Following the same procedure, we see that B and C will also conjunct in 10 days. Since A overtakes B in 10 days and B overtakes C in 10 days, all three must obviously conjunct at the end of 10 days; Johannes, however, unnecessarily calculates the time for mobiles A and C, which must also conjunct in 10 days. When these 10 days are added to the 20 days that have already passed, we must conclude that 30 days is required from one conjunction to the next—as was already shown in Chapter 21 (see above).

In a second example, Johannes falls into serious confusion and error when he overlooks, or ignores, the relative positions of the quicker and slower mobiles. He now assumes that the three mobiles have been separated for only 12 days since their initial conjunction. First he locates their present positions on the circle (in 12 days A will have traversed $12 \cdot 9 = 108$ parts and be in the 48th part of the circle [i.e., $\frac{108}{60} = 1 + \frac{48}{60}$]; B will have moved through $12 \cdot 7 = 84$ parts and arrive at the 24th part of the circle [i.e., $\frac{84}{60} = 1 + \frac{24}{60}$]; and C will have arrived in the zero, or 60th, part, since it will have traversed $12 \cdot 5 = 60$ parts). For each pair of relationships, we are now told to subtract the place of the slower from the place of the quicker body. Subtracting B from A, or the second from the first mobile, Johannes insists that we obtain 24, which, when divided by $V_A - V_B = 9 - 7 = 2$, results in 12, signifying that A and B will conjunct in 12 days. But this is false, since A and B cannot be interpreted to be separated by 24 parts, as indicated by Johannes's subtraction of 24 from 48. Rather, they are separated by 36 parts, since the measurement must be made counterclockwise from the 24th part to the 48th part of a circle divided into 60 equal parts. Since $\frac{36}{2} = 18$, we see that A will overtake and conjunct with B in 18, not 12, days.

By measuring the distance of separation clockwise, instead of counterclockwise, from the slower to the quicker mobile, i.e., from B to A, Johannes has seriously blundered. Since the mobiles are moving in a clockwise direction, the necessity of measuring counterclockwise is evident, since A, the quicker, must overtake B, the slower, before they can conjunct. This serious error[36] is re-

---

36. If Oresme was familiar with this par-   ticular chapter of the *Quadripartitum*, he de-

peated consistently in the subsequent calculations. Taking next A and C, or the first and third mobiles, he arrives at 48 parts as their distance of separation, which, when divided by their difference in daily motions—namely, 4—yields 12 days as the time of conjunction. But had he measured the distance of separation counterclockwise from C to A, their separation would have been $60 - 48 = 12$ parts, which, after division by 4, would have yielded 3 days as the time of next conjunction. From Chapter 21, however, we know that A and C will conjunct every 15 days (see above, p. 90), so that in $3 + 15 = 18$ days A and C will conjunct. But since A will also conjunct with B in 18 days, it is obvious that A, B, and C will conjunct simultaneously in 18 days. Although this should have sufficed, Johannes proceeds with superfluous calculations for B and C (just as he did in the first example of this chapter) and commits the very same error when he takes 24 parts as their measure of separation, which, when divided by 2, their difference in daily motions, yields 12 days as their time of next conjunction. Here again, had he taken the separation as $60 - 24 = 36$ parts and divided by 2, the time of next conjunction would have been 18, not 12, days.

After determining that 12 days is the time of next conjunction for each of the three pairs of mobiles, Johannes now subtracts 12 from 30 and says that 18 days must elapse before the next conjunction. Although correct, it does not follow from the results already obtained, since Johannes had previously shown that the three mobiles would conjunct in 12 days. Ironically, had he been content at the outset to subtract 12 from 30 (and 20 from 30 in the first example of this chapter), and omit the erroneous calculations for the mobiles taken two at a time, he would have obtained the correct answer. In light of the initial data, this is all that would have been required.

In Chapter 24, Johannes continues to concern himself with the determination of the time of conjunction of three mobiles whose daily motions are, as before, 9 parts for A, 7 for B, and 5 for C. But now, instead of giving the time during which each mobile has been in motion since the initial conjunction, he assigns arbitrary distances, expressed in parts of the circle, that each mobile is assumed to have traversed since the initial conjunction. Thus if A has already moved

---

tected the error and consciously avoided this pitfall in the *De commensurabilitate*, pt. I, prop. 9 (I.342–45); this problem did not arise in the *Ad pauca respicientes*) by insisting that the measurements be made counterclockwise from slower to quicker mobile (also see above, p. 15 and n. 17 of chap. 2). Such a blunder is even more surprising when we realize that in Chapters 24 and 28 (see below, pp. 94 and 97 and nn. 38 and 47; also App. 1) Johannes actually reverses the positions of quicker and slower mobiles, thereby obtaining different results for otherwise identical data. Because of this, he did not repeat his error, but apparently failed to see the bearing of all this on his faulty calculations in Chapter 23.

through 8 parts, B through 13, and C through 17, how much time must elapse before they conjunct simultaneously?[37]

To find the time of conjunction for any two of these mobiles, divide the distance that separates them presently by the difference in their daily motions. Since A and B are separated by $13 - 8 = 5$ parts, division by 2, the difference in their daily motions, gives $2\frac{1}{2}$ days as the time in which A and B will conjunct. Using the same procedure, we see that A and C will conjunct in $2\frac{1}{4}$ days, and B and C in 2 days (chap. 24, lines 20–22). But where on the circle will these three conjunctions occur? Multiplying the daily motion of A by its time of conjunction, we get $9 \cdot 2\frac{1}{2} = 22\frac{1}{2}$, to which we add 8, the distance it has already moved, and thus determine that A will conjunct at $30\frac{1}{2}$ parts of the circle, as will B (i.e., $7 \cdot 2\frac{1}{2} = 17\frac{1}{2} + 13 = 30\frac{1}{2}$).[38] Similarly, A and C will conjunct in $28\frac{1}{4}$ parts of the circle at the end of $2\frac{1}{4}$ days; and B and C will be in the 27th part of the circle when they conjunct in 2 days.

Although we now have the times and places of the next conjunction for the three mobiles taken two at a time, the objective is to determine the time of the next conjunction of all three mobiles simultaneously. In fact, this was already determined in Chapter 21 for three mobiles with the very same daily motions. There we saw that A and B, and B and C will conjunct every 30 days, and A and C every 15 days, so that all three must conjunct in 30 days after departure from their initial point of conjunction. At this point, Johannes says (chap. 24, lines 34–37) that if all three once conjuncted, they will conjunct again and again in the same time; but if they had never conjuncted in this time, they would never do so in the future. However, because it was assumed that A, B, and C had commenced their motions from conjunction, we would expect them to conjunct every 30 days. But on the basis of the data supplied, and contrary to the initial assumption, the three mobiles could not have started from conjunction. If A moves through 9 parts of the circle every day, and B moves through 7, and C through 5, *and* they start from simultaneous conjunction, A could not arrive in the 8th part of the circle when B reaches the 13th and C the 17th. Indeed,

37. Oresme has no counterpart to Chapter 24 anywhere in his treatises under discussion here.

38. Johannes also considers the case where the distances already traversed by A and B are the reverse of those given above. If A has traversed 13 parts and B only 8, then A and B have already conjuncted $2\frac{1}{2}$ days ago, for just as in the first situation, $(13-8)/(V_A - V_B) = \frac{5}{2} = 2\frac{1}{2}$ days. Obviously, the place of this conjunction must differ from $30\frac{1}{2}$. Multiply-ing V_A, or 9, by $2\frac{1}{2}$ gives $22\frac{1}{2}$, the number of parts traversed by A since its conjunction $2\frac{1}{2}$ days before; but since it is now in the 13th part of a circle divided into 60 equal parts numbered clockwise from 0 to 60, $22\frac{1}{2}$ must be subtracted from $13 + 60 = 73$, which leaves $50\frac{1}{2}$, signifying that $2\frac{1}{2}$ days earlier A and B had conjuncted in $50\frac{1}{2}$ parts of circle. Johannes also calculates the same result for B ($7 \cdot 2\frac{1}{2} = 17\frac{1}{2}$, so that $8 + 60 - 17\frac{1}{2} = 50\frac{1}{2}$).

Johannes seems to have been fully aware of this, for in the next chapter, 25, he denies that A, B, and C could conjunct every 30 days if the times of conjunction for the mobiles taken two at a time are $2\frac{1}{2}$, $2\frac{1}{4}$, and 2 days, respectively (see chap. 25, lines 20–25). And so he concludes that "if they never ceased to be moved in the past, and time had no beginning, they were never in conjunction in the same point of the circle, for which reason the hypothesis given before [i.e., in Chapters 23 and 24] that these three mobiles would be conjuncted some-time was assuming the impossible."[39]

At the conclusion of Chapter 24 we find the only other reference to incom-mensurable motions other than what appears in Chapter 14. Apparently draw-ing implications from the material summarized in the preceding paragraph, Johannes insists that no one has ever demonstrated that seven planets, or even three, must conjunct in the same point of the celestial sphere. But if the seven planets had once conjuncted simultaneously in the same point, and if their motions are commensurable, they would necessarily conjunct again in that very same point. However, Johannes is unconcerned as to whether or not these planetary motions are commensurable, since astronomers are satisfied if, when they predict a conjunction, what appears to the senses can be interpreted as a conjunction.[40] Although the senses are unable to inform us of many things (see chap. 24, lines 50–53), astronomers are content to rely on them.

---

39. Chap. 25, lines 25–28, in App. 1 below. Johannes has presented an indirect method for showing that three mobiles with com-mensurable motions may never conjunct. After assuming that they start from conjunc-tion and assigning them uniform daily mo-tions that necessitate a simultaneous con-junction every 30 days, he then arbitrarily assigns them places on the circle that would have made an initial conjunction impossible. In this way, the assumption of an initial con-junction is rendered absurd and impossible. Nowhere does Oresme utilize such a method for demonstrating the same thing. In the *Ad pauca respicientes*, Part 1, Proposition VII (Grant, *Oresme PPAP*, pp. 398–401) and in *De commensurabilitate*, Part I, Proposition 12, he presents cases involving three mobiles with mutually commensurable velocities that could never conjunct simultaneously through an infinite past and future. In both exam-ples, two of the mobiles are initially in con-junction while the third precedes them by some part of the circle. He then shows that

since mobiles A and B share no common points of conjunction with B and C, no simultaneous conjunction of three bodies is possible. Not only does Oresme's direct dem-onstration contrast sharply with the indi-rect method used by Johannes de Muris, but it is more graphic and lucid.

40. This attitude stands in sharp contrast with Oresme's attitude. Oresme would agree with Johannes that one cannot formally dem-onstrate whether or not planets will con-junct (see above, pp. 69–70). Like Johannes, he stressed many times the fact that astrono-mers are satisfied with approximate results (for example, in *Ad pauca respicientes*, pt. 2, prop. XX [Grant, *Oresme PPAP*, p. 428, lines 263–65] and *De commensurabilitate*, I.45–49). Oresme, however, was deeply con-cerned about whether or not the celestial motions are commensurable and sought *probable* answers rather than formal demon-strative knowledge of this (see above, pp. 72–76, and especially n. 113). By his own statement, Johannes disclaims any interest in

Since the substance of Chapter 25[41] is summairzed above in the description of Chapter 24, let us move on to Chapter 26. Although, in Chapters 12 and 13, Johannes flatly asserted the number of times two mobiles would conjunct in the course of a revolution (see above, n. 24 of this chapter)—i.e., the number of conjunctions that occur between two successive conjunctions in the same point—he devotes Chapter 26 to this problem, restricting himself to conjunctions that occur in parts of a circle represented by an integer and a fraction. Proceeding as always by example, he assumes that the two mobiles move on a circle divided into 60 equal parts and that, after starting from conjunction, A moves through 7 and B through 2 parts daily. On the basis of reasoning in previous chapters, they will conjunct in the 24th part of the circle.[42] If they should conjunct $\frac{2}{5}$ of the way through the 24th part—i.e., at $24\frac{2}{5}$ parts of the circle[43]—they will conjunct only in fifths of the circle and conjunct five times in the course of a revolution. If, however, they had conjuncted in $27\frac{9}{20}$ of the circle, they would conjunct in every 20th part of the circle and conjunct a total of 20 times during a whole revolution. The same reasoning can be extended to three or more mobiles. "Therefore, the reduction of fractions of the circle to prime [numbers] is the basis of this rule" (chap. 26, lines 16–17). Applying this to celestial motions, Johannes remarks that by this rule the number of conjunctions of the superior planets and other celestial relationships can be determined for any revolution (chap. 26, lines 15–16). Although he does not justify the claims in his examples, he does state that the denominator of the fraction will indicate the number of conjunctions in a revolution.[44]

The final chapter, 28,[45] has two parts. The first is concerned with two con-

---

(Note 40 continued)
the problem of whether or not the celestial motions are commensurable (chap. 24, line 45), and he does this because in astronomy only approximations are used (chap. 24, lines 45–47), thereby obviating the need for a decision or further discussion. Cf. also the remarks of Averroes quoted below on p. 107.

41. It is worth noting that the substance of Chapter 25, lines 2–5, is considered in the *Ad pauca respicientes*, pt. 1, prop. I (Grant, *Oresme PPAP*, pp. 386–88, lines 55–64).

42. Johannes omits the calculations but, using the method of Chapter 13, we see that A and B will first conjunct in 12 days $(60 / (V_A - V_B) = \frac{60}{5} = 12)$. But in 12 days, A will have traversed $12 \cdot 7 = 84$ parts and be in the 24th part of the circle; and B will

have traversed $12 \cdot 2 = 24$ parts and so conjunct with A in the 24th part.

43. By suddenly shifting to $24\frac{2}{5}$ parts of the circle, Johannes, perhaps unwittingly, renders superfluous the previous data which had produced a conjunction in the 24th part, since those data could not produce a conjunction in $24\frac{2}{5}$ parts of the circle.

44. Although framed in a wholly different context, Oresme's discussion in the *De commensurabilitate*, Part I, Proposition 22 (more specifically, I.750–74) deals with essentially the same problem and includes the same rule (I.762–63). For a comparison between Theodosius and Oresme, see above, n. 8 of this chapter.

45. Although Chapter 27 is concerned with the motion of two mobiles moving in the same direction, it has been omitted because

trary motions, as in Chapter 12 above, and the second with two retrograde motions.[46]

Assuming from Chapter 24 that A traverses 9 parts daily and B, 7 parts, Johannes arbitrarily locates A in the 8th part and B in the 18th part of the circle. If they begin to move in opposite directions, when will they first conjunct? He divides the sum of their daily motions, $9 + 7 = 16$, by the difference in their places, or $18 - 8 = 10$, and determines that they will first conjunct in $\frac{10}{16} = \frac{5}{8}$ of a day. This conjunction will occur at $13\frac{5}{8}$ parts of the circle (for A we get $9 \cdot \frac{5}{8} = 5\frac{5}{8} + 8 = 13\frac{5}{8}$; and for B, $7 \cdot \frac{5}{8} = 4\frac{3}{8}$, so that $18 - 4\frac{3}{8} = 13\frac{5}{8}$.[47]

If A were placed at the 18th part and its motion taken as retrograde, while B is located in the 8th part moving with a progressive or clockwise motion, they would, of course, conjunct in the same time as in the previous example—namely, $\frac{5}{8}$ of a day—but not in the same place. Under these conditions, conjunction will occur at $12\frac{3}{8}$ parts of the circle (for A this is obtained as follows: $18 - 5\frac{5}{8} = 12\frac{3}{8}$; and for B, $8 + 4\frac{3}{8} = 12\frac{3}{8}$).

In concluding Chapter 28, Johannes takes both motions to be retrograde and finds the time and place of their first conjunction by the same method used earlier for progressive motions (see description of Chapter 13, above). If the quicker mobile is now at part 8 and the slower at part 18, their conjunction will already have occurred, since their motions are retrograde and their present positions indicate that A, the quicker, must have recently passed B, the slower; however, if A is at part 18 and B at part 8, the conjunction will occur sometime in the future (i.e., when A gains 50 parts over B).[48]

From the preceding summary and notes, it is apparent that in his *Quadripartitum numerorum* Johannes de Muris had not only devoted chapters to the subject of circular and celestial commensurability, but among them had also enunciated and exemplified certain propositions very similar to, and sometimes nearly identical with, some of those in the *Ad pauca respicientes* and *De commensurabilitate*. Did Johannes de Muris influence Oresme, his junior contempo-

---

it assumes that A moves faster than B during the first half of the time and that their velocities are reversed during the second half of the time. No proposition of this kind is found in Oresme's treatises, perhaps because it could not represent, or be applied to, celestial motions. It has no genuine relevance to the subject matter of this volume. Johannes concludes that under such conditions the mobiles must reach a given terminus at the same instant, whether their motions be rectilinear or circular. In BN 7190 (see below, p. 359), Chapter 27 appears on fols. 78v–79r.

46. Oresme has no proposition corresponding to Chapter 28 of the *Quadripartitum*.

47. The number representing the place of A, or 8, must be added to $5\frac{5}{8}$, the distance A travels in $\frac{5}{8}$ of a day, because A is moving clockwise on a circle that is divided into 60 parts numbered clockwise from 0 to 60; but since B moves in a counterclockwise, or retrograde, direction, its daily motion, or $4\frac{3}{8}$, must be subtracted from its original place.

48. For the detailed calculations concerning each of these cases, see the text and translation in App. 1, below.

rary? Or was it the other way around? Did they arrive at these similar propositions independently? If so, who has priority? Or, perhaps, did both draw upon
a common source, as yet unknown? As of now, no categorical answer to these
questions is possible, and the last question need not even concern us, since I
presently know of no authors other than Johannes de Muris and Oresme who
considered problems of this kind.[49] Hence our reflections here must necessarily
be confined to the first four questions.

Although Johannes de Muris completed his *Quadripartitum numerorum* in
1343, no priority arguments can be made from this date alone, since no usefully
precise date of composition can be proposed for Oresme's *Ad pauca respicientes*,
probably the earliest of his extended discussions on this subject. Even if it were
true that the *De proportionibus proportionum* and *De commensurabilitate* were
written in or after 1351,[50] the lack of a *terminus post quem* for the *Ad pauca
respicientes* leaves open the possibility that it may have been written much
earlier, even earlier than 1343. And if Oresme wrote the *De proportionibus proportionum* prior to his *Algorismus proportionum*,[51] then one, two, or all three of
the relevant treatises—i.e., *Ad pauca respicientes*, *De proportionibus proportionum*, and *De commensurabilitate*—could have been written prior to the *Quadripartitum*.

Despite the lack of a *terminus post quem* for the *Ad pauca respicientes*, the
crucial treatise in any priority argument concerning Oresme and Johannes de
Muris, there is reason to conjecture that it was written after the relevant chapters
in the *Quadripartitum*. One bit of tantalizing evidence is a remark made by the
scribe who copied the *De proportionibus proportionum* and the *Ad pauca respicientes* in codex Bibliothèque Nationale, fonds latin, 16621. There, on folio 93v,
we are told that "A treatise *On Ratios* by Oresme follows with a certain astrological treatise taken [or had] by him [i.e., Oresme] from [de] Muris...."[52] Since
the *Ad pauca respicientes* is the "certain astrological treatise" which follows the
De proportionibus (*On Ratios*), we appear to have here an explicit declaration
that, in some sense or other, Oresme derived the *Ad pauca* from Johannes de

49. While it is true that the five propositions described above from Theodosius's *De
diebus et noctibus* are related to certain propositions in Oresme's treatises (see above,
pp. 79–84), they seem wholly unrelated to
the problems and situations presented in the
Quadripartitum, and for this reason Theodosius does not properly qualify as a precursor
of Johannes de Muris despite the fact that
some influence or inspiration cannot be categorically excluded. It is quite possible, and

perhaps even likely, that genuine precursors
or predecessors may come to light.
50. See above, p. 5 and Grant, *Oresme
PPAP*, pp. 13–14, 76–79.
51. See above, p. 5.
52. "Sequitur tractatus proportionum
Orem cum quodam tractatu astrologico ad
pauca aspicientes habito ab illo de Muris...."
—Quoted from Grant, *Oresme PPAP*, p. 125;
see also p. 126.

Muris. As the source of the *Ad pauca* did the scribe have in mind the chapters summarized above from Book 4 of the *Quadripartitum*?[53] Or perhaps some other treatise by Johannes de Muris?[54] Unfortunately, since we lack the approximate year in the fourteenth century[55] when this statement was written and are wholly ignorant of the scribe and his reliability, this statement cannot be properly evaluated or interpreted. Nevertheless, the *Quadripartitum numerorum* represents at least one treatise by Johannes de Muris which contains some chapters and examples that are closely related to some of the propositions, methods, and ideas in the *Ad pauca respicientes*, thus rendering plausible the scribe's claim that Oresme had drawn upon Johannes de Muris. Perhaps the more lucid, succinct, and formal presentation in the *Ad pauca respicientes* of the material it has in common with the *Quadripartitum* indicates that Oresme had read Johannes de Muris and then recast some of the latter's examples into a more mature and theoretical format, to which he added many additional and quite different consequences and propositions.

That priority of composition may lie with Johannes de Muris receives a modest degree of reinforcement from the fact that Oresme entered the College of Navarre as a bachelor of theology in 1348. To have composed the *Ad pauca respicientes* in or before 1343 would seem to imply that Oresme had already received his Master of Arts degree, for it is unlikely that such a treatise would have been written prior to the attainment of that degree. But if Oresme was already a Master in or before 1343, and was actively teaching at the University of Paris, it is then difficult to explain a delay of at least five years before he commenced theological studies in 1348 at Navarre.[56] However, because such

53. In attempting to evaluate the statement quoted in the immediately preceding note, I remarked (Grant, *Oresme PPAP*, p. 126) that the introduction of the name of Johannes de Muris would seem unwarranted since to my knowledge he was not associated in any manner with the subject matter of the *Ad pauca respicientes*. Obviously, this opinion must now be abandoned.

54. Possibly the treatise, or part of a treatise, promised at the conclusion of Chapter 14 (line 12). See above, n. 30 of this chapter.

55. The treatises in BN 16621 were copied in the fourteenth century (see Grant, *Oresme PPAP*, p. 126).

56. Although Oresme's approximate birthyear was estimated to fall between 1320 and 1325 (see above, p. 3), this is a conjecture

based on the belief that he was anywhere between 31 and 36 years of age when he received his doctorate around 1356 (this too is a conjecture, but a reasonable one; see above, p. 3 and Grant, *Oresme PPAP*, p. 5). If Oresme had been born in 1320 and had composed the *Ad pauca respicientes* sometime between 1340 and 1343, but before Johannes de Muris had written his chapters in the *Quadripartitum*, he would have been between 20 and 23 years of age. This is possible, since attainment of the Master of Arts degree at the age of 20 or 21 was quite common. But such an early date for Oresme's Master's degree requires that we explain the five or more years delay before he entered Navarre for theological studies. This is the most telling argument against supposing that Oresme was born before 1320, for the earlier

delays may have happened, perhaps more frequently than we suppose, Oresme's possible priority cannot be precluded. But the weight of evidence, such as it is, inclines me to the belief that Johannes de Muris wrote before Oresme on the subject of celestial commensurability and incommensurability.

Acceptance of the priority of Johannes de Muris does not, however, permit us to conclude that Oresme was aware of the chapters in the *Quadripartitum*, or had read them, when he wrote the *Ad pauca respicientes*.[57] Perhaps both

----

(Note 56 continued)

we set his birthyear, the greater does the temporal gap become between the time he would normally have received his Master's degree and the time (1348) that he entered the College of Navarre to begin theological studies. Since even a five-year lag is difficult to explain, it seems preferable to guess at a birthyear of 1325 and acquisition of a Master's degree sometime around 1346, thus leaving only a two-year gap, a time that might have been spent as a teaching master prior to commencement of full-scale theological studies. If we assume these dates and argue plausibly that Oresme did not write the *Ad pauca* until after he had received his Master's degree, then Oresme clearly could not have written the *Ad pauca respicientes* before Johannes de Muris had written the relevant chapters in the *Quadripartitum*.

57. It seems most unlikely that Oresme would publicly have dared to utter a claim to originality for the *De commensurabilitate* without it being true in some important sense, for he submitted it for correction to Fellows at the University of Paris (Prol. 45–48) and knew it would be read by many Parisian scholars, some of whom would probably have known the content of the *Quadripartitum* or other similar earlier treatise. Furthermore, if Johannes de Muris had written before Oresme and if we assume that Oresme was truthful in stating that the *De commensurabilitate* was an original treatise (Prol.41–44), we must suppose either that at the time of its composition he had not read, or otherwise known, the *Quadripartitum* (and *a fortiori* would have been ignorant of its content when he wrote the earlier *Ad pauca respicien-*

tes), or that he was familiar with it and his claim to originality is thus qualified, but nonetheless important.

That Oresme was more than likely aware of the *Quadripartitum* at the time he wrote the *De commensurabilitate* can be argued from a probable reference to the former treatise made in the *De proportionibus proportionum*, chap. 4, prop. VI, where he says: "It has indeed been assumed that you would know how to find any number of means in continuous proportionality between two given lines. Now Euclid teaches how to find only one mean, ... but Reverend Master Johannes de Muris has, I believe, shown how to find any number of them."—Grant, *Oresme PPAP*, p, 299. This is probably a reference to the *Quadripartitum*, bk. 1, chap. 20, which, in Latin MS Vienna 4770, fol. 199v, bears the title "Quot vis proportiones continuas invenire." Since the *De commensurabilitate* cites the *De proportionibus* by title (see II.201), and was therefore written after it, it would seem that Oresme had some first-hand acquaintance with the *Quadripartitum* at the time he composed the *De commensurabilitate*. But it remains an open question whether or not he knew the *Quadripartitum* when he composed the *Ad pauca* sometime earlier. Indeed, his claim to originality in the *De commensurabilitate* may derive from the fact that the *Ad pauca* was an original treatise written before the *Quadripartitum*, so that the *De commensurabilitate* is also original because it contains the substance of the *Ad pauca*, revised and greatly expanded.

But if, on the contrary, the *Quadripartitum* preceded the *Ad pauca* and Oresme had util-

authors drew from some as yet unknown common source.[58] But it is futile to multiply hypotheses and conjectures on a matter which must, as of now, be inconclusive. For our purposes, however, it will be assumed that not only did Johannes de Muris write his chapters in Book 4 of the *Quadripartitum* before Oresme composed any of his treatises on celestial commensurability, but also that Oresme was familiar with them. On the basis of these assumptions and with our earlier comparison of Theodosius and Oresme in mind, we can now determine the sense and extent of Oresme's claim to originality in Prol.42–45 (also see above, p. 78) and thereby arrive at his minimum possible contribution when this is projected against the total background of what has come to light thus far.

From the above summary of the chapters in the *Quadripartitum* and the notes connecting relevant propositions in Oresme's treatises, it is obvious from examples offered by Johannes de Muris that Oresme could have learned how to determine the time and place of first conjunctions for two and three mobiles, as well as how to calculate the number of conjunctions in a period of revolution. In Chapter 14 he could have read that when two mobiles move on incommensurable circles with commensurable speeds, they will never conjunct again in the point from which they started simultaneously. Furthermore, and of special importance, he would have seen that Johannes applied these kinematic examples and consequences directly to celestial motions.

Obviously, then, Oresme could have absorbed from Johannes de Muris a rather solid basis for his own work, a basis that could have served as a springboard for his own treatises. If this happened, what are we to make of Oresme's claim in the *De commensurabilitate* that "if another has set out the more fundamental principles [or elements found in this book], I have yet to see them" (Prol.44–45)? Under these circumstances, he could hardly have intended to proclaim full and complete originality (see above, n. 57 of this chapter). But if we concede the priority of the *Quadripartitum* and grant that Oresme had read the pertinent chapters, what elements or aspects of his treatises on celestial commensurability carry the stamp of originality? First and foremost would be the fact that Oresme had written independent and special treatises on this subject[59] which far surpassed in extent, subtlety, and content the few chapters in the

---

ized it, then his declaration of originality must be a qualified, but nonetheless, significant and meaningful claim. The sense of this claim must now be considered.

58. Perhaps even from some work by Campanus of Novara, the thirteenth-century editor of Euclid's *Elements*, who is cited by Jerome Cardano as the initial source of his acquaintance with problems of circular commensurability and incommensurability. See below, p. 158; and for the text of Cardano's statement see n. 186.

59. Besides Oresme's *Ad pauca respicientes* and *De commensurabilitate*, I know of no other separate treatises devoted to the same subject matter.

Quadripartitum. Moreover, these treatises are structured formally on the model of Euclid's *Elements*, with definitions and propositions, all systematically organized. With Oresme, the subject very nearly attained the status of a formal discipline worthy of study and analysis which stands in stark contrast with the few concrete and specific examples offered by Johannes de Muris to illustrate the application of numbers and fractions to rectilinear and circular motions, where rarely, if at all, general rules or procedures are formulated.

Perhaps more important than this is the emphasis on incommensurable motions and the application of the consequences of such motions to celestial motions and astrology. Only in Chapter 14, and incidentally in Chapter 24, does Johannes offer any discussion of incommensurable motions, restricting himself to the consequence—not very clearly formulated—that celestial configurations will never repeat twice in the same way. Indeed Oresme, and Johannes de Muris himself, could have derived as much from an earlier statement by Duns Scotus (see below, p. 119). The subsequent treatment and discussion of incommensurable circular and celestial motions by Oresme are so vastly superior in scope, quality, and imaginative speculation to anything that has come to light thus far that his claim to originality could rest on these aspects alone. Oresme's conviction that the celestial motions are probably incommensurable, a conviction that was mathematically demonstrated in the *De proportionibus proportionum*,[60] assumed in the *Ad pauca respicientes*,[61] and argued for implicitly in the *De commensurabilitate*,[62] finds no counterpart in the *Quadripartitum*, and no similar discussion earlier than Oresme's has yet come to my attention. In fact, whether or not the celestial motions are commensurable is dismissed by Johannes de Muris as inconsequential on grounds that astronomers are satisfied with approximations (chap. 24, lines 45–47; see also above, n. 40 of this chapter). Finally, only Oresme was to employ the consequences of celestial incommensurability against astrology and determininism as embodied in such concepts as the Great Year and exact cyclical repetition and prediction. None of these ideas are even hinted at in the *Quadripartitum*, although had Johannes de Muris been interested, he could have formulated such consequences from Chapters 14 and 24 (see below, App. 1), just as Duns Scotus had done in a preliminary and modest way some years earlier (see below, pp. 120–22).

In these different senses, then, Oresme is original and whatever debt he may have owed to Johannes de Muris seems overshadowed by the vastly superior intellectual products that he formulated and fashioned from those possible

---

60. See above, chap. 2, n. 113 for a summary of this demonstration and its application to celestial motions.

61. As a special case of Part I, Supposition II. See Grant, *Oresme PPAP*, pp. 384–86, lines 36–38.

62. See above, pp. 72–76.

starting points. And yet apart from Theodosius, whose relevant *De diebus et noctibus* was probably not available, only Johannes de Muris has emerged thus far as a potentially genuine precursor to Oresme. Despite their very different objectives, attitudes, and even techniques, it may have been Johannes de Muris who first generated and set in motion an interest in the problem of celestial commensurability and incommensurability that was to preoccupy Oresme in many treatises and that would remain an intense and abiding interest all his life.

The propositions and chapters summarized from the works of Theodosius and Johannes de Muris represent thus far all that is available in a mathematical form on the subject of celestial commensurability and incommensurability prior to Oresme (even the priority of Johannes de Muris being uncertain). Now, however, we must investigate other possible influences and traditions embedded in cosmological, philosophical, and astrological contexts.

Explicitly or implicitly, Greek and Roman authors seem to have assumed and believed that the celestial motions were related by rational ratios—i.e., were commensurable. This is the consequence of a widespread belief in the general uniformity of nature and more particularly in the regular and uniform repetition of celestial configurations and events, a belief which made possible the Pythagorean conviction that the celestial motions were related as the principal musical concordances,[63] and rendered plausible the widely held ancient credence in the existence of a Great, or Perfect, Year. It is this very widespread acceptance of the concept of a Great Year that reveals how firmly rooted was the belief in the commensurability of the celestial motions.

A definition that would have surely satisfied most adherents of a Great Year is given by Cicero when he tells us that "On the diverse motions of the planets the mathematicians have based what they call the Great Year, which is completed when the sun, moon, and five planets having all finished their courses have returned to the same positions relative to one another. The length of this period is hotly debated, but it must necessarily be a fixed and definite time."[64]

---

63. For the doctrine of celestial harmony see above, chap. 2, n. 28.

64. *De natura deorum* 2. 20. 51–52, trans. H. Rackham, Loeb Classical Library (London and New York, 1933), p. 173. This and six other references to the Great Year in the works of Cicero are discussed by P. R. Coleman-Norton, "Cicero's Doctrine of the Great Year," *Laval théologique et philosophique*, vol. 3 (1947), pp. 293–302. Included in this article are quite a few references to other authors, both Greek and Roman, who discussed or mentioned the Great Year.

Among these is Censorinus, who, in Chapter 18. 11 of his *De die natali* (written in 238 A.D.), not only includes a definition of the Great Year that is virtually the same as Cicero's, but also various estimates of its duration by Greek authors. Because it contains other interesting information, I quote it in full from William Maude's translation:

"There is also a year which Aristotle calls Perfect, rather than Great, which is formed by the revolution of the sun, of the moon and of the five planets, when they all come at the same time to the celestial point from

Although many accepted it as a periodic astronomical occurrence, they refrained from deriving deterministic consequences from it.[65] No such restraint was exercised by certain Stoics, however, who concluded that the Great Year entailed an exact and identical substantive and sequential repetition of all celestial

---

(Note 64 continued)

which they started together. This year has a great winter called by the Greeks the Inundation and by the Latins the Deluge; it has also a summer which the Greeks call the Conflagration of the world. The world is supposed to have been by turns deluged or on fire at each of these epochs. According to the opinion of Aristarchus this year was composed of 2484 solar years; according to Orestes of Dyrrachium, it was 5552 years; according to Heraclitus and Linus it was 10,800; according to Dion it was 10,884; according to Orpheus it was 10,020 years; and according to Cassandrus it was 3,600,000 years. Others have thought it infinite; and that it would never recur."—Censorinus, *De die natale* (*The Natal Day*), trans. William Maude (New York, 1900), p. 23. (The Latin text of this passage is in *Censorini De die natali liber*, ed. Friedrich Hultsch [Leipzig, 1867], p. 39.)

The inclusion by Censorinus of Aristotle among proponents of a Great Year was probably based on *Meteorologica* 1. 14. 352a28–30, where, in considering physical changes on the earth, Aristotle remarks: "...we must take the cause of all these changes to be that, just as winter occurs in the seasons of the year, so in determined periods there comes a great winter of a great year and with it excess of rain."—Trans. E. W. Webster, *The Works of Aristotle* (Oxford, 1931), vol. 3. This passage from Censorinus and Chapter 18 as a whole are evaluated briefly by Coleman-Norton, "Cicero's Doctrine of the Great Year," p. 295, n. 1.

Although the doctrine of the Great Year was apparently adumbrated by pre-Socratics, the first full and definite account of it in Greek was given by Plato in *Timaeus* 39D (see Coleman-Norton, "Cicero's Doctrine," p. 294). Coleman-Norton observes (p. 295, n. 1) that "One difficulty in deciding about

references to the Great Year is whether is meant the accordance only of lunar months and solar years in one period of whole numbers or the return of all the planets (including the sun and the moon) exactly to their same respective positions whence these first started on their revolving courses." This difficulty is of no consequence here, since, in the *De commensurabilitate* (pt. I, prop. 22, [I.814–24]), Oresme would identify as a Great Year any periodic return of sun and moon, and for that matter any two or more planets, to the same respective positions from which they started. Oresme's conception of a Great Year was as broad as could be expected, for it even included the case of a single celestial body with two or more simultaneous motions if, in equal time intervals, those motions always brought it back to the same point from which it started. Any arbitrary point or set of positions could be taken as the beginning and end points of a Great Year (see above, p. 33 and n. 51).

An extensive discussion of the Great Year in antiquity is given by Pierre Duhem in *Le Système du monde*, vol. 1, pp. 65–85, 275–97. Duhem also briefly discusses Indian and Islamic descriptions of the Great Year (vol. 2, pp. 214–23), as well as Hebrew views (vol. 5, pp. 223–26).

65. Among this group, Plato and Aristotle are most notable. For Plato, see Charles Mugler, *La Physique de Platon*, vol. 35, *Études et Commentaires* (Paris, 1960), p. 13. Although Aristotle insisted that the celestial motions govern sublunar changes (*Meteorologica* 1. 2. 339a21–33), he did allow for the contingency of events in *Physics* 2. 5 and 6 and in *De interpretatione*, chap. 9 (see *Averroes on Aristotle's "De generatione et corruptione," Middle Commentary and Epitome*, trans. Samuel Kurland [Cambridge, Mass., 1958], p. 228, n. 69).

configurations and terrestrial events. For them, Socrates and Plato, as well as every other individual, would return in each Great Year and do precisely what they had done in every preceding Great Year.[66] For opponents of this doctrine of individual return, an attack aimed at the Great Year would have served to undermine the former belief, which depended on the regular recurrence of Great Years. One obvious means of achieving this would have been to propose

---

66. The relationship between the conflagration, Great Year, and individual return in Stoic doctrine is summarized briefly by E. Vernon Arnold (*Roman Stoicism* [1911; reissue, London, 1958], p. 193):

"Upon the conflagration will follow the reconstruction of the world (παλιγγενεσία, *renovatio*), which will lead again to a conflagration; the period between one conflagration and the next being termed a 'great year' (περίοδος, *magnus annus*). The conception of the 'great year' was borrowed by the Stoics from the Pythagoreans, and leads us back ultimately to astronomical calculations; for a great year is the period at the end of which sun, moon and planets all return to their original stations. The phenomena of the sky recur in each new period in the same way as before; and hence we readily infer that all the phenomena of the universe, including the lives of individuals, will recur and take their course again. Although this doctrine appears only slightly connected with the general Stoic system, it was an accepted part of it...." (Also see the quotations from Nemesius and Origen in the Commentary, n. 55 for III.311–12.) It seems that Zeno of Citium, Chrysippus, and Cleanthes were Stoics who subscribed to this doctrine (see Duhem, *Le Système du monde*, vol. 1, pp. 279–80), although in the first of two passages quoted from Origen (the Commentary, n. 55 for III.311–12) we are told that the Stoics allowed for minute, but indistinguishable, variations between successive Great Years (see also the Commentary, fn. 8, p. 353). In his discussion of the Great Year, Duhem (ibid., vol. 1, pp. 79–80) includes a passage from Simplicius's commentary on the *Physics* of Aristotle in which Simplicius attributes to

the Pythagoreans the doctrine of numerically identical return in successive world cycles. Without proposing any textual evidence, Duhem (ibid., vol. 1, p. 278) subsequently ascribed to Plato the doctrine of individual return in successive Great Years.

In his discussion of cyclical recurrence, Aristotle distinguished between sequences that are *numerically* the same—i.e., identical in every way—and those that are the same *only in species*.

"In consequence of this distinction, it is evident that those things, whose 'substance' —that which is undergoing the process—is imperishable, will be numerically, as well as specifically, the same in their recurrence: for the character of the process is determined by the character of that which undergoes it. Those things, on the other hand, whose 'substance' is perishable (not imperishable) must 'return upon themselves' in the sense that what recurs, though specifically the same, is not the same numerically. That is why, when Water comes-to-be from Air and Air from Water, the Air is the same 'specifically', not 'numerically': and if these too recur numerically the same, at any rate this does not happen with things whose 'substance' come-to-be—whose 'substance' is such that it is essentially capable of not-being."—*De generatione et corruptione* 2. 11. 338b13–19 (trans. H. H. Joachim, *The Works of Aristotle* [Oxford, 1930], vol. 2).

Since the heavens are imperishable, they would produce sequences of numerically identical events (*De generatione* 2. 11. 337b 4–338b6), but individual men or animals, and individual perishable things generally, could not recur, for they form only rectilinear sequences (*De generatione* 2. 11. 338b8–12).

that some, or all, of the celestial motions might be incommensurable. With this assumption, if the planets started from some particular configuration, they could never again enter into the same relationship in the very same places (see II.290–302). As yet, however, I have failed to discover a single ancient who adopted precisely this tactic—not even among Christians, for whom the doctrine of cyclical and individual return was abhorrent.[67] But Alexander of Aphrodisias (fl. early 3d century A.D.), in an argument cited by Averroes in his *Epitome* of Aristotle's *De generatione et corruptione*, took an approach approximating this when he attacked the doctrine of individual return. Averroes tells us first that proponents of cyclical recurrence "argue that when the same arrangement in all of the parts of the sphere that prevailed at the time of Zaid's existence recurs, Zaid himself will return upon himself. And that is impossible as we have demonstrated." He then reports:

Alexander believes that the state and disposition of the spheres at any given time never revert individually. He maintains that if we assume all of the stars to be at a particular point in the sphere of the constellations, for example, in Ram, and then all of them, both the fast and the slow ones, begin to move, they need not necessarily all of them revert to the exact same point from which they began their movement, but the revolutions of some will be proportionate to those of others, so that, for example, when the sun completes one revolution the moon will have completed twelve. And there will be a similar relationship between the revolution of the sun and of each one of the stars. Then it should be possible for all of them to return to any one place, to any place you may postulate. But we find the exact opposite to take place. For the sun traverses its sphere in $365\frac{1}{4}$ days and the moon traverses its sphere in $27\frac{1}{2}$ days. When $27\frac{1}{2}$ days are multiplied [by twelve], they do not yield $365\frac{1}{4}$ days. Since this is so, and the efficient cause does not return upon itself numerically, and neither can the material cause do so, it becomes evident that it is impossible in any way whatsoever for the individual to recur. Now that is what we set out to prove.[68]

---

67. In *The City of God*, bk. 12, chap. 13 (14 in the modern Latin texts), Augustine reports and then attacks this doctrine (see below, the Commentary, fn. 6, p. 352), but fails to allude to, or even hint at, the approach I have suggested. Nor did Nemesius and Origen, who both report this popular ancient view (see below, the Commentary, n. 55 for III.311–12), invoke the possible incommensurability of celestial motions as a counterattack.

68. Translated from the Arabic by Kurland, *Averroes on Aristotle's "De generatione,"* pp. 137–38. Although Averroes does not cite the work in which Alexander for-

mulated this argument, and Kurland presents no reference, it is likely that it occurred in Alexander's lost commentary on Aristotle's *De generatione* (see F. E. Cranz, "Alexander Aphrodisiensis," *Catalogus translationum et commentariorum* [Washington, D.C., 1960], vol. 1, p. 79), where Alexander would have had occasion to raise this issue. Alexander seems to disagree with Aristotle on at least one point. Where Alexander denies the precise return of celestial bodies to previous configurations, Aristotle presumably would have affirmed such celestial return as part of his overall position that imperishable bodies do numerically return with cyclical regular-

While the details of this argument pose some difficulties,[69] of interest is Alexander's apparent denial of the repetition of celestial events—as represented by the example involving the return of sun and moon to the same place—because of an apparent lack of commensurability between the periods of revolution. This, in turn, serves as the basis for rejecting the doctrine of individual return. That the periods of revolution used in this example are not really incommensurable is quickly pointed out by Averroes, who, while he agrees with Alexander on the impossibility of individual return, rejects the latter's argument as inaccurate and ultimately incapable of verification. Here is what Averroes says:

> We might add to what we have already said that even though the revolution of the moon is not commensurate with that of the sun in days, it does not follow that they are not commensurate with one another at all. For it is possible that their common unit of measurement is a shorter time. But if that were so, the common measure would have to be one quarter of a day. To ascertain whether these revolutions of the stars are commensurable or not is most difficult or well nigh impossible, for that would have to be based upon a knowledge of the time of a single revolution in the case of each star as it is in truth. That is impossible because of the limited and approximated nature of our observation of these things. What we can ascertain in this matter is that they are approximately commensurate to one another, as the astronomers believe. Whatever the case may be, it is impossible for the individual to recur.[70]

Because it is impossible to determine the exact period of revolution for any celestial body, it becomes impossible to decide whether the periods of revolution of celestial bodies are commensurable or incommensurable. Thus Alexander's argument is inherently indeterminate. In this brief but important passage,

---

ity. However, from the context of the discussion Averroes is considering the doctrine of the return of one and the same perishable thing, as, for example, a man or animal, and on this point he and Alexander are in full accord with Aristotle (see above, n. 66 of this chapter).

69. All the planets are assumed to start from conjunction and to have integral ratios of circulations, as typified and exemplified by the 12 to 1 relationship between moon and sun. If we supposed all this *and* initial departure from a single point, it would follow necessarily that all the planets would, at some time in the future, arrive simultaneously in the same celestial point (see *De commensurabilitate*, pt. I, prop. 14). Nevertheless,

Alexander seems to deny this, saying that they need not revert to the very same point. He says that they could conjunct in any point whatever, which is impossible if the conditions described above are assumed. Thus the consequences that were drawn by Alexander on the assumption of integral ratios of circulations are false. Because the remarks by Alexander were probably made in his lost commentary on Aristotle's *De generatione* (see immediately preceding note), we cannot determine whether these difficulties are found in Alexander's discussion or whether Averroes has misinterpreted or misrepresented Alexander's position.

70. Trans. Kurland, *Averroes on Aristotle's "De generatione,"* p. 138.

Averroes offers the most basic and obvious reason why astronomers would have found it unprofitable to argue about the commensurability or incommensurability of celestial motions. The problem was simply insoluble as well as irrelevant, since all astronomers were perfectly aware that their observations were necessarily approximate. As Averroes tells us, they assumed that periods of revolution and celestial velocities were approximately commensurate, and no doubt rounded off their observations accordingly.[71] Lacking appropriate criteria and seeing the irrelevance of a decision, technical astronomers did not discuss this issue, and I believe we would look in vain for a consideration of this problem in their treatises.[72] The question whether or not the celestial motions are commensurable would be apt to turn up—if it appeared at all—in general cosmological discussions, in arguments concerning the theoretical validity of the foundations of astrology, and in considerations of a Great Year as well as in arguments about the truth of the philosophical and physical consequences that were drawn from it. Should discussions about the probability or reality of celestial commensurability have occurred in these contexts, it is likely that their substance and variety would have been quite similar to what appears in Part III of Oresme's *De commensurabilitate*. We have seen this to be true in the passage cited just above (p. 107) from Averroes' *Epitome* of Aristotle's *De generatione*. The reasonableness of this conjecture is reinforced by a passage in Book 1, Chapter 2 of the *Tetrabiblos*, or *Quadripartitum* as it was known in the Middle Ages, where Ptolemy says:

For in general, besides the fact that every science that deals with the quality of its subject-matter is conjectural and not to be absolutely affirmed, particularly one which is composed of many unlike elements, it is furthermore true that the ancient configura-

---

71. Cf. Oresme, *De commensurabilitate*, I.45–49, 210–14, and n. 40 of this chapter.

72. The problem is not even raised in such works as Ptolemy's *Almagest*, Geminus's *Elementa astronomiae*, Cleomedes' *De motu circulari corporum caelestium*, Theon of Smyrna's *Liber de astronomia*, nor, as far as I am aware, in any other more or less technical astronomical treatises. Why, then, did Theodosius raise the issue in his *De diebus et noctibus*, a treatise concerned with astronomical problems (see above, pp. 79–86)? Perhaps because he was aware of discrepancies in the values offered for the length of the solar year (see above pp. 79–83), he first showed how the intervals of time for these specific values might be computed before all

solar events repeated exactly as before. For the sake of completeness, he then sought to demonstrate that solar events would never repeat if the length of the year were incommensurable to a day—i.e., if the solar year was of an irrational length (cf. Paul Tannery, *Recherches sur l'histoire de l'astronomie ancienne* [Paris, 1893], pp. 42–43). But technical astronomers apparently chose not to entertain this latter possibility, choosing instead to offer precise values for the length of the year while, at the same time, recognizing these values as only approximate (see, for example, Ptolemy's *Almagest*, bk. 3, chap. 1 ["On the Year's Magnitude"], trans. R. Catesby Taliaferro, vol. 16, Great Books of the Western World [Chicago, 1952], p. 83).

tions of the planets, upon the basis of which we attach to similar aspects of our own day the effects observed by the ancients in theirs, can be more or less similar to the modern aspects, and that, too, at long intervals, but not identical, since the exact return of all the heavenly bodies and the earth to the same positions, unless one holds vain opinions of his ability to comprehend and know the incomprehensible, either takes place not at all or at least not in the period of time that falls within the experience of man; so that for this reason predictions sometimes fail, because of the disparity of the examples on which they are based.[73]

From the uncertainty expressed in this passage, Ptolemy's great authority could have been invoked for or against the doctrine of a Great Year. Either the planets do not all return to their same positions or, if they do, the period in which this occurs is too great for any generation of men to determine it. Although no mention is made here of celestial commensurability or incommensurability, had Ptolemy pursued the matter, our problem would almost certainly have arisen, and in this instance in an astrological context.

Of the three authors just mentioned, only Ptolemy's remarks could have been influential, for they were known in a medieval Latin translation. Alexander's commentary on the *De generatione*, from which Averroes probably drew his paraphrase of Alexander's argument, is lost without any extant Latin translation, and the *Epitome* of the *De generatione* by Averroes was translated from Hebrew into Latin only after the lifetime of Oresme—perhaps as late as the sixteenth century. Thus only the vague statement of Ptolemy and not the direct discussion of commensurability by Averroes could have been known in the Latin Middle Ages.

Turning now to the Latin Middle Ages, we shall see that the concepts of a Great Year and exact return played a considerable role in the few discussions of celestial commensurability that have thus far come to light. That Christians in the later Middle Ages were troubled by the doctrine that all things would return in successive Great Years is revealed by the fact that Etienne Tempier, bishop of Paris, found it necessary to condemn this opinion in 1277. Among the 219 articles, or propositions, condemned in that year, article 6, which could thereafter be held by Christians in Tempier's ecclesiastical domain only at the price of excommunication, reads: "That when all the celestial bodies have returned to the same point—which will happen in 36,000 years—the same effects

73. The substance of this passage in the medieval Latin translation made from the Arabic is in substantial agreement with the modern Greek text of Ptolemy's *Tetrabiblos*, trans. F. E. Robbins, Loeb Classical Library (London and Cambridge, Mass, 1940), pp. 15–17. For a medieval Latin version of this passage see *Liber Ptholomei quattuor tractatuum cum Centiloquio eiusdem Ptholomei et commento Haly*.... (Venice: Erhard Ratdolt, 1484), bk. 1, chap. 2, sig. a3r, col. 2. Because he was undoubtedly familiar with Ptolemy's *Quadripartitum*, Oresme was probably well aware of this passage.

now in operation will be repeated."[74] Although they make no mention of it, it is possible that both Henry Bate of Malines and Duns Scotus, whose views will be described below, were cognizant of, and influenced by, article 6; despite a belief in an astronomical Great Year, Henry Bate makes no mention of the identical return of all things, while Scotus employs the possible incommensurability of celestial motions as grounds for attacking the Great Year of 36,000 years and the frequently drawn consequence of exact cyclical return, which he calls contrary to faith.[75] Oresme may also have had article 6 in mind when he attacked as one of "many errors about philosophy and faith," the Great Year of 36,000 years and its consequence that celestial aspects and configurations will repeat in each Great Year.[76] To refute this, he demonstrated mathematically that "when two motions of celestial bodies have been proposed, it is probable that they would be incommensurable."[77]

Despite St. Augustine's unequivocal repudiation of the Great Year and the cyclical return of all things (see his explanation of Solomon's famous remark, below in the Commentary, fn. 6, p. 352) and the obvious fact that numerous Christian writers must have been fully aware of its potential danger to their faith and dogma, it is possible that article 6 was the more immediate stimulus that motivated Scotus and Oresme, and perhaps others, to supplement and reinforce explicit theological condemnation with scientific and mathematical argument.[78] Perhaps this partially explains why, in the *De proportionibus proportionum*, Oresme formulated detailed probability arguments concerning commensurability and incommensurability; or perhaps it explains why, after having formulated such probability arguments, he applied them to celestial motions.[79]

74. "Quod redeuntibus corporibus celestibus omnibus in idem punctum, quod fit in XXX sex milibus annorum, redibunt idem effectus, qui sunt modo."—H. Denifle and E. Chatelain, *Chartularium Universitatis Parisiensis*, vol. 1, p. 544. See also Grant, *Oresme PPAP*, p. 376.

75. See below, n. 98 of this chapter.

76. *De proportionibus proportionum*, IV. 606–9 in Grant, *Oresme PPAP*, p. 307 (Latin text, p. 306). In the *Ad pauca respicientes*, Oresme refers to Great Years of 36,000 and 15,000 years as "foolishness" (ibid., p. 383). In the *De commensurabilitate*, he speaks of a Great Year of 36,000 years for the sun and eighth sphere (fixed stars) only, and one of much greater duration involving all the planets and the eighth sphere (see *De commensurabilitate*, I.818–21). These Great Years

are assumed to occur only if the celestial motions are commensurable, which Oresme did not really believe.

77. *De proportionibus proportionum*, IV. 573–74 in Grant, *Oresme PPAP*, p. 305 (Latin text, p. 304). For a discussion of this passage and the mathematical demonstration which supports it, see above, chap. 2, n. 113.

78. The only Latin discussions of celestial commensurability or incommensurability known thus far postdate the Condemnation of 1277. Although this fact may be significant, no useful inferences may be drawn from it.

79. See above, chap. 2, n. 113. Oresme may also have been motivated by a desire to free King Charles V from the grip of his court astrologers (see Grant, *Oresme PPAP*, pp. 61–64).

But now let us describe and evaluate the contributions of Henry Bate and Duns Scotus.

In 1281, Henry Bate of Malines criticized an anonymous translator of Abraham ibn Ezra's *Book on the Universe or World* (*Liber de mundo vel seculo* composed in 1147, according to the author's own statement on fol. 80r, col. 2, in the edition cited below in n. 80). Although neither the original language nor the language into which the translation was made are known with certainty, this translation may have served subsequently as the basis for Bate's own translation of this same treatise into Latin.[80] A passage in ibn Ezra's text lies at the

80. It is possible that the translator referred to by Henry Bate (see below, this note and nn. 83, 84) was the Jew, Hagin. According to the colophon of MS français N° 24276, Hagin translated Abraham ibn Ezra's *Principium sapientiae* from Hebrew into French in 1273 while in the house of Henry Bate (the colophon is quoted by Duhem, *Le Système du monde*, vol. 4, pp. 27–28 and by Raphael Levy, *The Astrological Works of Abraham ibn Ezra* [Baltimore, 1927], p. 21). Although specific and convincing evidence is lacking, it is on this slender basis that Hagin is believed to have translated from Hebrew into French the *Liber de mundo vel seculo* (see Duhem, *Le Système*, p. 28), as well as other of ibn Ezra's astrological treatises. Whatever the merit of these conjectures, it was from French versions that subsequent Latin translations were made by Peter of Abano in 1293 and Arnold of Quinquempoix at the beginning of the fourteenth century (see Raphael Levy and Francisco Cantera, eds., *The Beginning of Wisdom: An Astrological Treatise by Abraham ibn Ezra* [Baltimore, 1939], p. 14). Perhaps Henry Bate's translations of ibn Ezra's works, made in 1281 and 1291, were also made from French, despite an explicit statement to the contrary in the colophon of Bate's translation of the *Liber de mundo vel seculo* cited and discussed in the next paragraphs.

The edition from which I shall cite all the relevant passages is the same as that used by Duhem in his *Le Système*, vol. 4, p. 27, and vol. 8, p. 445, n. 2, and bears the long title:

Abrahe Avenaris Judei Astrologi peritissimi in re iudiciali opera ab excellentissimo Philosopho Petro de Abano post accuratam castigationem in latinum traducta: Introductorium quod dicitur principium sapientiae; Liber rationum; Liber nativitatum et revolutionum earum; Liber interrogationum; Liber electionum; Liber luminarium et est de cognitione diei cretici seu de cognitione cause crisis; Liber coniunctionum planetarum et revolutionum annorum mundi qui dicitur de mundo vel seculo; Tractatus insuper quidam particulares eiusdem Abrahe; Liber de consuetudinibus in iudiciis astrorum et est centiloquium Bethen breve admodum; Eiusdem de horis planetarum. Colophon: Ex officina Petri Liechtenstein, Venetiis, anno domini 1507.

Although all the translations mentioned in the title are attributed to Peter of Abano, the colophon to the *Liber de mundo vel seculo*, on fol. 85r, col. 1, ascribes this particular translation to Henry Bate of Malines in the year 1281: "Explicit liber de mundo vel seculo completus die Lune hore post festum beati luce hora diei quasi 10 anno domini 1281 inceptus in Leodio et perfectus in Machilinia translatus a magistro Henrico Bate de Hebreo in Latinum."

The inclusion of this translation by Henry Bate is explained by the fact that Peter of Abano did not make a translation of the *Liber de mundo vel seculo* (see Lynn Thorndike, "The Latin Translations of the Astrological Tracts of Abraham Avenezra," *Isis*, vol. 35 [1944], p. 294). Although the explicit states that Henry translated this work from

root of the criticism. After reporting that the Persians held a doctrine of periodic celestial return,[81] he denies its truth and argues against the plausibility of such a conception:

But if someone should say that because of this every [period of] 75 years is similar to preceding periods [of 75 years]—since the planets and their participations are uniform —here is the response: Having realized that this cannot be [achieved] by means of a [single] ratio because an ascendant is found with a disposition proportional to it and the ratio of one to the other is perpetually uniform or equal—even if the world were eternal—you can consider this matter carefully. For Saturn has many diverse relationships both with the sun and with the planets; and similarly with the fixed stars, which are moved through one degree every 70 years, so that Saturn certainly will not have the

------

(Note 80 continued)
Hebrew into Latin, Raphael Levy insists that "Nothing warrants crediting Bate with a knowledge of Semitics and his translations of astrological works of Abraham ibn Ezra seem based on Hagin's intermediary French version" (*Astrological Works*, p. 30). In this view Levy sides with Duhem (*Le Système*, vol. 4, p. 27; vol. 8, p. 446) and opposes Lynn Thorndike (see Levy, p. 25), who, in his *History of Magic and Experimental Science* (vol. 2, p. 928), had accepted this colophon at face value, and continued to maintain this position in his later 1944 *Isis* article (p. 293). But the problem is even more complicated. In a postscript to MS *B* (BN 7281) of the *De commensurabilitate* (the Latin text is quoted below in its entirety on pp. 164–65), written in the hand of the scribe who copied the treatise, we are told that Henry Bate translated the *Liber de mundo* from Arabic to Latin ("...sed eius translator de arabico in latinum, Henricus Bate,..." [fol. 273r]). That there was a translation from Arabic is indicated in the text of the *Liber de mundo*, where we read (Abraham ibn Ezra, *Liber de mundo vel seculo* [Venice, 1507], fol. 78r, col. 2): "Inquit translator hec est itaque sermo Avenare secundum quod iacet in arabico...." Is Bate telling us, in effect, that the anonymous translator had translated ibn Ezra's text from Arabic? Or is he informing us that the anonymous translator rendered ibn Ezra's original text into Arabic? Or, and

this seems unlikely, was he claiming that his own translation into Latin was made from Arabic—an interpretation which may have been placed upon this statement by the author of the postscript in MS *B* of the *De commensurabilitate*.

On the basis of such conflicting evidence— and despite Levy's (unsupported) claim that Bate knew no Semitic language—it seems advisable to allow that Bate's Latin translation of Abraham ibn Ezra's *Liber de mundo vel seculo* may have been made from French, Hebrew, or Arabic. It is not even certain that the translation of "the translator," whom Bate repeatedly attacks, was the basis of Bate's Latin translation, which may have been rendered from the original language (Hebrew or Arabic).

81. "Dicunt autem Persarum sapientes quod semper advertendum est ad partes firdarie que quidem revertuntur secundum circulationem in quibuslibet annis 75..." (ibn Ezra, *Liber de mundo*, fol. 80r, col. 1). The substance of these few lines appears in the postscript to MS *B* of the *De commensurabilitate* (see below, p. 164). Furthermore, this passage and almost all others which I shall quote below from ibn Ezra's *Liber de mundo vel seculo* were translated into French by Duhem in *Le Système du monde* (vol. 8, pp. 445–46), who, however, included only a few lines of Latin text and offered no explanatory or interpretive comment.

same ratio [to the sun or the planets] that it has to one of the fixed stars, i.e., up to 25,200 years. But it is not necessary to prolong this discussion....

For this reason, the nature of one man cannot be assimilated to the nature of another, since no [celestial] orb remains in one mode; nor will there ever be a point of an hour whose ratio could return because there was not any like it, nor will there be. And the wise arithmeticians have come to recognize this.[82]

Thus Abraham ibn Ezra appears to dispute claims of cyclical regularity in celestial motions on the grounds that every celestial body has quite a number of different simultaneous, and seemingly irreconcilable (incommensurable?), relationships with all other celestial bodies. Furthermore, he seems to allude to, and deny, the doctrine that one and the same individual can return; or perhaps he is simply denying that identical natures can recur since the heavenly bodies will never exactly repeat previous configurations. It is apparently at this juncture in the text that the anonymous translator intruded in an effort to save the doctrine of the Great Year by imposing a more favorable interpretation on this passage.

The translator says: Although the multiplication of a number could increase to infinity, the revolutions of the celestial bodies are necessarily finite, as has been demonstrated with certitude in another part of philosophy. It is necessary, therefore, that similar [celestial] configurations should sometimes return, even though the [interval of] time is incomprehensible to us because of the enormity of these intervals. Perhaps this is what the author [i.e., Abraham ibn Ezra] means here.[83]

---

82. "Porro si loquens aliquis dicat quod ob hoc esse deberent quilibet 75 anni similes precedentibus cum planete, et ipsorum participationes sunt uniformes, responsio hoc est: scito quod hoc esse non potest secundum viam proportionis quod invenitur unum ascendens cum habitudine proportionali ad ipsum, et sit proportio unius ad alterum uniformis seu equalis in perpetuum etiam si mundus semper duraret. Et hanc quidem rem perpendere potes. Saturnus enim multas habet diversitates tum ex parte solis tum ex parte planetarum; item et ex parte stellarum superiorum que moventur in quibuslibet 70 annis per unum gradum quo circa proportionem non habebit utique quam habuit ad superiorum unam usque ad 25,200 [*corr. ex* 25,000] annos. Et non est quidem necesse in hac sermone protelare...."

"Quapropter esse non potest nativitas hominis que assimiletur nature alterius tamquam sibi non enim est orbis stans secundum

unum modum nec umquam erit punctus hore quin revertatur [*corr. ex* removetur] proportio quod [*corr. ex* que] ⟨non⟩ fuit sicut illa neque erit. Et sapientes [*corr. ex* aspicientes] quidem arismeticii [*corr. ex* ananetici] hoc noverunt."—ibn Ezra, *Liber de mundo*, fols. 80r, col. 2–80v, col. 1. The omission indicated at the end of the first paragraph consists of approximately six lines of text that are unintelligible to me, but which seem concerned with the construction or formation of astrological houses. They appear to be unrelated to the basic position adopted here by Abraham ibn Ezra, and, interestingly, are also omitted in the postscript to the *De commensurabilitate* found in MS *B*, where this entire passage from Abraham ibn Ezra is quoted (see below, pp. 164–65). The five corrections and one emendation made above were derived from the version quoted in *B*.

83. "Dicit translator: Quamquam multiplicatio numeri possit crescere in infinitum,

In this obscure passage, it is suggested by the anonymous translator that although ibn Ezra denied cyclical regularity, perhaps this was simply because he was concerned only with relatively short periods but would have accepted cyclical regularity over periods of extraordinary length. This interpretation aroused the wrath of Henry Bate for, while the latter was in basic agreement with the anonymous translator concerning cyclical regularity, he was annoyed with his seemingly timid and hesitant approach to the passage. Here is his reaction:

It is not the case however, etc. I do not know why the translator has defiled the parchment by interjecting his words into the text and showing that he knows mathematics (metaphysics?). Now because of the manifold diversity of the motions of celestial bodies, it must not be thought that whatever the number of revolutions it would be possible for them not to meet or communicate, as is the case with incommensurable lines, which, in the tenth book of the *Elements*, Euclid calls irrationals, or surds, since they lack a common measure. As the Philosopher [i.e., Aristotle] testifies in the twelfth book of the *Metaphysics*, "all things are ordered together;" and the Commentator [i.e., Averroes] says about this that "all the actions of celestial bodies in their mutual relationships in the regulation of the world are as the action of freemen in the regulation of a house." Thus even a modicum of thought about this makes it evident that if there must be order or association between some things, this order or association must be [realized] excellently in divine things. It is, therefore, absurd that the motions of the superior [i.e., celestial] bodies be thought to be irrational, or surd. And this [harmony or rationality] is what Pythagoras and other ancients wished to signify by worldly music; and on this matter, Plato says similar things in the *Timaeus* and elsewhere; and Chalcidius also, along with an infinite number of other philosophers. But [now] let us return to the text.[84]

(Note 83 continued)

revolutiones tamen corporum celestium finite sunt secundum necessario quemadmodum in alia parte philosophie demonstrari habet cum certitudine. Quapropter necessarium est consimiles interdum redire constellationes licet incomprehensibile sit a nobis tempus huiusmodi revolutionum propter intervallorum immensitatem et hoc forsan est quod hic innuit auctor iste."—ibn Ezra, *Liber de mundo*, fol. 80v, col. 1. This entire passage, and, with the exception of a few lines, the long passage in the next note are also repeated in the postscript in MS *B* (see below, p. 164).

The few lines that were omitted probably explain why the author of the postscript mistakenly attributed the Latin passage quoted in this note to Henry Bate rather than to the anonymous translator whom Bate was to criticize. The omitted lines introduce the quotation in note 84 and read: "Non est autem, et cetera. Nescio quare hic translator deturpavit pergamenum ponendo se in textu et ostendendo se scire ?mathematicam [*corr. ex* metaphysicam]." Obviously, Bate is speaking harshly of the anonymous translator who dared write these lines.

84. "Non est autem, et cetera. Nescio quare hic translator deturpavit pergamenum ponendo se in textu et ostendendo se scire ?mathematicam [*corr. ex* metaphysicam]. Non est autem opinandum quod propter multiplicem diversitatem corporum motuum

Here we have enunciated an emphatic denial that celestial motions can be incommensurable, for they are not related as are incommensurable lines in Book 10 of Euclid's *Elements*. Since the anonymous translator did not specifically mention celestial incommensurability, was Henry Bate reacting to some other text in which celestial incommensurability was proclaimed? As yet, no answer can be offered. For lack of information not even its influence, if any, can be assessed. Despite ignorance on these matters, however, the text was

---

celestium possibile sit ipsos in revolutionibus quibuslibet non convenire seu communicare quemadmodum est de lineis [*corr. ex libris*] incommunicantibus quas in decimo [*hab.* 10] elementorum Euclides [*corr. ex* Eulides] vocat irrationales sive surdas propter inpotentiam communicandi. 'Omnia namque coordinata sunt' ut testatur Philosophus duodecimo [*corr. ex* 10] *Metaphysice* super quo dicit Commentator quod 'actiones omnes corporum celestium in communicatione eorum adinvicem in constitutione mundi sicut sunt actio liberorum in constitutione domus.' Palam autem est etiam modicum consideranti circa hoc quod si inter aliqua debet esse ordo seu communicatio excellenter esse debet in divinis. Quare absurdum esset oppinari motus corporum superiorum in irrationales sive surdos, et hoc est quod Pictagoras [*hab.* Pictago] et alii antiqui per musicam mundanam innuere voluerunt. De qua similiter Plato in *Thymeo* et alibi necnon et Calcidius cum aliis philosophis infinitis. Sed redeamus ad textum."—ibn Ezra, *Liber de mundo*, fol. 80v, cols. 1–2.

I shall now cite the brief quotation from Aristotle's *Metaphysics*, as well as the lines immediately following, all of which served as the point of departure for the remarks quoted from Averroes, the Commentator:

"And all things are ordered together somehow, but not all alike—both fishes and fowls and plants; and the world is not such that one thing has nothing to do with another, but they are connected. For all are ordered together to one end, but it is as in a house, where the freemen are least at liberty to act at random, but all things or most things are already ordained for them, while the slaves

and the animals do little for the common good, and for the most part live at random; for this is the sort of principle that constitutes the nature of each."—*Metaphysica* 12. 10. 1075a15–23 (trans. W. D. Ross, *The Works of Aristotle* [Oxford, 1908], vol. 8).

In the Juntine edition of the *Metaphysica* (see bibliography under Averroes, vol. 8), the first line of Aristotle's passage is rendered (in the Renaissance version of Cardinal Bessarion) as: "Cuncta autem coordinata quodam modo sunt..." (vol. 8, fol. 337r, col. 2). The quotation from Averroes, which forms part of his commentary on the Aristotelian passage quoted above, appears in Text 52 of Book 12 of the *Metaphysica* (fol. 338r, col. 1 in vol. 8 of the Juntine edition) and is in virtual agreement with Bate's version, except that in the Juntine edition "omnes" is omitted and "est sicut" replaces "sicut sunt." In this brief passage, Averroes means to say that just as freemen are not at liberty to do as they please in operating and maintaining their households, but must order their actions for the common good of all who dwell therein (this is contrasted with slaves, whose actions lack this order and purpose), so also the celestial bodies are not free to act in any manner whatever, but must act to produce a given order.

The reference to Plato's *Timaeus* is, perhaps, to 35B–36B, where the World-Soul is divided into harmonic intervals. Since Chalcidius commented on these sections of the *Timaeus*, it is probably these comments that were intended (see *Timaeus a Calcidio translatus* (ed. Waszink), commentaries 28–55, pp. 78–103).

available to a wide reading audience since it was embedded in a translation of
a work of Abraham ibn Ezra, who was himself a very popular author. Thus
where the intrinsically significant passage quoted above from Averroes' *Epitome*
on the *De generatione et corruptione* (see pp. 106–7) was probably untrans-
lated and presumably unknown in the Middle Ages, Henry Bate's sentiments—
hostile though they were to the doctrine of celestial incommensurability—
might have provoked further consideration of this problem, perhaps even evok-
ing the later discussion by Duns Scotus.

In turning now to John Duns Scotus (ca. 1266–1308), we shall consider a
brief, but directly relevant, and highly significant passage, which represents the
first clear and explicit statement of the consequences of celestial incommensurability enunciated in the Latin West prior to the writings of Johannes de Muris
and Nicole Oresme. The locus of Scotus's treatment is his commentary on
Book 4 of the *Sentences* of Peter Lombard (*Quaestiones in quartum librum Sententiarum*) of which there are two major published versions, the *Reportata parisiensia*, probably composed in 1302–3 at the University of Paris, and the *Opus
oxoniense*, or *Ordinatio*, a much more complete account that was prepared at
Oxford, apparently no earlier than 1305.[85] Despite the possibility that Oresme

85. The conjectural dates of composition
for the two versions are given by Carl Balić,
O.F.M., in his critical study *Les Commentaires de Jean Duns Scot sur les quatre livres
des sentences* (Louvain, 1927), pp. 242–43,
244. The texts of these two works, from
which I shall quote, were published in the
Vivès edition, the title of which is: *Joannis
Duns Scoti... Opera omnia editio nova juxta
editionem Waddingi XII tomos continentem a
patribus Franciscanis de observantia accurate
recognita*, 26 vols. (Paris: Luis Vivès, 1891–
95). Although other versions, both published
and unpublished, are extant, only in recent
years has any genuine effort been made to
establish critical texts and eliminate much
extraneous material added by later Scotist
commentators and editors. In order to convey to the reader the meaning of the terms
Reportata and *Ordinatio*, as well as to give
some inkling of the difficulties that bedevil
editors who cope with Scotist texts, I offer
the following summary by Allan Wolter,
O.F.M. (*Duns Scotus Philosophical Writings*,
a selection edited and translated by Allan
Wolter, O.F.M. [London: Nelson, 1962],

pp. xvii–xviii):

"Scotus commented on the *Sentences* of
Peter Lombard at least twice, once at Oxford
and again at Paris. From the seventeenth
century down to our own, these two commentaries were referred to respectively as the
Opus oxoniense (or *Ordinatio*) and the *Opus
parisiense* (or *Reportata parisiensia*). Modern
research has not only revealed the existence
of other unedited reports of these lectures on
the *Sentences* but has rediscovered the meaning of the terms *ordinatio* and *reportatio*. The
original lecture of a master or bachelor as
copied down by one of his students, or some
scribe, is known as a *reportatio*. If such a
'reported version' was later checked by the
teacher himself, it is referred to as a *reportatio examinata*. In many cases the author
would revise his original lectures before presenting them for final publication. This last
redaction or finished product is known as an
ordinatio, inasmuch as it represents the final
draft as ordered or arranged by the author
himself. As applied to Scotus's *Commentary
on the Sentences*, the *Ordinatio*, to which the
redactor of the Assisi manuscript, used by

may have been influenced by Scotus's Parisian *Reportata* (see below, the Commentary, n. 55 for III.311–12), only the Oxford version, or *Ordinatio*, is quoted in the body of this narrative since it is the fuller account and may represent Scotus's final word on this problem. However, in view of the general meagerness of evidence that has thus far come to light on this topic, I have put into footnotes all the corresponding discussion from the earlier *Reportata parisiensia*.

In Book 4, Distinctio 43, Question 3, of his Oxford Sentence Commentary, Scotus considers the question "Whether nature could be an active cause of resurrection."[86] There are two parts to the question.[87] Because man consists of a body compounded from natural elements, the first step in any exact and identical resurrection of a human being must be the precise reconstitution of his bodily elements. The second step would be the reunion of that individual's intellective soul with his body. Upon completion of these two successive steps, the very same man would be reconstituted exactly as before. Scotus examines both parts of this question, but only the first concerns us here and it is considered in the broad sense of whether any corruptible thing—mixed or unmixed—can return, after corruption, exactly as it was before.[88]

Among a number of arguments that Scotus presents against this opinion, only one is relevant to our topic. It begins with a reference to Augustine's *City of*

---

us (Communalis 137), had access, is a revision of the Oxford lectures. Internal evidence suggests that some parts of the redaction antedate the Paris lectures, while for other portions Scotus made use of a *reportatio* of the latter. The *Opus oxoniense* as we have it in the Wadding and Vivès editions is not the pure *Ordinatio*, however, but contains other elements. One of the principal tasks of the Scotistic Commission under the direction of Carl Balić, O.F.M., at Rome is to reconstruct the text of the original *Ordinatio* and separate it from the major and minor additions taken from other writings of Duns Scotus. The enormity of this task can be realised from the fact that it was only after twelve years that Balić and his many collaborators published the first small fraction of the monumental *Ordinatio* in the two initial volumes of the critical Vatican edition of the *Opera omnia* of Scotus (Rome, 1950)."

86. "Utrum natura possit esse causa activa resurrectionis."—Vivès ed. vol. 20, p. 65, col. 2. The *Reportata parisiensia* substitutes

"efficiens" for "activa," but is otherwise in agreement (vol. 24, p. 508, col. 1).

87. These are: "Primo ergo videndum est, si natura possit reducere formam mixti eamdem numero. Secundo, si possit illi mixto dissoluto reunire animam intellectivam, ut sit idem homo."—Vivès ed., vol. 20, p. 66, col. 2. In the Paris version we are told: "igitur dupliciter potest intelligi quaestio, uno modo: An scilicet reparatio corporum habeat causam activam in natura; alio modo, an totaliter resurrectio hominis habeat totalem causam activam in natura."—Ibid., vol. 24, p. 509, col. 1. In the Parisian *Reportata*, Scotus explains that the complete resurrection of a man involves the reunion of body and soul (p. 509, col. 1). Hence the two versions are in substantial agreement.

88. This subquestion of Distinctio 43, Question 3, is expressed as follows: "Primum, si universaliter potest aliquod corruptibile redire idem numero...."—Vivès ed., vol. 20, p. 66, col. 2–p. 67, col. 1.

God, where the doctrine of individual return is described and rejected: "...in the *City of God*, [Book] 12, Chapter 13,[89] Augustine cites the opinion of certain philosophers who say that the same things return through fixed periods of time. It is set down by them that after a Great Year, that is after 36,000 years, all things would return the same as they were."[90]

Scotus then explains the rationale of these philosophers: "Their reason was that since the same cause returns, the same effect will also return. Furthermore, all celestial bodies will return to the same positions, for by assuming what Ptolemy says in the *Almagest* [namely] that the starry sky is moved one degree in one hundred years in a direction opposite that of the daily motion, it follows that this motion of the [fixed] stars from west to east will be completed in 36,000 years."[91]

---

89. In modern editions, the chapter to which Scotus refers is numbered 14 rather than 13. See below, the Commentary, fn. 6, p. 352.

90. For the Latin text, see below, the Commentary, fn. 7, p. 352; the corresponding Latin passage in the Parisian *Reportata* is given below in the Commentary, fn. 5, p. 351.

91. "Ratio eorum erat, quia redeunte causa eadem, redibit idem effectus; tunc autem omnia corpora coelestia redibunt ad eumdem situm, quia supponendo illud Ptolemaei in *Almagesto*, quod coelum stellatum moveatur in centum annis uno gradu, contra motum diurnum, sequitur quod complebitur motus ille ejus ab occidente in orientem in triginta sex millibus annorum."—Vivès ed., vol. 20, p. 67, col. 1. This passage and the lines immediately preceding (the latter are quoted in the Commentary, fn. 7, p. 352) were translated into French by Duhem, *Le Système du monde*, vol. 8, p. 447 (the Latin text was omitted). Ptolemy gives the value of precession of the equinoxes as 1° in 100 years in the *Almagest*, bk. 7, chaps. 2 and 3 (see the translation by R. Catesby Taliaferro in Great Books of the Western World, vol. 16, pp. 227 and 232; for a further discussion, see Heath, *Aristarchus of Samos*, pp. 172–73, and Grant, *Oresme PPAP*, p. 429, n. 2).

Scotus explained in the earlier Parisian version: "The reason for this is as follows: Celestial bodies have a universal causal ef-

ficacy on all lower [or sublunar] generable and corruptible things whose difformity in disposition and position is the cause of the difformity in the kinds of effects [that they produce]. It follows, therefore, that the uniformity of disposition which the celestial bodies have when they have produced a body will be the cause of the uniformity of its effect. Therefore, if a certain body were generated now by virtue of celestial influence, and afterward it should be corrupted, it follows that at other times the power [or force] of the sky will bring back and renew the same identical effect when it will be in the [very same] disposition in which it was when that body was generated [previously]. Moreover, according to Ptolemy in the *Almagest*, the celestial bodies can return to such a disposition after 36,000 years." ("Ratio ad hoc talis est: Ex quo corpora coelestia habent universalem causalitatem super omnia ista inferiora generabilia et corruptibilia, et difformitas eorum in dispositione et situ est causa difformitatis in genere effectuum, sequitur quod uniformitas dispositionis quam habent, quando aliquod corpus produxerunt, erit causa uniformitatis illius effectus; igitur si aliquod corpus virtute influentiae coelestis generetur modo, et postea corrumpatur, sequitur quod alias virtus coeli, quando erit in dispositione in qua fuit, quando illud corpus generavit, eumdem effectum numero reducet et reparabit; ad talem autem dispositionem

After citing Augustine's arguments against cyclical return, all of which were rooted in appeals to Scripture,[92] Scotus introduces his own argument, which depends upon the possible incommensurability of the celestial motions.[93]

> This opinion [i.e., exact return] can also be disproved with respect to its cause, for if it could be proven that some celestial motion was incommensurable to another[94] and this can be proved if the magnitude [or circle of the sphere] on which one [celestial body] is located were assumed incommensurable to another magnitude [or circle of a sphere on which some other celestial body is located], *and* if it is assumed that these two magnitudes [or circles] move with equal speed[95]— then, I say, it follows that all the motions will never return to the same place. Nor is there anything about the incommensurability of celestial motions that is contrary to the continuity of continuous motion, because if two bodies were moved [with equal speeds], one on the side of a square and the other on the diagonal [of the same square], these two bodies would not return to uniformity [i.e., to the same positions], even if they should endure forever.[96]

---

possunt redire corpora coelestia post 36. millia annorum secundum Ptolemaeum *in Almagesto*."—Vivès ed., vol. 24, p. 509, cols. 1–2.) Ptolemy did not discuss the Great Year in the *Almagest*, but it seems that because of his doctrine of precession, in which the stars turn from west to east in 36,000 years, he was usually identified as an adherent of that cosmological doctrine.

92. Vivès ed., vol. 20, p. 67, cols. 1–2. In the Parisian *Reportata*, arguments from Aristotle are added (Vivès ed., vol. 24, p. 509, col. 2–p. 510, col. 1).

93. If Scotus drew the details or the substance of this argument from others, he gives no hint of it.

94. In the *De proportionibus proportionum*, Oresme offers a demonstration to show that this is *probably* true. See above, chap. 2, n. 113 for full discussion and references.

95. In Part 1, Proposition IV of the *Ad pauca respicientes*, Oresme includes, with greater clarity and explicitness, substantially the same conditions as one of four ways in which two bodies can be moved incommensurably on their respective circles: "And if they are moved through an eternity, mobiles are said to be moved incommensurably with respect to the center when they describe incommensurable angles in equal times. Now this can occur because the circumferences

are incommensurable on which the mobiles are moved with equal speed or commensurable [motions]...."—Grant, *Oresme PPAP*, p. 395 (Latin text, p. 394); see also ibid., p. 432, and below, the Commentary, n. 2 for I.24–31). Although he does not specify equality of curvilinear velocities of the mobiles, Johannes de Muris may have had this case in mind in Chapter 14 (lines 3–5) of Book 4, Tract 1, of the *Quadripartitum numerorum* (see below, App. 1, and above, nn. 26 and 28 of this chapter).

96. Utilizing two squares instead of one (see Fig. 14 reproduced here from Grant,

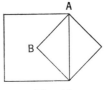

Fig. 14

Oresme PPAP, p. 399), Oresme includes this example in Part 1, Proposition V of the *Ad pauca respicientes*. "Furthermore, let there be two squares where the diagonal of one is the side of the other. Then let *A* and *B* be moved with equal speed from any angle on the figure, and never again would they be found in that angle or in another angle."—

But if, in the whole heavens, any [motion] should be found incommensurable to any other, this would require a great discussion of all the movements appropriate to epicycles and deferents."[97]

Scotus has enunciated an important conditional argument. If any two celestial motions should be incommensurable, the planets and fixed stars would never

---

(Note 96 continued)

Grant, *Oresme PPAP*, p. 399 (Latin text, p. 398). Both Johannes de Muris (*Quadripartitum numerorum*, bk. 4, tract 1, chap. 14) and Oresme would later illustrate the same point by relating two incommensurable circumferences as a diagonal to its side (see above, n. 26 of this chapter).

97. "Potest etiam illa opinio improbari quantum ad rationem ejus, quia si probetur aliquem motum coelestum alicui incommensurabilem, quod probari potest si ponatur magnitudo super quam fit esse incommensurabilis magnitudini, supposita hinc inde aequali velocitate; tunc, inquam, sequitur quod nunquam omnes motus redibunt ad idem; nec istud de incommensurabilitate motuum est contra continuitatem motus continui, quia si duo mobilia moverentur, unum super costam quadrati, illud super diametrum, motus illi essent incommensurabiles, nec in perpetuum, si durarent, redirent ad uniformitatem; sed istud requireret magnam discussionem singulorum motuum congruentium epicyclis et deferentibus, si aliquis possit alicui inveniri incommensurabilis in toto coelo."—Vivès ed., vol. 20, p. 67, col. 2). I have made slight alterations in punctuation. Duhem also translated this passage, but again omitted the Latin text (see *Le Système du monde*, vol. 8, pp. 447–48).

The corresponding passage in the *Reportata parisiensia* is briefer and less intelligible: "Besides, some celestial motions, or celestial bodies, are, perhaps, improportional and incommensurable since the magnitudes on which they are located can be incommensurable and improportional, as is evident *from the tenth book of Euclid on the side and diagonal of a square*. Thus motions made on its side and diagonal cannot be commensurable because the magnitudes cannot be commensurable, and, consequently, they could not be measured by one and the same measure. And so they cannot return to the same disposition as before, and therefore the same thing cannot recur numerically the same." ("Praeterea, aliqui forte motus coelestes, vel corporum coelestium, sunt improportionabiles et incommensurabiles, quia magnitudines super quas sunt, possunt esse incommensurabiles et improportionabiles, ut patet *ex 10. Euclidis de costa et diametro quadrati*. Unde motus factus super costam et ejus diametrum, non possunt esse commensurabiles, quia magnitudines non possunt esse commensurabiles, et per consequens non possent mensurari aliqua una et eadem mensura, et ita non possunt redire ad eamdem dispositionem quam prius, et ita nec eadem numero."—Vivès ed., vol. 24, p. 510, cols. 1–2.)

Although a proof of the incommensurability of the side and diagonal of a square was interpolated as Proposition 117 of Euclid's tenth book of the *Elements* (see I. L. Heiberg, *Euclidis Elementa*, vol. 3, containing Book 10 [Leipzig, 1886], App. 27, pp. 408–11, and Heath, *Euclid's Elements*, vol. 3, p. 2), it does not appear in *Euc.-Campanus*, the most widely used medieval edition of the *Elements*, nor is it found in the major medieval translations and versions of the *Elements* by Gerard of Cremona and Adelard of Bath. (I have examined the following manuscripts: Gerard of Cremona's translation in BN 7216; Adelard Version I in Bodleian Library, D'Orville MS 70; Adelard Version II in Bodleian Library, Auct. F. 5.28; Adelard Version III in Oxford, Balliol College, MS 257. For the basis of the threefold division of Adelard versions, see Marshall Clagett, "The Medieval Latin Translations from the Arabic of

return to identical positions in fixed intervals of time, and a Great Year could not occur. But he concedes that a "great discussion" would be required to demonstrate whether such incommensurability exists, and he wisely refrains from claiming it as true, or even probable. Nevertheless, he rejects the doctrine

---

the *Elements* of Euclid, with Special Emphasis on the Versions of Adelard of Bath," *Isis*, vol. 44 (1953), pp. 18–25.) If Scotus did not have access to a version that contained this interpolated proof, then either he was mistaken, or—and this is more likely—perhaps he only meant that by use of Book 10 one could demonstrate the incommensurability of side and diagonal.

Immediately following the incommensurability discussion in the Oxford version, but not in the Parisian *Reportata*, Scotus asserts that Thabit ibn Qurra had disproved Ptolemy's claim that the fixed stars would return to their former positions in 36,000 years. From Scotus's description of Thabit's position, it would seem that we have here a reference to the latter's trepidation theory. According to Scotus, Thabit's theory would rule out cyclical repetitions of the fixed stars and planets in 36,000 years. Scotus insists that if the motion in Thabit's theory were completed in such a period of time as to render impossible the return of the celestial bodies to the same positions which they held at the start of their motions, it would follow that exact cyclical return would also be impossible in Thabit's interpretation.

"Praeterea, illud fundamentum acceptum a Ptolemaeo improbatur per Thebit, qui probat coelum stellatum non sic moveri ab occidente in orientem, quia tunc aliquando foret in principio Cancri noni coeli stella illa, quae alias fuit in principio Capricorni noni coeli, et ideo ponit motum octavi coeli, vel stellati, esse in quibusdam parvis circulis descriptis super principium Arietis et Librae noni coeli; et quod ille est quidam motus accessus et recessus, secundum quod principium Arietis mobilis in suo circulo ascendit, et per oppositum caput Librae mobilis in suo circulo descendit, et alias e converso,

capite Arietis descendente, caput Librae ascendit, et sic moventur stellae in octavo coelo secundum longitudinem et latitudinem simul. Si ergo ille motus probaretur in aliquo tempore compleri, in quo tempore non possent omnes orbes inferiores redire ad eumdem situm, quem habuerunt in principio illius motus, sequeretur propositum."—Vivès ed., vol. 20, pp. 67, col. 2–68, col. 1.

Thabit's trepidation theory was included in his *De motu octave sphere*, presumably translated by Gerard of Cremona in the twelfth century, and, in modern times, edited twice by Francis Carmody and three times by J. M. Millás Vallicrosa. These editions are cited by Otto Neugebauer in his English translation of Thabit's brief tract ("Thâbit ben Qurra *On the Solar Year* and *On the Motion of the Eighth Sphere*," *Proceedings of the American Philosophical Society*, vol. 106, no. 3, (June, 1962), p. 264, n. 1). For a discussion of Thabit's theory, see J. L. E. Dreyer, *A History of Astronomy from Thales to Kepler*, 2d ed. (New York, 1953), pp. 276–77. By adding this paragraph about Thabit's trepidation theory to the Oxford version, Scotus wished to show that if Thabit was correct and Ptolemy wrong, then the latter's cyclical returns of 36,000 years could be rejected without recourse to any argument involving possible celestial incommensurability. However, according to Thabit's theory, it might yet be possible that after a period of oscillation the planets would return to the same positions that they had at the start of the period of trepidation, in which event periodic return would be possible. Though Scotus does not say, one would presumably attack this in the same manner as the theory of 36,000-year cycles—i.e., by assuming celestial incommensurability.

of exact and fixed cyclical return, not only because it was contrary to faith (it would have destroyed the Christian belief in a unique creation), but also because cyclical return was itself a doubtful assumption, so that *a fortiori* the inference drawn from it—that all things will inevitably return—must also be doubtful.[98]

Obviously, here was a source that Oresme may have known, and from which he could have drawn the fundamental idea that an effective attack against the doctrine of the Great Year ought to be based upon an assumption of celestial incommensurability. Oresme, who attacked the Great Year on precisely these grounds, may have drawn inspiration to explore this subject by a reading of Scotus—or by reading some other author who, in commenting upon Book 4, Distinctio 43, may have repeated, or elaborated upon, Scotus's arguments,[99] or

---

98. In the *Opus oxoniense*, he says: "The first [opinion], concerning the return of all things, is wholly improbable, at the very least because it is contrary to faith. Nor is the reason for it very efficacious, since the antecedent, concerning a return of the heavens, is doubtful, and the inference [drawn from it] is doubtful." ("Prima enim de reditu omnium est omnino improbabilis, saltem quia est contra fidem; nec pro ea est ratio efficax, quia illa de reditu coeli, et dubium antecedens supponit, et dubia est illatio."—Vivès ed., vol. 20, p. 74, col. 1.) The doubtful inference is that all things return because the celestial bodies return to identical positions at fixed intervals.

In the earlier *Reportata parisiensia*, Scotus invoked only the argument that cyclical return was heretical: "Respondeo igitur ad quaestionem primam, quod prima opinio circa eam, quod omnia scilicet redeant eadem numero secundum circulum, non est tenenda, quia est haeretica."—Vivès ed., vol. 24, p. 515, col. 2.

Although Scotus was convinced that exact cyclical return was an unwarranted inference from the Great Year, the theory of which he also rejected, he was careful to distinguish it from the doctrine that in nature the same cause acting on the same matter or thing would always produce an identical effect. This latter concept, which was independent of, and unrelated to, cyclical regularity, rep-

resented a basic medieval belief in the uniform effect of causal action. "Tertia opinio est media, quae ponit quod, etsi non omnia possunt redire eadem numero per actionem naturae, tamen aliquid potest redire idem numero. Pro hac opinione arguitur primo per illud Philisophi, 8. Metaph. 3. cap. *Si agens idem, et materia eadem, effectus erit idem*, quia non assignat diversitatem effectus possibilem, nisi propter diversitatem efficientis vel materiae."—Vivès ed., vol. 20, pp. 71, col. 2–72, col. 1. (The italics are in the edition cited.)

In defense of his opinion, Scotus offers an example with further explanatory comment: "...et ideo potest teneri tertia, quia non apparet quare non possit natura aliquid idem numero reducere. Si enim ubi est continua actio agentis naturalis respectu producti, sicut est in Sole respectu radii, si ponatur in primo instanti producere radium, et in tempore habito conservare in ultimo instanti, puta *B*, erit ibi idem radius, cum non dependeat illa identitas radii in secundo instanti ad seipsum in primo instanti, vel ab existentia ejus in tempore intermedio, quia sine illa existentia posset idem numero fuisse productum in eodem instanti; sequitur ergo, quod illo *esse* intermedio destructo, posset idem *esse* in utroque extremo...."—Vivès ed., p. 74, cols. 1–2.

99. In light of Scotus's fame, it seems reasonable to assume that some later Sentence

utilized them elsewhere—and eventually formulated his important mathematical demonstration in the *De proportionibus proportionum* that the celestial motions are indeed probably incommensurable. All this is possible, of course, but if it were true that Oresme received the seminal idea ultimately from Duns

commentators would have evaluated or reconsidered his incommensurability argument, or, at the very least, have utilized it against the doctrine of the Great Year and its consequence of exact return. Furthermore, since Book 4, Distinctio 43 was a logical place for the celestial incommensurability argument to appear, it is possible that Scotus was not the first to formulate this argument in a Sentence Commentary. Whatever the merits and actualities of these conjectures, an examination of ten discussions of Book 4, Distinctio 43 has revealed only one commentator who cites the incommensurability argument against the Great Year—and this was approximately one century after the death of Oresme.

In his *Reportata* on Book 4 of Scotus's Sentence Commentary, Petrus Tataretus (he was rector at the University of Paris in 1490) repeats Scotus's description of the Great Year (*D. Petri Tatareti...reportata, in quartum librum sententiarum Ioannis Duns Scoti...* [Venice, 1583], p. 389, col. 1), and then reports the argument on celestial incommensurability (p. 389, col. 2):

"This opinion [concerning the Great Year and cyclical return] can be disproved. Consequently, he [i.e., Scotus] disproves this opinion of the philosophers by the true reason, after he has disproved it by [appeal to] authorities. And his argument is formed as follows: speaking naturally, it is impossible that all motions return to the same point, therefore not all will return to the same place after 36,000 years. The antecedent is proved: some motions are mutually incommensurable, therefore they cannot return to the same point. The antecedent [of this last argument] is [now] proved: some magnitudes are mutually incommensurable, therefore the motions that occur on these magnitudes with equal speeds will never re-

turn to the same point. An example [of this is]: I take the magnitude of a side and the magnitude of a diagonal, [and] these two magnitudes are incommensurable." ("Potest etiam illa opinio improbari. Consequenter improbat istam opinionem philosophorum per veram rationem postquam improbavit per auctoritates. Et formatur sic ratio: impossibile est, naturaliter loquendo, omnes motus in idem punctum redire, ergo non omnia redibunt de hinc ad triginta sex mille annos. Antecedens probatur: aliqui sunt motus incommensurabiles adinvicem ergo illi non possunt redire in idem punctum. Antecedens probatur: aliquae sunt magnitudines incommensurabiles adinvicem ergo motus qui fuerunt supra illas magnitudines aeque veloces nunquam redirent in idem punctum. Exemplum: magnitudinem coste capio et magnitudinem diametri, illae duae magnitudines sunt incommensurabiles.")

Peter then summarizes and explains Scotus's description of Thabit ibn Qurra's trepidation theory and the latter's rejection of the Great Year of 36,000 years, all of which indicates that Peter probably had the Oxford version, since the paragraph on Thabit does not appear in the Parisian *Reportata*. In the quotation above, the "proof" of the second antecedent has no necessary connection with the antecedent itself, and shows that Peter did not understand the argument. Nothing more is said about celestial incommensurability and it is fair to say that its inclusion was prompted by little more than a desire for completeness.

Of the other nine authors, only four discuss the Great Year of 36,000 years—Petrus de Tarantasia (later Pope Innocent V, who died in 1276; see p. 406, col. 1), Richard of Middleton (d. ca. 1308; fol. 259r, col. 1–259v, col. 1), Petrus Aureoli (d. 1322; p. 196, col. 2), and John Major (1469–1550; fol.

Scotus,[100] it would be equally true to say that the two treatises on celestial commensurability and incommensurability which Oresme wrote bear as much resemblance to the brief relevant passage in Scotus's Sentence Commentary as the oak tree bears a physical resemblance to the acorn.

Oresme and After

Against this rather slender and meager background, it comes as quite a surprise to find that Oresme devoted two complete treatises (the *Ad pauca respicientes* and the later *De commensurabilitate*) to the topic of circular and celestial incommensurability and also found occasions to introduce it into a number of other works.[101] Advocating the probability of celestial incommen-

---

(Note 99 continued)
191r, cols. 1–2) (page and folio references are to the editions cited in the bibliography). Another, St. Thomas Aquinas (1225–1274), mentions a Great Year of unspecified duration after which human souls will return to human bodies (see p. 1063, col. 2 of edition cited in bibliography). These authors, however, do not mention, or even hint at, celestial incommensurability, but rest content to condemn the doctrine of cyclical return as heretical. The four remaining authors— Franciscus de Mayronis (d. after 1328), Petrus de Palude (d. 1342), Robert Holkot (d. 1349), and William of Ockham (d. ca. 1349–50)—mention neither the Great Year nor celestial incommensurability. (The editions that I examined are in the Manuscripta series [lists 5, 16, and 19] of Microfilms of Rare and Out-of-Print Books produced by St. Louis University. For Mayronis's Sentence Commentary, see List 5, item 34; for de Palude, see List 19, item 15; for Holkot, see List 19, item 19; and for Ockham, List 16, item 41.)

In light of the potential importance of discussions of Book 4, Distinctio 43, it is regrettable that the Sentence Commentary which Oresme is known to have written has not yet been identified, and is, perhaps, irretrievably lost. There is, of course, a very large number of extant Sentence Commentaries—F. Stegmüller lists approximately

1,400 in his *Repertorium commentariorum in sententias Petri Lombardi* (2 vols.; Würzburg, 1947), to which Victorin Doucet, O.F.M., adds even more in his *Commentaires sur les sentences, supplément au répertoire de M. Frédéric Stegmueller* (Quaracchi, Florence, 1954). It would come as no surprise to find others who discussed or used the celestial incommensurability argument against the Great Year, especially among Scotists. But on the basis of the present sampling of Sentence Commentaries, it seems reasonable to conclude that this particular argument did not become a usual feature, or integral part, of the arguments and counterarguments regularly incorporated into Book 4, Distinctio 43.

100. The evidence suggesting a link between Scotus and Oresme consists of: (1) the attack against the Great Year utilizing possible celestial incommensurability; (2) the definition of incommensurable circular motions (see above, n. 95 of this chapter); and (3) the example of two independent and incommensurable motions on the side and diagonal of the same square (see above, n. 96 of this chapter). It is not unreasonable to suppose that Oresme drew these particular ideas from an earlier, quite general discussion of celestial incommensurability, and perhaps even from the Sentence Commentary of Duns Scotus.

101. The relationships between the con-

surability[102] was not, for Oresme, a mere scholastic exercise. He saw in it a means by which he could reasonably demonstrate that celestial effects were inherently unpredictable and thereby weaken the foundations of astrology and strike a blow at the astrologers, who had aroused his deep concern by virtue of their considerable influence on the king of France, Charles V.[103] But although his ideas were not without influence, as we shall see below, the intense interest and fascination which he had for this subject seem to have died with him. As of now, it would be fair to say that he had no genuine successors.

As might be expected, a number of Parisian scholastics were to utilize certain of Oresme's conclusions, and some would even cite him by name. In this group we can include Henry of Hesse (Heinrich von Langenstein; d. 1397), Marsilius of Inghen (d. 1396), Pierre d'Ailly (1350–1420), and Jean Gerson (1363–1429). Although not a Parisian scholastic, we should add to this group the Bolognese physician and astrologer, John de Fundis, who, in 1451, wrote a hostile commentary on Oresme's *Ad pauca respicientes*,[104] the only critique concerning either of Oresme's two treatises on celestial incommensurability known thus far.[105]

---

tent and propositions of these two treatises have been noted in many places in this volume and in my edition of the *Ad pauca respicientes* in Grant, *Oresme PPAP*; however, see especially my discussion of their connections with the *De proportionibus proportionum*, ibid., pp. 72–80.

102. See III.457–66 of the *De commensurabilitate*, and above, chap. 2, n. 113.

103. See Grant, *Oresme PPAP*, p. 64.

104. It is not an attack on the *De commensurabilitate*, as Thorndike conjectured (*History of Magic*, vol. 4, p. 236, n. 12).

105. Despite its obvious connections with Oresme, the anonymous *Questio de proportione dyametri quadrati ad costam ejusdem* is omitted from consideration here since it has already been discussed in Grant, *Oresme PPAP*, pp. 77–78, n. 101, and again in this volume in Chapter 2, nn. 59 and 97. It is plausible to conjecture that this treatise was written either by Oresme himself, or by someone who extracted a number of different Oresmian works, one of which must have been the *Quaestiones super geometriam Euclidis*. Although it seems improbable, we cannot exclude the possibility that the anony-

mous *Questio* was written prior to Oresme's treatises on celestial commensurability and that it was Oresme, especially in his *Quaestiones super geometriam Euclidis*, who drew upon the anonymous *Questio*.

Brief mention must also be made of another anonymous treatise, this one a compendium of metaphysics and natural philosophy in six books, described by Thorndike, *History of Magic*, vol. 3, pp. 568–84. In the course of his summary discussion of its contents, Thorndike observes that its author rejected the Great Year. "Not only is it against human liberty of action, but since the moon does not fit into the solar year, it is impossible that after the great year of Plato everything should be the same again. Thus he approaches Oresme's argument from the incommensurability of the celestial movements."— Ibid., p. 582. In this long work of 293 folios (it occupies all the folios of Bibliothèque Nationale, fonds latin, 6752), the discussions described by Thorndike are found on folios 13 and 16r. If we are to judge from the nature of its diverse contents, this work probably derives from the latter part of the fourteenth century, and from its references to the Pari-

In his *Tractatus de reductione effectuum particularium in causas universales*, Henry of Hesse remarks that one cannot determine how many of 120 theoretically possible conjunctions[106] of the seven planets are realizable naturally, since "it is impossible to know whether the motions of all the planets are mutually commensurable, or even how many of their velocities have a common measure, as Master Nicolas Oresme has shown."[107] Perhaps Henry had in mind a statement in Part 2, Proposition XIX of the *Ad pauca respicientes*, where Oresme

---

(Note 105 continued)

sian condemnation of 1277 (ibid., pp. 568–69 and n. 3), we may conjecture that it was written at Paris. Thus it is possible that Oresme's influence is operative here. But the uncertainty of the treatise's date of composition and the fact that I have not had opportunity to examine personally the relevant discussions obviates my further consideration of it at this time.

106. That the seven planets could theoretically conjunct from the beginning to the end of the world (see next note) only 120 times had already been computed and formulated much earlier, and this information is even found in Oresme's *De configurationibus qualitatum*. For a discussion of the combinations used in arriving at the figure 120, see Clagett, *Nicole Oresme and Medieval Geometry*, pp. 444–47.

107. This appears in the fifth proposition of Chapter 24, which is brief enough to quote in full.

"Quinta propositio. Quot septem planetarum sunt naturaliter combinationes possibiles impossibile est homini investigare vel hominis investigatione inveniri. Apparet quia impossibile est scire an omnium planetarum motus sunt sibi invicem commensurabiles vel saltem quot illorum velocitates commensurantur, sicut deducit magister nycholaus oresme. Confirmatur, quia ad hoc oportet, necessario sciri totum spacium temporis inter inicium cursus nature et eius finem, quia alias semper nesciret an illud tempus [*corr. ex* totum spacium temporis] sufficeret vel non pro illis centum xx combinationibus coniunctionalibus perficiendum. Etiam supposito quod motus omnium essent

commensurabiles quot [*corr. ex* quod] tamen tempus est totaliter ignotum ergo, et cetera."

I have used Clagett's text of this passage (the textual corrections are his) as it appears in his *Nicole Oresme and Medieval Geometry*, p. 120, n. 19. The last sentence does not appear in Clagett's text, but is added here. The text is based primarily on British Museum, Sloane MS 2156, fol. 128v, col. 2, but Bibliothèque Nationale, fonds latin, MS 2831, fol. 114r was also used. Duhem, who designates this as the fourth proposition of Chapter 24, translated it into French without inclusion of the Latin text (*Le Système du monde*, vol. 8, pp. 483–84). That part of Henry's text about the possible number of conjunctions is omitted from Duhem's translation, no doubt because he used BN 14580 in addition to BN 2831.

In this proposition, Henry is concerned with whether or not all of the possible 120 conjunctions could occur in the time that must elapse between the creation and end of the world. He concludes that this is impossible to know because it cannot be determined whether the planetary motions are mutually commensurable—a precondition for these conjunctions to take place. Moreover, the time that will elapse between the creation of the world and its end is unknown, so that we cannot know whether there is sufficient time for all 120 conjunctions to take place. Finally, even assuming the commensurability of the planetary motions, which would guarantee the theoretical possibility of the occurrence of the 120 conjunctions, we could never know whether the existence of the world would be of adequate duration to permit the occurrence of all 120.

says that "one does not know if the motions are commensurable or not; and who is ignorant of the antecedent is necessarily ignorant of the consequent."[108] In another treatise, Henry seems to draw upon Oresme once again when he says that astrology cannot be based upon the same recurrent celestial phenomena, since these cannot repeat exactly "because of the variety and incommensurability of the superior [celestial] motions."[109]

Although Marsilius of Inghen makes no mention of Oresme, the substance of his remarks on celestial incommensurability indicates at least some familiarity with the latter's ideas on that subject. In Book 8, Question 3 of his *Questiones super octo libros physicorum*, Marsilius considers "whether a new action could arise from an eternal and immutable mover."[110] He divides the question into three parts, the second of which considers "how new effects arise from diverse immutable movers."[111] In the second conclusion of his response to this second part, Marsilius begins first to explore the consequences of celestial commensurability and then, in a third conclusion, the effects of incommensurability. He explains:

If the motions in the sky were mutually comparable, no disposition, aspect, conjunction, or anything of this kind would occur but that a similar disposition would have occurred an infinite number of other times—assuming that the world is eternal. This is proved: because if there never was a similar disposition or one that did not occur an infinite number of times, it follows that these motions are not mutually comparable, and thus are transformed [or made the opposite of mutually comparable].

Since all lower things depend upon superior things [i.e., on celestial bodies], it follows from this that there is never an effect on these lower things but that a like effect, according to species,[112] has not occurred an infinite number of times. And so, by assuming the commensurability of celestial motions, there cannot be, speaking naturally, any new effect according to species.[113]

---

108. Grant, *Oresme PPAP*, p. 427 (Latin text, p. 426, lines 245–46). Immediately before this passage, Oresme says that even if the celestial motions were commensurable, their precise ratios are indeterminable because our senses would be inadequate for such a task (ibid.).

109. "...propter motuum superiorum varietatem et incommensurabilitatem."—*Tractatus contra astrologos coniunctionistas*, in *Studien zu den astrologischen Schriften des Heinrich von Langenstein*, ed. Hubert Pruckner (Leipzig, 1933), p. 159. This is a standard theme in the *Ad pauca respicientes* (see, for example, Grant, *Oresme PPAP*, pp. 424–27) and *De commensurabilitate* (see, for example,

III.409–15).

110. "Queritur tertio utrum a motore eterno et immutabili possit provenire actio nova."—*Questiones subtilissime Johannis Marcilii Inguen super octo libros physicorum secundum nominalium viam* (Lyon, 1518; reprinted Frankfurt-a.-M., 1964), fol. 80v, col. 2.

111. "Secundo qualiter a diversis motoribus immutabilibus possunt provenire novi effectus."—Ibid., fol. 81r, col. 1.

112. For the distinction between the repetition of an effect according to species and the recurrence of an effect that is numerically the same, see the quotation from Aristotle's *De generatione* in this chapter, n. 66.

113. "Secunda conclusio est ista: quod si

But new effects can be produced, says Marsilius in the third proposition of his response:

If the celestial motions are mutually incommensurable, then in any instant whatever the sky is in such a disposition that never before was it in the same or a like disposition; nor will it be in the future. For by assuming that it could be, it follows that these motions are mutually commensurable, which is the opposite of what was assumed. Now it is clear from this that, by assuming incommensurability of the celestial motions, these lower [or sublunar] things are always in such a disposition that never before were they in a like disposition, nor would they be ever after. And so it is possible that some new effect could occur according to species, which no like effect has preceded.[114]

And now Marsilius raises the crucial question:

And if one should inquire whether or not the celestial motions are mutually relatable [or commensurable], the reply is that this cannot be known by experience. For a very small part, as [for example] one-thousandth of an imperceptible part, can make two quantities commensurable or incommensurable into infinity. And thus it is obvious how diverse effects can arise from an unmoved mover when an eternal motion mediates because of a difformity of the parts of the mobile.[115]

(Note 113 continued)
motus celi sint adinvicem comparabiles nunquam est aliqua dispositio in celo vel aspectus aut coniunctio, vel aliquod huiusmodi, quin alias infinities fuerat dispositio similis supposito quod fuerit mundus ab eterno. Probatur: quia si nunquam fuerat dispositio similis aut non infinities, sequitur quod illi motus non sunt adinvicem comparabiles, ita quod ista convertuntur.

"Ex isto sequitur quod cum omnia ista inferiora dependeant a superioribus quod nunquam est aliquis effectus in istis inferioribus quin infinities fuerit effectus consimilis sibi secundum speciem. Et sic supposita commensurabilitate motuum celestium non potest esse naturaliter loquendo aliquis effectus novus secundum speciem."—*Questiones super octo libros physicorum*, fol. 81r, col. 2. In the *De commensurabilitate*, Oresme says much the same thing (see III.406–8), and in the *Ad pauca respicientes* it could easily be inferred from Part 1, Propositions, II, III, and VI (Grant, *Oresme PPAP*, pp. 390–95, 398–99).

114. "Tertia conclusio est ista: quod si

motus celorum sunt adinvicem incommensurabiles, tunc in quolibet instanti celum est in tali dispositione quod nunquam ante fuerat in dispositione tali vel consimili nec in [*corr. ex* im] posterum erit; quia dato quod sit sequitur quod omnes illi motus adinvicem sunt commensurabiles quod est oppositum positi. Ex quo patet quod hoc supposito quod motus celi sint incommensurabiles quod semper ista inferiora sunt in tali dispositione quod nunquam ante fuerunt in consimili dispositione, nec unquam postea fuerunt. Et ideo possibile est quod fiat aliquis effectus novus secundum speciem cuius nunquam similis precessit."—*Questiones super octo libros physicorum*, fol. 81r, col. 2. The same ideas are expressed in much greater detail in the *Ad pauca respicientes*, pt. 2, props. XVII and XVIII (see Grant, *Oresme PPAP*, pp. 422–25), the *De commensurabilitate*, pt. II, prop. 12 (II.307–23), the *Questiones super de celo*, and *Le Livre du ciel et du monde* (see above, chap. 2, n. 90; for Oresme's explanation of such unique events, see chap. 2, n. 89).

115. "Et si queratur an motus celorum sint adinvicem comparabiles vel non, res-

But diverse effects arise not only because the celestial sphere is a mobile whose parts—namely, the planets—differ from one another, but also "in the second place because of the diverse velocities of the mobiles [i.e., planets], their irregularity and incommensurability."[116] Thus while denying the possibility of determining by experience or observation whether or not the planetary motions are commensurable, Marsilius argues that if there are unique and diverse effects in sublunar things one reason for their production would be the incommensurability of the celestial motions.

As a final illustration of how new effects might arise from immutable and eternal movers, Marsilius offers the following example:

In the third place let this be proved. Assuming that the celestial motions are mutually incommensurable and that there could be an eternal wheel to which a motor [or mover] could be applied [but which is] incapable of moving it because of an obstacle, then because of the incommensurability of the celestial motions, it could happen that some sphere or some conjunction could occur that would be nearer to that wheel than any other that ever happened through an eternal time. And because of its nearness, it might happen that the obstacle [to the wheel's motion] is removed and the wheel is moved. But it is true, as already stated, that this would be impossible unless there was an eternal mediating motion preceding [this event].[117]

---

pondetur quod hoc non potest sciri per experientiam nam in infinitum modica pars ut millesima unius partis insensibilis [*corr. ex insenbilis*] potest reddere duas quantitates commensurabiles vel incommensurabiles. Et sic patet qualiter a motore immobili possunt provenire diversi effectus mediante motu eterno propter difformitatem mobilis quantum ad suas partes."—*Questiones super octo libros physicorum*, fol. 81r, col. 2. The argument concerning inability to determine whether or not two quantities are commensurable when very minute alterations are made is probably derived from III.12–17 and 34–36 of the *De commensurabilitate*, even to the extent of specifying the minute alteration as "one-thousandth" of a part. Apparently it also appealed to Pierre d'Ailly, who plagiarized it from Oresme and included it, along with much of Part III, in his *Tractatus contra astronomos*, which was published in *Joannis Gersonii opera omnia*, col. 799 (see below, this chapter, for a further discussion of Pierre d'Ailly).

116. "Secundo propter diversitatem velo-

citatum diversorum mobilium et irregularitatem [*corr. ex* irraritatem] et propter incommensurabilitatem."—*Questiones super octo libros physicorum*, fol. 81r, col. 2. Although Oresme did, of course, explain the production of diverse effects by celestial incommensurability, he also sought to account for celestial incommensurability in the *Ad pauca respicientes* by noting that in the heavens "there are many circles, latitudes, distances, eccentricities and many motions and diversities."—Grant, *Oresme PPAP*, p. 423.

117. "Tertio probatur: supposito quod motus celi sint incommensurabiles adinvicem, tunc posito quod esset una rota perpetua cui motor esset applicatus non potens ipsam movere propter impedimentum, et propter incommensurabilitatem motuum celestium potest provenire quod aliqua sphera aut aliqua coniunctio esset propinquior illi rote quam unquam fuit aliqua alia toto tempore eterno. Et ex approximatione posset contingere quod impedimentum removeretur et moveretur rota. Verum est

On the basis of passages in Oresme's works cited in the preceding footnotes to Marsilius's arguments, we see that Marsilius's fundamental points and examples have close counterparts in Oresme's treatises. When we also consider that Marsilius was Oresme's junior contemporary at the University of Paris and was probably familiar with the works of so eminent a scholar,[118] it seems plausible to conjecture that Oresme was the source of his isolated discussion of celestial incommensurability.[119]

But if, for lack of more concrete and specific evidence, it cannot be stated categorically that Marsilius of Inghen drew his arguments directly from Oresme, no such qualification is in order for Pierre d'Ailly's thoughts about celestial incommensurability—they are all from Oresme, plagiarized from Part III of the *De commensurabilitate*. After attacking astrologers on various grounds through the first seven chapters of his *Tractatus contra astronomos*, d'Ailly sought to impress his readers with some powerful climactic arguments, and could think of nothing more fitting than Oresme's dialogue in Part III of the *De commensurabilitate*, a work which he could have known as a student at the College of Navarre, where a few years before his entry Oresme had been Grand Master.[120]

(Note 117 continued)
tamen sicut dictum est quod hoc est impossibile nisi mediante motu eterno precedente."—*Questiones super octo libros physicorum*, fols. 81r, col. 2–81v, col. 1. Marsilius's example has affinities with, and is perhaps derived from, Oresme's discussion of a perpetually moving wheel in *Le Livre du ciel et du monde*, bk. 1, chap. 29, 46a–46c, described above in chap. 2, n. 59. There only a conjunction can initiate motion of wheel *d* (see Fig. 4 in n. 59), which will be in perpetual motion ever after. Perhaps with this in mind, Marsilius tells us that a conjunction might occur close enough to the wheel and in some manner or other remove the retarding obstacle, so that it is set into motion. If his example is patterned after Oresme's, the latter's description has lost something in the reformulation. Despite its opaqueness, it is easy to see that Marsilius's example could have been derived from Oresme's.

118. Marsilius gave his inaugural lecture as Master of Arts in 1362 and thereafter served the University of Paris for about twenty years, leaving Paris sometime before, but very near to, 1382 (see George Sarton,

Introduction to the History of Science, vol. 3, pt. 2, p. 1435), the very year in which Oresme died. It seems reasonably certain that by 1362 Oresme was well established as a reputable scholar at Paris (see Grant, *Oresme PPAP*, p. 6).

119. Although Marsilius wrote *Questiones...super quattuor libros sententarium* (2 vols. [Strasbourg, 1501]), I have not found in it any discussion of celestial incommensurability. The omission of such a discussion does not in any way exclude Duns Scotus as a potential influence on Marsilius. But Scotus does not qualify as a likely candidate; other than the fact that both discussed celestial incommensurability in the context of the production of new effects, none of Marsilius's arguments were formulated by Scotus, whereas most of them seem to have a recognizable counterpart in the works of Oresme. Nor can the chapters described above from the *Quadripartitum* have played any role in shaping Marsilius's arguments, since Johannes de Muris did not consider the production of new effects.

120. D'Ailly entered the College of Navarre in 1368 at the age of eighteen (see

And so without acknowledgment of any kind and with only minor rearrangements, Pierre d'Ailly constructed almost the whole of chapters eight and nine (and a small section of the final chapter, ten) of his *Tractatus contra astronomos* from passages and arguments in Part III of the *De commensurabilitate*.[121] So

---

Coopland, *Nicole Oresme and the Astrologers*, p. 11). Oresme was Grand Master of Navarre from 1356 to ca. 1362.

121. To convey the extent of d'Ailly's plagiarism, I shall list first the column number of the *Tractatus contra astronomos* as it was printed in *Joannis Gersonii opera omnia*, vol. 1, pt. 2, and then give, in order of appearance, all of the passages plagiarized in that column from Part III of the *De commensurabilitate*. Despite some alteration of word order and more than occasional word substitutions, there is no doubt that d'Ailly plagiarized all of the passages listed below substantially as Oresme wrote them.

chap. 8, col. 793:	III.65–77, 81–94, 95, 103, 106–7, 112–13
cols. 793–94:	III.114–34
col. 794:	III.139–47, 150–64, 168–73, 185–88, 202–11
cols. 794–95:	III.223–31
col. 795:	III.253–55, 232–52, 256–61, 262–69, 270–71
cols. 795–96:	III.272–79
col. 796:	III.288, 294–95, 298–99, 305–7, 309–11, 311–12
chap. 9, col. 796:	III.331–51
col. 797:	III.351–53, 357–59, 353–56, 375–76
cols. 797–98:	III.381–419
col. 798:	III.421–24, 425–38, 448–52, 457–66
chap. 10, col. 799:	III.10–23, 27–28, 29–31
col. 802:	III.40–46
col. 803:	III.444–48

Pierre d'Ailly also found portions of Oresme's *Tractatus contra astrologos* worthy of plagiarism. Coopland, who detected this, tells us (*Nicole Oresme and the Astrologers*, p. 11) that d'Ailly "...was responsible, at a date apparently after 1409, for two tracts entitled *De Falsis Prophetis*...; the second is, in effect, a treatise against divination, in

which he covers very familiar ground at wearisome length, without either perceptible addition to knowledge or originality in method. At a late stage in his discussion he proceeds, with the words *Primo videbitur quod sit divinatio*, to consider questions similar to those raised in the *Tractatus*, and, so doing, annexes a good deal of Oresme's first chapter, a scrap of the second, much of the fourth and fifth, and part of the seventh.... Yet in this *Tractatus de Falsis Prophetis*, although he freely names St. Basil, St. Augustine, St. Thomas and others, when quoting their words, Oresme becomes *quidam doctor*, and, with still larger vagueness, *quidam doctores*, and this in spite of the fact that the Cardinal embodies in his Tract not merely paraphrased statements but long verbatim extracts with references included." (Both treatises are printed in Coopland, *Nicole Oresme and the Astrologers*, pp. 123–48.)

Obviously, d'Ailly could operate at different levels of plagiarism: in the *De falsis prophetis* he deliberately omits Oresme's name, but usually concedes that the lifted passages were composed by someone else; in his *Tractatus contra astronomos*, however, he wishes to be taken as the author of his plagiarisms, as he also does in his famous *Ymago mundi* of 1410 where passages from Oresme's *Traitié de l'espere* were translated verbatim into Latin without acknowledgment (as a consequence Oresme would indirectly exert some influence on Christopher Columbus's cosmological ideas; see Edmond Buron, *Ymago mundi de Pierre d'Ailly* [Paris, 1930], pp. 5–37). For brief mention of other plagiarisms by d'Ailly, see Coopland, *Nicole Oresme and the Astrologers*, p. 11, and Duhem, *Le Système du monde*, vol. 8, p. 492 (in Duhem's discussion, Roger Bacon is the victim).

successful was he that Jean Gerson ranked him along with Oresme as two authors who had argued that the question of celestial commensurability or incommensurability could be answered only in terms of probability, not certainty.[122]

Other authors who implied, or are alleged to have mentioned, celestial incommensurability might be discussed here,[123] but let us turn instead to three

122. "*Ninth Proposition.* That the heavens have commensurable or incommensurable motions of the signs, and moreover that the motions of the planets have dominion over this or that nation, is absolutely uncertain.

"*Comment.* As experience teaches, those who wish to report certitude where only rhetorical probability can be had fall into error, as Master Nicholas Oresme and, after him, Monsignor Peter, Cardinal of Cambrai, have shown, deriving from this one root of the difficulty of astrological judgments. And perhaps it is for this reason that the length of the solar year does not yet seem to have been found, nor the root [or basis] from which it proceeds. Besides, the variety [of opinions] of those who assign these or those dominions of the signs to this or that part of the earth shows sufficiently that there is uncertitude."

("*Propositio Nona.* Coelum habere commensurabiles vel incommensurabiles motus signorum; insuper et planetarum huic vel illi genti dominari, prorsus incertum esse.

("*Commentum.* Erraverunt, ut experientia docuit, certitudinem afferre volentes ubi solam posse haberi rhetoricam probabilitatem deducit magister Nicolaus Oresme et post eum Dominus Petrus, Cardinalis Camaracensis, sumens exinde radicem unam difficultatis astrologicorum judiciorum. Et inde forsan est quod nondum videtur reperta praecisa quantitas anni solaris, aut ex qua radice procedit. Ceterum varietas assignantium huic vel illi plagae terrae talia vel talia signorum dominia, satis incertitudinem esse monstrat."—*Trilogium astrologiae theologizatae* in *Joannis Gersonii opera omnia*, vol. 1, pt. 2, col. 193.)

Omitting the Latin text, Duhem translated this proposition into French in *Le Système du monde*, vol. 8, pp. 454–55.

123. Although no discussion of celestial incommensurability by Regiomontanus (1436–76) has come to my attention, he deserves first consideration here because it is reasonably certain that he knew Oresme's *De commensurabilitate* directly. For reasons that are unknown to me, he himself copied, or had copied, all definitions in Part I and the enunciations only of all the propositions in Parts I and II. A copy of this version containing the title of the treatise but no mention of Oresme appears on pp. 69–72 of Codex H 67 of the Stadtbibliothek in Schweinfurt. The codex was put into final form in 1575 by Johannes Praetorius, a student of Tycho Brahe. In addition to the *De commensurabilitate*, H 67 contains a number of other treatises copied by Regiomontanus which include only the enunciations of propositions. (See Ernst Zinner, *Leben und Wirken des Johannes Müller von Königsberg genannt Regiomontanus* (Munich, 1938), pp. 130–31, 258; I am indebted to Marshall Clagett for drawing my attention to the relevance of Codex H 67 and for directing me to Zinner's discussion of it.)

Of the authors who will be mentioned here, only Nicholas of Cusa (1401–64) says anything of direct relevance (however, see below, p. 159, n. 190, for a brief summary of a discussion by Blasius of Parma). In 1436, Cusa wrote the *Reparatio calendarii*, an appeal for calendar reform which he presented to the Council of Basel (for a summary of his proposed reform see Ferdinand Kaltenbrunner, "Die Vorgeschichte der Gregorianischen Kalenderreform," in *Sitzungsberichte der Kaiserlichen Akademie der Wissenschaften*, phil.-hist. Klasse, vol. 82 [Vienna, 1876], pp. 342–50). In it he reports that Thabit ibn Qurra denied the possibility of a precise trop-

Italian authors, Paul of Venice, John de Fundis, and Jerome Cardano, who, for quite different reasons, were to focus upon this topic.

The locus of Paul of Venice's (d. 1429) brief examination of this topic occurs where frequent discussions of it might have been expected, namely, in com-

---

ical year because of the equinoctial points, which are always in motion, moving alternately forward and backward (this is Thabit's influential trepidation theory; see above, n. 97 of this chapter). But Thabit and Indian astronomers (the latter are cited from reports of Abraham ibn Ezra) held that the time of the sun's motion with respect to a fixed star is constant. ("Thebith etiam qui ob diversitatem experientiarum imaginatus est motum accessus et recessus diversus est ab omnibus illis dicens annum non posse constare ex revolutione ad aequinoctium quia illa aequinoctialia puncta instabilia asserit accedere et recedere, sed dicit annum ex motu solis a stello fixa ad eandem constare, uti Indi annum mensurant secundum Abraam avenzre [i.e., Abraham ibn Ezra]... et loquuntur de anno revolutionis de stella ad stellam eandem...."—*Nicolai Cusae Cardinalis Opera*, 3 vols. [Paris, 1514; reprinted Frankfurt-am-Main, 1962], vol. 2, fol. 23r.) Continuing, Cusa says:

"But since in the present discussion, our concern is not with the [fixed] stars in Ram of the eighth sphere, but with the equinoctial point, then, so far, the year remains doubtful because some have been compelled to say that every supercelestial motion is incommensurable to human reason and occurs in a certain irrational ratio that has an incomprehensible and unnameable root because, given any human measure whatever that measures a motion closely, a closer measure can always be given. And so they say that a superior motion is as comprehensible to the human mind as a circle is squarable by the same [human] mind and [just as] an acute contingent angle is attainable by a rectilineal angle. Furthermore, Peter of Abano, in his *Treatise on the Eighth Sphere* refutes Thabit's opinion of the forward and backward motions [of the equinoctial points]; and simi-

larly [refutes the opinion] of al-Battani. And when he [i.e., Peter of Abano] sought why this motion was found to be slower in the time of Abracis [i.e., Hipparchus] and Ptolemy than in this time, he replies to this that the motion in all the heavens is unknown. Thabit posited [this discrepancy to arise] in the imperfection of the human reason and the limitation of instruments. Alpetragius [i.e., al-Bitruji], following the opinion of Aristotle, refutes the opinions of Ptolemy and all others concerning eccentrics and epicycles, and saves the appearances by assuming all concentric spheres."

("Unde cum consideratio nostra non sit de ariete stellato octavae spherae quo ad praesens negotium, sed de puncto aequinoctiali; tunc annus adeo remanet dubius quod quidam compulsi fuerunt dicere omnem motum supercaelestium incommensurabilem esse rationi humanae et in quadam irrationabili proportione habente surdam et innominabilem radicem cadere, quia quacumque mensura humana propinque motum mensurans, dabilis est semper propinquior. Et ita dicunt motum superiorem per humanum ingenium compraehensibilem sicut circulus per idem ingenium est quadrabilis et angulus acutus contingentiae attingibilis per rectilinealem. Petrus etiam de Ebano in *Tractatu de octava sphera* opinionem accessus et recessus Thebith repraehendit; similiter et Abbategni. Et cum quaereretur ab eo cur tempore Abracis et Ptolemaei ille motus tardior inventus sit quam tempore isto, respondet ad haec motum in caelis omnibus incognitum esse. Thebith in imperfectione humanae rationis et instrumentorum impedimentum posuit. Alpetragius, opinionem Aristotelis sequens, Ptolemei et aliorum omnium opiniones eccentricorum et epicyclorum confutat, apparentias salvans etiam ponendo spheras omnes concentricas."—*Ibid.*, fols. 23r–v.)

(Note 123 continued)

Without adopting any position, Cusa simply presents a number of explanations, each purporting to account for our ignorance of the length of the tropical year. Thus (1) it might possibly be the consequence of the incommensurability of the celestial motions, or (2), as Peter of Abano would have it, the consequence of our total ignorance of all celestial motions, or (3), as Thabit explains, the imprecision might arise from human imperfection or inadequate instruments. In commenting on his translation of this passage, Duhem implies, without justification, that Cusa definitely accepted celestial incommensurability and then suggests that Cusa may have derived this idea from the *De commensurabilitate*, which he attributed provisionally to Pierre d'Ailly, not Oresme (see *Le Système du monde*, vol. 10, pp. 311–13, and App. 2, below). In fact, on the basis of this passage, Duhem attributed to Cusa a general astronomical skepticism (ibid., p. 312). But near the end of the *Reparatio calendarii*, Cusa says: "Although from all these human experiences [or observations], no certitude can be had from past to future positions of the planetary motions, except as the order and regularity of the motions that have been found permit future positions to be anticipated." ("Licet ex his omnibus humanis experimentis particulariter captis, nulla ex motu luminarium de praeterito certitudo ad futuras habeatur, nisi quantum ordo et regularitas inventorum motuum praesumi de futuris permittit."—*Nicolai Cusae Opera*, vol. 2, fol. 29r.) Thus it is by no means obvious that Cusa is a complete skeptic on astronomical matters, for he concedes that certain judgments can be made about future planetary positions provided that one has truly discovered motions whose order and regularity allow for this—a possibility he does not exclude.

Brief mention must now be made of two authors who are said to have considered or discussed celestial incommensurability, but the sources of information are either mistaken or misleading. In *Le Système du monde*,

vol. 10, p. 312, Duhem says that Peter of Abano, along with Duns Scotus, discussed this topic and concluded that we cannot decide whether or not they are commensurable. He then refers the reader to a discussion in Part 5, Chapter 14, sections 2 and 3, of Volume 8 of *Le Système*, which is devoted to Oresme and d'Ailly and where Peter of Abano is not even mentioned; nor, indeed, is there any such discussion in *Le Système*, vol. 4, pp. 229–63, a section devoted solely to Peter of Abano's astronomical and astrological thought. Perhaps Duhem had in mind Cusa's reference to Peter in the passage quoted above in this note. If so, this must be rejected, since Cusa mentions only that Peter thought men were ignorant of all celestial motions and does not state or imply anything about celestial commensurability. If Peter really thought what Duhem ascribes to him, it becomes difficult to explain why Peter believed in a Great Year, a belief that implies celestial commensurability (Duhem discusses Peter's belief in the Great Year in *Le Système*, vol. 4, pp. 238–39, 261).

The claim for Christopher Clavius (1537–1612) is made by Antonius Hiquaeus, a seventeenth-century Spanish commentator on Duns Scotus. In commenting on the passage from Distinctio 43, Question 3, of Scotus's Sentence Commentary (the Latin text is quoted above at the beginning of n. 97 of this chapter), Hiquaeus remarks: "The Doctor [i.e., Scotus] says that this argument requires a longer discussion. Clavius furnishes a demonstration of it in the first chapter of *On the Sphere*, [called] *On the Periods of the Celestial Motions*, where he proves that these [celestial] motions are incommensurable." ("Hanc rationem dicit Doctor requirere longiorem discussionem, ejus demonstrationem subjicit Clavius *in cap. primum sphaerae, de periodis motuum coelestium*, ubi probat hos motus esse incommensurabiles."—*Joannis Duns Scoti Opera omnia*, Vivès ed., vol. 20, p. 69, col. 2.) But neither in the first chapter, nor elsewhere in his commentary on Sacrobosco's *De sphera* does Clavius even men-

mentaries on the very last paragraph of Aristotle's *De generatione* (2. 11. 338b13–19), where the question of numerically identical cyclical return is considered.[124] Paul's discussion is brief enough to speak for itself:

It must be noted that in this matter there are two extreme opinions. The first is that of Plato,[125] who says that in the revolution of a Great Year all inferior things [i.e., all terrestrial things] present and past are the same at their return not only as to species, but [also] as to number [i.e., as individuals][126] with the same dispositions as they now have or have had before. Or this opinion is based on this: when a Great Year has revolved, i.e., with the motion of the eighth sphere completed, the supercelestial bodies will be simultaneously in the same disposition in which they were at the beginning of the same Great Year, and then another Great Year occurs which is alike in all things with respect to the first. And because, according to a common conception of philosophers, the motions of all inferior things are caused and regulated by the motions of the superior bodies, as Aristotle says in the first book of the *Meteorologica*,[127] it is necessary that this lower world be continued [and preserved] by the superior motions so that its every power would be governed from above. And so both specifically and individually, things will necessarily return in the second Great Year.

In the first place, I argue against the foundation of this opinion because according to Ptolemy the eighth sphere is moved only one degree in one hundred years and as a consequence, according to him, its motion is completed in 36,000 years. The consequence is obvious since by multiplying the degrees of the starry orb, which are 360, by one hundred, there will result 36,000; and so much is the number of years assigned to

---

tion celestial incommensurability, to say nothing of "proving" it. In fact, he offers numerous estimates for the length of the Great Year, which he calls the Platonic Year, and remarks that some have said that all things would return in the same order in which they now occur. ("Quod quidem spacium, seu tempus, appellari solet annus Platonicus. Hoc enim intervallo sydera omnia ad eundem situm reditura autumant; immo quidam volunt, tunc omnia, quaecunque, nunc in mundo sunt, eodem ordine esse reditura, quo nunc cernuntur."—*Christopher Clavii Bambergensis ex societate Iesu in Sphaeram Joannis de Sacro Bosco commentarium* [Rome, 1570], chap. 1, "De periodis motuum caelestium," p. 77.) If Clavius had "proved" the incommensurability of the celestial motions in this treatise, it is difficult to believe that he would not also have explicitly rejected the Great Year.

Finally, I cite Pico della Mirandola (1463–

96), who, in his *Disputationes adversus astrologiam divinatricem* (ed. Eugenio Garin, 2 vols. [Florence, 1946 and 1952]), refers to Oresme's repudiation of the Great Year in the latter's *De proportionibus proportionum* (the text of the passage is given in Grant, *Oresme PPAP*, pp. 119n–120n).

124. The Aristotelian text is quoted above, p. 105, n. 66; the commentaries on this section by Alexander of Aphrodisias and Averroes are summarized above on pp. 106–7.

125. The second opinion (omitted here) is Aristotle's and follows the very lengthy evaluation of this first opinion, not all of which I have included.

126. While it is true that Plato discussed the Great Year in *Timaeus* 39D, there is no evidence that he held the doctrine of identical numerical return (see above, p. 104, n. 64 and p. 105, n. 66).

127. *Meteorologica* 1. 2.

the completion of the motion of the eighth sphere, as we find in the book *De proprie-tatibus elementorum*. Moreover, one can show that at the end of such a time all the planets will not be at the beginning of Aries, from which, according to Ptolemy, every motion of the superior bodies begins. This can be shown by the famous Alphonsine tables and [by the tables] of the same Ptolemy on the motions of the superior bodies. Therefore, at the end of such a year all the superior bodies will not be in the same disposition in which they were at the beginning of the same year, which is contrary to the very foundation of this opinion. Furthermore, this fundamental argument, which was accepted by Ptolemy, was rejected by Thabit, who showed that the motion of the eighth sphere, which is from west to east because of the forward and backward motion of the starry sky, cannot be made in 36,000 years, but by the Alphonsine tables it can be shown that this motion can be completed in a time slightly less than 49,000 years.

Besides, some celestial motions are mutually incommensurable, namely, epicycles and deferents.[128] Therefore mobiles that begin to be moved [simultaneously] from some point will never return simultaneously to the same point. This consequence is obvious because celestial motions are uniform. Therefore since the sides and diagonal [of a square] are mutually incommensurable, mobiles which begin to be moved from the same point will never return [simultaneously] to the same point if they are moved with continuous uniformity. For this reason Ptolemy, at the beginning of the *Quadri-partitum*, and Haly, his commentator, say that it will never happen that a constellation will be alike in all things with respect to a constellation that preceded it.[129]

128. Without further elaboration, it is by no means clear in precisely what sense Paul understood this incommensurability between epicycles and deferents. Are the epicycles and deferents of a given planet incommensurable? Or is the deferent of one incommensurable to the deferent of another; and the epicycle of one incommensurable to a corresponding epicycle of another? From what follows, however, it is apparent that the resultant motions of some planets are simply incommensurable—however this may occur—thereby preventing successive conjunctions in the very same point.

129. "Notandum quod in hac materia due fuerunt extreme opiniones quarum prima fuit Platonis dicentis quod in revolutione magni anni revertentur omnia inferiora presentia et preterita non solum eadem specie, sed numero cum eadem dispositione omnino quale nunc habent aut prius habuerunt. Aut fundatur hic opinio super hoc: quia revoluto magno anno, id est completo motu octave spere, erunt simul corpora superceles-

tia in eadem dispositione in qua fuerunt in principio eiusdem magni anni et tunc fiet alius magnus annus per omnia consimilis primo anno signato. Et quia secundum communem conceptum philosophorum motus omnium inferiorum causantur et regulantur a motibus superiorum corporum secundum illud Aristotelem primo *Metheorum*, necesse est mundum hunc inferiorem continuari superioribus lationibus [*corr. ex* lotionibus], ut omnis virtus eius gubernetur inde. Ideo omnia hec inferiora necessario consimiliter tam specifice quam individualiter in secundo magno anno redibunt.

"Contra istam opinionem similiter arguo primo contra fundamentum eius quia secundum Ptolemeum octava spera movetur in centum annis tantum uno gradu et per consequens secundum eum motus iste completur in triginta sex milibus annorum. Patet consequentia quia multiplicando gradus orbis stellati qui sunt 360 [*corr. ex* 36] per centum resultabunt triginta sex milia; et tantus numerus annorum assignatur ad completionem

Almost all the substantive parts of Paul's discussion could have been derived from either Duns Scotus or Oresme, while Johannes de Muris could have provided all the material on incommensurability (but not the discussion on the Great Year). Indeed, by the time Paul wrote, such discussions could have been available to him from many later writers who might have included brief treatments of the same problem, which were in turn derived from Scotus, Oresme, or Johannes de Muris. Perhaps the only significance of Paul's discussion is that it appeared in a commentary on the *De generatione*, so that we may confidently expect that others proposed similar arguments in the same context.

In John de Fundis (fl. 15th century) we see how an avowed astrologer[130] reacted to Oresme's fundamental attack against his discipline. In his hostile commentary on the *Ad pauca respicientes*,[131] de Fundis commented on most of

---

motus octave spere, ut habetur in libro *De proprietatibus elementorum.* Constat autem quod in fine tanti temporis non erunt omnes planete in principio Arietis a quo incipit omnis motus superiorum, secundum Ptolemeum. Quod quidem ostendetur per tabulas famosas Alphonsi et eiusdem Ptolomei de motibus corporum superiorum. Igitur in fine talis anni non erunt omnia corpora superiora in eadem dispositione in qua fuerunt in principio eiusdem anni quod est contrarium fundamento opinionis. Item illud fundamentum acceptum a Ptolemeo reprobatur per Thebit ostendentem motum octave spere, quod est ab occidente ad oriens propter accessum et recessum ipsius celi stellati non esse talem in triginta sex milibus annorum sed per tabulas Alphonsi probatur motus ille compleri in tempore modicum minori quam sint 49 milia annorum.

"Preterea, aliqui motus celi sunt invicem incommensurabiles, videlicet epiciclorum et deferentium. Igitur mobilia incipientia moveri ab aliquo puncto nunquam simul redibunt ad eundem punctum. Patet consequentia ex quo: motus celi est uniformis. Unde quia motus coste et dyametri sunt invicem incommensurabiles mobilia incipientia moveri ab eodem puncto nunquam redire simul ad eundem punctum si continue uniformis moverentur. Quapropter dicit Ptolemeus in principio *Quadripartiti* et Haly, commentator suus, quod constellatio per omnia consi-

milis alicui constellationi que prefuit nunquam eveniet."—*Expositio super libros de generatione et corruptione Aristotelis* (Bologna, 1498), fols. 100v, col.2–101r, col.1. This section was also translated into French by Duhem in *Le Système du monde*, vol. 10, pp. 386–87. Duhem observes that despite Paul's seemingly deficient understanding, his argument against the Great Year is reminiscent of Oresme.

130. Sometime around 1432, de Fundis published a *Nova theorica planetarum*, which, according to Olaf Pedersen, was little more than a reissue of the very widely used *Theorica planetarum*, written in the thirteenth century ("Theorica Planetarum-Literature," *Classica et Mediaevalia*, vol. 23 [1962], p. 229). John de Fundis's pretensions as an astronomer cannot be determined from this work.

131. Its full title is *Tractatus reprobationis eorum que scripsit Nicolaus Orrem in suo libello intitulato de proportionalitate motuum celestium contra astrologos et sacram astrorum scientiam, compilatus per Iohannem Lauratium de Fundis.* I know of only one copy, which is found in Bibliothèque Nationale, fonds latin, MS 10271, fols. 63r–153v. In this long work, only folios 63r–75r constitute the commentary on Oresme's *Ad pauca respicientes*, while the remainder consists of long refutations of numerous criticisms of astrology. In the explicit and colophon on

the first part, omitting only the first supposition,[132] the last part of the second supposition,[133] and Propositions VIII and IX.[134] Of the second part, however, only the introductory material and Propositions I, XIX, and XX were commented upon, and, strictly speaking, the last two mentioned propositions "are not propositions at all, but climactic pronouncements repudiating astrological prediction on the basis of the technical propositions wholly omitted in John's commentary."[135] That it is a hostile commentary is made perfectly evident when de Fundis introduces every comment with the words *Glosa reprobationis*, a gloss of rejection or repudiation.

De Fundis sought to show that not only are the celestial motions commensurable, but even if they were not, celestial conjunctions and aspects would repeat, contrary to Oresme's propositions in the *Ad pauca respicientes*. In his criticisms, he reveals a complete lack of understanding of Oresme's objectives and, together with numerous errors in his version of Oresme's text, the result is not a happy one.

That the celestial motions might be incommensurable is simply denied as contrary to nature: "The motions of the sky have been ordained commensurable by nature because, otherwise, nature itself would be imperfect. Therefore, when these [motions] are multiplied a certain number of times, they return to the same place. In his *Almagest* Ptolemy demonstrated this proposition, which he [i.e., Oresme] ignored. Here is a skilled sophist."[136] Categorical as his claim

---

(Note 131 continued)
folio 153v, the work is dated 1451:

"Explicit tractatus de reprobatione eorum que scripsit Nicholaus Orrem ut supra; necnon de reprobatione eorum que ab aliis multis obiecta sunt contra astrologos et sacram astrorum scientiam subiuncto quoque in hac doctrina modo et ordine iudicandi tam in nativitatibus quam in revolutionibus annorum compilatus per artium et medicine doctorem Iohannem Lauratium de Fundis Bononie commemorantem anno domini 1451, die 30 Octobris."

That part of the title of Oresme's work which is cited by de Fundis as *De proportionalitate motuum celestium* may have been seen by the latter in the colophon of the *Ad pauca respicientes* in BN 7378 A, fol. 17v. It has no special claim as the title of the work and differs from titles given in other manuscripts of the same work (see Grant, *Oresme PPAP*, p. 380). Because none of these titles

could be shown to have originated with Oresme, I chose to call it *Ad pauca respicientes*, after the opening words of the treatise.

132. This can be found in Grant, *Oresme PPAP*, p. 384, lines 33–35.

133. Ibid., p. 386, lines 39–44.

134. Ibid., pp. 400–402.

135. Ibid., p. 64, n. 82. In this note, I stated that in my subsequent edition of Oresme's *De commensurabilitate* de Fundis's treatise would be edited as an appendix. A more careful assessment of the treatise, however, made it clear that such attention was unwarranted. Only his major arguments against Oresme will be summarized here.

136. This gloss is to Oresme's second supposition of Part 1 (see Grant, *Oresme PPAP*, pp. 384–86, lines 36–38), which I shall quote in the version given by John de Fundis (BN 10271, fol. 66r):

"Auctoris intellectus: 'Propositis multis quantitatibus quarum proportio [*corr. ex*

was, de Fundis apparently realized that his opponents might, with equal dogma-
tism and validity, assert the "natural" incommensurability of the celestial motions. Seemingly in anticipation of such a counter claim, he formulated a more
basic attack, which, in effect, was an attempt to maintain that even if celestial
motions were incommensurable, celestial configurations and aspects would
nevertheless repeat, so that accurate astrological prediction would yet be possible.

Oresme's mistake, de Fundis insists, was to assume that the zodiac is infinitely divisible, when, in fact, astrologers and astronomers deal with the zodiac
as if it were only divisible into finite parts. This is brought out initially in de
Fundis's commentary on Part 1, Proposition IV of the *Ad pauca respicientes*,
which says that two mobiles moving incommensurably with respect to a common center will never conjunct twice in the same point.[137] In his *Glosa reprobationis*, he responds:

> It suffices here, indeed, that they be conjuncted in a degree, minute, second, or third,
> and not in an infinite division as Oresme subtilized when he said "there are an infinite
> number of points in which such mobiles, disposed in this way, have conjuncted, and
> the number of points in which they will conjunct are [also] infinite [in number],[138]
> and so on. And if such conjunctions are not true in simple motions they are never-
> theless true in [terms of] a multiplicity [of motions], as was declared above.[139]

---

propositio] est ignota possibile est dubium et
verisimile aliquam alicui incommensurabilem
esse.' Et est suppositio. *Glosa reprobationis.*
Motus celi commensurabiles sunt a natura
ordinate quia aliter natura ipsa ?imperficien-
ter. Ergo aliquotiens hiis replicatis in idem
reddunt quam propositionem Ptolomeus in
libro *Almagesti* demonstravit quam quidem
ignoravit. Hic est expertus sophista."

I do not know what demonstration in
Ptolemy's *Almagest* John de Fundis had in
mind. This is not the only passage where
Oresme is labeled a sophist. Somewhat later
on, Part 1, Proposition IX of the *Ad pauca
respicientes*, which Oresme called a "very
wonderful" demonstration ("valde pulcra";
see Grant, *Oresme PPAP*, p. 402, line 203),
is characterized by de Fundis as having "only
sophistical beauty, without a tincture of
reason" ("...sed solum sophisticam pulchritudinem sine rationis mixtura...."—fol. 72r
of BN 10271).

137. See Grant, *Oresme PPAP*, pp. 394–96.

De Fundis's version of this proposition (BN
10271, fol. 70r), includes only lines 122–32
of my edition and differs somewhat from my
text.

138. Here de Fundis is actually quoting the
enunciation of Part 1, Proposition V of the
Ad pauca respicientes (see Grant, *Oresme
PPAP*, p. 396).

139. "Sufficit quidem hic ut in gradu,
minuto, secundo, vel tertio, coniungantur et
non in infinita divisione ut subtilizavit Orrem cum dixit 'infinita puncta sunt quibus
fuerunt coniuncta talia mobilia sic disposita
et infinita quibus erunt coniuncta,' et cetera.
Que quidem coniunctio et si veritatem non
habeat in motuum simplicitate tamen in mul-
tiplicitate veritatem optinet ut superius est
declaratum."—BN 10271, fol. 70v. We see
here an attempt by de Fundis to further safe-
guard his argument by insisting that even if
Oresme's propositions were true for the
simple and few motions that he considered,
they would not hold true for the many mo-

De Fundis's point is simple and direct. If the zodiac is divided only finitely, and if the planets are accordingly moved with incommensurable motions, as Oresme would have it, they would nonetheless return to one and the same degree, or the same minute, or the same second, etc., and enter into conjunction repeatedly, even though this might not occur in the very same point within a particular degree, minute, or second. That is, they would conjunct in the same minute of arc, for example, but never in the very same point of that particular arc. Nevertheless, if the divisions are made finite, and extend no farther than, let us say, seconds or thirds, conjunctions would indeed repeat in such relatively large subdivisions of the zodiac.[140]

In this way, Oresme is accused of oversubtlety when he insists on infinite divisibility of every minute, second, third, etc. After all, says de Fundis, somewhat later on: "It follows from this that however many times things may have conjuncted when moved unequally with respect to their center, it is possible, probable, and necessary that they should be conjuncted again, as [stated] above. Furthermore, no demonstrations can be made in these matters except by experience, that is, partly with instruments of calculation and sight, and partly by true knowledge of the effects of a place."[141]

De Fundis reiterates his fundamental argument in commenting on Part 1,

---

(Note 139 continued)
tions that really obtain in the heavens. Thus de Fundis appears to be arguing that as more motions are taken into account the true state of celestial affairs becomes more simplified and the results of the planetary motions approximate what they would be if the celestial motions were commensurable.

140. In the *De commensurabilitate*, Part I, Proposition 25 Oresme readily concedes that "An astronomer is satisfied if he knows that a conjunction is in a certain degree, or minute, or second, and so forth, even though he does not know in what [precise] point of a particular minute it occurs; or, he is content when an error is visually undetectable through use of an instrument"—(I.879–82). But if the celestial motions are incommensurable, then by Part II, Proposition 4 of the *De commensurabilitate*, Oresme would insist that, however small an arc is taken, conjunctions must necessarily occur there in the future, but never twice in the same point. Thus Oresme recognized that if the celestial motions were incommensurable and if, as a

result, no conjunction could possibly occur twice in the same point, astronomers would nevertheless be compelled to ignore what falls below the threshold of observation and conclude that conjunctions do occur in the same point if they are observed within the same minute of arc—even though the precise minute of arc is indeterminable. Oresme could easily concede this, since he was in no way concerned with the practice of astronomers, but was only interested in precise punctual aspects (in support of this, examine the references in n. 145, below), a fact of crucial importance but completely ignored by John de Fundis.

141. "Ex quo sequitur quod mota inequaliter respectu centri quotienscumque coniungentur in puncto aliquo possibile est, verisimile, et necessarium ut vice alia coniungantur ut supra. Nulle autem in istis demonstrationes fieri possunt nisi ex experientia ex parte, scilicet calculi atque visionis cum instrumentis, et vera ex parte loci effectuum noticia."—BN 10271, fol. 71r.

Proposition V of the *Ad pauca respicientes*. After quoting the text of two successive examples offered by Oresme to show that if two mobiles move with incommensurable speeds they will never meet again in any point from which they may be assumed to have started their motions,[142] he says:

Although one could respond to this by the statements made above, nevertheless, in order that the solution might be made more explicit, it should be said that an astrologer [or astronomer] does not consider the division of the zodiac into infinity. Moreover, in an astrologer's [or astronomer's] consideration of this, the division of the zodiac has only a finite number of degrees, minutes, seconds, or thirds; and this suffices for him. Besides, if a division of the zodiac were taken to infinity, such a division is rather more in the understanding than it is real, and among astrologers [and astronomers] no thought has been given to this [kind of intellectual, but unreal division].[143]

These criticisms form the basis of John de Fundis's attack on Oresme's *Ad pauca respicientes*[144] and the many consequences that were derived from the

142. These examples are quoted on folio 71v of BN 10271 and appear in Grant, *Oresme PPAP*, pp. 396, 398, lines 150–59.

143. "Et quamquam per dicta superiora possit ad hoc responderi, tamen ut expressius habeatur solutio, dicatur quod astrologus non considerat divisionem zodyaci in infinitum. Habet autem ipse in numero finito gradus, minuta, et secunda; sive tertia, circa que in sua speculatione versatur et sibi sufficit hoc. Preterea divisio zodyaci et si restat in infinitum est talis divisio potius per intellectum quam sit realis de qua apud astrologos consideratio nulla facta est."—BN 10271, fol. 71v. I have translated "astrologus" as "astrologer" and "astronomer," since in the Middle Ages this term was frequently used for practitioners of both arts (see Shlomo Pines, "The Semantic Distinction Between the Terms *Astronomy* and *Astrology* According to Al-Bīrūnī," *Isis*, vol. 55 [1964], p. 344, col. 1). Moreover, it is apparent that what de Fundis says here is meant to apply to both.

144. Much of the remaining portions of the commentary consists largely of purely astrological arguments. In the parts that are relevant to the subject matter discussed here, the arguments rely, for the most part, on themes that I have already described.

Occasionally, de Fundis's knowledge of Oresme's text was faulty, resulting in unjust accusations against Oresme. For example, as his version of the last part of Part 1, Proposition III of the *Ad pauca respicientes*, de Fundis has: "Item [*corr. ex* idem] si prescise sol cursum suum faceret in uno anno, mars in duobus commensurabiliter, numquam nec in uno loco coniungentur."—BN 10271, fol. 68v. ("Thus if the sun should complete its path in one year and Mars in two years commensurably, they will never be conjuncted, not [even] in one place.") But had de Fundis replaced *nec* with *nisi*, as the text requires (see Grant, *Oresme PPAP*, p. 394, line 102), he would not have concluded that "what he [i.e., Oresme] inferred was false, [namely] that Mars and the sun would never be in conjunction; for indeed they would be conjuncted, and in the same place..." ("falsum esset id quod intulit quod mars et sol numquam coniungerentur; ymo coniungantur et in eodem loco...."—BN 10271, fol. 68v). In truth, de Fundis has actually adopted Oresme's true conclusion, which is that "if the sun completed its path in exactly one year and Mars in two years, they would never be conjuncted except in one place."—Grant, *Oresme PPAP*, p. 395; Latin text, p. 394.

assumption that the celestial motions are probably incommensurable. In evaluating them, it must be said that they are utterly irrelevant, since they rely on the actual practices of astronomers and astrologers, who, as everyone was well aware, depended upon observations and tables that were necessarily approximations. Oresme stressed this point frequently.[145] But it was not with approximations that Oresme was concerned. Only exact punctual relationships on circular continua, and the necessary consequence of this, namely, infinite divisibility, play any role in these two major treatises. From this standpoint, if we assumed, as did many astrologers, that the heavens governed and influenced terrestrial events and that celestial bodies have uniform and regular motions, and that the same cause can produce only the same effect, then, if the celestial motions are incommensurable and exact punctual relationships are the only concern, Oresme's many arguments concerning the non-repetitivity of celestial configurations and aspects and the various consequences deriving therefrom should have threatened the foundations of astrology and caused astrologers the gravest apprehension.[146] In the absence of conclusive proof, it was equally plausible to assume that the celestial motions were incommensurable rather than commensurable. But even if the celestial motions were commensurable, Oresme had derived consequences that were at variance with the data and assumptions of astrologers. To counter all this, as did John de Fundis, by denying categorically the incommensurability of these motions, or by denying infinite divisibility because it was an impractical and inconvenient assumption in the calculations of astronomers and astrologers, is to reveal total ignorance of Oresme's objectives and the nature of his onslaught against astrology and its assumptions and predictions.

The last of the post-Oresmian authors who will be discussed here is by far the most significant. Jerome Cardano (1501–76), in his *Opus novum de proportionibus*,[147] a diverse collection of 233 propositions on physics and mathematics, devotes six propositions (Propositions 47–52 inclusive) to mobiles moving

145. In the *Ad pauca respicientes*, see pt. 1, supp. III (Grant, *Oresme PPAP*, p. 386, lines 45–50) and pt. 2, prop. XX (ibid., p. 428, lines 263–64); and in the *De commensurabilitate*, see I.45–49, 210–17, and 879–82.

146. These conditions are substantially the same as those adopted by Oresme in the enunciation of Part 2, Proposition XIX of the *Ad pauca respicientes* (Grant, *Oresme PPAP*, pp. 424–27) where, despite such concessions to astrology, he insists that prediction of future dispositions is impossible if the celestial motions are incommensurable..

147. The full title is: *Hieronymi Cardani Mediolanensis civisque Bononiensis, philosophi, medici, et mathematici clarissimi Opus novum de proportionibus numerorum, motuum, ponderum, sonorum, aliarumque rerum mensurandarum, non solum geometrico stabilitum, sed etiam variis experimentis et observationibus rerum in natura solerti demonstratione illustratum ad multiplices usus accommodatum et in V libros digestum. Praeterea Artis magnae, sive de regulis algebraicis liber unus.... Item De Aliza regula liber, hoc est algebraicae....* Basel, [1570].

on circles, especially emphasizing their times and places of conjunction. Although the basic substance of all these propositions can be found in the *De commensurabilitate*, Cardano's emphasis and presentation are strikingly different from that of Oresme. Nowhere, for example, does Cardano apply any of his propositions to celestial motions, a fact that distinguishes Cardano from all those whom we have mentioned in this chapter. Since Cardano restricted the scope of his propositions, our comparison must perforce be confined to the purely kinematic aspects of their respective propositions. But even such a comparison reveals great contrast between the two treatises. In these few propositions, Cardano leaves the strong impression that he sought to incorporate as much Euclidean proportionality terminology, and even more recent algebraic terminology, as was possible, and in so doing present the appearance of a formidable exercise in mathematical manipulation and ingenuity. This may explain certain obscurities of expression as well as the tedious multiplication and extension of examples beyond reason, with no other apparent purpose than to prolong the parade of mathematical terminology and in this way impress the reader by sheer cumulative impact.

Cardano's desire to replace the older Latin Euclidean terminology, much of it derived through Latin translations from the Arabic, with Latin transliterations of Greek terms is also much in evidence. Substitutions of this kind, while not affecting the essential arguments, would undoubtedly have given the appearance of greater mathematical elegance, profundity, and modernity. Terms like *analogae, omiologa, rhete, alogae*, defined at the beginning of the treatise[148] and used wherever possible in our six propositions, displaced their older Latin equivalents *continua proportio, similis, rationalis*, and *irrationalis*, respectively.

The differences between the two treatises seem exaggerated by yet another factor. Specific material and ideas that were scattered over many separate propositions in the *De commensurabilitate* are conflated, rearranged, and given different emphasis in Cardano's propositions, to such an extent as to make his propositions appear quite independent of those formulated by Oresme. And since these few propositions are infiltrated with Greek terminology and expressed with a mathematical brevity often bordering on obscurity, the two treatises might appear unrelated to a casual reader. Indeed, when we take into account Cardano's own explicit statement at the end of Proposition 52 that the basic idea for these propositions came from Campanus of Novara, Oresme's possible influence seems improbable. But all the differences cited above are but

---

148. Ibid., pp. 2 and 3, Definitions 9, 19, and 22. In Definition 19, brevity of expression and avoidance of a term used by non-Greeks are given as the reasons for replacing *continua proportio* by *analogae*. ("Quanti-

tates que in continua sunt proportione analogae vocantur. Dictum est hoc ad fugiendum nomen barbarum, etiam ut breviter tamen possemus sententiam explicare.")

mere appearances, for we cannot lightly set aside the fundamental and indisputable fact that almost all of Cardano's ideas and thoughts on this subject can be found in Oresme's *De commensurabilitate*, not in embryonic and ill-defined form, but concretely and specifically. On the basis of the summary and analysis of Cardano's six propositions to follow, it seems more plausible to conjecture that, directly or through intermediaries, his ultimate debt is to Oresme and the *De commensurabilitate* rather than to the chapters by Johannes de Muris in the *Quadripartitum numerorum* (a crucial argument for this conjecture is given below, n. 175 of this chapter).

In the first of these propositions, Proposition 47, Cardano includes material that can be found in two of Oresme's propositions in Part I of the *De commensurabilitate*. He tells us in the enunciation of the proposition that "If two mobiles are moved uniformly with their proper motions on the same circle,[149] the product of their respective times of circulation will be equal to the product of the difference between their times of circulation and the time of their first conjunction."[150] Let A and B represent the mobiles, and f and g the times in which A and B complete one circulation, respectively, where, furthermore, $f > g$ and $f - g = h$; finally, if A and B are now in conjunction in some point b,[151] let k be the time it would take before they conjunct in some other point d. It is Cardano's objective to demonstrate that $f \cdot g = h \cdot k$.

It is assumed that when A completes one circulation, B will have reached only as far as c, so that arc cdb is the difference between arc bc and the distance

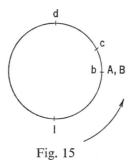

Fig. 15

149. In Part I, Proposition 6 (I.304–5), Oresme says that it does not matter whether the mobiles are assumed to move on one circle or several.

150. "Si duo mobilia aequaliter in eodem circulo iuxta proprios motus moveantur, productum temporis circuituum invicem erit aequale producto differentiae temporum circuitus ductae in tempus coniunctionis pri-

mae."—Cardano, *Opus novum de proportionibus*, p. 36, but mistakenly numbered 46.

151. In his figure (ibid., p. 36), which I have altered in Fig. 15, Cardano lets b represent both a mobile and the initial point of conjunction. Since I have designated both mobiles by capital letters, b will be used to represent the point of conjunction.

represented by the whole circumference of the circle. Relating the distances and times, Cardano says that the Circumference of the circle: arc $bc = g:f$ and the Circumference of the circle: arc $cdb = g:h$ and, finally, that arc bc:arc $cdb = f:h$.[152]

Cardano's next move is rather unclear. After assigning k as the time in which A and B will conjunct in d, he says that k, f, g and h, all temporal quantities, are homologous (*omiologa*)—i.e., "quantities that have a similar discontinuous proportion"[153]—just as the relationships between distances. Cardano seems to be saying that the sorts of relationships that can be had between these four temporal quantities will correspond to the relationships between the distances that have been assigned previously. Thus the relationships between $k, f, g,$ and h should correspond to those between the $\dfrac{\text{circumference of the circle} \cdot \text{arc } bc,}{\text{arc } cdb}$ i.e., between the circumference of the circle and arc bc, and their difference, arc cdb.[154]

Continuing with the relationships between the four temporal quantities, Cardano says "let $s = f \cdot g$, [and] I say that dividing s by h gives k, the time of the first conjunction."[155] His purpose, then, is to demonstrate that if $s = f \cdot g$, then $s/h = k$, from which it must follow that $f \cdot g = h \cdot k$.

---

152. The Latin text for the material summarized in this paragraph is as follows: "Dum itaque *A* perficit circulum, *B* perveniat in *c*, igitur *cdb* est differentia quae superanda est, et proportio circuli ad *bc* ut *g* ad *f*, quare reliqui ad reliquum ut residui ad residuum, scilicet circuli ad *cdb* ut *g* ad *h* et *bc* ad *cdb* ut *f* ad *h*."—*Opus novum de proportionibus*, p. 36.

153. This is a translation of Cardano's definition of homologous quantities (Definition 22), which reads: "Quantitates quae similem habent proportionem non continuatam, omiologae appellantur."—Ibid., p. 3.

154. Here is the Latin text that has been interpreted in this paragraph. "Coniungantur igitur in *k* tempore, eruntque *k f g h* omiologa, ut productum ex circulo in *bc* diviso per certam quantitatem et cum circulo et *bc* et *cdb* differentia."—Ibid., p. 36. At this point, Cardano seems to be anticipating the relationship that he will derive at the end of this proposition (see below), namely, that $(o \cdot p) / q = n$, where o is one circulation and represents the circumference of the circle; p is the distance traversed by B while A traverses o, i.e., $p = $ arc bc; q is arc cdb, the difference between the whole circle and arc bc; and, finally, n is the total distance traversed by B from its conjunction in b to its next conjunction in d.

Observe that Cardano chose to use the term *omiologa* (for his definiton, see above, n. 153), a transliteration of the Greek term ὁμόλογα, meaning "similar" or "corresponding," instead of the usual Latin term *similis*. Cardano frequently preferred a Latin transliteration of a Greek term rather than a Latin translation. If Cardano used Zamberti's translation from the Greek of Euclid's *Elements*, he could have found ὁμόλογα placed in the margin alongside Book 5, Definition 11, where it is translated as "similis" in the definition itself. In *Euc.-Campanus*, which also includes Zamberti's translation, see p. 113; and also Heath, *Euclid's Elements*, vol. 2, p. 134. This definition does not appear in Campanus's version of the *Elements*.

155. "Et sit *s* productum ex *f* in *g*, dico quod divisa *s* per *h* exibit *k*, tempus coniunc-

Next Cardano explains that when A and B conjunct in d, one of the mobiles must have traversed one more complete circle than the other.[156] He calls this extra circle, or whole circumference, o, and designates m and n as the total distances traversed by A and B, respectively, in moving from their first conjunction to their very next conjunction in d. Since A is the quicker moving body, it will have traversed one more circulation than B when they conjunct in d, so that $m - n = o$; and while A traverses o, B will traverse a certain lesser distance, say, p, where $q = o - p = $ arc $cdlb$. (Note that Cardano now includes l in Fig. 15, so that arc cdb is now cited as arc $cdlb$.) Now in equal times, A moves through m and B traverses n, whence it follows that $S_A:S_B = m:n$ and $S_A:S_B = o:p$ (where S is distance), so that $m:n = o:p$. Furthermore, since $o = m - n$ and $q = o - p$, it follows from Euclid 5. 19 that $m:o = o:q$ and $m:o = n:p$.[157] Therefore, $n:p = o:q$ and $(o \cdot p)/q = n$, where n is the total distance traversed by B from one conjunction to the next. Or, expressed in the temporal quantities used above, $(f \cdot g)/h = k$.[158] It is now obvious, concludes Cardano, that

(Note 155 continued)

tionis primae."—*Opus novum de proportionibus*, p. 36. In Part I, Proposition 6 of the *De commensurabilitate*, Oresme shows how to find the time between two successive conjunctions (see I.273–74 for the enunciation). The rule that Oresme gives at the conclusion of Proposition 6 (see I.307–9 and above, p. 13) is essentially the same as what has been quoted here from Cardano, although the latter will seek to derive the time exclusively from distance relationships.

156. "Sit itaque d locus coniunctionis, dico igitur quod differentia spatii pertransiti ab A et a B in reditu ex coniunctione prima ad d est unus circulus completus. Non enim possunt esse plures, nam sequeretur quod A aliquando pertransisset B et sic non esset prima coniunctio; nec potest esse minus, nam sic cum A et B sint in d ultra perfectas circulationes uterque eorum pertransivit arcum bc, igitur nullo modo differentia potest esse minor circulo, neque maior, ut declaratum est. Igitur est unus circulus ad unguem." —*Opus novum de proportionibus*, pp. 36–37. Oresme says much the same thing in I.290–92, 300–302.

157. Euclid 5. 19 reads: "If, as a whole is to a whole, so is a part subtracted to a part

subtracted, the remainder will also be to the remainder as whole to whole."—Heath, *Euclid's Elements*, vol. 2, p. 174; Zamberti has the same thing in *Euc.-Campanus*, p. 128. In Cardano's proposition, Euclid 5.19 is applied as follows: $o:q = (m-n):(o-p)$, where $m:o = n:p$ (since $m:n = o:p$), so that $m:o = o:q = n:p$.

158. The text of the demonstration described in this paragraph up to this point is as follows: "Hoc declarato ponatur m spatium compositum ex circulis pertransitis ab A cum spatio bd, et n [*corr. ex* etenim] spatium quod pertransit B a coniunctione in a, ad coniunctionem primam in d. Et erit ex demonstratis horum differentia circulus qui vocetur o; et sit p spatium quod pertransit B in tempore eodem in quo A pertransit o; et sit q differentia o et p, que in circulo est $cdlb$. Quia igitur in eodem tempore A pertransit m et B n, erit m ad n ut A ad B; et eadem ratione A ad B ut o ad p. Igitur ex undecima quinti Euclidis, m ad n ut o ad p; quare cum o sit differentia m et n, et q differentia o et p, erit ex decimanona quinti Euclidis, m ad o ut o ad q. Et ita circulus est analogus inter spatium pertransitum a motore velociori et inter differentiam spatii quae accidit, dum velocior motor pertransit circulum, id est quod circulus *acd*

$k:f = g:h.$[159] Curiously, he fails to draw the final consequence that $f \cdot g = h \cdot k$, which was claimed in the enunciation and was presumably the objective of the proof.

Proposition 48 illustrates how Cardano could complicate and unduly prolong the discussion of a quite simple proposition. In the enunciation he asserts that "If three mobiles [or bodies] depart from the same point and conjunct two by two in commensurable times, these three mobiles will conjunct again in a time *that has been produced from the denominator of the division of the greater time by the lesser [multiplied?] into the lesser [time], or by the numerator into the greater [time].*"[160] That part of the enunciation which I have italicized seems rather unintelligible. Directly after the enunciation, and without any general or formal discussion of the kind found in Proposition 47—where no examples were offered—Cardano fortunately "proves" his proposition by example and in so doing clearly reveals his meaning, a meaning completely obscured in the enunciation.

"Let there be three mobiles: A, which completes a circulation in 2 years, B in 5 years, and C in 7 years. I say that they will first return [to conjunction] in [a time equal to] the number produced by [the multiplication of] 7, 5, and 2, which are prime numbers; and this number will be 70 years."[161] By multiplying the three times of circulation, Cardano obtains the least common multiple, which indicates the time required before the three mobiles will conjunct again in the same point.[162] It is puzzling that Cardano chose to express himself so obscurely

---

est analogus inter *cdlb*, et circulos pertransitos ab A cum portione *bd*. Revertor igitur ad propositum. Cum sit m ad o ut o ad q; et m ad o ut n ad p, ex sextadecima quinti Euclidis, erit ex undecima eiusdem n ad p ut o ad q. Quare ex sextadecima sexti *Elementorum* ducto o, id est circulo, seu maiore numero, in p, spatium pertransitum a B; seu ducto f in g et diviso per q, differentiam spatiorum, seu per h, exibit n, seu spatium quod pertransit B ab una coniunctione ad aliam, quod erat demonstrandum."—*Opus novum de proportionibus*, p. 37.

A method for determining the total distance traversed and the place of the next conjunction is given by Oresme in Part I, Proposition 8 of the *De commensurabilitate* (see above, p. 14). The mode of expression and the procedure differ markedly from that of Cardano.

159. "Ex hoc patet quod proportio temporis coniunctionis ad tempus tardioris motus circuitionis est veluti temporis circuitus velocioris motoris ad differentiam temporis motus tardioris et velocioris motoris in uno circuitu."—*Opus novum de proportionibus*, p. 37.

160. "Si tria mobilia ex eodem puncto discedant, fuerintque duorum ac duorum coniunctiones in temporibus commensis illa tria mobilia denuo coniungentur in tempore *producto ex denominatore divisionis temporis maioris per minus in minus, aut numeratore in maius.*"—Ibid.

161. "Sint tria mobilia: A quod circuat in duobus annis, B in quinque, C in septem. Dico quod primum redibunt in numero producto ex septem quinque et duobus, qui sunt numeri primi, et erit ille numerus septuaginta annorum."—Ibid.

162. In demonstrating the very same proposition, Oresme discusses, in general, direct,

in the enunciation and so plainly and directly in the body of the proposition.

In an effort to demonstrate the truth of the example, Cardano offers an "indirect proof" showing that 70 years must necessarily be correct, since a time less than 70 years would not produce a conjunction in the same point. In succession, he shows that intervals of 35, 17, $14\frac{1}{2}$, and $10^{1/2}$ years are not times in which successive conjunctions of the three mobiles can occur in the same point.[163] On the basis of these four negative examples, he concludes that "they will not meet before 70 years."[164]

Two corollaries are now drawn which seem to say much the same thing. The first proclaims that the three mobiles cannot conjunct in any point other than the one in which they have already conjuncted,[165] from which it follows—and

---

(Note 162 continued)
and uncomplicated terms, the finding of a least common multiple. See *De commensurabilitate*, pt. I, prop. 15, I.579–615 and the Commentary, n. 9 for these lines. A least common multiple was also utilized for the same purpose by Johannes de Muris in the *Quadripartitum numerorum*, bk. 4, tract 1, chap. 21, lines 11–13 and chap. 22, lines 6–11 (see below, App. 1 and above, pp. 90–91).

163. To convey something of Cardano's prolixity and apparent desire to interject and employ Euclidean jargon wherever and as often as possible, I shall quote only the example for 17 years:

"Now if you should say that *A*, *B*, and *C* are conjuncted in 17 years, a number that is not numbered [or measured] by any of these times, they will [each] be carried through [a certain number of] complete circulations and [then] $\frac{1}{2}$ of *A* will remain, $\frac{2}{5}$ of *B*, and $\frac{3}{7}$ of *C* [i.e., $A = \frac{17}{2} = 8\frac{1}{2}$; $B = \frac{17}{5} = 3\frac{2}{5}$; and $C = \frac{17}{7} = 2\frac{3}{7}$]. In order that they meet in the same point after they have completed their circulations, it will be necessary that these [fractional] parts be equal. Therefore, ratios 7:3, 5:2, and 2:1 are equal, so that by alternating (*permutando*) [ratios 7:3 and 5:2], 3:2 must be as 7:5. But 7 and 5 are mutually prime numbers and therefore, by what is said in the seventh [book] of the *Elements*, in their lowest ratio. Therefore, 3 and 2 are not in the same ratio [as 7 and 5]." ("Quod si dicas *A*, *B*, *C* coniungi in decemseptem annis

numero non numerato ab aliquo illorum temporum, auferantur perfectae circulationes et remanebunt dimidium ex *A*, duae quintae ex *B*, tres septimae ex *C*. Igitur oportebit ut hae portiones sint aequales ut post perfectas circulationes in idem punctum conveniant. Ergo $\frac{1}{2}$ et $\frac{2}{5}$ et $\frac{3}{7}$ aequivalebunt, quare proportio 7 ad 3, et 5 ad 2, et 2 ad 1 est una, quare permutando 3 ad 2 ut 7 ad 5. Sed 7 et 5 sunt contra se primi, ergo in sua proportione minimi per dicta in septimo *Elementorum*. Ergo tria et duo non sunt in eadem proportione."—*Opus novum de proportionibus*, p. 38.)

By presenting a series of negative examples of this kind, Cardano was seemingly convinced that he was demonstrating his proposition indirectly. Obviously he was mistaken, since all the possibilities were not exhausted. If Oresme's *De commensurabilitate* influenced Cardano at all, it is possible that two brief statements in Part I, Proposition 5 (I.237–45 and 258–62) may have triggered Cardano's parade of negative examples. Oresme, however, reveals an awareness that examples do not constitute a formal proof when he says: "But all these things can be shown more readily and in briefer compass by an example rather than by formal demonstration" (I.596–97).

164. "Igitur non conveniunt ante septuaginta annos."—*Opus novum de proportionibus*, p. 38.

165. "Ex hoc sequitur quod nullibi con-

this is the second corollary—that through all eternity they cannot meet anywhere other than in their initial point of conjunction.[166]

Reducing the number of mobiles to two, Cardano, in Proposition 49, assumes once again that the mobiles depart simultaneously from conjunction in a given point. Let the mobiles be A and B and assume that after departure from some initial point of conjunction, they will conjunct again in another point, c, whose distance from the initial point of conjunction is given. If the time of one circulation is known for one of the mobiles, Cardano says that he can determine the time required for the other to complete one circulation, as well as the total time that will elapse between its departure from the initial point of conjunction and its subsequent conjunction in c.[167] Since the formal discussion is followed immediately by two examples which parallel the formal presentation closely and are themselves almost identical in form, it will be convenient to summarize this proposition solely in terms of the first example.

If A and E be the two mobiles, Cardano will find both the time in which E will complete one circulation and the time in which it will conjunct with A in c. Certain data are assumed for A. Let $f = 4$, the number of integral circulations that A will complete prior to conjunction with E in c; $b = 2\frac{1}{2}$ years, the time required for A to make one complete circulation; finally, let c be distant $\frac{4}{5}$ of

---

venient praeterquam in eodem puncto, scilicet quo ab initio coniuncti fuerunt."—Ibid.

166. "Sequitur denuo ex propositione ipsa repetita et primo corrolario, quod nullibi alibi convenient quam in dato primo puncto in quo coniuncti fuerant ab initio etiam usque in aeternum."—Ibid. While Oresme does not explicitly state that three mobiles will not meet elsewhere, he does demonstrate, in Part I, Proposition 14 of the *De commensurabilitate*, that the three mobiles will of necessity conjunct repeatedly in the same point from whence they started; and since the conditions described in this proposition are such as to allow for repeated conjunctions in only that particular point, it would follow that conjunctions cannot occur elsewhere. Indeed, although in Part I, Proposition 17 more than one point of conjunction is involved, it is nevertheless a finite number, so that conjunctions of all three mobiles could not occur in any other points (see the Commentary, n. 11 for I.647–77).

Cardano concludes Proposition 48 with yet another example which he subsequently

confirms by a negative example. Here the mobiles A, B, and C complete single circulations in $2\frac{1}{2}$, $3\frac{1}{3}$, and $4\frac{1}{4}$ years, respectively. Cardano shows that all three will conjunct in the same point every 170 years, but not before. That conjunctions cannot occur before 170 years is shown by assuming a period of 33 years and then demonstrating that this is impossible. See *Opus novum de proportionibus*, pp. 38–39.

167. The enunciation says: "When the time of one circulation has been given for one mobile, and the distance [of the next conjunction given], to find from this the circulation of one mobile, which, after departing from the same point, meets with the other mobile in a given point after any number of circulations whatever; also to find the time of circulation." ("Proposito mobilis in circulo circuitus tempore, dataque ratione distantiae ab illo mobilis circuitum invenire quod ex eodem puncto discedens cum alio mobili in dato puncto conveniat sub quocunque numero circuituum; tempus quoque coniunctionis."—Ibid., p. 39.)

the circle from A's previous point of conjunction, so that if all motions and distances are taken clockwise, arc $ac = \frac{4}{5}$.

The product of $f \cdot b = 4 \cdot 2\frac{1}{2} = 10$ years, the time required for A to arrive at its initial point of conjunction—namely, a—after precisely four circulations.

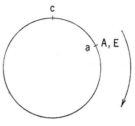

Fig. 16

But it must reach c, $\frac{4}{5}$ of the circle away. Now if h represents the additional time in which A will reach c, then $b:h =$ circumference of the circle: arc ac. Inserting the values given above, we get $2\frac{1}{2}:h = 1:\frac{4}{5}$, so that $h = \frac{4}{5} \cdot \frac{5}{2} = 2$ years. If we add 2 years to 10 years, A will travel $4\frac{4}{5}$ circulations in 12 years.

To obtain the time in which E will complete one circulation, Cardano divides 12 by $\frac{4}{5}$. If in 12 years, A will arrive in c to conjunct with E, it is obvious that E must also have been in motion for 12 years before it conjunct with A in c; and since it is assumed that E will move only $\frac{4}{5}$ of the circle in that time, 15 years are required for E to complete one circulation (i.e., $12/\frac{4}{5} = 15$). However, E need not move through one entire circulation, but only through $\frac{4}{5}$ of the circle, which will require only 12 years (i.e., $\frac{4}{5} \cdot 15 = 12$). In this manner Cardano determines that E takes 15 years to complete one circulation and 12 years to conjunct with A in point c.[168]

168. "Exemplum primi in re paulo obscuriore: sit f 4 et b $2\frac{1}{2}$ et ac $\frac{4}{5}$. Ducemus 4 in $2\frac{1}{2}$ fit 10; adde $\frac{4}{5}$ b [corr. ex 6], quod est 2 fit 12; divide per $\frac{4}{5}$, seu multiplica per $\frac{5}{4}$, quod idem est, fit 15 circuitus E. In quatuor ergo circuitibus et $\frac{4}{5}$ qui sunt duodecim anni perveniet A ad c; et in duodecim annis E perveniet ad c, nam 12 sunt $\frac{4}{5}$ ipsius 15."—Ibid., p. 40.

Although the formal part of this proposition and the two examples that follow it are presented by Cardano in a manner that differs considerably from Oresme, the basis for his manipulations can be easily derived from the *De commensurabilitate*, Part I,

Proposition 6, where Oresme determines the time between two successive conjunctions in different points (see above, p. 13). Indeed, the same thing can also be determined by use of Part I, Proposition 11 (I.423–40). Oresme did not devote a special proposition, or part of one, to the trivial problem of determining the time of one circulation for one mobile after having assumed it for another.

The second example which Cardano offers differs only slightly from the first, for instead of E traversing only part of its circle before conjunction in c, it now makes more than one circulation. The objectives, however, are identical with those of the first example—

In Proposition 50, Cardano considers incommensurable relationships for the first time, where the distance between successive conjunctions is always equal and incommensurable to the whole circle. In assuming at the outset that "all portions of circulations are repeated in the same times,"[169] he asserts, in effect, that the distances separating successive conjunctions are always equal.

Let A and B be in conjunction in point a[170] and assume that they will conjunct successively in points c, d, e, f, g, h, k, l, etc. Since the distances between immediately successive conjunctions are always equal, it follows that arcs $ac = cd = de = ef = fg = gh = hk = kl$, etc. Furthermore, let arc ac be incommensurable to the whole circle. Cardano seeks to demonstrate that if arc ac is incommensurable to the whole circle, then arc ak is also incommensurable to the circle and arcs ak and kc are mutually incommensurable (see Fig. 17).[171] For convenience, the proof will be presented in outline form.

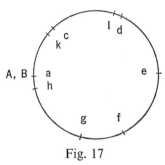

Fig. 17

---

namely, to determine the time of one circulation for E and the time between its successive conjunctions, both of which are achieved from data assumed for mobile A. For this superfluous example, see *Opus novum de proportionibus*, p. 40. In both examples, Cardano found it necessary, or at least desirable, to compute separately the time required for E to conjunct with A in c, after he had already calculated it for A. Obviously, the time of conjunction must be the same for both mobiles. When faced with the same situation, Oresme was fully aware that his result applied equally to both mobiles (cf. pt. I, prop. 6) and would no doubt have thought it superfluous to make two separate calculations.

169. "Omnes circuituum portiones in eisdem [*corr. ex* eiusdem] temporibus repetuntur."—*Opus novum de proportionibus*, p. 40.

170. Once again, Cardano uses a to represent both a mobile and a point of conjunction. For convenience, I have altered Fig. 17 so that a represents the point of conjunction and A one of the two mobiles. A few lines below, and in n. 171, it will be observed that $ac=cd=de$, and so on. These arcal equalities are distorted in Cardano's figure and, therefore, in Fig. 17, which faithfully reproduces it (with the exception of the previously mentioned alteration).

171. I shall quote somewhat more than is directly relevant to the particular proof that follows: "Sint in circulo a, c, d, e, f, g [*corr. ex a, b, c*...]; A et B iuncta et in primo congressu iungantur in c, in secundo in d, in tertio in e, in quarto in f, in quinto in g, in sexto in h, in septimo in k, in octavo in l. Et sic deinceps cumque tempora sint aequalia, erunt et circuitus totidem numero et excessus aequales etiam ac, cd, de, ef, fg, gh, hk, kl. Et si aggregatum A, scilicet circulo-

1. Arc ak + whole circle are commensurable to ac. (Since arc ak + whole circle = arcs $ac + cd + de + ef + fg + gh + hk$, and since arcs $ac = cd = de = ef = fg = gh = hk$, it follows that arc ak + whole circle are related to arc ac as a number to a number [as 7 to 1, to be precise]; hence they are commensurable.)

2. Arc ak + whole circle are incommensurable to the circle. (This follows from the fact that while arc ak + whole circle are multiple and commensurable to arc ac, arc ac itself is incommensurable to the whole circle.)

3. Therefore arc ak + whole circle are incommensurable to arc ak. (By 2, we know that arc ak + whole circle are incommensurable to the circle; but since the circle is commensurable to itself, this can be true only if arc ak is incommensurable to the circle so that when it is added to the circle the sum is also incommensurable to the circle.)

4. And it also follows that arc ak is incommensurable to the circle. (See the reasoning in 3, above.)

5. Therefore arc ak will be incommensurable to arc ac. (Since arc ak is incommensurable to arc ak + whole circle and arc ak + whole circle are multiple and commensurable to arc ac, arcs ak and ac must be incommensurable.)

6. Consequently, arc ck will be incommensurable to arcs ak and ac. (Since arc ac is incommensurable to arc ak [by 5 above] and arc ac = arcs $ak + kc$, it follows by Euclid 10. 16 that arc ac is incommensurable to arc ck and that arcs ck and ak are mutually incommensurable.)[172]

7. And, finally, it follows that arc ck is incommensurable to arc ak + whole circle. (Since arc ck is incommensurable to arc ac [by 6 above] and arc ac is commensurable to arc ak + whole circle, therefore arc ck is incommensurable to arc ak + whole circle.)[173]

From all this, Cardano concludes[174] that if arc ac were commensurable to

---

(Note 171 continued)

rum et portionis, fuerit commensum circulo, et ita de *B*, erunt omnia commensa ad circulum et etiam inter se. Et si inter se aggregata, vel portiones erunt, et eodem modo reliqua.... Erunt tamen incommensa necessario, si partes fuerint incommensae toti. Ponatur *ac* incommensa toti circulo, dico quod *ak* etiam est incommensa toti circulo; et etiam *ak* et *kc*."—*Opus novum de proportionibus*, p. 40.

172. In the margin, Euclid 10. 14 is cited. But whether added by Cardano or the printer, it does not apply.

173. The seven parts into which I have arbitrarily divided the proof are described

by Cardano as follows: "Quia enim *ac* est incommensa circulo, et *ka* cum toto circulo semel est commensa *ac*, quia multiplex ei. Igitur cum circulus et *ak* dividantur in circulum et *ak*, et circulus sit incommensus circulo cum *ak*, erit aggregatum ex circulo et *ak* incommensum ipsi *ak*; et *ak* pariter incommensa circulo. Rursus quia *ak* est incommensa circulo cum *ak*, et circulus cum *ak* sit multiplex ad *ac*, erit *ak* incommensa *ac*, quare erit *ck* incommensa *ak* et *ac* et circulo addita *ak*."—*Opus novum de proportionibus*, pp. 40–41.

174. The text of the conclusion summarized in this paragraph extends almost to the end of the proposition (only a few additional

the circle, then all parts of the circle divided by successive points of conjunction, as well as all the distances between successive conjunctions added together, would also be commensurable to the whole circle. The truth of this depends on the fact that the distance separating any two successive conjunctions is always equal to arc *ac*. Indeed, if arc *ac* were commensurable to the circle, it follows that the circle plus any part of arc *ac* divided by points of conjunction would also be commensurable to the circle; and the relationships between any parts divided by points of conjunction (as, for example, arcs *ak* and *kc*) would be commensurable. But if arc *ac* is incommensurable to the circle, as in the proof above, all these relationships would be incommensurable, and the regular occurrence of conjunctions equidistant from one another would result in a division of the circle into smaller and smaller parts, all of which are mutually incommensurable. Thus the initial distance separating the first two conjunctions determines whether or not later conjunctions can occur in certain places. For example, if the distance separating the first two points of conjunction is $(\frac{1}{2})^{1/3}$ part of the circle, a distance incommensurable to the circle, and if *a* is the first point of conjunction, no subsequent conjunction can occur in points that are $\frac{1}{2}$ a circle distant from *a*, or $\frac{1}{4}$ part distant from *a*, or $\frac{1}{8}, \frac{1}{3}, \frac{1}{6}, \frac{1}{9}, \frac{1}{5},$ or $\frac{1}{10}$ part of the distance from *a*, since all of these are parts that are commensurable to the whole circle.[175]

---

examples are omitted here):

"Si ergo *ac* sit commensa circulo, erunt omnes portiones e genere numeri: et si potentia rhete erunt omnes, vel potentia rhete vel circulis detractis, ut *ak* et *al* recisa. Et *ac* sit potentia secunda rhete, id est radix cubica, erunt omnes *cd, de, ef,* potentia secunda rhete; et radices cubicae numeri, seu latera corporum rhete, *ak* vero et *al*, et huiusmodi in infinitum recisa potentia rhete.

"Ex hoc patet quod cum circulus possit dividi infinita genera quantitatum quae non sunt invicem commensae, cumque coniunctiones hae semper in eodem genere maneant quod infinita puncta et infinitis in speciebus quantitatum remanebunt in quibus *A* et *B* in perpetuum nunquam convenient. Velut si coniunctio prima fiat in ℞ cu. $\frac{1}{2}$ alicuius circuli, nunquam convenient, neque in medietate, neque in quarta parte, nec octava, nec tertia, nec sexta, nec nona, nec quinta, nec decima, et sic de singulis in genere commensarum toti circulo."—Ibid., p. 41.

The special medical symbol used here and in n. 178, below, is a close approximation to the algebraic symbol used by Cardano and others in which the letters *R* and *x* were joined together to represent the Latin term *radix* ("root").

175. The basis for this paragraph (the Latin text is given in n. 174) can be found in much greater detail and subtlety in the *De commensurabilitate*, pt. II, props. 2 and 4 (see above, pp. 40–41 and 43–47).

Because nothing remotely resembling these propositions is found in the chapters on celestial commensurability by Johannes de Muris, it is almost certain that the *Quadripartitum numerorum* was not the treatise given to Cardano by his father (see below, n. 185 of this chapter). Of all the authors and treatises mentioned in this chapter and known thus far, only Oresme's *De commensurabilitate* could have furnished the basis for Proposition 50 of Cardano's *Opus novum de proportionibus*.

In Proposition 51, Cardano seeks "to declare by example the operations that have been stated."[176] This consists of certain elaborations and clarifications of material in previous propositions. Thus, in citing Proposition 50, Cardano assumes that arc ac is $\sqrt{7}$, which is greater than 1 and therefore cannot be a fractional part of a circle. To make it a fractional part, Cardano assigns an arbitrary value of 10 to the circle so that $\sqrt{7}$ is divided by 10 and arc $ac = \sqrt{7}/10$. To avoid the use of irrationals, he squares this quantity and obtains $\frac{7}{100}$, which is a rational fractional part of the circle smaller than $\sqrt{7}/10$ but in the same geometric series. Similarly, if arc $ac = \sqrt[3]{16}/10$, cubing and reducing it gives $\frac{2}{125}$, which is also a rational fractional part of the circle but one that is smaller than $\sqrt[3]{16}/10$.[177] However, as successive conjunctions occur and the successive fractional parts are summed until they exceed unity, it becomes necessary to subtract the total number of integral circulations traversed by one of the mobiles from the total distance represented by the successive conjunctions. This will yield the position of the last conjunction in relation to the initial point of conjunction in a.[178]

Cardano now shows that mobiles A and B cannot conjunct elsewhere in less time than it took to conjunct in c after departure from their initial conjunction in a. He assumes, for example, that A and B require 10 years to conjunct in c after having departed from conjunction in a. After measuring the distance traversed with reference to point a, he finds that A will traverse 3 circulations plus arc ac prior to conjunction with B in c, while B completes 6 circulations plus arc ac. Since the mobiles move with unequal but uniform velocities, A will complete 3 circulations in 9 years, or 1 circulation in 3 years, plus $\frac{1}{3}$ of a circulation in 1 year, so that arc $ac = \frac{1}{3}$ of a circle. However, B makes $6\frac{1}{3}$ circulations in 10 years. To determine the time of one circulation for B, divide 10 by $6\frac{1}{3}$, which equals $1\frac{11}{19}$ years. Since B completes more circulations than A in the same

176. "Operationes dictas exemplo de- clarare."—*Opus novum de proportionibus*, p. 41.

177. Cardano's objective is unclear to me, but I suspect that we have here a faint and almost unrecognizable reflection of Part II, Proposition 7 of the *De commensurabilitate*, where Oresme divides the circle into different irrational geometric series and shows that three mobiles will conjunct in the same points if, when taken two at a time, the points of one division are also points of the other division or geometric series.

178. "Supponamus in circulo praedicto ac ℞ 7 constat quod esse non potest quia ℞ 7

est maior monade, ideo toto circulo. Quare non poterit esse pars circuli sed referetur ad quantitatem certam, velut quod circulus sit 10. Semper ergo dividemus ℞ 7, seu eam portionem, per 10 quantitatem circuli et exibit $\frac{7}{100}$. Et haec erit portio circuli. Et ita si portio sit ℞ cub. 16; dividemus ℞ cub. 16 per 10 exibit ℞ cub. $\frac{2}{125}$, et ita de aliis.

"Sed cum ex repetitione crescat portio illa donec exuperet monadem aut aliquem quemvis numerum, detracta monade aut numero circuituum habebit rationem recisi."—*Opus novum de proportionibus*, p. 41. For the second paragraph of this quotation, cf. Oresme, *De commensurabilitate*, I.328–32.

time, it has a greater velocity, from which it follows that no conjunction is possible until B has overtaken A; this event cannot occur until B makes at least one complete circulation after departing from any conjunction whatever. The minimum possible time for a conjunction must then be $1\frac{11}{19}$ years.[179]

At this juncture, Cardano offers a negative example of the same kind presented earlier in Proposition 48 (see the example in n. 163). In terms of the specific data presented in this proposition, he wishes to demonstrate that A and B could not conjunct in c in less than 10 years. If such a conjunction occurred in 4 years, B would have had to traverse $2\frac{1}{3}$ circulations to reach c. But if this were true and $1\frac{11}{19}$ years is required for one circulation, B could not reach c in 4 years, but only in 3.68 years (Cardano expresses this as $3\frac{13}{19}$ years). In fact, as Cardano shows, B must traverse $2\frac{8}{15}$ circulations—not $2\frac{1}{3}$—in order to reach c in 4 years (i.e., $4/1\frac{11}{19} = 2\frac{8}{15}$). However, $\frac{8}{15}$ does not equal $\frac{1}{3}$, which is the distance that B must travel beyond a to reach c. It is obvious that 4 years cannot be the time in which A and B conjunct in c.[180]

Now if the conjunctions of A and B are repeated regularly, then however many times they may conjunct elsewhere on the circle, it is necessary that they conjunct again in a immediately before they conjunct again in c. If this were not true, then assume that they conjunct in point e, rather than a, immediately before conjuncting again in c. In Fig. 18, which is reproduced exactly as it

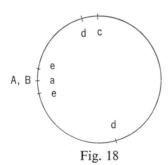

Fig. 18

179. The text summarized in the paragraph above reads: "Quod vero non contingat coniungi in alio loco, neque tempore sit ut A B iungantur in c. Et sit revolutio A triplex integra et B sexcuplex, et tempus totum decem annorum, ita ut ac sit tertia pars circuitus et A circuitus tres anni. Et quia circuitus B sunt sex cum tertia, dividemus decem per $6\frac{1}{3}$ exit $1\frac{11}{19}$. Dico quod non prius, neque in alio puncto."—Ibid., pp. 41–42.

180. The text for the material summarized in this paragraph reads: "Si enim primum in eodem puncto et, gratia exempli, in quatuor annis congruit enim et B dicamus quod peragrit [*corr. ex* peregrit] duas revolutiones cum tertia. Hoc enim est necessarium si debet pervenire ad c. Et erunt anni tres et $\frac{13}{19}$, non ergo anni quatuor. Cum enim tempora diversa dividuntur per numeros habentes proportionem, erunt qui prodeunt numeri in eadem ratione. Diviso ergo 10 per $1\frac{11}{19}$ exit $6\frac{1}{3}$; et diviso 4 per $1\frac{11}{19}$ exit $2\frac{8}{15}$, igitur $6\frac{1}{3}$ ad $2\frac{8}{15}$ ut 10 ad 4, igitur $\frac{8}{15}$ non potest esse aequale $\frac{1}{3}$."—Ibid., p. 42.

appears in the text, *e* is marked twice, since it may fall within or without arc *ac*. Since the same argument would apply in either situation, Cardano considers only the case where *e* falls inside arc *ac*. Should this happen, arc *ec* would equal arc *ac*, because it is assumed that the distance separating any two successive conjunctions is always equal (see the summary of Proposition 50 above, pp. 151–53). But this signifies that the part, arc *ec*, would equal the whole, arc *ac*, which is impossible.[181]

Another counterargument is now introduced against this position. As before, he assumes that immediately before they conjunct again in *c*, *A* and *B* will not conjunct in *a*, but in some other point, this time *d*, which, as with *e* before, might fall within or without arc *ac*. He considers only the case where *d* falls within arc *ac* (see Fig. 18; if *d* fell outside arc *ac*, the same argument would apply *mutatis mutandis*). By Proposition 50, then, arc *dc* must be a submultiple part of arc *ac*, so that arc *dc* : arc *ac* is expressible as a ratio of prime numbers. But, says Cardano, it was not assumed that arcs *dc* and *ac* are related as prime numbers.[182] The remainder of the proposition is devoted to a trivial exemplification of this point.[183] Thus, in the example, *A* and *B* were assumed to meet in *c* in 10 years and arc *ac* was said to be $\frac{1}{3}$ of the circle. After departing initially from *a* and conjuncting in *c* in 10 years, they would meet again in *a* in 30 years and in *c* in 40 years. If anyone had assumed 40 years as the time of conjunction

---

181. Here is the text on which the discussion in this paragraph is based: "Si enim per praecedentem repetuntur, ergo non possunt redire donec iterum coniungantur in ipso *a*. Si enim aliter sit ut ex *e*, igitur *ec* est aequalis *ac*, pars toti, quod contingere non potest."—Ibid., p. 42.

182. "Sin vero coniunctio fiat in *d*, igitur per praecedentem *dc* est pars *ac* submultiplex quomodolibet, quare non fuerunt assumpti primi numeri."—Ibid.

One again Cardano makes explicit what Oresme clearly implies but did not deem worthy of elaboration. In Part I, Proposition 10 of the *De commensurabilitate*, Oresme notes that where velocities are commensurable, the points of conjunction are distributed at equidistant intervals around the circle. However, the order of conjunctions may be such that points are omitted in regular sequence. When this occurs, any conjunction located between two points in which successive conjunctions have occurred must

produce relationships between the parts of the circle involving the intermediate point and the two points where successive conjunctions have occurred that are expressible as prime numbers, or reducible to prime numbers. See above, pp. 16–18.

183. Here is the text of the example: "Veluti in exemplo constituimus quod *A* et *B* conveniunt in *c* in decem annis et *ac* est tertia pars circuitus, ergo in triginta annis conveniunt in *a* et in quadraginta rursus in *c*. Si ergo quis assumpsisset quadraginta annos ab initio pro congressu et divisisset per $1\frac{11}{19}$ exiret $25\frac{1}{3}$; et si per 3 exiret $13\frac{1}{3}$ et manifestum est quod uterque numerus potest dividi per eundem numerum, utpote 4, et exit numerus cum eadem parte, scilicet $6\frac{1}{3}$ et $3\frac{1}{3}$. Ergo convenient ante, non ergo assumpsisti minimos in ea proportione. Illi autem nequaquam amplius dividi non possunt eodem modo."—*Opus novum de proportionibus*, p. 42.

in c, he would have divided 40 years by $1\frac{11}{19}$ years (the time it takes B to make one circulation) and obtained $25\frac{1}{3}$ circulations for B in the course of the 40 years; and if he divided by 3 years, he would have obtained $13\frac{1}{3}$ circulations for A. But it is evident that these numbers, namely, $25\frac{1}{3}$ and $13\frac{1}{3}$, are both divisible by the same number, 4, and yield quotients of $6\frac{1}{7}$ and $3\frac{1}{3}$, respectively. Therefore, the person who posited 40 years before conjunction in c did not assume, as he should have, that $25\frac{1}{3}$ and $13\frac{1}{3}$ are prime to each other. However, $6\frac{1}{7}$ and $3\frac{1}{3}$ are not further divisible in such a manner as to have the same part, from which Cardano infers that conjunction occurs in c before 40 years—namely, in 10 years, and not before.

The final proposition of the sequence, Proposition 52, is devoted exclusively to incommensurable relationships; its brevity permits citation in full.[184]

Three mobiles in conjunction in the same point will never meet in any point if they meet two at a time in parts that are incommensurable to each other.

Let A, B, and C be in conjunction, and let A and B conjunct again first in d and B and C first in e; and let arcs ad and ae be incommensurable (see Fig. 19). I say that

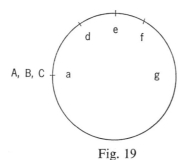

Fig. 19

A, B, and C will never meet in any point—not in the first point [in which they conjuncted] and not in any other. If [this is] not [true], let them meet in f. Therefore, in time g they will have made an integral number of revolutions and, in addition, a part of a revolution, arc af. Now since this [total distance, which equals an integral number of revolutions plus arc af,] consists of [a certain number of successive] conjunctions of

184. "Tria mobilia coniuncta in eodem puncto quorum duo et duo conveniant in partibus incommensis inter se in perpetuum in nullo unquam puncto convenient.

"Sint A, B, C iuncta; et primo iungantur A et B iterum in d, et B et C in e; et sint ad, ae incommensae. Dico quod A, B, C nunquam convenient in aliquo puncto, seu primo seu alio a primo. Si non, conveniant in f. Erunt ergo in g tempore revolutiones integrae et portio af insuper. Et quia hae constituuntur per congressus B cum A et sunt spatia ad; et B cum C et sunt spatia ef, igitur spatium af erit ex genere quantitatis ad et ae, per quinquagesimam. Harum ergo erunt commensae, quod est contra suppositum."—Ibid., pp. 42–43.

B with *A* [each of which is] equal to arc *ad*, and of [a certain number of successive] conjunctions of *B* with *C* [each of which is] equal to arc *ae*, then, by the fiftieth proposition, arc *af* will be constituted of the same kind of quantity as are arcs *ad* and *ae*. Therefore, these [three] quantities will be [mutually] commensurable, which is contrary to the assumption [that they are incommensurable]."[185]

Cardano concludes Proposition 52 and the six propositions devoted to the kinematics of commensurable and incommensurable circular motions with an admission that "the basis (*principium*) for these propositions was furnished by Campanus of Novara, the expositor [or interpreter] of Euclid, in a certain little unedited [or unpublished?] book which came to me through the diligence of my father, Fazio."[186] If this is true, Cardano's statement would be of major significance, since such a treatise by Campanus might also have served as a basic source for both Johannes de Muris and Oresme. But, as yet, I am not aware of any complete or even partial treatise attributed, rightly or wrongly, to Campanus of Novara where the commensurability and incommensurability of circular or celestial motions was formally or informally treated.[187] Under these circumstances, we must, for the present, attempt to identify the most likely source of Cardano's propositions from among the extant and known works where this subject matter was treated. As candidates, we have only the works of Theodosius of Bithynia, Johannes de Muris, and Nicole Oresme. Theodosius, however, cannot be considered seriously, since the five propositions in the *De diebus et noctibus* (summarized above) are not at all like Cardano's propositions,

185. Despite some obvious differences, the basis for this proposition, and much more, is found in Oresme's *De commensurabilitate*, Part II, Propositions 7 and 8. See above, pp. 48–52. Nothing like it appears in the chapters from Johannes de Muris's *Quadripartitum numerorum*, cited below in Appendix 1, thus reinforcing the conjecture that de Muris did not influence Cardano, whereas Oresme may well be the source of these propositions (see also n. 175 of this chapter).

186. "Et harum propositionum principium est traditum a Campano Novariensi, Euclidis expositore, in quodam libello non edito qui diligentia patris mei Facii ad me pervenit."—*Opus novum de proportionibus*, p. 43.

187. Since it is obvious that Oresme's *De commensurabilitate* was more likely to have influenced Cardano than Johannes de Muris's *Quadripartitum* (see above, nn. 175,

185; and below, this paragraph), perhaps Cardano had a manuscript codex containing an anonymous copy of Oresme's *De commensurabilitate* that followed immediately after a treatise ascribed explicitly to Campanus. If so, Cardano may then have assumed that the anonymous copy of the *De commensurabilitate* was also by Campanus. Confusions of this kind were not uncommon in the Middle Ages. Or, perhaps Cardano received from his father a manuscript of the *De commensurabilitate* that had been wrongly attributed to Campanus, just as MS *P* in this edition of the *De commensurabilitate* was mistakenly ascribed to Jordanus de Nemore (see below, p. 168). If, however, Campanus did actually compose such a treatise, it may turn up in the future, whereupon the sources of de Muris, Oresme, and Cardano will have to be reexamined and reevaluated.

and even if they were the treatise seems not to have been available in Latin translation prior to 1591 (see above, n. 14 of this chapter). And despite a more than casual resemblance between the chapters of Johannes de Muris and the propositions of Cardano, the conclusions based on incommensurable velocities by Cardano could not have been derived from the *Quadripartitum* (see above, nn 175 and 185 of this chapter). This leaves only Oresme's *De commensurabilitate*, which contains all the substantive ideas and consequences of Cardano's propositions;[188] although specific evidence is lacking to demonstrate convincingly Cardano's dependence on Oresme, the striking agreement between them lends reasonable credence to the conjecture that Oresme—not Campanus, at least not until we have a candidate for Cardano's reference (see also n. 187)— was the source of Cardano's propositions.

Before concluding this extended discussion of Cardano's six relevant propositions, we must ask why they were included at all in the *Opus novum de proportionibus*. Their presence is probably owing to the fact that, in this mélange of 233 propositions,[189] there are many devoted to traditional problems of rectilinear and circular motion. Since Cardano was already aware of the treatise which he ascribes to Campanus, perhaps he decided to reformulate all or some of those propositions in his own language and style and include them as representative problems of bodies moving in perfect circular and uniform motion. They were obviously suitable as additional problems on circular motion, of which there are a goodly number scattered throughout the treatise, since they were illustrative of the application of mathematical proportionality to physical problems—and this was one of the major objectives of the book.[190] But if it

188. Since this cannot be said of Oresme's *Ad pauca respicientes*, which was written even earlier than the *De commensurabilitate*, it is a much less plausible candidate to represent Oresme's possible influence on Cardano. For this reason, it has not been discussed in this section.

189. The enunciations of all 233 are listed immediately after the preface in a "Table of Propositions" (*Tabula propositionum*) extending over a series of unnumbered pages. Even a cursory glance reveals that the propositions are very varied and seemingly in no particular order or sequence.

190. Blasius of Parma (d. 1416), on a much-reduced scale and for a somewhat different reason than Cardano, also found occasion to employ the incommensurability of linear and circular motions in a lengthy

discussion of the conditions required for a ratio to obtain between two quantities. In his *Questions on the Treatise of Proportions of Thomas Bradwardine* (*Questiones super tractatu de proportionibus magistri Thome Berduerdini*; while this title accurately reflects the purpose of the treatise, it is a composite formed from terms in the incipit and explicit of MS Vat. lat. 3012, fols. 137r–164r), Blasius argues:

"There is some local rectilinear motion to which [another] rectilinear motion cannot be proportioned. Let two distances be taken that are related to each other as the diagonal and the side of a square; and I assume that two mobiles are moved over these distances with motions [or velocities] that are proportional to these distances. Then this is proved: just as a distance is to a distance in length, so

was Oresme's *De commensurabilitate* which Cardano actually had in his possession, then it can be said that he selected for inclusion only the more elementary and less interesting propositions, and even these were rendered with far less clarity and concision than in their original Oresmian versions.

(Note 190 continued)

is a motion to a motion in velocity in the case premissed here [i.e., $S_2:S_1 = V_2:V_1$, where $T_2 = T_1$ and S, V, and T represent distance, velocity, and time, in that order]. But in this case, the distances are not proportional in a rational ratio; therefore, neither are the motions in velocity comparable by a rational ratio. But another conclusion follows from this: not every circular motion is comparable to another circular motion. This is obvious because if the circumferences of two circles are mutually related as the diagonal and side of the same square and [if] two mobiles are moved with two proportional motions over these distances, then what was proposed above applies also in this case. From these [two] conclusions, there comes this conclusion: on the assumption of the eternity of motions of two bodies now in conjunction, it is possible that they will not conjunct again. Let these things be declared, since two planets (*astra*) now in conjunction on different circles can be taken. Now they begin to be moved and are perpetually moved with motions related mutually as the diagonal and side of the same square. What has been proposed is now obvious, for if they were conjuncted in some time [or other], then they would be conjuncted. But as a time is to a time, so is a distance to a distance, and a motion [or velocity] to a motion [or velocity]; but since [in this example] the time is equal to the time, it might be concluded from this that these motions would be comparable. Where indeed you maintain that the motion is the mobile itself, you could say that these two local motions are not mutually comparable, neither according to the common notion of proportion nor according to another. Let this be declared: that a man and an ass are taken and moved through an hour traversing a dis-

tance of one foot. But a man and an ass are not mutually comparable according to any ratio; but a man and an ass made these two motions, therefore, etc. [the two motions are not comparable]."

("Et hec erit quinta conclusio in ordine. Aliquis est motus localis rectus cui motui recto rectus non proportionatur, declaratur. Capiantur duo spacia se habentia invicem ut dyameter et costa eiusdem quadrati. Et volo quod duo mobilia moveantur super hec motibus proportionabilibus spaciis istis. Tunc probatur sic: sicut spacium ad spacium in longitudine ita motus ad motum in velocitate casu premisso. Sed ex casu ista spacia non sunt proportionalia proportione rationali; ergo nec motus in velocitate sunt comparabiles tali proportione. Sed ex hoc sequitur alia conclusio: non omnis motus circularis est comparabilis alteri motui circulari. Patet quia capiantur due circumferentie duorum circulorum se habentes invicem ut diameter et costa eiusdem quadrati et moveantur duo mobilia duobus motibus proportionalibus istis spaciis et tunc in isto casu patet propositum ut supra. Ex istis conclusionibus stat ista conclusio: possibile est quod aliqua duo corpora nunc esse coniuncta que supposita eternitate motuum ipsorum et non coniungentur de cetero. Declarantur, quia capiatur duo astra in diversis circulis que nunc sunt coniuncta et incipiant moveri et sic perpetue moveantur motibus se invicem habentibus ut dyameter et costa eiusdem quadrati. Et tunc patet propositum quia si coniungerentur in aliquo tempore, coniungerentur. Sed sicut tempus ad tempus, ita spacium ad spacium, et motus ad motum [*corr. ex* motus] et cum tempus et sit equale tempori quia sibi ipsi concluderetur quod isti motus essent comparabiles. Ubi vero tu tenens quod motus est ipsum mobile tu posses dicere quod hii duo motus

Since no others who dealt with this subject are known to me, I terminate this summary of ideas and materials relevant to the development and history of the concept of the commensurability and incommensurability of circular and celestial motions. One incontrovertible conclusion emerges. On the basis of the evidence presented here, Oresme's works easily stand as the most significant contributions to this subject. Despite direct antecedents, his treatment of the subject seems to have been original and a few later authors who had occasion to mention celestial incommensurability have cited Oresme as their authority (e.g., Henry of Hesse, Pierre d'Ailly, who failed to mention him but plagiarized his work, Jean Gerson, and John de Fundis). It is plausible to expect that in the future other treatises, or parts of treatises, will be found in which these problems are considered. But in light of the evidence presently available, it is a subject that, in its most highly developed and original expression, justly deserves to be associated with the name of Nicole Oresme.

---

locales non sunt comparabiles invicem, nec secundum proportionem communiter dictum nec secundum aliam. Declaratur: capiantur homo et asinus et moveantur per horam pertranseunto spacium pedale. Tunc sic homo et asinus non sunt invicem comparabiles secundum aliquam proportionem; et homo et asinus sunt isti duo motus, ergo, etc."— Vat. lat. 3012, fol. 138v, col. 1.)

Blasius reveals a lack of understanding (e.g., he declares that the two bodies in motion on the incommensurable circles will never conjunct, instead of restricting his claim to points in which they have already conjuncted; Oresme demonstrates this in Part II, Proposition 1 of the *De commensurabilitate*). And his chief interest in the incommensurability of circular motions lies only in whether or not they are really comparable, and even this concern forms but a small part of the larger question of the conditions under which ratios can be formed. The passage is of interest only because it reveals another instance in which the incommensurability of circular motions is mentioned and utilized.

4

Sigla and Descriptions
of Manuscripts

The seven manuscripts described below are those that I used in establishing
the text for the present edition of the *De commensurabilitate*. Their relationships have already been discussed (see above, under "Editorial Procedures"),
and their sigla are listed here in alphabetical order according to the abbreviation that I have assigned to each manuscript.

1. *A* = Vatican, Biblioteca Vaticana, Latin MS 4082, fols. 97v–108v.[1]
   Date: November 11, 1401.[2]
   This codex contains twenty-two treatises which are listed by B. Boncompagni, "Catalogo de' lavori di Andalo di Negro" in *Bullettino di Bibliografia
e di storia delle scienze matematiche e fisiche*, vol. 7 (Rome, 1874), p. 366, n. 1.[3]
The *De commensurabilitate*, which is the eighth treatise, is neatly written in
double columns with the enunciations of the propositions of Parts I and II
set off in large letters. A total of eighteen figures, some incomplete, are scattered through the margins of the first two parts of the treatise extending over
folios 98v–105r. Oresme's *Algorismus proportionum* occupies folios 109r–113v
following immediately after the *De commensurabilitate* and, like the latter,
also copied by Peter de Fita, but some months earlier, on March 8, 1401.

2. *B* = Paris, Bibliothèque Nationale, fonds latin, 7281, fols. 259r–273r.[4]
   Date: 15th century (perhaps in, or after, 1420).

---

1. Titles and colophons for this and all
other manuscripts can be found in the variant readings on pp. 172 and 322, respectively.

2. This exact date was furnished in the
colophon (see variants, p. 322) by the scribe,
Peter de Fita.

3. Boncompagni derived his list from the
unpublished longhand Vatican *Inventarium*.
Additional information on Vat. lat. 4082 is
furnished by Lynn Thorndike in "Notes

upon Some Medieval Latin Astronomical,
Astrological and Mathematical Manuscripts
at the Vatican," *Isis*, vol. 47 (1956), pp. 397–
401.

4. V. P. Zoubov's Russian translation of
the *De commensurabilitate* cited above, p. xix,
was based almost wholly on MS *B* (he also
lists MSS *A*, *F*, *L*, and *U*, but makes no
mention of *P* and *R*).

Nineteen treatises are listed for this codex (of 280 folios) in the *Catalogus codicum manuscriptorum Bibliothecae Regiae*, pt. 3, vol. 4 (Paris, 1744), p. 334, col. 1, where it is dated as fifteenth century. This date is probably derived from at least one of two statements written in the same hand, the first of which occurs at the top of folio 235r and reads "Credo quod compositor huius libri fuit magister [at this point a name, perhaps that of Joh. de Troyes[5] is crossed out] quia hoc tempore, 1420, fuit Parisius ?volentissimus," while the other appears at the end of a treatise, on folio 241r, and says "anno Christi domini nostri Jesu 1420." The treatise with which these remarks are associated is an anonymous astronomical work that appears to have been composed in, or slightly later than, 1420,[6] so that it would have been copied in, or after, 1420. The scribe who wrote these remarks also appears to have copied the *De commensurabilitate* (the resemblance of the hands seems close, especially that between the *De commensurabilitate* and folio 241r, where the second of the two statements quoted above is found). If so, the likelihood is great that the *De commensurabilitate* was copied in the fifteenth century, sometime in, or after, 1420. But if they are not by the same scribe, the time when the *De commensurabilitate* was copied could range anywhere from the latter part of the fourteenth century to sometime in the fifteenth century.

The contents of BN 7281 are exclusively astronomical.[7] A scribe other than the copier of the anonymous astronomical work and the *De commensurabilitate* seems to have copied everything up to and including folio 232r. This particular copy of the *De commensurabilitate* is not always readily legible. Furthermore, the scribe omitted the accompanying letters from most of the twelve figures that are found in the text and margins. For the most part, the figures themselves are incomplete, consisting of two or three concentric circles.

Immediately after the end of the *De commensurabilitate*, and in the same

5. Thorndike and Kibre, rightly it seems, suggest this name in their *A Catalogue of Incipits of Mediaeval Scientific Writings in Latin*, rev. ed. (Cambridge, Mass., 1963), col. 200, under the incipit: "Centrum est cuius ab initio circumductio...," where they give the composition date as ca. 1420 and list only this manuscript (BN 7281, fols. 235r–241r).

6. This is further reinforced by a few statements on folio 238r that concern calculations of the motion of the eighth sphere made for the period 1420–1520. See also n. 5, above.

7. Perhaps because of this, the *De commensurabilitate* represents Oresme's only work. In their respective codices, all other manuscripts, including *R*, are either immediately preceded (*R* and *U*) or followed (*A*, *F*, *L*, and *P*) by complete versions of Oresme's *Algorismus proportionum*. Basic differences between *B* and the other five complete manuscripts are described above on pp. xv–xvii.

hand, is a postscript that may have been added by the scribe of *B* or sometime earlier by another scribe who made the version of the *De commensura-*
bilitate on which *B* was ultimately based. I deem it improbable that this
postscript was added by Oresme, since it is lacking in all other manuscripts
used in this edition. Here is the full text of it from folio 273r:

Cum igitur scriberem hec ad memoriam rediit id quod dicit Abraham Avenesre
in libro *De seculo* loquens de firidariis quod revertuntur secundum circulationem
in quibuslibet 75 annis. Sic dicit: "Si quis dicat ob hoc deberent esse similes quilibet 75 anni precedentibus cum planete et sue participationes sunt similes et uniformes, responsio hoc est: scito quod hoc esse non potest secundum viam proportionis quod inveniatur unum ascendens cum habitudine proportionali ad ipsum, et
sit proportio unius ad alterum uniformis seu equalis in perpetuum et si mundus
semper duraret. Et hanc quidem rem perpendere potes. Saturnus multas habet
diversitates tam ex parte solis quam ex parte planetarum superiorum que moventur quibuslibet 70 annis per 1 grada quo circa proportionem non habebit quam
prius habuit ad superiorem unam usque 25,200 annos. Et non est necesse in hoc
protelare sermone et cetera."

Deinde dicit: "Quapropter esse non potest nativitas hominis que assimiletur
nativitati alterius tamquam sibi, non enim est orbis stans secundum unum modum
nec umquam erit punctus hore quin revertatur proportio quod non fuit sicut illa
nec erit. Et sapientes arismetici hoc noverunt."

Non plus dicit de hoc, sed eius translator de arabico in latinum, Henricus Bate,
magnus in quadrivio, dicit consequenter: "Quamquam multiplicatio numeri crescere possit in infinitum revolutiones tamen celestium corporum necesse est finitas
esse secundum speram, quemadmodum in alia parte philosophie demonstrari habet cum certitudine. Quapropter necessarium est consimiles interdum redire constellationes licet incomprehensibile sit nobis tempus huius revolutionis propter
intervallorum immensitatem et hoc forsan est quod iste auctor [*corr. ex* actor]
innuit. Non est autem opinandum quod propter multiplicem diversitatem motuum
corporum divinorum possibile sit ipsos in revolutionibus quibuslibet non convenire seu non communicare quemadmodum est de lineis incommunicantibus quas
decimo elementorum Euclidis vocat irrationales seu surdas propter impotentiam
communicandi. Omnia namque coordinata sunt ut testatur Aristoteles duodecimo
Metaphysice, super quo dicit Commentator quod omnes actiones corporum superiorum in communicatione eorum adinvicem in constitutione mundi sunt sicut
actio liberorum [*corr. ex* lignorum; *see above, chap. 3, n. 84, where this passage is*
quoted by Henry Bate] in constitutione domus. Palam autem est esse modicum
consideranti circa hoc quod si inter aliqua debet esse ordo vel ?communicatio
excellenter et proprie debet esse in divinis. Quare absurdum est opinari motus
corporum superiorum irrationales et incommunicantes et surdos et hoc est quod
Pyctagoras et alii antiqui per mundanam musicam innuere voluerunt. De qua similiter Plato in *Thymeo* et alibi loquitur necnon et Calcidius cum aliis philosophis
infinitis." Non plus de hoc, sed credo quod de hac materia ipse plus tractet in suo

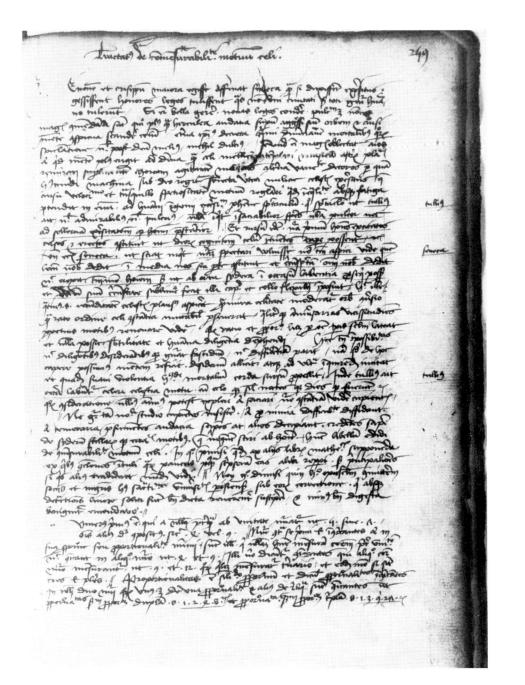

Plate 1. MS *B* = Paris, Bibliothèque Nationale, fonds latin, 7281, fol. 259r (*De commensurabilitate*, title and beginning of treatise).

Plate 2. MS *F* = Paris, Bibliothèque de l'Arsenal, 522, fol. 114v (*De commensurabilitate*, Part I, extending over Propositions 20 to 22).

Plate 3. MS *U* = Utrecht, Universiteits Bibliotheek, 725, fol. 185r (*De commensurabilitate*, conclusion of Part I).

Plate 4. MS *A* = Vatican, Biblioteca Vaticana, Latin MS 4082, fol. 103r (*De commensurabilitate*, Part II, extending over Propositions 1 to 4).

Plate 5. MS *P* = Cambridge, Magdalene College, Bibliotheca Pepysiana 2329, fol. 122r (*De commensurabilitate*, Part II, extending over Propositions 9 to 11).

Plate 6. MS *L* = Florence, Biblioteca Medicea Laurenziana, Ashburnham 210, fol. 168v (*De commensurabilitate*, end of Part II and beginning of Part III).

Plate 7. MS *R* = Vatican, Biblioteca Vaticana, Latin MS 4275, fol. 99v (*De commensurabilitate*, the conclusion of Arithmetic's oration and the beginning of Geometry's).

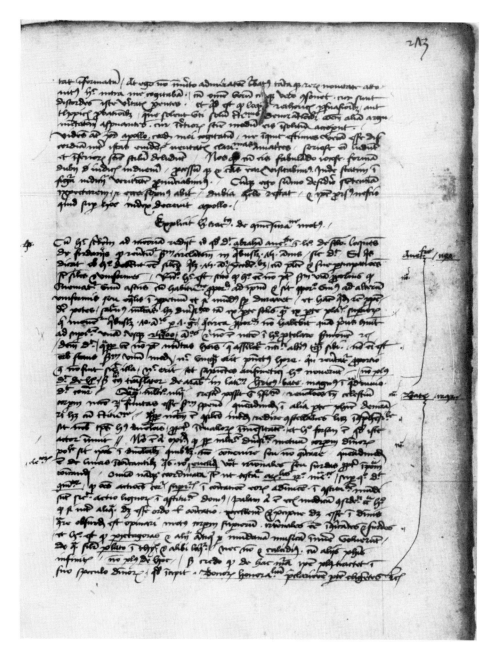

Plate 8. MS *B* = Paris, Bibliothèque Nationale, fonds latin, 7281, fol. 273r (*De commensurabilitate*, conclusion of Part III and postscript).

speculo divinorum quod incipit: "Bonorum honorabilium preclariorem partem eligentes, et cetera."[8]

3. *F* = Paris, Bibliothèque de l'Arsenal, 522, fols. 110r–121r, col. 1.

Date: Perhaps between 1395 and 1398.

The incipits and explicits of the thirteen treatises of this codex of 187 folios in double columns are cited by Henry Martin, *Catalogue des manuscrits de la Bibliothèque de l'Arsenal* (Paris, 1885), vol. 1, pp. 370–72. Although most of the works in this codex are attributed to Oresme, only the *De commensurabilitate, Algorismus proportionum, De configurationibus qualitatum,* and *Tractatus de communicaccione ydiomatum* are genuine.

According to a statement on folio 187v, this codex was in the possession of the library of the College of Navarre,[9] where Oresme studied theology (see above, p. 3). All the works in it were copied by a single scribe, Johannes Monachus, whose name appears in the colophons of three of the treatises. In one of these colophons, that which follows the *Liber de anima* (fols. 33r–56v) of Pierre d'Ailly, the latter is called Bishop of Puy en Velay (fol. 56v),[10] an office he held between 1395 and 1398 before becoming bishop of Cambrai in 1398. If Johannes Monachus formulated the entire colophon

---

8. The authorship of this passage and its bearing on the *De commensurabilitate* are discussed on pp. xvi–xvii. The Latin quotations from Henry Bate's translation of Abraham ibn Ezra's *Liber de mundo vel seculo* are repeated in Chapter 3, notes 82–84, where corrections have been made from MS *B*; for translations of these passages, see pp. 112–14; for a full discussion of Bate's translation of ibn Ezra's work, see chap. 3, n. 80.

If, as we are told in the closing lines of the postscript, Henry Bate debated these matters further in his very lengthy *Speculum divinorum*, such a discussion does not appear in the first two parts of it (there are 23 in all) which have been published in G. Wallerand, ed., *Henri Bate de Malines Speculum divinorum et quorundam naturalium*, Étude critique et texte inédit, fasc. 1, pts. 1–2, *Les Philosophes belges*, vol. 11 (Louvain, 1931). (On p. 16, n. 17, Wallerand quotes almost all of what remains of this postscript after the quotations from Abraham ibn Ezra and Henry Bate have been eliminated.) Part 22, which treats of the celestial motions, is the most likely place for such a discussion. For

the contents of this unpublished part, see vol. 1 of the new edition of Bate's *Speculum divinorum et quorundam naturalium* by E. Van de Vyver in *Philosophes médiévaux*, vol. 4 (Louvain and Paris, 1960), pp. 42–43. This volume contains only the Proemium and first part, although in 1953, Van de Vyver edited Part 6 for his doctoral dissertation at the University of Louvain.

9. "Iste liber est librarie parve Artistarum in regali collegio Navarre." It is also quoted in Martin, *Catalogue des manuscrits*, p. 372.

10. "Explicit liber de anima inchoatus et compilatus per reverendum in Christo patrem magistrum Petrum de Ally, sacre theologie professorem, Anyciensem episcopum. Suessionensis diocesis oriundum, scriptus per manum Johannis Monachi, ejusdem diocesis oriundi. Deo gracias."— Ibid., p. 371. Drawing his information from Salembier, *Petrus ab Alliaco* (1886), pp. xiii, 146, Thorndike reports that the *De anima* was composed by d'Ailly in 1372, while a student at Paris (*History of Magic*, vol. 3, p. 746).

himself—that is, if he did not merely copy from his exemplar the statement about d'Ailly's episcopal office in Puy—then it seems plausible to assume that he did indeed copy this manuscript while d'Ailly was bishop of Puy. The basis of this assumption is Johannes Monachus's biographical interjection that he himself originated from the same diocese in which d'Ailly was bishop. Even if Johannes no longer lived in this diocese—indeed, by his own statement he lived in Paris during the time he copied at least one of the works in the codex[11]—it seems reasonable to suppose that he would have known the name of the incumbent bishop of his natal diocese at the time he composed the colophon. Since Johannes also copied the *De commensurabilitate*, it too would have been copied sometime around 1398. If, however, Johannes was merely copying the information about d'Ailly from his exemplary copy, then the period 1395–98 may only serve as a *terminus post quem* and the codex may even have been copied long after that, but surely no later than the fifteenth century.

In describing the separate works of Arsenal 522, Martin not only failed to identify the *De commensurabilitate* as a separate work, but conflated it with Part 1 of the *Algorismus proportionum* as the tenth work in the codex under the latter title. As an aid in explaining Martin's confusions and errors, I quote his description of the tenth and eleventh treatises in Arsenal 522. "10°.—Fol. 110. Nicolas Oresme: 'Algorismus proporcionum.'—Commencement: 'Zenonem et Crisippum majora egisse affirmat Seneca quam si duxissent exercitus, gessissent honores, leges tulissent....'—Fin: '...determinatum est sufficienter. Explicit algorismus proporcionum.' 11°.—Fol. 122v°. Proportions géométriques.—Commencement: 'Est autem istarum regularum de algorismo proporcionum....'"[12]

The root of the difficulty lies in the complete anonymity of the *De commensurabilitate* in this codex. It has neither title nor author on its opening page and it lacks a colophon at the end. While it was obvious to Martin

11. On folio 168v, in the colophon to the *Tractatus de proportionibus velocitatum in motibus*, mistakenly attributed to Oresme, Johannes Monachus informs us that he copied this treatise at Paris ("...compilatus per magistrum egregium Nicolaum Oresme, scriptus Parisius per manum Johannis Monachi, Suessionensis diocesis, scriptus in vigilia sancti Pauli." Quoted also in Martin, *Catalogue des manuscrits*, p. 372). In 1411, a "Johannes Monachus" was listed among the arts faculty at the College of Navarre and since Arsenal 522 was once in possession of the library of that college, we probably have here the very scribe who copied the codex under discussion. See James F. McCue, ed., "The Treatise *De Proporcionibus velocitatum in motibus* attributed to Nicholas Oresme" (Ph.D. dissertation, University of Wisconsin, 1961), p. xxiii; on pp. iv–v, McCue shows conclusively that it was not Oresme, but Symon de Castello who composed the *Tractatus de proportionibus velocitatum in motibus*.

12. Martin, *Catalogue des manuscrits*, p. 372.

that a distinct and separate treatise had begun on folio 110r, he could not determine by simple inspection the actual point at which it concluded because the prologue of the *Algorismus proportionum* begins immediately after, as if it were but the next paragraph in a work that is continuous from folio 110r. Indeed there is no break until folio 122v, where we read "... determinatum est sufficienter. Explicit algorismus proporcionum," which is the explicit of Part 1 of the *Algorismus proportionum*. Martin, therefore, made the understandable error of concluding that folios 110r–122v constituted a single work bearing the title *Algorismus proportionum*. From this error, he fell quite naturally into another by interpreting as an independent treatise—he lists it as the eleventh—what is only Part 2 of the *Algorismus proportionum*, namely, what he calls "proportions géométriques" with the opening line "Est autem istarum regularum de algorismo proporcionum...." This second part of Oresme's *Algorismus* terminates on folio 123r, where it is immediately followed by Part 3 of that same treatise. The three-parted *Algorismus proportionum* concludes, finally, on folio 126r.[13]

Throughout its length, MS *F* is easily readable and carefully written. It contains many decorative initial letters, quite unusual for a work of this kind. Although many blank spaces were left for diagrams and figures, none were inserted.

4. *L* = Florence, Biblioteca Medicea Laurenziana, Ashburnham 210, fols. 159r–172r, col. 1.

Date: Early 15th century.[14]

This codex of 184 folios in double columns contains thirteen works that are listed with incipits and explicits by C. Paoli, *I Codici Ashburnhamiani della R. Biblioteca Mediceo-Laurenziana di Firenze*, vol. 1, fasc. 3 (Rome, 1891), pp. 225–27.[15] All but the last work on folios 178r–183r were copied

---

13. For an edition of all three parts, see Maximilian Curtze, ed., *Der "Algorismus proportionum" des Nicolaus Oresme* (Berlin, 1868). For an edition which includes the Prologue (this is lacking in Curtze's edition) and Part 1, see Edward Grant, "The Mathematical Theory of Proportionality of Nicole Oresme" (Ph.D. dissertation, University of Wisconsin, 1957), pp. 331–39. Both Prologue and Part 1 are translated by E. Grant in "Part I of Nicole Oresme's *Algorismus proportionum*," *Isis*, vol. 56 (1965), pp. 327–41.

14. This approximate date is based on a statement by the scribe following immediately after Oresme's *De configurationibus*

qualitatum, on fol. 129v, col. 1: "Et sic est finis huius tractatus quem... [*linea scr. et del.*] scribi feci Parisius anno Domini 1410." Since the same scribe also copied the *De commensurabilitate*, it too was probably completed sometime around 1410. On the last folio of the codex, 184r, this statement appears: "1473. Iste liber est mei Bernardini Cepolle de Verona."

15. This is Volume 8 in the series Ministero della Pubblica Istruzione, *Indici e cataloghi*. For an earlier description of Ashburnham 210 see L. V. Delisle, "Notice sur des manuscrits du fonds Libri conservés à la Laurentienne à Florence," *Notices et extraits des*

by the same scribe, and all but two of the treatises are either by Oresme or
Henry of Hesse. The *De commensurabilitate* is clearly and neatly written
with many initial letters, some of which are quite ornate. The enunciations
of the propositions in the first two parts are written in large, bold letters
with each enunciation commencing with a large initial letter. Twelve complete figures with letters are drawn in the margins of scattered folios of the
first two parts of the work. Oresme's *Algorismus proportionum* follows immediately after on folios 172r, col. 1–177v.

5. *P* = Cambridge, Magdalene College, Bibliotheca Pepysiana 2329, fols. 111v–
   128r.

   Date: Early 15th century.[16]

   A complete and thorough description of this codex is given by M. R.
   James, *A Descriptive Catalogue of the Manuscripts of the Library of Peterhouse* (Cambridge, 1899), pp. 353–56.[17] The *De commensurabilitate*, beautifully written in double columns, includes many initial letters but no diagrams, although blank spaces were left for their eventual insertion. This
   version is incomplete with everything in the text before "refert" in I.62
   omitted, although folio 111r was left blank for possible later inclusion of this
   material. Of the manuscripts used in this edition, only MS *P* attributes the
   De commensurabilitate to someone other than Oresme. In the colophon (see
   the variants for III.481), Jordanus de Nemore is mistakenly cited as the

---

(Note 15 continued)
*manuscrits de la Bibliothèque Nationale et
autres bibliothèques*, vol. 32, pt. 1 (Paris,
1886), pp. 49–51.

16. On folio 45r, in the colophon to the
Arithmetica of Jordanus de Nemore, the
scribe gives his name and the date of completion: "Explicit...totus liber de elementis
Arismetrice artis magistri Jordani de Nemore
scripta Parisius per manus Servatii Tomlinger de Bavaria anno domini millesimo
quadricentesimo septimo octava die post
festum penthecoustes finitus liber est iste."
Thus we learn that Servatius Tomlinger of
Bavaria, the scribe, completed the copying
of the *Arithmetica* in 1407. Since it is obvious
by inspection that Servatius also copied the
De commensurabilitate, it too was probably
completed sometime around 1407. (At the
termination of Oresme's *De proportionibus
proportionum* on folio 110v, and a few lines
before the beginning of the *De commensura-*

bilitate, Servatius asserts that he is the scribe
of the *De proportionibus* ["...Scriptus per
me Servatium"]; see Grant, *Oresme PPAP*,
pp. 127–28.)

17. Because this Magdalene College manuscript formerly belonged to Peterhouse
College, it is described in James's Peterhouse
catalogue, where it has the number 277. "It
was given to Peterhouse in 1472, seen there
by Leland, borrowed in 1556 by Dr. John
Dee, and was in Pepys' possession in 1697
when the 'Catalogi Manuscriptorum Angliae
et Hiberniae' were published. The contents
appear in that work (Vol. II, p. 208, 9) cat-
alogued as separate items under nos. 6767–
76, 6778, 6780–84." James, *Catalogue of the
Library of Peterhouse*, p. 353. Its temporary
possession by John Dee necessitated a brief
repetition of its contents in James's later
*Lists of Manuscripts Formerly Owned by Dr.
John Dee* (Oxford University Press, 1921),
p. 25, no. 91.

author. Two other works of Oresme are also found here. The *De proportioni-bus proportionum* immediately precedes the *De commensurabilitate* on folios 93v–110v (see Grant, *Oresme PPAP*, pp. 127–28), and an anonymous version of the *Algorismus proportionum* follows immediately after on folios 128r–131r.

6. *R* = Vatican, Biblioteca Vaticana, Latin MS 4275, fols. 96r–101r.
Date: 14th or 15th century.[18]
Of all the manuscripts used in this edition, *R* alone is a partial version, containing only Part III. In this codex of 127 folios, *R* is preceded by a complete, but anonymous, copy of Oresme's *Algorismus proportionum* (fols. 90r–96r) and followed by a complete copy of Oresme's *De proportionibus proportionum* (fols. 102r–127r; see Grant, *Oresme PPAP*, pp. 128–29). Writ-ten in a single column in a clear, neat hand, this anonymous version bears the title *Pulchra disputatio: si omnes motus celi sint invicem commensurabiles an non* (fol. 96r).

7. *U* = Utrecht, Universiteits Bibliotheek, 725, fols. 172r–193v.
Date: 15th century.[19]
The contents of this codex of 246 folios are listed briefly, without incipits or explicits, by P. A. Tiele in *Catalogus codicum manu scriptorum Bibliothecae Universitatis Rheno-Trajectinae* (Utrecht, 1887), vol. 1, pp. 190–91.[20] The scribe of MS *U* had not only a rather unattractive handwriting, but was also subject to careless omissions, some of considerable length.[21] Because of numerous alternative readings supplied by the scribe in the margins, he undoubtedly had before him at least two copies of the *De commensurabilitate*. Perhaps this is why he provided two different titles—one at the beginning and another in the colophon (see variants below, pp. 172 and 322). How-ever, the work is correctly attributed to Oresme. Four figures are also in-cluded in the margins.

---

18. So dated by Thorndike, *History of Magic*, vol. 3, p. 400, n. 8.

19. This date is supplied by Tiele (see below).

20. A second volume was completed by A. Hulshof in 1909. For a recent and more accurate identification of the works included within folios 201v–246r, see L. Thorndike, "A Summary Catalogue of Reproductions 296–383 of Medieval Manuscripts Collected by Lynn Thorndike," *Medievalia et Humanistica*, fasc. 13 (1960), p. 89, no. 324.

21. For example, these lines are omitted: III. 202–61, 266–70, 279–93, 295–96, 300–17.

De commensurabilitate
vel incommensurabilitate
motuum celi

Tractatus de commensurabilitate
vel incommensurabilitate motuum celi

[Proemium]

"Zenomen et Crisippum maiora egisse," affirmat Seneca, "quam si
duxissent exercitus, gessissent honores, leges tulissent: quas non uni
civitati, sed toti humano generi tulerunt." Si enim bella gerere, novas leges
condere pulchrum est nonne magis commendandi sunt qui plusquam
Herculea audacia superum agressi sunt orbem et ausi mente conspicua
scandere celum eterna ipsius decreta primi pronunciare mortalibus;
quorum speculatione nihil post deum melius, nihil dulcius. Quid enim
magis oblectat animos, aut quid mentem plus erigit ad divina quam celi
mellicam contemplari musicam astrorum; remirari lucentem choream
agitantem modulos alterna varietate decoros per quam huius mundi machina sub deo regitur subiecta virtuti militie celestis exercitus que cursu
velocissimo et tranquillo speciosam diversitatem motuum regulari quadam
inequalitate absque fatigatione protendit in evum ad exercitium humani
ingenii perhemne.

Spectaculum "quo spectaculo," ut ait Tullius, "nihil admirabilius, nihil
pulchrius"; "nulla," inquit, "insatiabilior species, nulla pulchrior nec ad
solertiam et exercitationem hominum prestantior." Et rursum idem natura
"primum homines excitatos celsos et rectos constituit ut deorum cognitio-
nem celum intuentes capere possent." Unde etiam Seneca "ut scias,"
inquit, "naturam spectari voluisse non tantum aspici, vide quem locum

---

1 *ante* Tractatus *hab. LU* incipit
1–2 Tractatus de...celi *om. AFP*
2 vel incommensurabilitate *om. BLU* /
celi: celestium editus a reverendo philosopho Nycolas Oresme de commensurabilitate motuum *L* supercelestium Magistri Nicolai Orem *U*
Prol.4–I. 62 Zenonem....parum *om. P*
5 gessissent *om. U* / tulisset *A*
6 humano generi: genere humano *B* /
enim: vero *F*
7 commendandi: ?*F* / plusquam: ?*A*
8 Herculea: herculis *A* / audacia: iudicia *F*

9 primi: pre *A*
10 nihil post deum: post Deum nihil *U*
11 aut *om. L*
12 *ante* remirari *hab. B* planetarum / remirari: rimari *L* demirari *U* / *post* remirari *hab. B* super / lucentem: lucem
U / choream: choreis *U*
13 modulos *obs. B*
15 et: atque *U* / transquillo *BFL* / speciosam: speciositatem *B* / diversitatem *om.*
B
16 *ante* -gatione *in* fatigatione *mg. hab. U*
nota super vero [*the next word is illegi-*

Treatise on the Commensurability
or Incommensurability of the Celestial Motions

[Prologue]

Seneca says that "Zeno and Chrysippus performed greater things than if they had led armies, held public offices, [and] brought laws not merely to one city but to the whole human race."[1] Now if to wage wars [and] to establish new laws is wonderful, ought they not to be praised even more who, with a boldness greater than that of Hercules, have approached the upper orb and, with clear minds, were the first to ascend to the sky and pronounce its eternal laws to mortals. After God [Himself], nothing could be better or sweeter than the contemplation of these things. For what could be a greater delight to souls, or more readily elevate the mind to divine things, than to contemplate the melodious music of the stars in the heavens; to wonder at their brilliant dance which maintains, with varied turns, the appropriate rhythms by means of which the fabric of this world is regulated under God and subjected to the band of celestial bodies which extend into eternity, for the perennial exercise of the human mind, a brilliant diversity of motions [moving] effortlessly with a certain regular inequality through the swiftest and [most] tranquil course.

A spectacle, but, as Tully says, "nothing can be more marvelous or more beautiful than this spectacle."[2] "No sight," he says, "satisfies more, nothing is more beautiful or more excellent for the ingenuity and exercise of men."[3] And, again, he says that nature "first raised men tall and upright so that they might be able to behold the sky and so gain a knowledge of the gods."[4] Thus Seneca also says "in order that you may know that nature wishes to be observed

---

ble] nota quo exemplari fuerat corruptissimum ?immo fore potius / fatigatione: fatiga B / protendit: procedit A / evum: eis A / exercium A

16–17 exercitium humani ingenii: humani ingenii exercisium B

18 ait Tullius tr. B / post Tullius mg. hab. B Tullius

19 ante nulla¹ mg. hab. U mirabilior vel amabilior vel ?infra amabilior aut ?in-

scrutabilior

20 post et¹ add. LU ad / exersitationem B existationem F / ante hominum hab. B que / rursus U

21 excitatos: exercitatos A exitatos F

22 carpere AF / posset A / post possent add. B et cetera / etiam ?A et L / sciat BF

23 post non scr. et del. U tamen / tantum: tamen A

nobis dederit: in media nos sui parte constituit et circumspectum omni
25 nobis dedit nec erexit tantummodo hominem sed...ut ab ortu sidera in
occasum latentia prosequi posset et vultum suum circumferre sublime fecit
illi caput et collo flexibili imposui" hec ille.

Quatinus scilicet iocunditatem celestis plausus aspiceret quam mira
celeritate moderatur orbis conversio quam rato ordine celi constantia
30 mutabilis perseverat, qualitercumque anniversarias vicissitudines perpetuis
motibus renovare videtur, quorum ratio et proportio licet per omne temporis seculum lateant et nulla possint subtilitate aut humana diligentia
deprehendi. Hec tamen impossibilitas nec generat fastidium nec diffiden-
tiam parit. Nam quidem de hoc capere possumus mentem reficit et allicit
35 atque ad ultra inquirendum incitat et quadam suavi violentia mortalium
corda sursum propellit. Inde Tullius ait: "'cetera labuntur celeri celestia
motu/cum celoque simul noctesque diesque feruntur,' quorum contemplatione nullius animus potest expleri nature constantiam videre cupientes."

Ne igitur tam nobili studio cupientes insistere aut pro nimia difficultate
40 diffidant aut temeraria presumentes audacia seipsos et alios decipient
credentes sapere de siderum motibus que sciri nequeunt ab homine, hunc
libellum edidi de commensurabilitate motuum celi in quo premisi quedam
ex aliis libris mathematicis supponenda ex quibus conclusiones intuli
quarum paucas postquam scripseram alibi reperi. Sed principaliores si
45 alius tradidit nundum vidi. Non ergo dimisi quin hoc opusculum committerem sociis et magistris huius sacratissime universitatis Parisiensis sub
eorum correctione qui absque detractionis livore soliti sunt benedicta
reverenter suscipere et minus bene digesta emendare benigne.

---

24 dederit: dedit *ABU* / circumspectum:
circumspectionem L
25 *post* dedit *mg. hab.* B Seneca
26 occasu *A* / labentia *ABU* lambentia L /
posset et: ut *A* / vultum: ?*B* / suum: simul *U* / circumfere *A*
27 colli *A*
28 scilicet *om. A* / iocunditatem celestis
plausus: celestis plausus iocunditatem *U*
29 *ante* quam *hab. U* vel / rato: raro *A* /
constantia: ?*F*
30 qualiterque *BLU*
31 renovare: removere *A* revolvere *L* /
quorum: quarum *U* / et: atque *ALU*
31–32 temporis: tempus *A*
32 lateat *ABL* latea *F* / possit *ABFL* / subtilitate: sublimitate *A* / aut: et *B*
33 *ante* generat *add.* B diligentibus deside-

rantibus quam
34 quidem: quod *FL* / et: desiderium *B*
35 atque *om. A* / *ante* ultra *hab. A* quoque /
post suavi *add.* B huiusmodi
36 *ante* Inde *add.* F et / Inde: unde *L* / Tullius ait *tr. ALU* / *post* ait *mg. hab.* B
Tullius / celeri: celi *U*
37 fuerunt *BF* finitur *U* / *ante* quorum *add.*
B et
37–38 contemplatione: consideratione *B*
38 potest: poterit *L* / expleri *om. A* / *post*
expleri *hab.* B aut sacrari / naturam *A*
39 Ne: nec *A* / igitur: quo *U* / cupientes insistere *tr. F* / pro: pre *AU*
40 temeraria presumentes *tr. L* / et: ac *B*
40–41 audacia...credentes *om. A*
41 de *om. A* / siderum: sydereis *U* / *ante*
motibus *hab.* B stellarumque ?ac alia-

and not merely to be thought about, see the place it has given us: it has put us in the center and made us part of it giving us a view of all the things around us; not only did nature bring forth man but...so that he might follow the stars from their risings to their settings and turn his face [with the whole revolving universe], it made a head for him which could be raised and placed it on a flexible neck "[5]

Although man could now view the delights of the sky and applaud the manner in which the revolution of an orb is regulated by an extraordinary swiftness, and the way the variable uniformity of the sky persists in a fixed order despite the annual changes which are repeated because of its perpetual motions, the reason and ratio of these things lie hidden through every period of time and can be discovered by no human diligence or subtlety. This impossibility, however, does not produce disgust or distrust. For we realize that it refreshes and attracts the mind stimulating it to further inquiry, and, by a certain pleasant intensity, incites the thought of mortals upwards. Thus, Tully says: "'Swiftly the other heavenly bodies glide, / All day and night travelling with the sky,' and no one who loves to contemplate the uniformity of nature can ever be tired of gazing at them."[6]

Therefore, lest those who are eager to pursue so noble a study either despair before such great difficulty, or, believing that they know things about the motions of the stars which cannot be known by man, deceive themselves and others with rash impudence, I have written this little book on the commensurability of the celestial motions. In this book I have set forth certain things assumed from other mathematical books [and] from these have inferred [a number of] conclusions, a few of which I discovered after I had written [on this subject] elsewhere.[7] But if another has set out the more fundamental principles [or elements found in this book], I have yet to see them. For this reason I did not release this little book without [first] submitting it for correction to the Fellows and Masters of the most sacred University of Paris, who are accustomed to receive respectfully, without malicious slander, things that are well put, and to alter, in a kindly way, things not adequately formulated.[8]

---

rum / motibus *om. A*
44 *ante* alibi *add. B* eas
45 *ante* alius *add. B* quis / tradiderit *BLU* / hoc: illud *F*
45–46 *ante* committerem *hab. F* dimitterem sive / committerem: commentarem *A*

46 *post* sociis *hab. L* meis / sub: ?experner *A*
47 correctionem *A* / detractionis: detectionis *A* / benedicta: bene constructa *A*
48 digesta: dicta *A* / emendare benigne: benigniter emendare *B*

[Prima pars]

Numerus primus est qui a nullo preter quam ab unitate numeratur ut 5 sive 7; omnis alius dicitur compositus sicut 4 vel 9.

Numeri contra se primi, vel incommunicantes aut in sua proportione seu proportionalitate minimi, sunt qui nullam habent mensuram communem preter unitatem, nec communicant in aliquo numero, ut 4 et 9. Illi vero qui aliquo communi numero mensurantur dicuntur communicantes, ut 9 et 12 quorum quilibet mensuratur ternario. Et eodem modo si fuerint tres, vel plures.

Proportionalitas est similitudo proportionum; et dicuntur proportionalitates incommunicantes quando nulli duo numeri quorum unus est de una proportionalitate et alter de reliqua sunt communicantes, sicut proportionalitas secundum proportionem duplam, scilicet 1, 2, 4, 8, et cetera; proportionalitas secundum proportionem triplam, scilicet 1, 3, 9, 27, et cetera. Ille vero sunt communicantes quarum aliquis numerus unius cum aliquo numero alterius communicat sicut proportionalitas secundum proportionem duplam et proportionalitas secundum quadruplam.

Quantitates dicuntur commensurabiles quarum est aliqua mensura communis vel quarum proportio est sicut numeri ad numerum, ut si una est bipedalis et altera trium pedum; incommensurabiles sunt quarum nulla est communis mensura nec ipsarum proportio est sicut numerorum, sicut sunt dyameter et costa quadrati quarum proportio est medietas duple, que solum in continuis et numquam in numeris reperitur.

---

1 [Prima pars] *om. ABF* finito prohemio incipit prima pars *L* suppositiones *U*
3 sicut: ut *L* / vel: sive *F*
4 vel: sive *F* aut *L* / aut: sive *F*
5 seu: sive *F* / seu proportionalitate *om. U* / *post* proportionalitate *hab. F* numeri / minimi sunt *tr. F* / habent *om. U*
6 preter: nisi propter *F* / Illi: qui *U*
7 qui *om. U* / qui…communicantes: dicuntur communicantes qui aliquo communi numero mensurantur *B* / communi numero *tr. U*
8 mensuratur: numeratur *U* / *ante* ternario *add. U* a / fuerint: sint *B*
9 vel: sive *F*
12 alter: alius *B* / sicut: ut *B* sed *L*
13 scilicet: dicendo *FLU* dicendi *A*
14 cetera *om. AFLU* / scilicet: dicendo *AFL*
15 et cetera *om. FU* / Ille: illi *ABL* / Ille…communicantes: communicantes

[Part I]

A prime number is one which is numbered [or measured] by unity only, as 5 or 7; every other number is said to be composite, as 4 or 9.

Numbers which are prime to one another, or incommunicant, or in their least ratio or proportionality, are those which have no common measure other than unity; nor are they divisible into any number, as 4 and 9. But those which are measured by a common number are said to be communicant, as 9 and 12, each of which is measured by 3. The same applies if there are three or more [numbers].

A proportionality is an equality of ratios. Proportionalities are said to be incommunicant when no two numbers—taking one from each proportionality —are communicant, as [for example] a proportionality consisting of double ratios, namely, 1, 2, 4, 8,..., and a proportionality consisting of triple ratios, namely, 1, 3, 9, 27,.... Those [proportionalities] are [called] communicant when any number from one communicates with any number from the other proportionality, as is the case with double and quadruple proportionalities.

Quantities [or magnitudes] are said to be commensurable which have some common measure, or which have a ratio of a number to a number; as [for example] if one [quantity] is two feet and the other three feet. Those [quantities] are incommensurable which have no common measure and do not constitute a ratio of numbers, as is the case where the ratio between the diagonal and the side of a square is half of a double;* but this is found only in continuous quantities and never in numbers.[1]

---

* I.e., square root of two.

---

vero sunt ille *U* / quarum: quorum *BFL*
16 sicut: ut *B*
17 proportionem...secundum *om. F* / *ante* quadruplam *add. BU* proportionem
18 est aliqua *tr. L*
19 vel: aut *F* / numerum: numeros *F*

20 est: esset *FLU* / trium pedum: tripedalis *B* / *ante* sunt *hab. F* enim / sunt *om.* *U* / quarum: quibus *BFU*
21 ipsorum *F* / numerorum sicut *om. F* / *ante* sicut[2] *hab. L* et
22 sunt *om. AL*

Commensurabilitatem et incommensurabilitatem motuum circularium
25 accipio penes quantitatem angulorum descriptorum circa centrum aut
centra, sive in respectu circulationum, quod idem est, ita quod illa moventur commensurabiliter que in temporibus equalibus describunt angulos
commensurabiles circa centrum sive que in temporibus commensurabilibus
suas circulationes perficiunt. Et circulationes sunt incommensurabiles que
30 in temporibus incommensurabilibus fuerint complete, et quibus descri-
buntur temporibus equalibus anguli incommensurabiles circa centrum.
Unde secundum huiusmodi mensuram accipiuntur et fuerint coniunctiones, oppositiones, et aspectus, motusque omnes qui ab astronomis assig-
nantur in celo, quoniam proportio velocitatum secundum proportionem
35 linearum circularium a mobilibus descriptarum. Non facit ad propositum
sive debeat taliter accipi sive non. Incommensurabilitas autem potest in
omni genere continuorum reperiri ac in eis in quibus ymaginatur continuitas sive secundum extensionem sive secundum intensionem. Est namque
magnitudo incommensurabilis magnitudini, angulus angulo, motus motui,
40 velocitas velocitati, tempus tempori, proportio proportioni, gradus gradui,
et vox voci, et ita de similibus.

Circulationem voco unius mobilis circulariter de aliquo puncto ad
eundem reditionem. Revolutionem voco plurium mobilium de aliquo statu
reditionem ad statum vel aspectum omnino consimilem.

45 Intentio in hoc libello est loqui de precisis et punctualibus aspectibus
mobilium circulariter motorum, et non de aspectibus prope punctum de
quibus communiter intendunt astrologi qui non curant nisi quod non sit
sensibilis defectus quamvis modicus error imperceptibilis multiplicatus
per magnum tempus notabilem defectum efficiat.

50 Voco ergo pro nunc coniunctionem aliquorum mobilium quando eorum
centra sunt in eadem linea egrediente a centro [mundi]; tunc est coniunctio
corporalis aut saltem in eadem superficie vel circulo transeunte per polos

24 *post* motuum *hab. B* scilicet
25 aut: sive *F*
26 centra: celorum *A* / idem est *obs. B* / ita
quod: itaque *U* / illa: ista *A*
27 *post* commensurabiliter *scr. et del. L*
?continue / que: cum *A*
29 *post* perficiunt *add. B* et complent
30 in *om. L* / incommensurabilibus: commensurabilibus *U* / complete *om. AFLU*
32 accipiuntur: summuntur *F*
32–33 coniunctiones oppositiones *tr. F*
33 *ante* aspectus *hab. B* ceteri / motusque:
que *A* et motus *L* / qui *om. AF* que *BU*

34 *ante* secundum *hab. U* motuum
36 sive[2] *rep. F* / autem *om. L* / potest *om. F*
37 ac: ut *L* aut *U* / *ante* in[1] *scr. et del. U* ac
38 sive[1]: aut *F* / sive[2] *om. F* / sive[2] secundum intensionem *om. A* / intentionem
U
39 magnitudini *om. A* / *ante* angulus *add.*
FU et
40 *ante* velocitas *add. LU* et / *ante* proportio *add. U* et / *ante* gradus *add. U* et
42 de: ab *L*
43 eandem *F* / voco *om. U* vero *ABF* / de
om. A / statu *om. A*

I take the commensurability and incommensurability of circular motions in terms of the magnitude of the angles described around the center or centers, or in terms of the circulations, which is the same thing. Thus, things are moved commensurably when, in equal times, they describe commensurable angles around the center, or when they complete their circulations in commensurable times. Circulations are incommensurable when they are completed in incommensurable times, and when, in equal times, incommensurable angles are described around the center.[2] Accordingly, conjunctions, oppositions, aspects, and all the motions ascribed to the heavens by astronomers are to be measured in this way, since a ratio of velocities varies as a ratio of the circular lines described by the mobiles. Whether or not it ought to be taken in this way is not relevant to what is proposed here. Moreover, incommensurability can be found in every kind of continuous thing, and in all instances in which continuity is imaginable, either extensively or intensively. For a magnitude can be incommensurable to a magnitude, an angle to an angle, a motion to a motion, a speed to a speed, a time to a time, a ratio to a ratio, a degree to a degree, and a voice to a voice, and so on for any similar things.

The return of *one mobile* along a circular path from any point to that same point, I call a *circulation*. The return of *several mobiles* from any state to a wholly similar state, or aspect, I call a *revolution*.*

In this little book, my purpose is to consider *exact* and *punctual aspects* of mobiles that are moved circularly. I do not, however, propose to deal with aspects near a point which is usually the intention of astronomers, who care only that there be no sensible discrepancy—even though a minute, undetectable error would produce a perceptible discrepancy when multiplied over a long [period of] time.

For the present, I say that a conjunction occurs when the centers of any mobiles are on the same line drawn from the center [of the world]. At the very least, then, there is a physical conjunction either on the same surface, or

---

* The terms *circulatio* and *revolutio* are discussed above, chap. 2, n. 2.

---

44 reditionem *om. BL* / *ante* ad *mg. hab. B* intentio huius / vel *om. B* sive *F* veri *A* / aspectum *om. B* aspectus *A* / consimilem: similem *F* / *post* consimilem *hab. B* reversionem
45 *ante* in *add. B* autem / *ante* aspectibus *add. B* coniunctionibus et
46 motorum *om. AFLU*

47 astrologi: astronomi *F*
49 *ante* per *hab. A* et / magnum *om. ABFU* / notabilem defectum *tr. B* / defectum: effectum *A*
50 aliquorum: quorundam *F* / eorum: ipsorum *F*
51 egrediente: ?*B* / [mundi] *B*; *om. AFLU*
52 aut: sive *F* / vel: sive *F*

mundi sicut in eodem meridiano, vel per polos orbis signorum, quia possibile est quod duo planete sint in primo puncto, vel in prima linea, Arietis
55 et tamen non sint in eodem meridiano.

Suppono etiam quod motus celestes sint velocitate inequales, et de illis saltem est sermo; et quod quilibet unus motus sit continuus, perpetuus, et regularis quamvis ex pluribus motibus regularibus quandoque fiat motus irregularis. Dicendum est etiam de illis motibus qui fuerint ad eandem
60 partem quia de aliis poterit faciliter videri ex determinandis. Dicetur etiam de motibus istis ac si essent concentrici quoniam quantum ad propositum parum refert et etiam posterius videbitur in quo differt. Et quia necesse est omnes motus celestes esse invicem commensurabiles, aut quod aliqui eorum sint invicem incommensurabiles, ideo in prima parte huius operis
65 videbitur si sint commensurabiles quid inde sequatur; in secunda vero quid si sint incommensurabiles; in tertia inquiretur de supposito, scilicet si sint commensurabiles aut non. Sint ergo primo omnes commensurabiles.

Conclusio prima. Si fuerint quotlibet numeri ab unitate continua proportionalitate dispositi, nullus eorum numeratur ab aliquo primo numero nisi ab
70 *illo vel illis, si fuerint, qui numerant illum qui in illa proportionalitate immediate sequitur unitatem.*

Probatur per undecimam noni Euclidis que sic est: "Si numeris quotlibet ab unitate continua proportionalitate dispositis aliquis numerus primus ultimum numeret eum quoque qui unitatem sequitur numerare necesse
75 est." Sint ergo post unitatem continue dispositi *A, B, C, D*, et cetera, et sit *G* aliquis numerus primus. Ergo per conclusionem allegatam si *G* numerat *D*, similiter ipse *G* numerabit *A*. Ergo a destructione consequentis, si *G* non numerat *A* sequitur quod *G* non numerabit *D*. Et ita de quolibet alio in illo ordine et consimiliter de quolibet alio numero primo.

---

53 sicut: sint *A* / vel: sive *F* / per *om. A*
54 vel: et *B* sive *F* / in[2] *om. AB* / prima *om.*
    B
55 non *om. A*
56 Suppono: suppositio *A* / quod: pro *B* /
    celestes: celestium *B* / ante sint *hab.* *F*
    mobilium *et L* veloces / inequales: equales *F*
57 saltem est sermo: est sermo saltem *B* est
    sermo *L* / perpetuusque *U*
58 motibus *om. L* / quandoque fiat *tr. L*
59 est *om. LU*
60 poterit: potest *U*
61 quoniam: quia *B*
62 et[1] *om. B* / posterius videbitur *tr. B*

63 aut: a *P*
64 sint invicem *tr. A* sunt invicem *F* sint
    P / parte *om. L* / operis *om. L*
65 videtur *F* / *ante* commensurabiles *scr.*
    et del. F in / inde *om. L* / sequitur *U* /
    vero *om. B*
65–66 quid...imcommensurabiles *om. P*
66 quid *om. AL* / *post* quid *hab. B* sequatur / in *om. U* et *AFLP* / scilicet *om. A*
67 si *om. F* an *L* / sint[1]: aut *BL* an *APU*
    sive *F* / Sint ergo primo: primo ergo sint
    B dic ergo primo *A* / *ante* commensurabiles[2] *scr. et del. B* in / *post* commensurabiles *hab. F* et sic finis huius *et hab.*
    P et cetera

circle, passing through the poles of the world—i.e., on the same meridian—or passing through the poles of the sphere of the signs, since it is possible that two planets could be in the first point, or line, of Aries and yet not be on the same meridian.

I also assume that celestial motions have unequal speeds—our discussion is restricted to such motions only—and that any one motion is continuous, eternal, and regular, although sometimes an irregular motion can be produced from several regular motions. Furthermore, only motions in the same direction will be considered, since other kinds can be easily understood from what will be determined. Also, these motions will be discussed as if they were all concentric, since this has little bearing on what will be considered; later, however, the sense in which differences arise will be seen.[3] And [finally] since it is necessary that all celestial motions be either mutually commensurable, or that some be mutually incommensurable, we shall see, in the first part of this work, what [consequences] follow if they are [assumed] commensurable; and, in the second part, what follows if they are incommensurable; and in the third part, we shall investigate what has been assumed—that is, whether they are commensurable or not. To begin with, then, let them all be commensurable.

Proposition 1. If any numbers whatever are arranged in continuous proportionality beginning with unity, no prime number would number [or measure] any of them unless it numbers [or measures] the number immediately following unity.

This can be demonstrated by the eleventh proposition of the ninth book of Euclid which reads: "Should any numbers whatever be arranged in continuous proportionality beginning with unity, and if some prime number numbers [or measures] the last number, it is necessary that it also number the number which immediately follows unity." Thus, let *A*, *B*, *C*, and *D* be arranged in continuous proportionality after unity, and let *G* be a prime number. Then, by the proposition cited [above], if *G* measures *D* it will also measure *A*. Therefore, by denying the consequent, if *G* does not measure *A* it follows that *G* will not measure *D*. This applies to any other number in this series as well as to any other prime number.[4]

---

68 Conclusio prima *mg. hab. L post* numeri *om. P* prima conclusio *mg. hab. BF ante* si fuerint *et hab. U* propositio prima *et hab. A* prima

69 eorum: illorum *F* / ab: ex *F* / nisi: non *P*

70 illo: isto *P* / vel: sive *F* / si fuerint *om. U* / qui[1]: que *P*

70–71 immedietate *P*

72 Probatur: probatio *A* / que sic est: que est hoc *B* scilicet *U* / quotlibet: qui-libet *A*

73 primus: post *L*

74 eum quoque: cum quocumque *L* / qui: ab *A* / unitate *AP*

76 numerat: numeret *B* numeratur *L*

77 similiter *om. U* / ipse *om. BU* ipsa *L* / *G om. U* / numerabit: numerat *AF*

78 *ante* non[1] *add. U* etiam

79 in *om. P* / illo: numero *P* / consimiliter: similiter *FL* / *ante* primo *hab. P* ymo

80 Verbi gratia, sint post unitatem 6, deinde 36, postea 216, et sic in infinitum secundum proportionem sextuplam. Dico ergo quod nullus primus numerus numerat aliquem istorum nisi illi qui numerant 6, qui sunt 2 et 3. Nec per consequens aliquis numerus multiplex alicuius alterius numeri primi quoniam qui numerat numerantem numerat numeratum. Ergo
85 submultiplex numerat numeratum a suo multiplice.

Conclusio secunda. Si per ymaginationem aliquod continuum dividatur in aliquot partes et quelibet illarum in totidem et sic in infinitum in nullo puncto cadet diviso in quo caderet si divideretur secundum aliam proportionalitatem nisi numeri immediate sequentes unitatem illarum proportionalitatum sint
90 *communicantes.*

Hoc intelligitur de puncto dividente continuum in duas partes sicut est quilibet punctus signatus in linea recta et similiter quilibet signatus in linea circulari preter primo signatum qui non dividit in duo nisi adveniente secundo. Est igitur conclusio vera universaliter de continuo recto et de
95 circulari signato tamen primitus uno puncto qui non facit divisionem. Quo facto est eadem ratio utrobique.

Sit igitur *A* aliquod continuum dividendum in partes numeratas numero *D* et quelibet illarum in totidem et continue proportionaliter in infinitum. Rursumque ipsummet *A* dividatur in partes numeratas numero *E* et
100 quelibet illarum in totidem et sic in infinitum. Et sint numeri *D* et *E* incommunicantes.

Quoniam igitur *A* dividitur secundum numerum *D* in partes equales, omnis punctus huius sectionis dividit *A* in partes denominatas aliquo numero quia totum *A* est sicut numerus *D* et omnis numeri quelibet partes sunt denominate aliquo numero. Sed nullius numeri partes denominantur
105

---

80 6: g *F* / deinde *om. B* / 36: tres g *F* / postea *om. B* / 216: 21 g *F*
80–81 in infinitum: ultra *B*
81 sexduplam *P* / ergo *om. L* / quot *P*
81–82 primus numerus ?*A*; *tr. B* primus *LU*
82 istorum: illorum *L* / illi *om. B* / 6: g *F* / qui: que *A*
83 alterius numeri *tr. U*
83–84 alicuius…primi *om. F*
84 quoniam: quia *B* / qui: que *A* / *post* numera- *in* numeratum *scr. et del. A* -ntem
84–85 Ergo submultiplex numerat: quid est multiplex *U*
85 numeras *P* / multiplice: multiplici *L* submultiplice *U* / *post* multiplice *hab. F* et

86 sic finis est huius *et hab. P* et cetera
86 Conclusio secunda *om. FP* propositio secunda *U* secunda *mg. hab. AB ante* Si per / ymaginem *P / ante* aliquod *hab. F* videlicet / aliquot *L* / dividitur *U*
87 aliquod *P / post* aliquot *scr. et del. A* continuum
88 cadet: caderet *A* / aliam: illam *L* / proportionalitatem: proportionem *U*
89 illarum *om. U* istarum *A* / proportionalitatum ?*AB*; *om. U* / sint: sit *A* sicut *P*
91 *ante* Hoc *hab. U* et / Hoc: illud *F* / continuum *om. P* / duas: d *A*
92 quilibet[1] *obs. B* / punctus *om. U* / signatus: figuratus *F* / et *om. F* / similiter: super *P* / quilibet[2] *om. B* quelibet *P* /

For example, assume the sequence 1, 6, 36, 216, and so on to infinity in accordance with a sextuple ratio. I say, then, that no prime number numbers [or measures] any of these numbers except those [prime numbers] which number 6—namely, 2 and 3. As a consequence, no number [in the series] is the multiple of any other prime number, since that which numbers [or measures] the measuring [or first] number [also] numbers [or measures, all the numbers in the series measured by the first number]. Therefore, it is through its multiple that a submultiple measures the terms in the series.

Proposition 2. If, in the imagination, any continuum were divided into aliquot parts, and any of these into aliquot parts, and so on ad infinitum, no point in this division would coincide with any point of another division—if the continuum were [also] divided by another proportionality—unless the numbers following immediately after unity in both proportionalities were communicant.

This must be understood with respect to a point dividing the continuum into two parts, as [for example] any point in a straight line; and, similarly, in a circular line a second point must be assigned since the first point cannot divide the circle into two parts. This proposition is, therefore, universally true for both rectilinear and circular continua despite the fact that [in the circular continuum] the first point assigned will not make the [required] division; but after the second point has been assigned the same reasoning applies.

Let *A* be any continuum divided into parts equal to number *D*, and any of these parts into just as many parts, and so on to infinity in continuous proportionality. Now once again, let *A* be divided into *E* parts, and any of these into just as many parts, and so on to infinity. And assume that the numbers *D* and *E* are incommunicant [or prime to each other].

Then, since *A* is divided into *D* equal parts every point of this division divides *A* into parts denominated by some number, since *A*, the whole, is as number *D* and any parts of every number are [themselves] denominated by some number. But no number has its parts denominated by any number other

post quilibet² *mg. hab. U figuram*
93 circulari: articulari *P* / preter *om. P* / primum *U* / qui: que *P*
94 *ante* Est hab. *P* et / conclusio: quo *P* / universaliter: similiter *P* / de² *om. AU*
96 eadem ratio: ad eandem rem *P*
97 A aliquod: dividendo *L* / dividentem *P*
98 D: e *L* / illarum: istarum *AB* istorum *F* / *post* et² *add. B* ita / in infinitum *om. B*
99 rursum *B* / ipsum *BU*

100 illarum: istarum *AP* / in infinitum: continue proportionales *B* / numeri *om. U* unum *L* / et³ *om. A*
100–101 *post* incommunicantes *mg. hab. B* nota
102 Quoniam: quando *F* / igitur *om. U* / A *om. L* / dividitur *om. FP* / *post* dividitur *scr. et del. B* in / *ante* equales *hab. P* e
103 A: d *L*
104 *ante* numeri *scr. et del. U* numerus
105 sunt denominate *tr. U*

aliquo numero nisi illo vel illis qui numerant illum totum vel ipsomet toto vel aliquo sibi submultiplice ut satis patet ex tricensimaoctava septimi Euclidis. Ergo nullus punctus huius divisionis secat *A* nisi in partes denominatas ipso *D* vel aliquo numero qui numeret ipsum *D*.

110 Et eodem modo probabitur quod nullus punctus divisionis ipsius *A* in partes numeratas numero *E* dividit ipsum *A* nisi in partes denominatas numero *E* vel aliquo numero qui numeret ipsum *E*. Sed nullus numerus qui numeret *D* numerat *E* quia positi sunt incommunicantes ergo nullus numerus qui denominat partes divisionis que sit per *D* denominat partes
115 divisionis que sit per *E*. Et eodem modo partes unius totius habent eandem denominationem igitur nulle partes prime divisionis sunt partes secundum divisionem secundam. Ergo nullus punctus secans secundum primam divisionem est aliquis punctus secans secundum divisionem secundam. Ergo si aliquod continuum dividatur duabus divisionibus secundum numeros
120 incommunicantes nullus punctus unius divisionis est punctus alterius divisionis.

Sint ergo due proportionalitates *B* et *C* quarum numeri immediate sequentes unitatem sint *D* et *E*. Ergo si *B* et *C* sint communicantes in aliquo, sicut in *G* numero, sequitur quod numerus *G* numerat aliquos in utraque
125 proportionalitate per diffinitionem proportionalitatum communicantium. Ergo per conclusionem precedentem ipse *G* numerat *D* et *E*. Ergo si *B* et *C* proportionalitates sunt communicantes *D* et *E* numeri erunt communicantes. Ergo a destructione consequentis, si *D* et *E* sint incommunicantes, *B* et *C* erunt incommunicantes; ergo per diffinitionem proportionalitatum
130 incommunicantium quilibet numerus unius proportionalitatis est incommunicans cuilibet numero alterius proportionalitatis.

Et iam probatum est quod sectiones que fuerint secundum numeros incommunicantes in nullo puncto conveniunt. Ergo si *A* divideretur in infinitum secundum unam proportionalitatem in nullo puncto fieret divisio

---

106 illo: isto *FP* / vel¹: sive *F* / numerat *L* / illum: illud *F* / vel²: nisi *ABFLP*
107 *ante* vel *hab.* *U* toto / vel: sive *F* / submultiplici *BFLP* multiplici *A* / ex tricensimaoctava septimi: per Aristotelem 38 *P*
108 punctus *om.* *AFP* / punctus huius divisionis: huius divisionis punctus *U* / *ante* secat *scr. et del.* *B* q
109 vel: sive *F*
110 Et *om.* *ABU* / probatur *B* / ipsius *om.* *B*
111 dividet *L* / *ante* denominatas *scr. et del.* *L* numeratas numero E

112 vel: sive *F* / numerus: ?numeret *B*
113 numeret: numeres *P* / numerat: numeret *P* / quia: qui *PU*
114 denominat¹: denominet *L*
114–15 denominat²...per *om.* *L*
115 divisionis *om.* *A* / *ante* Et *hab.* *U* et cetera / modo *om.* *BLU* / unius: huius *L* / habet *AB*
116 denominationes *U* / nulle: ulle *A*
116–17 secundum divisionem secundam: divisionis secunde *B*
117–18 primam divisionem *tr.* *B*
118 divisionem secundam *tr.* *FPU* secun-

than those which number [or measure] the whole—[i.e.] either by the whole [number] itself, or by some [number] submultiple to it, which is evident by the thirty-eighth proposition of the seventh book of Euclid.[5] Consequently, no point of this division cuts A into parts unless denominated by D itself or by some number which numbers [or measures] D.

In the same way it will be shown that no point of division of A divides it into parts equal to number E, unless it be divided into parts denominated by number E [itself], or by some number which measures E. But no number that measures D [also] measures E, since they were assumed incommunicant [or prime to each other]. Therefore, no number representing parts of the division made by D can represent parts of the division made by E. And in the same manner, the parts of one whole have the same denomination so that no parts of the first division are parts of the second division. Thus, no point dividing [the continuum] in the first division is a point dividing [it] in the second division. For this reason, if any continuum were divided by two [distinct] divisions represented by incommunicant [or mutually prime] numbers, no point of one division is [also] a point of the other division.

Let there be two proportionalities B and C whose numbers immediately following unity are D and E. Now if B and C are communicant with respect to some number, say, G, it follows, by the definition of communicant proportionalities, that number G numbers [or measures] any number in each proportionality. Hence, by the preceding proposition, G numbers D and E, so that if B and C are communicant proportionalities numbers D and E will be communicant. Then, by denying the consequent, if D and E are incommunicant [it follows that] B and C will be incommunicant; and by the definition of incommunicant proportionalities any number of one proportionality is incommunicant to any number of the other proportionality.

It has now been demonstrated that divisions will not meet in any point if they are made with incommunicant numbers. If A were divided into infinity according to one [particular] proportionality, no point of this division would

---

dam L

119 *post* si *mg. hab. B* conclusio
122 due proportionalitates *tr. U*
123 sint[2]: sunt P
124 sicut: sint A / numerus G *tr. U* / utrumque P eadem L
125 proportionalitate: proportione L / communicantium *om. A*
126 ipse *om. AB* / enumerat F / *post C hab.* F et

127 proportionalitates sunt *tr. B* / sint U
128 *ante* incommunicantes *hab.* L numeri / incommunicantes: communicantes AP
129 *ante* diffinitionem *hab.* U 1
130 quelibet P / proportionalitatis *om. A*
132 numeros *om. B*
133 dividetur AL
134 unam: aliam ?dividi L / *post* in *mg. hab.* B conclusio / divisio: communicatio B
134–135 in...proportionalitatem *om. L*

135 in quo fieret si divideretur in infinitum secundum aliam proportionalitatem nisi numeri immediate sequentes unitatem istarum proportionalitatum communicarent. Verbi gratia si aliquod continuum dividatur in 2, deinde in 4, postea in 8, et sic in infinitum nullibi cadet divisio ubi caderet si divideretur in 3, deinde in 9, postea in 27, et sic in infinitum secundum pro-
140 portionalitatem triplam.

Unde patet quod si esset taliter facta divisio in infinitum secundum proportionalitatem triplam vel etiam duplam, nihil restaret dividendum et tamen continget ymaginari infinita puncta in quibus nulla cecidit divisio. Sed si duarum proportionalitatum numeri immediate sequentes unitatem
145 essent communicantes, tunc sectiones essent communicantes que fierent secundum huiusmodi proportionalitates sicut si continuum divideretur secundum proportionalitatem triplam, deinde secundum proportionalitatem sextuplam.

Conclusio tertia. Dividendo continuum per fractiones phisicas quantum-
150 *libet impossible est prescindere partem seu partes aliquotas seu denominatas aliquo numero primo aut sibi multiplici preter 2, 3, et 5.*

Quoniam omne quod sic dividitur est sicut unitas et numerus partium prime divisionis est sicut 60, et numerus partium secunde divisionis est 3,600 [et numerus partium tertie divisionis est 216,000] et sic continue
155 secundum proportionalitatem sexagintuplam. Sed immediate sequens unitatem in ista proportionalitate, videlicet 60, non numeratur ab aliquo primo numero nisi ab istis tribus, scilicet 2, 3, et 5. Ergo per primam conclusionem nec aliquis numerus huius proportionalitatis numeratur ab aliquo primo numero nisi ab istis [tribus] 2, 3, et 5, ergo quilibet numerus
160 huius proportionalitatis est incommunicans cuilibet primo numero nisi tribus predictis.

---

135 fieret: fiet *P* / in infinitum *om. U*
136 nisi: ubi *U*
137 divideretur *U*
138 *ante* postea *hab. L* et / postea: deinde *FU* / *post* infinitum *hab. B* secundum proportionem duplam / nullibi: nullicubi *FLPU* nulliuscubi *A*
139 *ante* in[1] *hab. A* primo *et scr. et del. B* per / postea: deinde *BU* / in infinitum: ultra *B*
139–40 proportionalitatem: proportionem *AB*
141 *ante* facta *hab. A* iam
141–42 Unde…triplam *om. B*
142 duplam: quadruplam *U* / restaret divi-

dendum: referret dividendo *U*
143 infinita: in ?summis *L* / cecidit: cadet *L*
144 *ante* Sed *mg. hab. B* ?nota
145 essent[1] *om. L* / *post* tunc *hab. B* et / tunc…communicantes *om. AP* / *ante* que *hab. P* sicut
147 proportionalitatem: proportionem *L*
147–48 proportionalitatem: proportionem *L*
148 *post* sextuplam *hab. P* et cetera *et B* divisionis per sectiones communicarent / *infra* sextuplam *mg. hab. B* figuram
149 Conclusio tertia *om. FP* propositio tertia *U* tertia *mg. hab. AB ante* dividendo / *ante* dividendo *hab. F* si / dividendo: si

coincide with any point of a second division [of this same A] unless the numbers following immediately after unity were communicant in both proportionalities. For example, if any continuum were divided into two [parts], then into 4, then 8, and so to infinity, nowhere would the division be the same if the continuum were [subsequently] divided into 3 [parts], then into 9, then 27, and so on to infinity in accordance with a triple proportionality.

It is obvious, however, that if such a division into infinity were made in accordance with either a triple or double proportionality, nothing would remain to be divided—and yet infinite points can be imagined in which no division fell. But if, in two proportionalities, the numbers immediately following unity should be communicant, then the divisions made according to such proportionalities would [also] be communicant—as if [for example] a continuum were divided first by a triple, and then by a sextuple proportionality.

Proposition 3. However much a continuum is divided by physical [i.e., sexagesimal] fractions, it is impossible to divide [and arrive at] a part or aliquot parts denominated by any prime number—or multiple of any prime number— other than 2, 3, and 5.

Since every thing which is divided in this way serves as a unit, the number of parts of the first division is 60, of the second division 3,600, of the third division 216,000, and so on, always in accordance with a sexagintuple proportionality.* But in this proportionality, that which directly follows unity, namely, 60, is not numbered [or measured] by any prime number except by these three—i.e., 2, 3, and 5. By the first proposition, then, no number in this proportionality is numbered by any prime numbers except 2, 3, and 5, so that any number of this proportionality is incommunicant to any prime number except the three just mentioned.

---

* That is, by 60^n, where $n = 1, 2, 3, 4, \ldots$.

---

videndo *L* / phisicas *obs. B*
149–50 quantumlibet: quamlibet *P*
150 precindere *A* / seu[1]: vel *B* / seu[2]: sive *F*
151 *ante* aliquo *add. B* ab / aut: sive *F* / preter: et *L* / 2: propter *F* / *ante* 3 *add. AFP* et
153 est[1] *om. U* / partium secunde divisionis: divisio secunde *A*
154 *ante* 3,600 *hab. A* sicut / [et…216,000] *om. AFLPU*
155 proportionalitatem sexagintuplam: proportionem sextuplam *L* / *post* Sed *hab. A* numerus

156 videlicet: scilicet *U*
157 primo numero *tr. B* / primo *om. A* / tribus *om. U* / scilicet *om. L* / *post* 2 *add. AFP* et / et *om. LU*
158 huius: alicuius *U*
158–60 numeratur…proportionalitatis *om. L*
159 primo *om. FP* / [tribus] *om. FPU* / *post* 2 *add. FP* et
159–60 numero…primo *om. A*
160 huius: unius *FP* / est *om. F* / incommunicans: communicans *P* / cuilibet: tribus *P*

Ergo, per precedentem conclusionem, omnes sectiones sunt incommunicantes quarum una fieret secundum aliquem numerum istius proportionalitatis et alia secundum aliquem primum numerum alium a predictis; ergo
nullus punctus unius sectionis esset punctus alterius sectionis. Ergo dividendo secundum proportionalitatem sexagintuplam numquam abscinderetur portio denominata aliquo primo numero alio a prefatis; igitur nec
aliquo multiplici alterius primi numeri quia tunc quotiens ille esset multiplex totidem partes denominate tali multiplici esset pars denominata illo
primo numero.

Verbi gratia, si posset abscindi $\frac{1}{14}$. Cum 14 sit duplex ad 7, ergo posset
abscindi $\frac{1}{7}$ quia $\frac{2}{14}$ sunt $\frac{1}{7}$, et sic de aliis. Ex quo patet quod per divisionem
secundum phisicas fractiones quantumcumque multiplicaretur numquam
abscinderetur $\frac{1}{7}$ pars rei divise taliter nec $\frac{1}{14}$, et cetera; et similiter nec $\frac{1}{11}$,
nec $\frac{1}{22}$, nec etiam $\frac{1}{13}$, nec $\frac{1}{26}$, et ita de aliis quia tunc aliquis numerus de
huiusmodi proportionalitate sexagintupla posset dividi per 7, vel per 11,
vel per 13, et cetera, sive numerari ab eodem quod est impossibile per
primam conclusionem quoniam nullus numerus habet partem aliquotam
vel partes aliquotas nisi denominatas ab ipsomet vel ab aliquo qui numerat ipsum.

Unde manifestum est quod si aliquod mobile circulariter pertransiret in
die aliquot gradus et partem seu partes gradus denominatas aliquo primo
numero vel multiplici alicui primo numero preter 2, 3, 5, numquam haberetur precisus motus ipsius per phisicas fractiones quantumcumque pro-

162 conclusionem *om. U* / ante sunt *add. B*
huius
163 fieret: fiet *U* / istius: huius *B*
163-64 istius...numerum *om. A*
164 primum *om. B* / a *om. A*
165 punctus[1] *om. FP* / unius *om. LP* nullius
F / ante sectionis *hab. BL* huius / sectionis esset *om. P*
166-67 abscinderet *A* abscindetur *L*
167 portio: proportio *BL* / primo *om. U* /
a: ?7 *P* / prefatis: prefato *A* predictis *L*
prelatis *P*
168 aliquo *om. U* alio *BFP* / post aliquo *hab.*
L d / alterius: alicuius *L* / ille esset:
esset iste *A* ille sunt *F* ille est *P*
169 esset: sunt *AP* essent *BL* / pars *rep. U*
partem *A* / post denominata *add. A* tali
multiplici est pars denominata / illo:
isto *P*
171 $\frac{1}{14}$: una quartadecima *AB*; *obs. P* una

14 *F* una quarta *L* / Cum 14 *om. P* /
post sit *hab. A* ?proportio / duplex: duplus *AL*
171-72 $\frac{1}{14}$...abscindi *om. U*
172 abscidi *F* / $\frac{1}{7}$[1]: una septima *BLU* una 7
AFP / $\frac{2}{14}$: due quartedecime *BFL* ?P
due 24 *A* / $\frac{1}{7}$[2]: una 7 *AFP* una septima
BL / ante Ex *hab. P* et
173 quantumcumque: quantumlibet *F* quan-
tumque *P* / multiplicetur *A*
174 abscideretur *F* / $\frac{1}{7}$: 7 *A* septima *BFLP* /
pars *om. U* / nec[1]: ut *L* / $\frac{1}{14}$: quartadecima *ABLP* 14 *F* / et cetera: et sic de
aliis *U* / et[2] *om. ABFPU* / similiter nec
tr. B / $\frac{1}{11}$: undecima *ABFL* nulla *P*
175 nec[1]: vel *AU* / $\frac{1}{22}$: vicensima secunda
BLP 22 *F* 13 *A* $\frac{1}{13}$ *U* / etiam *om. AU* /
$\frac{1}{13}$ *om. A* tertia decima *LP* 13 *BF* $\frac{2}{22}$
U / nec[3] *om. FP* vel *U* / $\frac{1}{26}$: vicensima
sexta *B* 26 *AF* 16 *P* $\frac{2}{26}$ *U* / et ita: et sic

Therefore, by the preceding proposition, all divisions are incommunicant whenever one of them is based on any number in this proportionality, and another is based on any prime number other than the three mentioned before, so that no point of one division is [also] a point of another division. Hence, in dividing a [continuum] by a sexagintuple proportionality, no section could ever be cut which is denominated by any prime number other than the aforementioned; therefore [it could] not [be denominated] by any [number] multiple to another prime number, for then there would be just as many parts denominated by such a multiple as the multiple contains the part denominated by that prime number.

For example, assume that $\frac{1}{14}$ of a part [of a continuum] could be divided. Since 14 is double 7, a $\frac{1}{7}$ part [of this continuum] could [also] be divided because $\frac{2}{14}$ equals $\frac{1}{7}$; and this applies to other divisions of this kind. It is clear from this that however far division by sexagesimal fractions is carried, it could never produce a $\frac{1}{7}$ part of the thing divided, nor a $\frac{1}{14}$ part, and so on; and, similarly, it could never produce parts such as $\frac{1}{11}$ and $\frac{1}{22}$, or $\frac{1}{13}$ and $\frac{1}{26}$, and so on, because [it would follow that] any number in a sexagintuple proportionality is divisible by 7, or 11, or 13, and so forth, or could be numbered by any of these, which is impossible by the first proposition since no number contains an aliquot part, or parts, unless [the part, or parts, is] denominated by [the number] itself, or by some [smaller] number which numbers [or measures] that number.

It is obvious, therefore, that if any mobile moving on a circle traverses in a day several degrees plus some part, or parts, of a degree denominated by any prime number—or multiple to any prime number—other than 2, 3, and 5, its exact motion could not be determined by means of sexagesimal fractions,

---

B et ista P ita U / ante quia mg. hab. B numerari

175–76 nec $\frac{1}{26}$…sexagintupla om. L

176 vel: sive F / 11: undecima F

177 vel: aut F / et cetera om. L / sive…eodem om. B / numerari: per multiplicem U / quod: quidem A

178 quoniam: quia B

179 vel¹: sive F seu L / ante vel² hab. P est / vel²: sive F / ab² om. L

181 quod om. P / pertransieret P

182–83 primo numero tr. L primo P

183 vel multiplici om. L sive multiplici F / alicui om. P aliquo FU / alicui primo numero om. B / primo numero tr. L primo gradu U / post preter hab. F propter

184 precisius A precise B precisi P / motus: modus F / ipsius: quamvis U / phisicas fractiones tr. B / quantumcumque: quantumlibet L quantum habet P quantumcumlibet U

184–85 procedetur P

185   cederetur nec etiam locus, nec tempus, nec coniunctiones eius cum alio
mobili, nec oppositiones, nec ceteri aspectus, nec aliquod tale punctualiter
et precise.

Cum igitur secundum divisionem tabularum totus circulus sit sicut sex
unitates quarum quelibet est duo signa et quarum quelibet dividitur in 60
190   gradus, et quilibet gradus in 60 minuta, et quodlibet minutum in 60 secunda, et quodlibet secundum in 60 tertia, et sic continue; et 6 non numeratur
ab aliquo primo numero nisi sit 2 aut 3; nec 60 numeratur ab aliquo primo
numero nisi ab istis, scilicet 2, 3, 5, sequitur ergo quod circulus totus numquam divideretur per divisionem tabularum in partes aliquotas denomina-
195   tas aliquo primo numero vel multiplici alicuius primi alterius quam 2, aut
3, aut 5. Nec posset abscindi per hoc pars aliquota secundum alium primum
numerum ut dictum est.

Quare si quilibet circulus celi ymaginaretur dividi per 17 partes que
vocarentur signa, et quodlibet tale signum per 17 gradus, et gradus per 17
200   minuta, et sic consequenter; et secundum hoc fierent tabule numquam
huiusmodi cum tabulis communibus quibus nunc utimur punctualiter
concordarent. Eodem modo si fieret divisio continue proportionalitatis
per 61, vel aliquem alium primum numerum.

Non igitur sequitur si motus celi sint commensurabiles quod per tabulas
205   factas possint commensurari precise vel equari quia possibile est quod
unum mobile pertranseat in die unum gradum precise et aliud [mobile]
gradum cum $\frac{1}{7}$ parte unius gradus, vel etiam in die $\frac{1}{13}$ partem unius gradus,
vel $\frac{1}{22}$ partem totius circuli, aut secundum aliquem alium numerum secundum quem non potest abscindi aliqua pars per divisionem tabularum com-
210   munium. Verumtamen astronomi, compositores tabularum, non inten-

---

185 locus *om.* L / nec²: aut *ABP* aut locus
sive F / nec³ *om.* P / eius *om.* *FLPU* /
cum: in L / *ante* alio *add.* *FLP* uno /
alio: vero U
185–86 eius…aspectus *om.* A
186 mobile F ?modali L motu U / ceteri *om.*
FLPU / nec³ aliquod tale *om.* B / aliquid U
187 et *om.* P / prescise L
188 sicut *om.* A divisus in U
189 *post* est *scr. et del.* A ?re / et *om.* BU
190 gradus¹: grada U / quodlibet: quotlibet
P / minutum: minuta U
191 *post* et² *hab.* U et cetera
191–92 6 non…aut 3 *om.* B
192 primo numero *tr.* AU / sit: sint A / aut:

vel L / nec 60: 60 non B / infra 60 *scr.
et del.* B 6
192–93 nisi…numero *om.* P / primo numero
tr. L
193 *post* nisi *hab.* A 2 aut 3 / 2, 3, 5: 2 et 3
et 5 B / ergo *om.* F
194 dividetur AU
195 primo numero *tr.* B / vel: sive F / primi
alterius: alterius numeri primi B / quam
om. L / aut: sive F
196 aut: sive F / possit L potest U / abscidi F
197 dictum est *tr.* L ?dictum U dictum fuit F
198 quelibet AB / celi *om.* A / ymaginetur A
199 gradus¹: grada B / gradus²: grada A
qualibet grada B
200 hoc: illud F / fieret B

however far one proceeded; nor, indeed, could its position, or time, or its conjunctions or oppositions with another mobile, or other aspects, or any such thing [be determined] punctually and exactly.

Since in a division made according to the [astronomical] tables, a whole circle is as six units any one of which is equal to two signs, and any sign is divisible into 60 degrees, any degree into 60 seconds, any second into 60 thirds, and so on; and since 6 is not measured by any prime number except 2 or 3, and 60 by no prime number except 2, 3, and 5, it follows that a whole circle could never be divided by a [sexagesimal or] tabular division into aliquot parts denominated by any prime number, or multiple of any prime number, other than 2, 3, or 5; nor, as stated already, could an aliquot part be isolated in this way by another prime number.

For this reason, if any circle of the sky were imagined as divided into 17 parts, called signs, and any such sign divided into 17 degrees, and any degree into 17 minutes, and so on, *and* if the tables were based on this [mode of division], they would never agree punctually with the common tables which we now use. The same could be said if a division of continuous proportionality were made by 61, or any other prime number.

Thus [even] if the celestial motions are commensurable, it does not follow that they could be measured exactly or equated by the tables that have been constructed, since it is possible that one mobile might traverse exactly 1 degree per day, while another traverses $1\frac{1}{7}$ of a degree, or even $\frac{1}{13}$ of a degree, per day; or it traverses $\frac{1}{22}$ part of the whole circle per day, or [traverses a distance represented] by some other number, any part of which cannot be divided in accordance with the divisions used in the common tables. However, the

---

201 *post* huiusmodi *hab. AL* tabule / cum *om. F*

202 *ante* concordarent *add. B* et precise / *ante* Eodem *add. P* et / divisio continue *tr. U* / proportionalitatis: proportionaliter *LU*

203 *post* 61 *hab. A* gradus / vel: seu *F* / *ante* aliquem *add. L* per / alium *om. AP*

204 quod *om. A* / per *rep. A*

204–5 per tabulas factas: tabulis factis *L* per tabulas *U*

205 vel: sive *F*

206 pertranseat in die: in die una pertransiret *U* / gradus *A* / alium *L* / *post* aliud

hab. *B* 1 / [mobile] *om. ABLPU*

207 gradum: gradus *AB* / $\frac{1}{7}$: septima *BFP* 7 *A* 61ª *L* / parte *om. U* / gradus *om. B* / vel: sive *F* / $\frac{1}{13}$: tertiamdecimam *BFLP* tertiam *A* / partem *om. U*

208 vel: sive *F* / $\frac{1}{22}$: vicensimam secundam *ABLP* 22 *F* $\frac{1}{12}$ *U* / partem *om. U* / alium *om. L*

208–9 numerum secundum: secundum *FP* terminum *U*

209 abscidi *F* abscindere *U*

210 astronomi: philosophi *AFLP* phisici *U* / compositorem *P*

210–11 intendebat *P*

debant talem precisionem quia per nullas tabulas unius proportionalitatis
posset haberi omnimoda precisio omnium motuum. Sed usi sunt divisione
secundum proportionalitatem sexagintuplam quia ipsa est ad eorum intentionem aptissima. Nihilominus in hoc libello, in quo loquendum est
215   magis mathematice, oportet uti fractionibus omnino precisis que vocantur
vulgares quia iam ostensum est quod alius modus non sufficit ad equan-
dum omnem velocitatem precise.

   *Conclusio quarta. Si duo mobilia nunc sint coniuncta necesse est ut alias in
puncto eodem coniungantur.*
220   Et semper est [in hac parte libri] intentio de motis commensurabiliter.
Sint itaque *A* et *B* coniuncta in puncto *g* quia idem est quam ad propositum
de coniunctione in puncto sicut in linea aut superficie. Et quia motus sunt
commensurabiles sequitur per quintam decimi [Euclidis] quod proportio
istorum duorum motuum est sicut [proportio] duorum numerorum. Sit
225   igitur motus ipsius *A* sicut numerus *C*, et motus *B* sicut numerus *D*. Ergo
in eodem tempore in quo *A* facit circulationes numeratas numero *C*, ipsum
B facit circulationes numeratas numero *D*, ergo in fine illius temporis
aliquot circulationes erunt precise complete ab ipso *A*, et similiter aliquot
ab ipso *B*; ergo quodlibet eorum erit ubi est modo. Ergo tunc erunt con-
230   iuncta in puncto in quo nunc sunt.

   Verbi gratia, sit velocitas ipsius *A* sicut 5, et velocitas ipsius *B* sicut 3.
Ergo, quando *A* fecerit 5 circulationes *B* fecerit 3, ergo tunc erunt ubi nunc
sunt. Et conformiter argueretur de [tempore] preterito et quod alias
fuerunt coniuncta.

211 *post* precisionem *hab. B* punctualiter
212 motuum: motum celi *B*
213 est *om. B*
214 *ante* aptissima *add. B* est / aptissima:
   ?approbata *L* / hoc: isto *F* / in quo *om.*
   U / loquendum: locutum *L*
215 *ante* oportet *hab. U* et
216–17 equandam *U*
217 *post* omnem *hab. P* in / *post* precise *add.*
   B et punctualiter *et add. F* et sic finis
   sit huius partis *et add. U* ergo et cetera
218 Conclusio quarta *om. FP* propositio
   quarta *U et ante* Si duo *mg. hab. B*
   quarta *et post* coniuncta *mg. hab. A*
   quarta / nunc: non *L* / *post* alias *mg.*
   hab. U nunc
219 puncto eodem *tr. L*
220 *ante* Et semper *mg. hab. B* de duobus
   mobilibus quarta huius de pluribus

duodecima huius de motis ?eccentricis
vicensima huius; de motis pluribus motibus habetur in vicensimasecunda
huius / Et semper…commensurabiliter
om. U / [in hac parte libri] *om. AFLP* /
motis: motibus *A* / commensurabiliter:
commensuratis *AL*
221 sit *A* / et *om. P* / puncto: principio *P* /
   quam: quantum *FLU*
222 de coniunctione *om. U* / puncto: principio *P* / sicut in: sicut *FP* aut *B* / aut:
   sive *F* / *post* aut *add. U* in / motus:
   modus *P*
223 sequitur…decimi: per quintam sequitur decimi *A* / [Euclidis] *om. AFLPU*
224 duorum[1] *om. L* / [proportio] *om. AFPU*
   istorum *L*
225 motus[1] *om. A* modus *P* / et motus *rep.*
   P / *infra* numerus D *mg. hab. B figuram*

astronomers who constructed the tables did not intend such precision, since precision for all the various types of motions cannot be obtained by any tables based on one proportionality. Instead, the astronomers have utilized a sexagintuple [or sexagesimal] proportionality because it is the most appropriate for their purpose. Nevertheless, in this little book, in which the discussion is more mathematical, it is necessary to use exact fractions, called "vulgar" fractions, since it has already been shown that no other way is adequate for equating every velocity exactly.

Proposition 4. If two mobiles are now in conjunction, it is necessary that they conjunct in the same point at other times.

In this part of the book the discussion is always about things moved commensurably. And so let A and B be in conjunction in point g, since, as far as what is considered here, a conjunction in a point is the same as on a line or surface. Since the motions are commensurable, it follows by the fifth [proposition] of the tenth [book] of Euclid that a ratio of these two motions is as a ratio of two numbers. Therefore, let A's motion be as number C, and B's motion as number D. Then, in the same time in which A makes C circulations, B makes D circulations, so that at the end of this time A and B will have completed an exact number of circulations and each of them will be where it is now. Consequently, they will be in conjunction in the point in which they are now.

For example, let A's velocity be as 5 and B's as 3. Then, when A will have made 5 circulations, B will have made 3, and they will be where they are now. A similar argument could be made for past time, namely, that they had conjuncted at other times.

226 in² *om. B* | A *om. A* | facit: sunt *L*
227 temporis: partis *A*
228 aliquot¹: aliquos *P* | erunt: essent *F* | precise complete *tr. B* | ipso *om. B* | et *om. B* | et similiter *tr. U*
229 ipso *om. B* | quotlibet *AB* | eorum: illorum tunc *B* | erit: et *A* | est modo *tr. U* | erunt: essent *F*
229–30 *ante* coniuncta *hab. L* tunc / coniuncta: coniuncter *P*
230 puncto: principio *P* | in quo: quo *B*

ubi *A*
231 sit *om. B* | A *om. L* | *ante* sicut¹ *hab. B* sit / ipsius² *om. AL* | sicut²: igitur *F*
232 fecerit²: fecit *PU* | erunt: essent *FP*
233 arguatur *U* | [tempore] *om. AFLPU* | *infra* preterito *mg. hab. U* figuram | et² *om. L* | quod: ?quoque *P* | post quod *hab. B* sit
234 fuerint *BFP* | post coniuncta *add. L* et cetera *et add. F* et sic finis esto huius ?presentis

235   *Conclusio quinta. Tempus invenire quando primitus coniungentur in puncto in quo nunc sunt.*

Compleat sicut prius ipsum *A* aliquot circulationes secundum numerum *C* quando *B* compleat circulationes secundum numerum *D*, et sint *C* et *D* in sua proportione minimi. Planum est autem quod *A* et *B* numquam
240   coniungentur ubi nunc sunt donec utrumque fecerit precise circulationes aliquot perfectas. Sed hoc erit quando *A* fecerit circulationes in numero *C* et ipsum *B* in numero *D*, et non prius quia tunc numeri circulationum [suarum] essent in alia proportione eo quod isti numeri *C* et *D* positi sunt minimi et ergo velocitates essent in alia proportione, quod est
245   contra positum.

Patet itaque quod tunc primo erunt iterum ubi nunc sunt quando *A* compleverit circulationes secundum numerum *C* et similiter *B* secundum numerum *D* qui sunt primi numeri proportionis velocitatum. Cum ergo velocitates sint commensurabiles sequitur quod est aliquod tempus quod
250   mensurat circulationes secundum primos numeros proportionis velocitatum. Sit ergo dies gratia exempli, quia idem est iudicium in proposito si est dies vel hora vel quodlibet aliud [tempus], ita quod *A* facit unam circulationem in diebus numeratis numero minori quando *B* facit unam in diebus numeratis numero maiori. Sed *A* facit circulationes secundum numerum *C*
255   antequam coniunguntur ubi nunc sunt. Et quamlibet circulationem facit in diebus numeratis numero *D*, ergo multiplicando *C* in *D* exit tempus quando primo coniungentur ubi nunc sunt.

Verbi gratia, compleat *A* 5 circulationes quando *B* complet 3, patet ergo quod si coniungentur ubi nunc sunt quando *B* fecerit pauciores circulatio-
260   nes, oporteret velocitates habere aliam proportionem quia nullus numerus minor quam 3 habet eandem proportionem ad aliquem numerum quam

235 Conclusio quinta *om. FP* propositio quinta *U et ante* Tempus *mg. hab. AB* quinta / invenire *tr. B post* sunt *in linea 236* / quando: in quo *A* quo *U* / primitus: primo *B*
236 puncto: principio *P* / in *om. B*
237 sicut: ut *U*
237–38 numerum…secundum *om. F*
238 et² *om. B*
239 *ante* in *hab. B* numeri / minimi: numeri *L* / numquam *om. A*
240 coniungeretur *U* / precise: precisio *P*
241 hoc: illud *F* / erat *B*
242 ipsum *om. B* / *ante* in numero *hab. L* et ipsum / prius: plus *A* / numeri: nu-

merum *A*
242–43 circulatio *F* circulationem *P* proportionum *U*
243 [suarum] *om. AFLPU* / essent: sunt *P* / eo: et *P*
244 velocitas *BP* / essent: ille esset *B* sunt *P* / proportione: proportionalitate *U*
246 erunt: essent *FL* / iterum *om. AF* / *ante* ubi *hab. L* ibi
246–47 quando…circulationes *om. L*
247 secundum numerum: numero *BFPU* / similiter B *tr. B* sit B *L*
248 qui: que *P* / primi *om. A* / *ante* numeri *hab. B* minimi / proportionis: proportionales *A*

Proposition 5. [How] to find the time when the two mobiles will first conjunct in the point in which they are now.

As before, let *A* complete *C* circulations when *B* completes *D* circulations, and [assume that] *C* and *D* are in their least ratio. It is evident, however, that *A* and *B* will never conjunct in their present position until each will have completed several exact circulations. But this will occur when *A* will have made *C* circulations, and *B*, *D* circulations—and not before since the numbers representing their circulations would then be in a ratio other than *C* to *D*, which were assumed to be in their least ratio, and, consequently, the velocities would be in another ratio, contrary to what was assumed.

It is obvious, however, that they will first return to where they are now when *A* will have completed *C* circulations and *B*, *D* circulations, which are the prime numbers of the ratio of velocities. Since the velocities are commensurable, it follows that there is a certain time which measures the circulations in terms of the prime numbers of the ratio of velocities. For example, let [the unit of time] be the day—the conclusion is the same whether a day, hour, or any other time unit is used—so that when *B* makes one circulation in a number of days represented by the greater number, *A* completes one circulation in a number of days represented by the smaller number. But before they are conjuncted in their present position, *A* makes *C* circulations. And since *A* makes a circulation in *D* days, the time in which they will be in conjunction where they are now is found by multiplying *C* by *D*.

For example, should *A* complete 5 circulations when *B* completes 3, it is obvious that if they are to be conjuncted where they are now with *B* making fewer [than 3] circulations, it would be necessary for the velocities to be related in another ratio because no number smaller than 3 bears the same

---

250 *ante* secundum *hab. B* illius / primos *om.* *L* / proportionis: proportionales *A*
251 sint *ALU* si *P* / iudicium in proposito: in proposito iudicium *U* / est: sit *L*
252 vel¹: aut *F* / *ante* vel² *add. B* vel annus / vel²: sive *F* / quodlibet *om. B* / [tempus] *om.* *AFPU* huiusmodi *B*
253 quando *L* quam *BF* scilicet D sed *U* / *post* unam *add. U* circulationem
253–54 minori...numero *om. A* / quando... maiori *om. P*
254 *post* maiori *add. U* scilicet C / *ante* numerum C *scr. et del. U* in *mg.* ?aliter

forte sit et B inter ?C circulationes facit et cetera
255–56 Et quamlibet...exit *om. P*
255–57 Et quamlibet...nunc sunt *om. A*
256 C in D: D in C B / exit: exibit *F*
258 complet 3: et *P*
259 coniungerentur *LPU* / fecerit: facit *U* / *post* fecerit *hab. P* 3
259–60 pauciores circulationes *tr. FP*
260 oportet *AU*
261 3: trius *F* / haberet *L* / *ante* quam *scr. et del. U* quod

ipse ternarius habet ad 5. Ergo primo coniungentur quando *B* fecerit 3 circulationes et similiter *A* fecerit 5. Et iterum patet quod velocitas *A* ad velocitatem *B* est sicut 5 ad 3, ergo *A* complet unam circulationem in tem-
265 pore quod se habet ad tempus circulationis ipsius *B* econverso, sicut 3 ad 5. Et sic *A* complet unam circulationem in 3 deibus et *B* in 5, et in fine 5 circulationum ipsius *A* ipsa coniungentur. Ergo oportet multiplicare 3 per 5 et exit numerus dierum quesitus, scilicet 15 dies. Vocetur itaque istud tempus peryodus revolutionis *A* et *B* quod tempus invenitur multiplicando
270 unum per aliud de minimis numeris proportionis velocitatum motuum. Quo tempore adimpleto coniunctiones et omnes aspectus quantum est ex parte talium mobilium rursum incipiunt fieri omnino sicut ante.

Conclusio sexta. Datis velocitatibus duorum mobilium nunc coniunctorum tempus prime coniunctionis sequentis reperire.

275 Velocitates date sunt quando proportio earum est data. Sint itaque ut prius *A* mobile velox et *B* tardum; et motus *A* sit sicut numerus *C*, motus vero ipsius *B* sicut numerus *D*. Ergo *A* pertransit partem circuli denomina-tam numero *D* quando *B* pertransit partem circuli denominatam numero *C* quia maior pars numeratur minore numero. Sed minimus numerus qui
280 habet partem secundum *C* et *D*, scilicet qui numeratur ipsis *C* et *D*, est ille qui fit ducendo *C* in *D* ut patet per tricensimamquartam septimi Euclidis. Ergo dividendus est circulus et signandus per illum numerum qui fit ex ductu *C* in *D*.

Verbi gratia, in exemplo iam posito, quoniam *A* pertransit $\frac{1}{3}$ circuli
285 quando *B* pertransit $\frac{1}{5}$, scilicet in una die, et primus numerus qui habet 3 et 5 est 15, ergo totus circulus erit sicut 15 sicut et tempus totius revolutionis est 15. Cum igitur *A* sit velocius, sequitur quod quando *A* lucratum fuerit super *B* unam circulationem tunc primo coniungentur et tunc *A* attinget *B*. Primo subtrahatur itaque illud quod *B* pertransit in die ab illo quod *A* per-

---

262 5: quinarium *B* / coniungeretur *P* / *post* quando *hab. A* ?fecit

263 similiter *om. B* / *post* 5 *add.* U circula-tiones / Et² *om. B*

264–65 unam...tempus *om. A*

265 circulationes *A* / ipsius *om. L* / sicut: sit *FP*

266 unam *om. AFLP* / *post* 5¹ *add. B* ?die-bus

267 circulationem *P* / *ante* ipsa *hab. B* tunc / coniungeretur *BP* / Ergo oportet *tr. B*

268 exit: exibit *A* ?*F* / 15 dies *tr. B* / vocatur *L*

269 tempus¹: tempora *A* / multiplicando:

270 aliud: reliquum *A* / *post* aliud *add. BFPU* et / velocitatis *FPU*

271 adimpletus *P*

271–72 et omnes...mobilium *om. A*

272 incipiunt: incidunt *P* / omnino: omnis *P* / *post* ante *hab. F* et cetera et cetera *et mg. hab. B figuram*

273 Conclusio sexta *om. FP* propositio sexta *U* sexta *mg. hab. B ante* Datis *et mg. hab. A post* mo- *in* mobilium / velocitatibus: velocibus *P*

274 tempus: tunc *L*

275 quandoque *L* / eorum *LPU* ipsarum *F* /

multitudo *P*

ratio to some number as 3 bears to 5. Hence, they will first conjunct when B will have made 3 circulations and A 5. But again it is clear that the velocity of A to the velocity of B is as 5 to 3, so that A completes one circulation in a time which bears an inverse ratio to the time of B's circulation—that is, as 3 to 5. Thus A completes one circulation in 3 days and B in 5, and at the end of 5 circulations of A they will be in conjunction. It is, therefore, necessary to multiply 3 by 5 to arrive at the number of days sought, namely, 15 days. Let this time be called the period of revolution of A and B because the time is found by multiplying together the least numbers of the ratio of the velocities of the motions. When this time has passed, the conjunctions and all [other] aspects undergone by such mobiles begin again exactly as before.

Proposition 6. [How] to find the time of the first conjunction that will follow a [present] conjunction of two mobiles whose velocities have been given.

Velocities are given when their ratios are given. And so, as before, let the mobiles be A, the quicker, and B, the slower; and let the motion of A be as number C and that of B as number D. Since the greater part [of a circle] is measured by the smaller number, A traverses part of a circle represented by D when B traverses part of a circle represented by C. But the smallest number which contains C and D as parts—that is, which is numbered [or measured] by C and D [separately]—is the product of their multiplication, as is shown by the thirty-fourth proposition of the seventh book of Euclid.[6] This number, the product of C and D, must divide and designate the circle.

For instance, referring to the example already presented, since A traverses $\frac{1}{3}$ of a circle in one day while B traverses $\frac{1}{5}$, and the prime number containing 3 and 5 is 15, the whole circle will be as 15, just as the time of a complete revolution is 15. However, because A is quicker, it follows that when A has gained one circulation over B it reaches B and they will be in conjunction at the first place. To begin with, then, the distance which B traverses in a day

sit U

276 *post* B *add.* B mobile / tardum: tardet P / sicut *om.* U

277 ipsius *om.* B / sicut: sit U

278 D quando...numero *om.* F / circuli *om.* APU

279 numeratur: denominatur AL / minori FP a minori U / *ante* qui hab. F est

280 scilicet: ?aut A / ipsis *om.* B / est *om.* FU et B / ille: iste A

281 per *om.* L / septimi: aut P

282 dividendus: ducendus A / illum: istum

FP

284 iam posito: dicto B / quoniam: quando L / pertransivit U / $\frac{1}{3}$: unam tertiam $ABLP$ unam triplam F

285 quando *om.* L / $\frac{1}{5}$: unam quintam $ABFLP$ / una die *tr.* L

285–86 3 et 5: tertiam et quintam BLP triplam et quintam F $\frac{1}{3}$ et $\frac{1}{5}$ U

286 erat U / sicut²: sic A / et *om.* B quod U

288 unam *om.* F / coniungeretur P

289 substratur F / illo: isto FP

289–90 ab...et *om.* L / A pertransit *tr.* B

290 transit in die et restabit spatium quod *A* lucratur super *B* in die. Sumatur
ergo istud superlucratum totiens quod valeat unum circulum et in fine
temporis in quo totum hoc erit superlucratum *A* et *B* primo coniungentur.

Et hoc fiet et invenietur dividendo totum circulum per numeratorem
illius excessus unius diei sive superlucrati de illis numeris a velociori super
295 tardius. Et idem est quod dividere denominatorem per numeratorem qui
numerant et denominant illud superlucratum in una die.

Verbi gratia, positum est quod *A* complet circulationem in 3 diebus et *B*
in 5, ergo in die *A* pertransibit $\frac{1}{3}$ circulationis, vel circuli, et *B* $\frac{1}{5}$. Ergo si
subtrahatur $\frac{1}{5}$ ab $\frac{1}{3}$ exibit illud quod *A* superlucratur in die et hoc est $\frac{2}{15}$. Et
300 cum in toto circulo sint 15, patet statim quod in fine 7 dierum cum dimidio
totus circulus erit superlucratus et quod *A* fecerit unam circulationem
plusquam *B* et tunc primo coniungentur. Unde si istius superlucrati
denominator dividatur per numeratorem eiusdem, videlicet 15 per 2, exit
tempus quesitum. Et nihil refert si, gratia exempli, ponantur moveri super
305 unum circulum vel super plures et etiam idem est si loco diei capiatur hora,
vel minutum, vel annus, aut quomodolibet.

Est ergo regula talis ad inveniendum propositum. Subtrahatur motus
unius a motu alterius et residuum habet numeratorem et denominatorem.
Dividatur itaque denominator per numeratorem et exibit tempus quesitum.

310 *Conclusio septima. Datis duobus motibus duorum mobilium, numerum
coniunctionum totius revolutionis invenire, videlicet quot erunt coniunctiones
quousque rursum incipiant fieri sicut ante et coniungentur in puncto in quo
nunc sunt.*

Cum enim motus sint regulares semper inter quaslibet duas coniunctio-
315 nes proximas est equalitas de tempore. Non oportet ergo nisi dividere

290 et *om.* *A* / restabit: remanebit *B* / lucra-
bitur *B*
291 totiens: totius *F*
292 in *om.* *BL* / totum hoc *tr.* *L* / erit *om.*
A erat *U* / primo coniungentur *tr.* *B*
293 hoc: illud *F*
294 illius excessus *tr.* *U* / superlucratum *A*
superlucratis *FP* / de illis numeris *om.*
A / velociori: velocitate *A*
295 est *om.* *A* / per numeratorem *om.* *P* /
post numeratorem *add.* *A* de illis numeris
296 denominant: denominatum *A*
297 compleat *U*
298 *ante* die *add.* *B* una / *ante* pertransibit
scr. et del. *A* complet circulationem /

pertransit *B* pertransit B *FPU* / $\frac{1}{3}$: unam
tertiam *ABFLP* namque pertransit ter-
tiam *U* / vel: sive *F* / $\frac{1}{5}$: unam quintam
BFLPU unam 5 *A*
299 $\frac{1}{5}$: una quinta *BFLP* sed una *rep.* *F*
una 5 *A* / ab: a *F* / $\frac{1}{3}$: una tertia *BFLP*
una 3 *A* / superlucrabitur *FU* / *ante* die
add. *B* una / $\frac{2}{15}$: due 15 *BF* due 15^e *AP*
due ?coniuncte ?deinde *L*
300 dimidio: dividere *P*
301 erit: exit *U* / quod *om.* *A* que *P*
302 coniungeretur *P*
303 denominatur *A* / dividat *L* / videlicet: ut
A scilicet *B* / exit: ?exibit *F*
304 ponantur moveri *tr.* *U*
305 vel: sive *F* / super *om.* *U* sunt *F* / etiam:

is subtracted from the distance traversed by *A* in one day and what remains is the distance gained by *A* over *B* in the course of a day. The distance gained is then taken as many times as required to make one circle, and at the end of the time in which the whole circle will have been gained, *A* and *B* will be in conjunction in the first place.

This is achieved and found by dividing the whole circle by the numerator of those numbers representing the excess gained in one day by the quicker over the slower. And this is the same as dividing the denominator by the numerator, which measures and denominates the gain made in a single day.

For example, it has been assumed that *A* completes a circulation in 3 days and *B* in 5, so that in one day *A* will traverse $\frac{1}{3}$ of a circulation, or circle, and *B*, $\frac{1}{5}$. If, then, $\frac{1}{5}$ is subtracted from $\frac{1}{3}$, the result is $\frac{2}{15}$ and represents what *A* has gained [over *B*] in a day. Since there are 15 [parts] in the whole circle, it is immediately evident that at the end of $7\frac{1}{2}$ days a whole circle will have been gained, and *A* will have made one more circulation than *B* and they will be in conjunction in the first place. The time [for this first conjunction] is found when the denominator of [the fraction representing] what has been gained [in a day] is divided by the numerator of the same [fraction], namely, 15 by 2. And it does not matter if it is assumed that the mobiles are moved, for example, on one circle or on several; and it is also the same if instead of a day, an hour, or minute, or year, or any other [unit] were taken.

Here is the rule for finding what has been proposed here. The motion of one [mobile] is subtracted from the motion of another and the remainder has a numerator and denominator. The time sought is produced by dividing the denominator by the numerator.

Proposition 7. [How] to find the number of conjunctions in a complete revolution when the motions of the two mobiles have been given; that is, how many conjunctions will occur before they will conjunct in the point where they are now and begin to repeat the previous cycle of conjunctions.

Now since the motions are regular, the time between any two successive conjunctions is always equal. Therefore, nothing more is required than to

---

ideo *L* / capiatur: sumatur *F*

306 vel¹: sive *F* / vel²: aut *F* / aut *om. P* sive
    F vel *L* / quomodolibet: aliquolibet *P*

307 regula: ?regulatur *L*

308 motu: motum *P* / et¹ *om. B*

309 itaque: igitur *L* / et *om. P* / *post* quesitum *hab. P* et cetera

310 Conclusio septima *om. AFP* propositio septima *U* septima *mg. hab. B ante*

Datis / duobus *om. L*

311 coniunctionum *om. ABU* / revolutionis: coniunctionis *U* / quot: quod *P* / erunt: essent *F*

312 rursus *A* / rursum incipiant *tr. B* / coniungantur *LPU* / in² *om. B*

314 enim: vero *F* / sunt *A* / inter: per *L*

315 equalitas: equaliter *AFP* equale *BU* / de tempore: tempus *B*

tempus totius revolutionis quod cognoscitur per quintam conclusionem
per tempus quod est de una coniunctione ad coniunctionem sibi proximam
seu proximo sequentem quod tempus habetur per conclusionem immediate
precedentem et exibit numerus quesitus.

320   Verbi gratia, in exemplo predicto, 15 erit numerus sive totum tempus
totius revolutionis A et B, et 7 cum dimidio est numerus seu tempus inter
duas coniunctiones invicem proximas. Dividatur igitur 15 per 7 cum
dimidio et venient 2 qui est numerus coniunctionum totius revolutionis
ipsorum A et B ita quod tertia coniunctio erit ubi fuit prima, et quarta ubi
325   fuit secunda, et quinta iterum ubi fuit prima, et sic consequenter.

Conclusio octava. Hiis duobus mobilibus nunc coniunctis locum prime
coniunctionis sequentis assignare.

Quia iam per sextam conclusionem habetur tempus huiusmodi coniunctionis non restat nisi videre quantum alterum istorum mobilium pertransit
330   in illo tempore. Deinde ab illo pertransito subtrahatur totus circulus quotiens poterit subtrahi, si est possibilis talis subtractio. Et hoc est idem quod
dividere totum circulum per illud pertransitum et habebitur propositum.

Verbi gratia, ex data velocitate in casu prius posito, patet quod B pertransit in die $\frac{1}{5}$ circuli; et quia tempus usque ad primam coniunctionem
335   sequentem est 7 dies cum dimidio per sextam conclusionem patet quod B
pertransit illo tempore $\frac{7}{5}$ circuli cum dimidia quinta, videlicet totum circulum et medietatem eiusdem. Ergo [sequens] coniunctio erit in puncto
opposito illi puncto in quo A et B nunc sunt coniuncta. Et idem invenientur operando per motum ipsius A.

316 cognoscitur: scitur *B* / quintam: 5 *A*
quamlibet *U* / conclusionem: coniunctionem *U*
317 sibi proximam *tr. U* proximam *B*
318 seu: vel *B* sive *F* / proxime *FU* / tempus *om. B* / haberetur *B* / ante immediate *scr. et del. U* coniunctionem
318–19 conclusionem immediate precedentem: sextam *B*
320 predicto: preaddito *FP* / 15 erit numerus: 15 de erit *B* erit numerus 15 *L* / erat
U / sive totum *om. B*
321 totius...tempus *om. F* / et 7 *om. A* / 7:
a *L* / numerus seu *om. B* nunc seu *A* /
tempus *om. A*
323 venient: provenient *B* veniant *F* veniunt *L* / numerus *om. B* / coniunctionis *A*
324 quod *om. FP* / fuit: erat *L* fuerit *P*

325 fuit¹: fuerit *BP* / quinta: 5 *A* / fuit² *om.*
AFP / prima *om. FP*
326 Conclusio octava *om. AFP* propositio
octava *U* octava *mg. hab. B ante* Hiis /
locus *P*
328 sextam: 6 *A* / conclusionem habetur:
haberetur *B* / huiusmodi: huius *AU*
329 non restat: nunc restet *U* / nisi videre
tr. A videre *U* / alterum istorum mobilium: istorum mobilium alterum *B* /
istorum: illorum *A*
330 illo¹: isto *P* / Deinde: enim *P* / illo²:
isto *A*
331 poterit: potuit *A* / subtrahi *om. B* / post
subtrahi *add. L* et / hoc *om. P* / idem
quod *om. A*
332 habebitur: habetur *AU* haberetur *B*
334 $\frac{1}{5}$: unam quintam *ABFLP* / circuli *om.*
F / et *om. U* / coniunctionem: conclu-

divide the time of a complete revolution, known by the fifth proposition, by the interval between one conjunction and the very next one following—and this time can be determined by the proposition immediately preceding. [In this way] the desired number is produced.

For example, following the previous illustration, 15 will be the number or total time of the revolution of A and B, and $7\frac{1}{2}$ is the number or time between two successive conjunctions. Then 15 should be divided by $7\frac{1}{2}$ to yield 2, which is the number of conjunctions in a complete revolution of A and B. Thus, the third conjunction will be [in the place] where the first was, the fourth where the second was, the fifth again where the first was, and so on in this order.

Proposition 8. [How] to determine the place of the first conjunction following the present conjunction of these two mobiles.

Since, by the sixth proposition, it has already been shown [how to find] the time required for such a conjunction, there only remains to be seen how great a distance either of these [two] mobiles traverses in that [very same] time, and then to subtract the whole circle from that distance as many times as possible. This is the same as dividing the whole circle by that distance so that what has been proposed will be found.

For example, from the velocity given in the case presented before, it is obvious that in one day B traverses $\frac{1}{5}$ of its circle. But since by the sixth proposition, $7\frac{1}{2}$ days represents the time elapsed before the first conjunction following [the present conjunction], it is evident that during that time B traverses $\frac{7}{5}$ of a circle plus $\frac{1}{2}$ of $\frac{1}{5}$ of a circle—that is, the whole circle plus half of it. Consequently, the [very next] conjunction following will occur in a point opposite the point in which A and B are now in conjunction. The same result will be obtained by operating with the motion of A.

---

sionem P
335 per *om.* P / sextam: 6 A / conclusionem: precedentem B
336 illo: isto AFP / $\frac{7}{5}$: 7 quintas $ABFP$ septem quintas L / circuli *om.* B / dimidia quinta: $\frac{1/2}{5}$ U dimidia 5 F / videlicet *om.* B videt P

337 medietate AP / eiusdem: eius B / [sequens] *om.* $AFLPU$ / erit: est L
337–38 puncto opposito *tr.* L
338 puncto *om.* B / in *om.* $BFPU$ / Et *om.* B
338–39 invenientur: invenitur F
339 A *om.* A

340   *Conclusio nona. Assignata distantia duorum mobilium locum et tempus
       prime coniunctionis sequentis dare.*

       Hec distantia signanda est secundum circuli portionem incipiendo a
       velociori ita quod mobile velocius ponatur retro. Unde si *A* precederet *B*
       per unum arcum parvum, ut per unum signum, tunc *B* diceretur ante *A* per
345   residuam circuli portionem, scilicet per undecim signa. Accipienda est
       igitur talis distantia signata per suum numerum et rursum capiendi sunt
       minimi numeri proportionis velocitatum mobilium et differentia eorum;
       igitur si illa differentia est numerus per quem signata est distantia dico
       quod *A* coniungetur cum *B* quando quodlibet eorum motuum fuit secun-
350   dum numeros date proportionis et in fine spatii ab istis numeris numerati.

       Si vero differentia minimorum numerorum proportionis velocitatum sit
       alius numerus quam ille per quem distantia est signata, tunc sicut ista
       differentia se habet ad illum numerum ita quilibet numerus proportionis
       velocitatum ad aliquem qui ostendit locum et tempus coniunctionis quesite
355   quod absque prolixitate demonstrationis patet evidenter exemplo.

       Verbi gratia, sit *A* velocius quam *B* in proportione 8 ad 3 quorum dif-
       ferentia est 5 et distent per 5 gradus. Patet ergo quod cum *A* pertransiverit
       8 gradus et *B* pertransiverit 3, tunc ibidem coniungentur. Si vero distent
       per 2 gradus sicut 5 se habet ad 2 ita 8 ad illum numerum graduum quem
360   pertransibit *A* antequam coniungantur, et est 3 gradus et $\frac{1}{5}$. Et similiter
       sicut 5 ad 2 ita 3 ad illum numerum quem pertransibit *B* antequam con-
       iungantur et est unus gradus et $\frac{1}{5}$. Sic ergo *A* transibit $\frac{16}{5}$ et *B* $\frac{6}{5}$ que se
       habent in proportione velocitatum posita.

340 Conclusio nona *om. AFP* propositio
    nona *U* nona *mg. hab. B ante* assignata
342 signanda est *tr. F* / circuli portionem
    tr. B
343 precederet: precedat *A* precedit *U*
344 unum[1] *om. B* / ut: id est *A* scilicet *B* et *L*
345 residua *P* / circuli portionem *tr. B* talis
    portionem *U*
345–46 est igitur *tr. L*
346 signata: signorum *U* / rursua *U*
347 minimi numeri *tr. AFU* / proportionis:
    proportionales *A* / *ante* velocitatum *scr.
    et del. U* proportionum / earum *F* isto-
    rum *A* illorum *B*
348 igitur si: si ergo *B* / illa: ista *AF* ita
    P / quem: quam *F*
349 coniungeretur *U* / quolibet *A* / eorum:
    illorum *L* / motum *BLU* / fuerit *L*
350 numeros: numerum *FP* / numerati: nu-

meratis *BFP*
351 vero: enim *L* / minimorum numerorum:
    numeratorum *U* / proportionum *P*
352 quam: quod *U* / ille: iste *A* / distantia
    est signata: signata est distantia *B* / *post*
    distantia *add. L* illa / sicut: ut *F* / ista:
    illa *F*
353 se habet *om. AFLPU* / quelibet *BP*
354 *post* aliquem *add. U* numerum / *ante*
    locum *hab. B* et
355 quod: quidem *A* / *ante* demonstratio-
    nis *hab. P* per / demonstratis *A*
356 8: octupla *A*
357 5[1]: quinta *P* / et distent per *om. A* /
    distant *L* / *post* distent *add. B* ab invi-
    cem / grada *B* / Patet: post *L* / quod *om.
    A* / A *om. A* / pertransiverit: pertransivit
    LP pertransierit *U*
358 8: ergo *P* / gradus *om. U* grada *B* / ibi-

*Proposition 9. [How] to determine the place and time of the first conjunction
following [a given conjunction] when a distance between the two mobiles has been
assigned.*

The distance which is to be assigned covers a section of the circle starting
from the quicker mobile which is assumed to be behind [the slower mobile].
Thus, if *A* precedes *B* by a small arc—let us say, by one sign—then *B* is said
to be ahead of *A* by the section of the circle remaining, namely, by eleven
signs. When such a distance has been designated by its number, it must be
considered along with the difference between the least numbers representing
the ratio of the velocities of the mobiles. Then, if that difference equals the
number designated as the distance, I say that *A* would be in conjunction with
B when their motions are represented by the numbers of the given ratio [of
velocities]; and this conjunction will occur when distances have been traversed
which are represented by those numbers.

If, however, the difference between the least numbers of the ratio of velocities
is a number other than that which designates the distance, this difference is
related to that number [designating the distance] just as any number of the
ratio of velocities is related to any number giving the place and time of the
conjunction which is sought. All this can be shown quite clearly by example
without resorting to a lengthy demonstration.

For example, let *A* be quicker than *B* and related to it in a ratio of 8 to 3,
where the difference is 5; and let them be separated from one another by 5
degrees. Clearly, then, when *A* will have traversed 8 degrees and *B*, 3 degrees,
they will be in conjunction in the same place. But if they are separated by 2
degrees, then as 5 is to 2 so is 8 to that number of degrees which *A* will traverse
before conjunction [with *B*]; and this is $3\frac{1}{5}$ degrees. Similarly, as 5 is to 2 so
is 3 to that number [of degrees] which *B* will traverse before conjunction [with
A]; and this is $1\frac{1}{5}$ degrees. Thus *A* will travel $\frac{16}{5}$ [degrees] and *B*, $\frac{6}{5}$ [degrees];
and these distances are related as the ratio of velocities assumed [at the outset].

---

dem: ibi *A* / coniungetur *A* / vero: non *FP*
/ distat *U* / *post* distent *add.* *B* invicem
359 gradus: grada *B* / sicut: ut *F* / se habet
om. AFLPU / ita *om.* *P* / graduum: gradum *B* / quem: quam *A* ?quia *F*
360 pertransivit *P* / *post* antequam *hab.* *U*
?pari / coniungentur *P* / 3: trius *F* /
grada *B* / $\frac{1}{5}$: una quinta *BFLP* una 5 *A*

360–62 Et similiter…et $\frac{1}{5}$ *om. BFPU*
361 quem: quam *A*
362 $\frac{1}{5}$: una quinta *L* / *A om. AFPU sed tr.* *B*
post transibit / transibit: pertransibit
AL / $\frac{16}{5}$: 16 quintas *B* quintas 5 *A* / et³
om. *A* / $\frac{6}{5}$: 6 quintas *B*
363 posita: proposita *AU* / *post* posita *hab.*
L et cetera

Conclusio decima. Numerum et seriem punctorum reperire in quibus
365 *umquam talia duo mobilia coniungentur.*

Ex conclusione septima patet numerus coniunctionum totius revolutionis et quia inter quaslibet duas proximas est equalitas de tempore propter regularitatem motuum statim sequitur quod quot sunt coniunctiones
in una revolutione tot sunt loca sive puncta coniuntionum equaliter dis-
370 tantia ab invicem. Atque dividentia circulum per equalia in quibus punctis
et non in aliis fient coniunctiones istorum mobilium. Ex quo patet quod
hec puncta distant ab invicem per portionem circuli commensurabilem
toti circulo secundum proportionem multiplicem, videlicet per partem
circuli aliquotam sicut si sint tria talia puncta quelibet duo proxima dis-
375 tant per $\frac{1}{3}$ circuli; et si quatuor quelibet duo proxima distant per $\frac{1}{4}$; si
quinque per $\frac{1}{5}$, et sic de aliis.

Sed quo ordine procedant coniunctiones per hec puncta invenietur per
octavam conclusionem que docet quantum unaqueque coniunctio distet
localiter ab illa que ultimo fuerat ante ipsam. Unde in circulo fit divisio
380 per equalia secundum aliquot circa puncta. Si coniunctio nunc est in uno
istorum punctorum non oportet quod coniunctio que immediate post erit
sit in puncto immediate sequenti, sed quandoque procedit ad tertium vel
ad quartum saltando ordinate et hoc diversimode in diversis secundum
variationem velocitatum, ut per conclusiones positas potest experiri et
385 etiam per sequentem in multis exemplis quorum unum proferatur in
medium.

Sit velocitas *A* sicut 12 et velocitas *B* sicut 5. Tunc per istam conclusionem et per sequentem invenietur quod numerus punctorum in quibus *A* et

364 Conclusio decima *om. FP* propositio *U* decima *mg. hab. B ante* conclusione (*linea 366) et post* re- in reperire *mg. hab. A* decima / reperire *om. FP sed hab. B* reperire *post* coniungentur (*linea 365)*
365 umquam *om. L* numquam *FP* / coniungentur: moventur *U*
366 conclusione septima *tr. B* conclusione 7 *A*
366–67 *ante* revolutionis *scr. et del. U* circulationis
367 et *om. U* / equalitas: equaliter *AFLP* equale *B* / de tempore: tempus *B*
368 regularitatem: regularem *P* / sunt coniunctiones *tr. B* coniunctiones *U*
369 loca sive puncta: puncta sive loca *F* loca vel puncta *B* / equaliter: equalitatis *L*
370 ab: ad *P*
371 et *om. A* / fient: fierent *A*
372 ab: ad *P* / invicem: invicis *L* / per portionem: proportionem *P*
373 multiplicem: multiplicitatem *P* / videlicet: videt *P* / post* partem *scr. et del. B* celi
375 per $\frac{1}{3}$...distant *om. FPU* / $\frac{1}{3}$: unam tertiam *ABL* / circuli et *om. B* / quelibet: quedem *A* / quelibet...distant *om. B* / *post* duo *add. A* puncto / $\frac{1}{4}$: unam quartam *BFLP* unam 4 *A* / si: scilicet *FU*
376 quinque: per $\frac{2}{4}$ *U* / $\frac{1}{5}$: quintam *AFLPU* unam quintam *B*
377 procedunt *F* precedant *P* / invenitur *L* scietur *B*
378 octavam: unam *P* / unaqueque: quelibet *B* / distat *L*

Proposition 10. [How] to find the number and series of points in which two such mobiles will always conjunct.

[The manner of determining] the number of conjunctions in a complete revolution was made clear in the seventh proposition. Since there is an equal interval of time between any two successive [conjunctions]—a consequence of the regularity of the motions—it follows directly that there are as many conjunctions in one revolution as there are places, or points, of conjunction equidistant from one another. And the circle must be divided into equal parts such that the conjunctions of these mobiles are made in these points [of division], and in no others. From this it is evident that these points are separated from one another by a sector of the circle commensurable to the whole circle in a multiple ratio—that is, by an aliquot part of the circle, as, for instance, if there are three such points, any two immediate neighbors are separated by $\frac{1}{3}$ of a circle; and if there are four points, any two immediate neighbors are separated by $\frac{1}{4}$ [of a circle]; if there are five points, they are separated by $\frac{1}{5}$ [of a circle], and so on for any number of points.

But the order in which the conjunctions proceed through these points will be found by the eighth proposition which shows by how much any particular conjunction is separated from the one immediately preceding. The division of a circle into equal parts is made according to the number of points [that are to be located] around it. If there is presently a conjunction in one of these points, it is not necessary that the conjunction occurring immediately after be in a point which is directly next to it; but sometimes the conjunction immediately following proceeds, by a regular jump, to a third or fourth point. And this happens in different ways under different circumstances depending upon the variation in velocities, as can be proven by the propositions [already] posited and also by following one example, among many, which is now set forth.

Let the velocity of A be as 12, and that of B as 5. Then, by this proposition and the next, it will be found that the number of points in which A and B

379 localiter *om. L* / illa: ista *P* / illa que *om.*
   B / ultimo fuerat: ultima proxima *B* / fit:
   sit *BF*
380 secundum aliquot: sicut autem *A*
381 post erit: sequitur *B*
382 sit *om. U* sic *FP* / *post* puncto *hab.*
   AFLPU istorum / *ante* sequenti *hab. B*
   etiam / proceditur *AFLPU* / tertiam

U 3 *A*
383 hoc *om. P*
384 variationem: nominationem *BFPU* / ut:
   et *L*
385 proferatur: perferatur *A*
387 *ante* Sit *add. B* verbi gratia
388 *post* et[1] *hab. B* etiam / invenitur *F* /
   quod: quam *P*

B umquam coniungentur est 7 et per octavam conclusionem reperitur
390 quod unaqueque coniunctio distat localiter ab ultimo preterita per $\frac{5}{7}$
circuli, ergo cum sint septem puncta circuli equaliter ab invicem distantia
erit coniunctio in uno, deinde in sexto ab illo, scilicet quatuor punctis
intermissis, deinde in sexto ab isto aliis quatuor intermissis, et sic semper.
Et in aliquo casu omitterentur 2, vel 3, et quandoque nullum sed sine saltu
395 coniunctiones fierent ordinate per hec puncta [et cetera].

*Conclusio undecima. Omnium duorum mobilium tot sunt coniunctiones in
una revolutione et tot puncta in quibus umquam possunt coniungi quota est
differentia minimorum numerorum proportionis velocitatum motuum.*

Et nunc suppono quod quodlibet [eorum] moveatur unico motu simplici
400 quia de pluribus motibus videbitur post. Sit itaque velocitas *A* sicut numerus *C* et velocitas *B* sicut numerus *D* qui numeri sint contra se primi,
vel in sua proportione minimi, quod idem est. Et sint [*A* et *B*] coniuncta.

Ergo quando *A* fecerit circulationes secundum numerum *C*, et *B* secundum numerum *D*, tunc primo erunt ubi nunc sunt et erit completa revolutio
405 tota, ut patet per quintam conclusionem. Sed tunc *A* lucratum fuerit super
B tot circulationes quota est differentia ipsius *C* super *D*, vel quot unitatibus *C* superat *D*, et quotienscumque *A* lucratur super *B* unam circulationem totiens coniunguntur per sextam conclusionem. Ergo in totali revolutione totiens coniungentur quot unitatibus *C* excedit *D*, qui *C* et *D* sunt
410 numeri minimi proportionis velocitatum. Et post revolutionem predictam

389 umquam coniungentur *tr.* B umquam *U* numquam coniungentur *F* nequam coniungentur *P* / octavam: 8 *A* / reperietur *A*

389–90 est…localiter *om. U*

390 localiter *om. L* / preterita: puncto *A* / $\frac{5}{7}$: 5/7as *A* 5 septimas *B* septimam *L*

391 septem *om. L*

392 erit: et *F* / sexto: 6 *A* / illo: isto *AP* / scilicet: alio *P*

392–93 punctis intermissis[1]: intermissis *PU* pretermissis *B*

393 *infra* intermissis[1] *mg. hab. U figuram* / deinde…intermissis *om. FP* / sexto: 6 *A* / aliis *om. B* alio *U* / *post* quatuor *hab. B* etiam / intermissis[2]: pretermissis *B*

394 Et[1] *om. L* / *ante* in *hab. B* aliquando / 2: 3 *A* / vel: aliquando *B* sive *F* / 3: 2 *A* tertia *P* / et[2] *om. B* / quando *F* / nulla *U* / saltum *FP* statu *L*

395 ordinate *om. BFPU* / hec: ista *F* / [et cetera] *om. AFLU* secundum ordinem *B*

396 Conclusio undecima *om. FP* Propositio undecima *U* undecima *mg. hab. AB ante* Omnium / *ante* sunt *scr. et del. U* tunc / in *om. F*

397 *post* tot *hab.* B sunt / umquam: numquam *FP* / possent *LU*

398 velocitatis *U*

399 nunc *om.* B / quodlibet: quelibet *A* / [eorum] *om. AFLPU* / moveatur: motatur *A* / unico: uno *L* / motu simplici *tr. F*

400 quia: et *B* / *ante* videbitur *mg. hab.* B in 22^a / post: postea *U* / itaque: igitur *FPU*

401 et *rep. A* / *post* sint *hab. U* etiam / *post* se *hab. FP* ipsum / primi *om. P*

402 vel: sive *F* / proportione: speciei *F* / *ante* quod *hab. A* et / Et *om. A* / [A et B] *om. AFLPU*

will always conjunct is 7; and by the eighth proposition it is found that any conjunction whatever is separated from its immediate predecessor by $\frac{5}{7}$ of the circle. Consequently, when there are seven equidistant points in a circle [and] a conjunction occurs in [any] one, the next conjunction occurs in the sixth point from it, namely, with four points omitted; and the next conjunction happens in the sixth point from that point, with four other points omitted, and so on endlessly. And in certain cases, two points might be omitted, or three points, and now and then no point is omitted so that the conjunctions occur without a jump, [i.e.,] in the sequential order of the points [and so on].

Proposition 11. The number of conjunctions of any two mobiles in one revolution, and the number of points in which they can always conjunct, is equal to the difference between the least numbers representing the ratio of their velocities.

And now I assume that any of these mobiles is moved with a single, simple motion since the movement of a mobile with several [simultaneous] motions will be considered later. And so, assume that the velocity of *A* is as number *C*, the velocity *B* as number *D*, and that these numbers are prime to one another, or, which is the same thing, are in their least ratio. [Finally] let *A* and *B* be in conjunction.

Therefore, when *A* has completed *C* circulations and *B*, *D* circulations, they will, for the first time, be in their present position and will have completed a whole revolution, which is evident by the fifth proposition. But then the number of circulations added [or made] by *A* over and above those made by *B* equals the difference between *C* and *D*, or the number of units by which *C* exceeds *D*; and, by the sixth proposition, as often as *A* gains one circulation over *B*, just so often will they be in conjunction. In a whole revolution, then, the number of times they will be in conjunction is equal to the number of units by which *C* exceeds *D*, where *C* and *D* are the least numbers of the ratio of velocities. After the aforesaid revolution, the conjunctions begin again as

---

403 *ante* B *hab. L* d
404 *post* primo *hab. B* et non ante / erunt: essent *F* / sint *B*
404–5 revolutio tota *tr. B*
405 quintam: 5 *A* / *ante* tunc *hab. L* quotiens / tunc: cum *B*
406 quotam *B* / *post* ipsius *add. B* numeri / super: similiter *P* / *post* super *add. B* numerum / vel: sive *F*
407 quotienscumque: quotiens *L* / super: si-

militer *P*
408 totiens: totius *L* / coniungitur *AF* / sextam: 6 *FP* / *post* totali *scr. et del. U* coniunctione
408–9 per sextam…coniungentur *om. A* / revolutione: coniunctione *FP*
409 coniunguntur *L* / sunt *om. A*
410 *post* proportionis *hab. B* istarum / post *om. U sed post* predictam *mg. hab.* forte deest post

coniunctiones reincipient fieri ut prius, et ut ante ergo tot sunt puncta in quibus possunt coniungi et tot coniunctiones in una revolutione quota est differentia predictorum numerorum.

Verbi gratia, in casu sepe posito, velocitas *A* est sicut 5 et velocitas *B* ut
415 3 quorum differentia est 2. Ergo in una revolutione *A* et *B* bis coniunguntur et in duobus locis. Et numquam alibi coniungentur quoniam cum *A* fecerit duas circulationes cum dimidia et *B* unam cum dimidia, tunc *A* lucratum fuerit super *B* unam circulationem et coniungentur in puncto opposito illi puncto in quo nunc sunt [coniuncta]. Et quando *A* fecerit 5
420 circulationes et *B* 3 iterum coniungentur ubi nunc sunt et *A* lucratum fuerit duas circulationes. Et resumetur cursus eorum sicut a principio et sic semper.

Recolligendo ergo ex predictis demonstrationibus fabricetur una ars sive regula generalis in hunc modum. Datis velocitatibus signando eas per
425 minimos numeros proportionis ipsarum, horum numerorum differentia ostendit quot locis *A* et *B* possunt coniungi; que loca vel puncta dividunt circulum totum per equalia, ut dictum fuit. Et multiplicando unum istorum numerorum per alterum habetur tempus totius revolutionis ut patet ex quinta conclusione; quod quidem tempus dividendum est per numerum
430 punctorum seu coniunctionum revolutionis et habetur tempus inter quaslibet duas coniunctiones proximas. Quo habito cum data velocitate invenietur spatium locale inter duas coniunctiones proximas et quo ordine procedunt coniunctiones per huiusmodi puncta.

Verbi gratia, sint velocitates sicut 8 et 3 ita quod *A* perficiat unam cir-
435 culationem in 3 diebus et *B* in 8. Igitur sunt quinque puncta equaliter

411 *ante* coniunctiones *mg. hab. U figuram* / *ante* ut¹ *hab. B* ubi nunc / ut¹: sicut *A* ubi *FP* / et: est *A*
412 et tot coniunctiones *om. A*
413 predictorum: punctorum *A*
414 A *om. L* / ut: est sicut *L*
415 *ante* 2 *hab. L* ut / bis coniunguntur: coniungentur bis *A* bis coniungantur *P*
416 coniungerentur *L* / quoniam *om. P* et *U*
417 duas circulationes *tr. A* duas revolutiones *P* / et *B*...dimidia *om. FL* / A *om. F*
418 fuit *U* / *post* et *hab. B* sic
419 opposito *om. A* / puncto: dicentur *B* / sunt *om. L* / [coniuncta] *om. AFLPU*
420 3: unam cum dimidia *L* / iterum coniungentur *tr. L* / *post* et² *add. B* tunc
421 fuerit: erit super B *B* / Et *om. F* / cursus earum *F* quivis *U* / principio: puncto *P*

423 *ante* ex predictis *mg. hab. F* ?regula / predictis: dictis *A* predictas *F* / demonstrationibus: monstrationibus *P* / fabricatur *U* / una *om. B*
424 sive: vel *B* / generalis: generaliter *A* / modum: medium *P* / signandos *L*
425 ipsarum: earum *U* / horum numerorum differentia: differentia horum numerorum *B* quorum numerorum differentia *U*
426 possint *A* / vel: sive *F*
427 *ante* circulum *hab. U* per / circulum totum: totum regulum *B* / ut: sicut *F* / *ante* fuit *hab. L* est / fuit: est *B* fuerit *U*
427–28 istorum: eorum *B*
428 alterum: alium *B* / *ante* habetur *hab. U* istorum / ut patet *om. B*
429 quod...est: et dividendo tempus *B* /

before; and just as before, the number of points of conjunction, as well as the number of conjunctions in one revolution, equals the difference between the numbers mentioned above.

For example, in the case already frequently used, the velocity of *A* is as 5 and that of *B* as 3, the difference being 2. In one revolution *A* and *B* will be in conjunction twice in two [separate] places; and they will never be conjoined elsewhere, for when *A* will have made $2\frac{1}{2}$ circulations and *B* $1\frac{1}{2}$, *A* will have gained one circulation over *B* and they will conjunct in a point opposite their present point of conjunction. Then, when *A* will have completed 5 circulations and *B*, 3, they will be in conjunction once again in their present position, with *A* having gained two circulations. Their course is then repeated from the beginning, and so on perpetually.

By recalling things said in earlier demonstrations, a law or general rule may be formulated in this way: By assigning least numbers to a ratio of two given velocities, the number of places in which *A* and *B* can be in conjunction is indicated by the difference between these numbers. And, as already stated, these places or points divide the whole circle into equal parts. Multiplying one of these numbers by the other gives the time of the whole revolution, as shown in the fifth proposition. This time must then be divided by the number of points or conjunctions in the revolution to find the time between any two successive conjunctions. When both the time between any two successive conjunctions and the given velocity [of either mobile] are known, the distance between two successive conjunctions could be found and from this [knowledge] the sequence of conjunctions through these points [can be determined].

For example, assume that the velocities are as 8 and 3 so that *A* completes one circulation in 3 days and *B* in 8. Hence there are 5 equidistant points in

---

ante dividendum *scr. et del. F* ?conclu-
dendum / dividendus *A*
430 seu: sive *F* / revolutionis: in illa revolutione *B* / et *om. B* / *post* habetur *scr. et
del. U* et habetur / tempus *om. P*
431 *ante* Quo *hab. F* et / *post* Quo *hab. A*
facto

431–32 habito…et quo *om. U*
433 *ante* coniunctiones *hab. U* cum
434 *ante* 8 *hab. F* sunt
434–35 circulationem: revolutionem *A*
435 3 diebus *tr. B* / Igitur *om. AP* ergo
BU / *infra* puncta equaliter *mg. hab. A*
figuram

distantia in quibus in perpetuum coniungentur. Et igitur per quintam conclusionem tempus totius revolutionis est 24 dies, quod tempus dividendum
est per 5 et venient 4 dies et $\frac{4}{5}$ quod est tempus inter duas coniunctiones
consequenter se habentes ita quod in perpetuum coniungentur de 4 diebus
440   et $\frac{4}{5}$ in 4 diebus et $\frac{4}{5}$. Quod adhuc aliter invenitur per sextam conclusionem.
　　　Et quoniam *B* in una die pertransit $\frac{1}{8}$ et *A* $\frac{1}{3}$ circuli, ideo multiplicando $\frac{1}{3}$
per 4 et $\frac{4}{5}$, quod est tempus inter duas coniunctiones, venit $1\frac{3}{5}$ quod est
spatium pertransitum ab *A* inter duas coniunctiones, videlicet una circu-
latio et $\frac{3}{5}$ unius a quo subtrahatur circulus et remanent $\frac{3}{5}$ quod est spatium
445   inter duas coniunctiones proximas. Et eodem modo venient $\frac{3}{5}$ quod est
spatium pertransitum ab ipso *B* in tali tempore si multiplicetur pertransitum ab ipso *B* in uno die per tempus predictum inter duas coniunctiones
proximas.
　　　Liquet itaque quod in hoc casu coniunctiones procedunt per hec puncta
450   saltando. Signentur itaque in circulo 5 puncta equidistantia *e, g, k, f, h*; et
nunc sint *A* et *B* coniuncta in puncto *e* et moveantur versus *g*. Igitur prima
coniunctio sequens erit in *f* et postea alia in *g*, et sic semper pretermissis

---

436 *post* quibus *hab. B* et non alibi / coniun-
　guntur *A* / Et *om. B* est *F* / quintam:
　5 *AU*
436–37 *post* conclusionem *mg. hab. U*:
　　　　Notanda
　Quia et 4 per 5 divisus exhibit quintam,
　scilicet $\frac{8}{4}\frac{4}{5}$ unius que quinta est tempus
　unius coniunctionis secundum hypote-
　sim premissam de proportionalitate mo-
　tuum *A* et *B*, scilicet $\frac{8}{24}$ et $\frac{3}{4}$. Quia post
　4 dies et $\frac{4}{5}$ unius diei movebitur *A* per
　totum circulum, scilicet $\frac{24}{24}$ et insuper
　per $\frac{8}{24}$ et per quatuor quintas $\frac{8}{24}$ sive,
　quod idem est, per quatuor quintas $\frac{5}{24}$
　et per quatuor quintas $\frac{3}{24}$ faciens precise
　$\frac{8}{24}$ et quatuor quintas $\frac{3}{24}$. Simul fac *A*
　?enim ultra circulum totum [*two illeg-
　ible words follow*] adhuc de secunda
　circulatione $\frac{12}{24}$ et insuper $\frac{8}{5}$ de $\frac{3}{24}$. Simi-
　liter movebitur *B* in eodem tempore, vi-
　delicet per 4 dies movebitur per $\frac{12}{24}$ pre-
　cise insuper per $\frac{8}{5}$ unius diei movebitur
　etiam per $\frac{4}{5}$ motus unius diei scilicet
　motus quatuor quintarum de $\frac{3}{24}$. Ergo
　erunt tunc ambo mobilia, scilicet *A* et
　B, ?pariter primo coniuncta, scilicet post

4 dies et $\frac{4}{5}$ unius diei. Sit itaque patet
quod in una revolutione continens 24
dies contingeret 5 coniunctiones ipsorum duorum mobilium in 5 certis punctis
in perpetuum ?videlicet
437 revolutiones *A* / est *om. F*
438 venient: provenient *B* / $\frac{4}{5}$: 4 quintas *B*
　5 *L* / tempus *om. L* / *ante* duas *hab. U*
　quaslibet
439 ita quod: itaque *P* / 4 diebus *tr. U*
440 et $\frac{4}{5}$ *om. A* et 4 ?quintis *B* et $\frac{24}{5}$ *L* in 5
　quartis etiam *F* / *ante* in *hab. P* ?quartis
　et hab. U ad vel / diebus: dies *F* / et^2:
　cum *U* / Quod: hoc idem *B* / sextam:
　6 *AU*
441 quoniam: quando *F* quandoque *P* / B
　in … circuli: A in una die pertransivit $\frac{1}{3}$
　circuli et B $\frac{1}{8}$ *U* / in: cum *L* / pertransi-
　bit *AP* / $\frac{1}{8}$: unam octavam *BFLP* unam
　8 *A* / multiplicando: multitudo *P*
442 4: quartam *A* / venit: provenit *B* / $1\frac{3}{5}$:
　unum et $\frac{3}{5}$ *BP* unum et $\frac{2}{5}$ *FL* unum $5\frac{3}{5}$ *A*
442–46 quod est … pertransitum *om. A*
443 ab A *om. B* a B *FPU* / videlicet: videt
　FP scilicet *B*
444 $\frac{3}{5}$^{1}: 3 quinte *BFP* 3 coniuncte *L* / remanet
　F / $\frac{3}{5}$^{2}: $\frac{2}{5}$ *F* / *ante* quod *hab. B* eius / *post*

which the mobiles will conjunct perpetually. By the fifth proposition, the time of the whole revolution is 24 days which must be divided by 5 to obtain $4\frac{4}{5}$ days, the time intervening between two successive conjunctions. The mobiles are so related that they will always conjunct every $4\frac{4}{5}$ days. This is found in yet another way by the sixth proposition.

Now since B traverses $\frac{1}{8}$ of a circle in one day and A, $\frac{1}{3}$, multiplying $\frac{1}{3}$ by $4\frac{4}{5}$—the time between two [successive] conjunctions—gives $1\frac{3}{5}$, which is the distance traversed by A between two [successive] conjunctions, i.e., one circulation plus $\frac{3}{5}$ of one circulation from which the [whole] circle is subtracted, leaving $\frac{3}{5}$, which is the distance between two successive conjunctions. In the same way, if the distance traversed by B in one day is multiplied by the aforementioned time [interval] between two successive conjunctions, $\frac{3}{5}$ will be obtained as the distance traversed by B in the stated time.

It follows, moreover, that in this case the conjunctions proceed through these points by jumps. Let e, g, k, f, and h be five equidistant points in a circle, and assume that A and B are now in conjunction in point e and are [then] moved toward g. The first conjunction following will be in f, and then another in g,

---

spatium *add.* B locale
445 venient: provenit *B* veniunt *L* venit *U*
446 spatium *om.* U / *post* spatium *add.* B locale / si *om.* P
447 una *A* duo *P* / inter: in *FP*
449 itaque *om.* A / hec: hoc *BU*
450 5: quinte *F* / *post* equidistantia *mg. hab.* U:

Notanda

Sic et in hiis numeris duobus, scilicet A 9 et B e, denominantur nam tempus totius revolutionis est 27, quod per 6 divisum exhibet sextam, scilicet 4 et $\frac{1}{2}$, que sextam est tempus unius coniunctionis. Quia dum movetur A cotidie per $\frac{9}{27}$ et B per $\frac{3}{27}$ patebit quod A per tempus illius sexte moveatur per $\frac{49}{27}$ et $\frac{1}{2}$ unius 27^e, B vero per idem tempus movebitur per $\frac{13}{27}$ et $\frac{1}{2}$ unius 27^e. Erunt igitur tunc primo ambo ipsa mobilia, scilicet post octavam circulationem factam per A, scilicet post 4 ?et $\frac{1}{2}$. Sit itaque patet quod erunt 6 puncta coniunctionum in una revolutione in quibus contingerent 6 coniunctiones [*illegible word*

follows] et cetera.

Similiter, patet posset de hiis duobus numeris, scilicet A 10 et B 3. Nam tempus totius revolutionis exibit 30 dies. Et foret $\frac{1}{3}$ pars 4 et $\frac{2}{3}$?itaque fecerint 7 coniunctiones in septem punctis in perpetuum.

Prima ?est ?tempus coniunctio erit in fine arcus 12$\frac{6}{7}$ de $\frac{3}{30}$.

Secunda coniunctio in fine arcus 24$\frac{4}{7}$ et $\frac{3}{30}$.

Tertia erit in fine arcus 36$\frac{6}{7}$ de $\frac{3}{30}$.

Quarta in fine arcus 41$\frac{1}{7}$ de $\frac{3}{30}$.

Quinta in fine arcus 63$\frac{3}{7}$ de $\frac{3}{30}$.

Sexta in fine arcus 75$\frac{3}{7}$ de $\frac{3}{30}$.

Ultima in fine arcus 90 sive in puncto primo perficiens A 10 revolutiones, B autem 3 revolutiones.
451 nunc: tunc *A* / puncta *A* punctis *P* / e et *om.* F / moveatur *BU* moveantur que *F* / primam *P*
452 in^1 *om.* FPU / et^1 *om.* AP / postea *om.* B / alia *om.* U / sic semper: et cetera U sic sepe A

duobus punctis. Et si velocitas *A* esset sicut 9 et velocitas *B* sicut 4 ita quod *A* transiret circulum in 4 diebus et *B* in 9, tunc prima sequens coniunctio fieret in *h* et alia in *f*, et sic retrocedendo vel antecedendo tribus intermissis; et eodem modo si velocitas *A* esset sicut 11 et velocitas *B* sicut 6.

Sed si velocitas *A* esset sicut 12 et velocitas *B* ut 7, tunc prima coniunctio sequens fieret in *k*, postea in *h*, et sic semper uno intermisso puncto. Et sic variis modis secundum nominationem velocitatum stante adhuc eadam differentia. Et adhuc fieret nominatio differentia variata sed de modo variandi regulam non inveni.

Ex ista conclusione [undecima] sequitur quod si aliquorum duorum mobilium celestium proportio in velocitate esset in aliqua principalium proportionum armonicarum que sunt in sonis dyapason, dyapente, dyatesseron, tonus, qui faciunt symphoniam sive consonantiam, numquam taliter mota coniungerentur nisi tantum in uno loco eo quod cuiuslibet talis proportionis minimi numeri differunt unitate. Sicut si medius motus Martis esset precise in duplo velocior medio motu solis, numquam fieret ipsorum media coniunctio nisi tantummodo in uno loco sive puncto. Et patet etiam quod si duorum motuum celestium fuerit proportio que in sonis vocatur dyessis vel semitonum minus, numquam fierit coniunctio nisi in altero 13 locorum seu punctorum eo quod differentia minimorum numerorum illius proportionis est 13.

Cum ergo non inveniatur in motibus celestibus quod ex duobus motibus oriatur constellatio que non posset fieri nisi in uno puncto celi, consequens est ut nulli duo motus celestes in velocitate teneant proportionem armonicam principalem. Et ergo si corpora celestia faciant consonantiam in

453 esset: est *AP* / sicut[1]: velut *A* / sicut[2]: ut *ALPU*
454–55 sequens coniunctio *tr. L*
455 vel: sive *F*
455–56 *ante* tribus intermissis *hab. B figuram*
456 et *om. BU* / esset: est *A* / velocitas[2] *om. B*
456–58 si...Sed *om. U*
457 *ante* sicut *hab. L* esset / 6: g *P*
458 si: sicut *P* / esset: est *A* / sicut: ut *B*
459 *ante* postea *add. AP* et / postea: alia *B* posita *U* / *ante* in[2] *scr. et del. U* ?tamquam / uno intermisso *tr. BU* / puncto: punctorum *B* / Et sic: sicut *FP*
460 *post* secundum *scr. et del. B* variatio-

nem / *ante* nominationem *hab. U* variam / nominationem: variationem *AL*
461 nominatio: variatio *AL* denominatio *U*
462 inveniri *B*
463 *post* ista *hab. A* enim / [undecima] *om. AB* nulla *P* enim vel 11 *F* et undecima *U* / aliquorum *om. B* aliquod *A*
464 celestium *A* ?*B*; *om. FP* / esset: est *A*
465 dyapason *rep. P*
465–66 *ante* dyatesseron *hab. U* et / dyatesseris *F*
466 tonus *om. L* / qui: que *L* / *post* faciunt *hab. L* in / sive: seu *L* / numquam: ?nequiquam *P*
467 coniungentur *A*
468 si *om. F*

and so on, always omitting two points. And if the velocity of *A* were as 9 and that of *B* as 4, so that *A* traverses the circle in 4 days and *B* in 9, the first conjunction following would occur in *h*, and another in *f*, and whether receding or advancing there will always be three [points] omitted. The same would apply if *A*'s velocity were as 11 and *B*'s as 6.

But if the velocity of *A* were as 12 and the velocity of *B* as 7, the first subsequent conjunction would occur in *k*, and then in *h*, and so on, with [only] one point always omitted. Although the difference in [each of] these various examples is constant, there are variations [in the number of points omitted] which depend upon the velocities assigned. Indeed, the difference between the velocities could also be varied, but I have found no rule for such a variation.[7]

It follows from this eleventh proposition that if the ratio of velocities of any two celestial mobiles were in any of the principal harmonic ratios in music, namely, the diapason, diapente, diatessaron, and tone, which make a concord or harmony, the mobiles will never conjunct except in one place only, since the least numbers of such a ratio differ only by a unit. As an example, if the mean motion of Mars were exactly twice the speed of the sun's mean motion, there would never be a middling conjunction of these two bodies for [they would conjunct] in only one place, or point. Furthermore, it is apparent that if the ratio of two celestial motions were what is called in music a diesis, or lesser semitone, there could never be a conjunction except in one of 13 places, or points, because the difference between the least numbers of this ratio is 13.

Since no configuration consisting of two motions is found to occur in only one point of the sky, [it follows] as a consequence that no two celestial motions have velocities related in a principal harmonic ratio. Therefore, if celestial bodies in motion produce a harmony, it is not necessary [to assume] that such

---

469 *ante* Martis *scr. et del.* U ?miratus / Martis: arcus *L* / *ante* esset *hab.* F si / esset: est *A* / in *om.* FPU / duplo: duplus *U* / medio motu *tr.* U / sol *A* / numquam: ?nequiquam *P*

470 ipsorum: ipsa *A* / media coniunctio *tr.* *P* / tantummodo *om.* AB / *ante* sive *hab.* B tantum / *ante* puncto *hab.* B uno / Et *om.* B

472 diessis *B* dieresis *F* dierisis *L* / vel: sive *F* / numquam: ?nequiquam *P*

473 *ante* nisi *hab.* B eorum / seu: vel B sive *F* / eo: et *P*

474 illius proportionis *tr.* F

475 in *om.* A / duabus *P*

476 possit *BFP* / uno *om.* A / celi B ?F tunc *A* tenetur *P*

477 in velocitate teneant: teneant in velocitate B

478 celestia: supercelestia *A* / consonantiam: ?P

movendo non oportet quod ex velocitatibus motuum proveniat huiusmodi
480 consonantia sed potest aliunde oriri ut postea videbitur per alias rationes.

  *Conclusio duodecima. Si fuerint mobilia plura duobus possibile est quod
numquam coniungentur simul plura quam duo.*

  Sint tria mobilia *A* et *B* et *C*. Igitur per precedentes conclusiones sunt
determinata puncta vel loca in quibus, et non alibi, possunt coniungi *A* et
485 *B*. Vocetur ergo quilibet talis punctus *d*. Consimiliter sunt aliquot certa
puncta quibus, et non alibi, possunt coniungi *B* et *C* quorum quilibet
vocetur *e*. Igitur si nullum *d* est *e*, quod est possibile, numquam hec tria
mobilia simul erunt coniuncta. Eodem modo si fuerint 4, vel 5, aut quotlibet potest contingere quod numquam coniungentur simul nisi duo quia
490 possibile est quod loca in quibus coniunguntur bina et bina sint incommunicabilia; et possibile est etiam quod sint in aliquibus communicabilia et
quod 6 mobilium 3 possint simul coniungi, aut 4 et non plura, aut 5, aut
omnia, et sic de aliis.

  Verbi gratia, sint *A* et *B* coniuncta in puncto *d* et *C* mobile precedat ipsa
495 per $\frac{1}{8}$ partem circuli, scilicet per unum signum cum dimidio. Et velocitas
A sit sicut 4, et velocitas *B* ut 2, et velocitas *C* ut 1; sitque *e* unus punctus
distans a puncto *d* per $\frac{1}{6}$ partem circuli, scilicet per duo signa; et *f* et *g* sint
duo alia puncta ita quod *e*, *f*, *g*, dividant circulum in tria equalia. Et sit

479 proveniat: perveniat *L* procedat *U*
480 *infra* per alias *mg. hab. A figuram* / alias
 rationes *tr. B*
481 Conclusio duodecima *om. FP* Propositio undecima *U* undecima *mg. hab. AB
 ante* Si
482 simul *om. FP* sic *L*
483 Sint: sicut *P* / mobilia *om. AFPU* / et[1]
 om. BLU / et[2] *om. BU* / sunt: sint *U*
484 determinata puncta *tr. B* / vel *om. B* sive
 F / loca *om. B* / *ante* in quibus *mg. hab.*
 B de 3 mobilibus / possunt coniungi *tr.*
 L possent coniungi *U*
485 quelibet *P* / Consimiliter: similiter *B* /
 sunt aliquot: sint aliqua *U*
485–86 certa puncta *tr. B* talia puncta *L*
486 *ante* puncta *scr. et del. B* aliqua / possent *U* / coniungi *om. A*
486–87 quorum quilibet vocetur: et vocetur
 quilibet eorum *B* quorum quidem vocetur *F*
487 nulla *P* / quod: ut *B* quodam modo *A* /
 possibile: impossibile *F* / hec *om. FU*
 illa *B*

487–88 quod...coniuncta. *om. P*
488 simul erunt *tr. B* simul essent *F* /
 coniuncta: iuncta *B* / vel: sive *F*
489 potest: possit *A* / *post* potest *add. U* quidem / quod: et *P* / simul *om. B*
490 in *om. B* / coniungentur *B* / *ante* bina[1]
 scr. et del. B fuit
490–91 incommunicabilia: incommensurabilia *B* communicabilia *P*
491 sint in aliquibus: in aliquibus sint *FP* /
 post aliquibus *hab. U* etiam
492 *ante* mobilium *hab. B* aut plurium / *ante*
 possint *hab. P* est / possent *L* / aut[1]: sive
 F / *post* aut[2] *hab. B* etiam / *post* aut[3]
 hab. B aut plura
492–93 *infra* aut[3] omnia *mg. hab. L figu-*
 ram
494 mobile *rep. P*
496 sicut: ut *L* / ut: sicut *U* / et[2] *om. U* / e
 om. A
497–99 per[1]....puncto d *om. U*
498 duo alia *tr. B* duo illa *A* / f: et *BFP* /
 Et: ut *L*

a harmony arises from the velocities of their motions, but perhaps it stems from some other source for other reasons, as will be seen later.

Proposition 12. If there were more than two mobiles, it is possible that no more than two will ever be in conjunction at the same time.

Let *A*, *B*, and *C* be three mobiles. By the preceding propositions, *A* and *B* can conjunct in only a fixed number of points, or places, and nowhere else; let any such point be called *d*. Similarly, *B* and *C* conjunct in a certain number of fixed points, and nowhere else; let any such point be called *e*. Then, if no

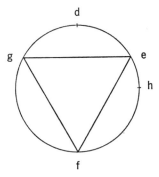

Fig. 20

With only trivial differences, this figure appears in MSS *A* (Vat. lat. 4082, fol. 100v, in the margin to the left of col. 1) and *L* (Florence, MS Ashburnham 210, fol. 162v, in the margin below col. 2).

d is an *e*, which is possible, these three mobiles will never be in conjunction at the same time. In a like manner, if there were four or five, or any number of mobiles, it can happen that only two will ever conjunct at the same time since it is possible that when the mobiles are taken two at a time, they do not share any common places of conjunction; but it is also possible that they could share some common points, and that [for example], of six mobiles there might be simultaneous conjunctions of three, or four and no more, or five, or all; and the same reasoning applies to other sets of mobiles.

For example, assume that *A* and *B* are in conjunction in point *d* [see Fig. 20], and mobile *C* precedes them by $\frac{1}{8}$ part of the circle—i.e., by $1\frac{1}{2}$ [zodiacal] signs. And let the velocity of *A* be as 4, that of *B* as 2, and *C* as 1; let point *e* be separated from point *d* by $\frac{1}{6}$ part of the circle—i.e., by two signs—and let *f* and *g* be two other points such that *e*, *f*, and *g* divide the circle into three

h alter punctus distans a puncto *d* per $\frac{1}{4}$ partem circuli, scilicet per tria
500 signa.

Rebus igitur sic dispositis demonstrabitur per nonam conclusionem
quod *A* et *C* primo coniungentur in puncto *e* et per eandem patebit etiam
quod *B* et *C* primo coniungentur in puncto *h* et per decimam, aut etiam
per undecimam, ostendetur quod *A* et *B* numquam coniungentur nisi in
505 puncto *d* et per eandem quod *B* et *C* numquam coniungentur nisi in puncto
h. Ex quo statim sequitur quod *A* et *B* et *C* numquam simul coniungentur
quin ymo per eandem apparebit quod *A* et *C* numquam coniungentur nisi
in altero trium punctorum *e, f, g*, ex quo iterum argueretur quod *A, B, C*
numquam simul coniungentur. Possunt etiam aliqua plura mobilia taliter
510 se habere quod 3, aut 4, aut quotlibet possunt simul coniungi et quid de
hoc contingat videbitur postea.

Conclusio tertiadecima. Omnium trium aut plurium mobilium que num-
quam simul coniungentur est certa distantia citra quam approximari non
possunt.

515 Sit *A* [mobile] velocius, et *B* medium, et *C* tardius; moveantur tali
commensuratione quod numquam ista tria coniungantur simul quod
possibile est ex precedenti conclusione. Dico ergo quod non possunt
appropinquari magis vel esse omnia in minori spatio quam aliquando dum
A velocius et *C*, quod est tardius, sunt coniuncta; quoniam si *A* et *C* sunt
520 coniuncta et *B* precedat, igitur immediate ante *A* et *B* magis distabant
quam nunc distent, et immediate post *B* et *C* magis distabunt quam nunc
distent. Si vero *B* sequatur, et *A* et *C* sint coniuncta, igitur econverso *A* et
B immediate post plus distabunt quam nunc distent; et *B* et *C* immediate

499 a puncto d: a puncto *B* d a puncto *P* /
partem *om. B* / *ante* circuli *scr. et del.*
B ?circuli
501 *ante* Rebus *mg. hab. A* figuram
502 A et C *om. U*
502–3 in puncto…coniungentur *om. A*
503 primo coniungentur *tr. FP* / *post* coniungentur *mg. hab. B* figuram / aut:
sive *F* / etiam *om. U*
504 per *om. ABL* / ostenditur *B*
504–5 numquam…B et C *om. U*
505 et¹ *om. P*
506 et¹ *om. B* / simul coniungentur *tr. L*
507 ymo: primo *U* / per eandem apparebit:
apparebit per eandem *L*
508 arguetur *ABU* / *ante* C *add. B* et
509 *ante* numquam *hab. F* A et B / simul
om. L

510 aut¹: sive *F* / possent *U* / et *om. A* /
quid: quidem *A*
511 postea: post *AFLP*
512 Conclusio tertiadecima *om. FP* Propositio tertiadecima *U* tertiadecima *mg.*
hab. AB ante Omnium / *ante* trium *hab.*
A aut / aut plurium *om. L* sive plurium
F
513 simul *om. A* / citra quam: extra *A*
513–14 approximari non possunt: non possunt approximari *B* coniungi non possunt *L*
515 [mobile] *om. AFLPU* / et¹ *om. BL* / moveatur *FLU*
516 numquam ista tria: ista tria numquam *U*
517 possibile est *tr. LU* / precedente *P* / conclusione *om. B*
518 magis *om. A* / vel: sive *F* / in *om. P* /

equal parts. [Finally] assume that h is another point separated from point d by $\frac{1}{4}$ part of the circle—i.e., by three signs.

Having now set out the data, we shall demonstrate, by [means of] the ninth proposition, that A and C will conjunct first in point e; and by the same [ninth] proposition it will be obvious that B and C will conjunct first in point h; and by the tenth, or even eleventh, proposition it will be shown that A and B will conjunct only in point d, and, by the same proposition, that B and C will conjunct only in point h. It follows immediately from this that A, B, and C will never conjunct simultaneously. Indeed, by the same [tenth] proposition it will be seen that A and C conjunct only in the other three points, e, f, and g, from which it could again be argued that A, B, and C will never conjunct simultaneously. When more mobiles are involved it can happen that 3, or 4, or any number of them can be in conjunction simultaneously. How this happens will be considered later.

Proposition 13. For any [set of] three or more mobiles never able to conjunct simultaneously, there is a fixed distance below which the mobiles cannot approximate to each other.

Let A be the quickest mobile, B a mobile with a middling speed, and C the slowest mobile; and assume that their motions are so related that these three will never conjunct simultaneously—and this is possible by the preceding proposition. I say that they are nearest one another, or encompassed within the smallest [possible] space, only when A, the quickest, and B, the slowest, are in conjunction. For if A and C are in conjunction and B precedes them, it follows that immediately before [conjunction] A and B were separated by a greater distance than now; and immediately after [conjunction] B and C will be separated by a greater distance than at the time of conjunction. If, however, B should follow and A and C are in conjunction, then, conversely, immediately after conjunction A and B will be separated by a greater distance than at the time of conjunction; and immediately before conjunction B and C were separated by a greater distance than at the time of conjunction. Thus if all

quam: quando *A* / aliquando: alii *L* / *ante* dum *mg. hab. B* scilicet

519 quod est *om. B* / sunt[1]: sint simul *L* super *P* / sunt[2]: sint *L*

520 distabat *A*

521 distent *om. B* / *ante* B *hab. A* hoc *et scr. et del.* a et

522 distent *om. A* distant *L* / *post* distent *hab. P* et immediate post B et C magis distabunt quam nunc distent

523 *ante* post *scr. et del. A* plus / *post:* ante *U* / plus: minus *U* / distabant *LPU* / distent *om. B* distant *ALP*

523–24 et B...distent *om. LP*

ante plus distabant quam nunc distent. Ergo si omnia tria sunt disiuncta
525 non erunt in tam parvo spatio quam cum *A* et *C* sunt coniuncta.

Quod si negetur, arguitur adhuc aliter. Et primo ponatur quod *A* precedat et *B* sequatur postea *C*. Ergo erant propinquiora quando *A* coniungebatur cum *B* et adhuc propinquiora quando *A* coniungebatur cum *C* vel
quando coniungetur cum ipso ut statim patet ex ordine velocitatum. Et
530 eodem modo potest argui in omnibus aliis combinationibus, possunt namque sex modis combinari seu ordinari: secundum, precedere, et sequi, ut
patet exemplo. Ergo non possunt esse in minori spatio quam quando *A* et
C sunt coniuncta.

Signentur itaque omnia puncta quibus *A* et *C* possunt coniungi sicut
535 docet decima conclusio. Et videatur in qualibet coniunctione totius revolutionis istorum trium mobilium quantum distabit *B* ab *A C* tempore cuiuslibet coniunctionis *A* et *C* et minor illarum est minima quam huiusmodi
tria mobilia possunt distare ita quod nulla minore possunt distare nec
amplius approximari, ut statim patet ex precedentibus. Et quelibet talis
540 distantia statim poterit haberi ex velocitatibus datis et proportione ipsorum.

Verbi gratia, in casu prius posito, quando *A* et *C* primo coniungentur in
e tunc *A* pertransiverit duo signa et *C*, quod movetur quadruplo tardius et
precedit per unum signum cum dimidio, pertransiverit dimidium signum et
545 *B*, quod movetur velocitate media et nunc est coniunctum cum *A* in *d*,

524 ante: post *U* / plus: minus *U* / distabunt
U / quam: quod *A* / distent *om.* *B*
distant *AF* / sint *FLP* / disiuncta *om.* *P*
525 erunt: essent *F* / tam: ita *L* tria *P*
526 arguetur *A* / alter *P* / *post* aliter *hab.* *B*
probatur / primo *om.* *U*
527 sequitur *AP* / *infra* sequatur *mg. hab.* *L*
figuram / ante postea *hab.* *L* et
527–28 quando…propinquiora *om.* *BFU*
quando A *P*
528 A *om.* *P* / cum² *om.* *B* / vel: sive *F*
529 coniungetur: iungetur *B* / cum *om.* *L* /
ut: tunc *P* / *post* ex *hab.* *F* eo / velocitatis *F* / Et *om.* *B*
530 *ante* possunt *hab.* *U* et cetera / possunt:
possit *P*
530–31 namque: enim *B*
531 modi *F* / seu: et *B* sive *F* / secundum:
sed *P*
532 non possunt esse: nunc sint *U* et *post*
nunc *scr. et del.* *U* possunt

534 itaque: ita quod *P* / C: B *A*
536 istorum trium: coniunctionis A et C *U* /
distabunt *A*
537 C et *om.* *A* / illarum: istarum *A* / quam:
qua *L*
538 *post* distare *scr. et del.* *B* nec amplius
approximari / ita…distare² *om.* *F* / minori *U* / possunt ?*B* posset *A*
539 quilibet *P*
540 statim *om.* *U* / habere *A* / *post* haberi
hab. *B* et sciri / velocitatibus datis *tr.* *B*
540–41 ipsorum: eorum *B* ipsarum *FP*
542 primo: post *FP*
543 pertransivit *FLU* pertransibit *A* / *post*
movetur *hab.* *B* in
544 precedat *AB* / pertransivit *ALU* per-
transeunt *FP*
545 movetur: movendo *A* / *post* et *hab.* *A*
?invicem est coniungentur / est coniunctum *tr.* *B*

three mobiles are separated, they will not be in as small a space as when *A* and *C* are in conjunction.

If this should be denied, yet another argument can be put forth. First, let it be assumed that *A* precedes, and *B* follows after *C*. Consequently, they were closer when *A* was in conjunction with *B*, and even closer when *A* was in conjunction with *C*, or when it will be in conjunction with *C*. This is immediately evident from the sequence of velocities. One can argue in the same way about all other combinations, since the mobiles can be combined or arranged in six sequences [see Fig. 21]—[i.e., where each mobile is] second [or in the

$$
\begin{array}{llll}
(1) & A \rightarrow & B \rightarrow & C \\
(2) & A \rightarrow & C \rightarrow & B \\
(3) & B \rightarrow & A \rightarrow & C \\
(4) & B \rightarrow & C \rightarrow & A \\
(5) & C \rightarrow & A \rightarrow & B \\
(6) & C \rightarrow & B \rightarrow & A
\end{array}
$$

Fig. 21

This figure represents the arrangement of mobiles exactly as they appear in MSS *A* (Vat. lat. 4082, fol. 100v, col. 2) and *L* (Florence, MS Ashburnham 210, fol. 163r, col. 2), although in *A* the mobiles are placed on six concentric arcs, whereas in *L* six parallel straight lines are used. The numbers in parentheses, ordering the six cases, and the arrows are editorial additions.

middle, then] precedes, [and then] follows, just as in the example. Therefore the three mobiles cannot be in a space smaller than when *A* and *C* are in conjunction.

And now, in accordance with the tenth proposition, assign all the points in which *A* and *C* can conjunct. And in any conjunction [during the course] of a whole revolution of these three mobiles, see how far *B* will be from *A* [and] *C* at the time of conjunction of *A* and *C*. The smallest of these distances is the least distance which can separate these three mobiles. This distance cannot be made smaller, nor can the mobiles approach closer, as is obvious by the preceding discussion. Any such [minimal] distance can be determined directly from the given velocities and their ratios.

For example, in the case considered before, when *A* and *C* first conjunct in *e*, *A* will have traversed 2 signs and *C*, which moves four times more slowly [than *A*] and precedes it by $1\frac{1}{2}$ signs, will have traversed $\frac{1}{2}$ of a sign, while *B*, moving with a middling velocity and presently in conjunction with *A* in

pertransiverit unum signum. Igitur *A* et *C* et *B* distabunt per unum signum. Item quando in sequenti coniunctione *A* et *C* erunt in *f*, tunc ab initio motus ipsum *A* pertransivit 18 signa et *B* 9, ergo hec tria distabunt per tria signa. Item quando postea *A* et *C* coniungentur in *g*, tunc a principio motus
550 *A* pertransivit 34 signa et *B* 17. Ergo subtrahendo [totum] circulum restat quod distabunt per 5 signa.

Item quando rursum *A* et *C* coniungentur in *e* tunc *A* pertransivit 50 signa et *B* 25, ergo distabunt per unum signum, sicut prius. Et erit completa revolutio totalis et erit initium alterius simul et coniunctiones re-
555 incipient fieri et procedere sicut ante. Unde patet quod in isto casu *A* et *B* et *C* non possunt fieri ita propinqua quin ad minus distent per unum signum et adhuc non ubilibet possunt tantum approximari sed in certis locis. Et eodem modo diceretur de pluribus mobilibus. Possibile est ergo quod aliqui 3 planete, vel 4, numquam coniungentur in eodem gradu, vel in eodem
560 minuto, et forte de aliquot possibile est quod non possunt approximari quin distent [ad minus] per 2 gradus, vel 3, si omnes commensurabiliter moveantur. Que de duobus mobilibus dicta sunt ad quotlibet possunt se extendere.

Conclusio quartadecima. Si plura mobilia nunc sint coniuncta necesse est
565 *ut in puncto eodem alias coniungantur.*

Sint *A* et *B* et *C* coniuncta in puncto *d*. Cum ergo velocitates sint commensurabiles, ut suppositum est, erunt sicut 3 numeri qui per ordinem sint *e*, *f*, *g*. Ergo quando *A* fecerit precise aliquot circulationes numeratas numero *E*, tunc *B* compleverit precise aliquot numeratas numero *F*; et
570 similiter *C* numero *G*. Ergo pro tunc quodlibet eorum erit primo regressum ad punctum *d* quod faciliter patet eodem modo sicut arguebatur de duobus [mobilibus] in quarta conclusione.

546 pertransivit *AU* pertransiverint *F* / A rep. *A* / et[1] *om. BL* / C et B: B *A* B et C *L* / distabant *A*
547 *ante* coniunctione *hab. B figuram* / erunt: essent *F*
548 9: cum *P*
549 signa *om. P* / coniungerentur *P* / a *rep. A*
550 A *om. A* / signa *om. AB* / Ergo: et B igitur *U* / subtrahendo: ?*B* / [totum] *om. AFLPU* / restant *AFP*
551 distabit *U* / 5: 9 *AB*
552 quando rursum *tr. A* quando rursus *U* / A[2] *om. FPU*
553 *ante* per *hab. B* ut primo / unum: ?*A* /

sicut prius *om. B* / prius: primo *U*
553–54 completa revolutio totalis: revolutio completa resolutio totalis *A* completa totalis ?revolutio *B*
554 simul: similis *A*
554–55 reincipient: incipient *B*
555 sicut: ut *F* / et[2] *om. B*
556 ita: in *P* / quin: quando *LU* quia *P*
557 non ubilibet *tr. L* / *post* locis *hab. B* tantum *et hab. L* et temporibus
558 eodem modo: similiter *L* / dicetur *U* / ergo: tamen *F*
558–59 aliqui: quaqua *U*
559 3 planete *tr. B* / vel[1]: sive *F* / coniungetur *U* / gradu *obs. B* igitur *U* / vel[2] *obs.*

d, will have traversed 1 sign. Thus, *A*, *C*, and *B* will be separated by 1 sign. When *A* and *C* will be in the next conjunction at *f*, *A* has traveled 18 signs from the start of its motion and *B*, 9, so that 3 signs separate the three mobiles. Again, when *A* and *C* conjunct afterward in *g*, *A* has moved [a distance of] 34 signs and *B*, 17, from the initial point of their motions. By subtracting the whole circle, we see that they will be separated by 5 signs.

Once again, when *A* and *C* will conjunct in *e*, *A* has traversed 50 signs and *B*, 25, and, as before, they will be separated by 1 sign. And now the whole revolution will have been completed and another commences, with the conjunctions repeating just as before. In this case *A*, *B*, and *C* cannot be closer than 1 sign, and, indeed, only in certain places can they be as close as that. In much the same way, what has been said is applicable to more [than three] mobiles. Therefore, if all the planets are moved with commensurable speeds, it is possible that any three or four planets might never be in conjunction in the same degree, or minute; and perhaps some planets are unable to approach closer than two degrees, or three degrees. What has been said about two mobiles can be extended to any number of mobiles.

Proposition 14. If several [i.e., three or more] mobiles are presently in conjunction, it is necessary that they conjunct in that same point at other times.

Let *A*, *B*, and *C* be in conjunction in point *d*. Since by assumption the speeds of *A*, *B*, and *C* are commensurable, they will be related as three numbers, *E*, *F*, and *G*, respectively. Thus when *A* will have made *E* circulations, *B* will have completed *F*, and *C*, *G* circulations. When this occurs any one of these mobiles will have returned to point *d*, which is easily shown in the same way as for two mobiles in the fourth proposition.

B sive *F* / in eodem² *om. U* scilicet gradu vel *B*

560 *post* possunt *hab. L* tantum / approximari: appropinquari *P*

561 [ad minus] *om. AFLPU* / gradus *om. B* grada *U* / vel: aut *F* / *post* vel *hab. L* per / *post* 3 *mg. hab. L figuram*

562 Que: quod *P* / *post* Que *hab. B* autem / *post* sunt *hab. B* potes / ad: aut *F* / possunt se *om. AFPU* voluere *B*

563 extendere: extendetur *U* / *post* extendere *hab. A* sequitur quartadecima conclusio et *infra mg. hab. A figuram*

564 Conclusio quartadecima *om. FP* Propositio quartadecima *U et ante* Si *mg. hab. B* quartadecima *et mg. hab. A* 14 / sunt *BLU*

565 puncto eodem *tr. B* / *post* coniungantur *mg. hab. L figuram*

566 et¹ *om. B* / et² *om. B* / in puncto *om. P* in *B* / *ante* sint *hab. B* eorum

567 suppositum est: supponitur *L* / *ante* erunt *scr. et del. U* et ?nunc / erunt: essent *F* / sic *A* / 3: est *P* / per ordinem: portionem *P*

568 sunt *A* / Ergo *om. P* / fecerit precise *tr. L*

569 E...numero F *om. U* / precise *om. B* / numero² *om. U* / et *om. B*

570 quodlibet eorum *om. U* / quotlibet *AF* / regressus *FP*

572 [mobilibus] *om. AFLPU* / conclusione: coniunctione *U*

Verbi gratia, sit velocitas *A* sicut 6, et velocitas *B* ut 5, et *C* ut 4. Igitur quando *A* compleverit 6 circulationes tunc iterum erit in puncto *d*; et
575 similiter *B* erit ibidem quando compleverit 5, et *C* quando fecerit 4. Et hec perficient in eodem tempore, igitur in fine illius temporis erunt omnia simul in *d* puncto. Et eodem modo, et per idem, diceretur de 4 mobilibus [vel 5, vel quotlibet] consimili ratione.

Conclusio quintadecima. [Tempus] quando hoc primo fiet invenire.

580 Iam patet ex precedenti quod hoc primo erit in fine circulationum numeratorum per numeros quibus velocitates signantur qui sunt *E*, *F*, *G*. Et quoniam proportio temporum in quibus fuerint circulationes est econtrario proportioni velocitatum, inveniantur 3 numeri qui se habeant proportionaliter econtrario istis tribus et in minimo illorum *A* faciet circulatio-
585 nem unam, et *C* in maximo et *B* in medio. Sint igitur isti numeri *H*, *K*, *L* et per tricensimamsextam septimi [Euclidis] inveniatur minimus ab eis numeratus, qui sit *M*. Dico ergo quod *M* est tempus in cuius termino *A*, *B*, *C* primitus coniungentur ubi nunc sunt; et unitas mensurans istum numerum *M* est maximum tempus numerans istas velocitates, vel maxima
590 mensura communis earum sicut dies vel annus.

Si ergo de ipso *M* accipiatur pars denominata numero maiori velocitatum, scilicet numero *E*, habebitur tempus in quo *A* facit unam circulationem, et si abscindatur pars denominata numero *F* veniet tempus in quo *B* facit unam circulationem et pars denominata numero *G* est tempus in quo
595 *C* complet unam circulationem; et proportio istarum partium est econver-

573 sicut: ut *B* / ante et[1] *scr. et del. F* circulationes / et[1] *om. B* / velocitas[2] *om. L* / B *om. P* / ut[1]: sic *A* sicut *F* / et[2]: velocitas *B*
573–75 Igitur…similiter *om. A*
574 complevit *U*
575 quando compleverit *om. P* cum compleverit *B* quando complevit *U* / quando[2]: cum *B* quandoque *P* / fecerit: compleverit *BU* / Et[2] *om. L* / hec: illud *F* hoc *P*
576 illius: istius *P* / erunt: essent *F* / omnia: illa 3 *B*
577 d puncto *tr. B* / Et[1] *om. B* / et[2] *om. L* / dicetur *U*
578 [vel 5 vel quotlibet] *om. AFLPU* / consimili: simili *BF* / post ratione *hab. P* et cetera
579 Conclusio quintadecima *om. FP* Propositio quintadecima *U et ante* [Tem-

pus] *mg. hab. B* quintadecima *et ante* quando *mg. hab. A* 15 / [Tempus] *om. AFLPU* / primo fiet: erit primo *A*
580 patet ex precedenti: ex precedenti patet *B* ex precedenti *L* / hoc *om. L* / primo *om. F*
580–81 numeratorum: numerare *B*
581 signatur *LP* / qui: que *P* / ante E *hab. U* quidem
583 proportionum *P*
583–84 se habeant proportionaliter: proportionaliter se habeant *L*
584 illorum: istorum *AU* / faciet: faceret *A*
584–85 circulationem unam *tr. AL* circulationem *F*
585 maxima *A* / et *C*…medio: B in medio et C in maximo *B* / sit *AP* / ante igitur *hab. F* sic / L *om. L*
586 *ante* septimi *hab. A* conclusionem / septimi: aut *P* / [Euclidis] *om. AFLPU* /

For example, let the velocity of *A* be as 6, that of *B* as 5, and *C* as 4. When *A* will have completed 6 circulations it will again be in point *d*; and the same may be said for *B* and *C* when they will have completed 5 and 4 circulations, respectively. Now these circulations are completed in the same time so that all the mobiles will be in point *d* at the end of this interval. By the same reasoning, and following the same procedure, one may say the same thing about four, five, or any number of mobiles.[8]

Proposition 15. [How] to find the time when [the several mobiles] first [conjunct again in their present point of conjunction].

From the preceding proposition it is already obvious that upon completion of the circulations numbered as *E*, *F*, and *G*, which represent the velocities, the mobiles will again be in their first point of conjunction. Now since the ratio of times in which any mobiles complete their circulations is reciprocally proportional to their ratio of velocities, three numbers can be found which bear a reciprocal proportional relationship to *E*, *F*, and *G*. The least, middle, and greatest of these numbers represent, respectively, the times in which *A*, *B*, and *C* complete one circulation. Let these numbers be *H*, *K*, and *L*. Now by the thirty-sixth [proposition] of the seventh [book] of Euclid, a least number [or common multiple] can be found that is numbered [or measured] by *H*, *K*, and *L*; let us call this number *M*. I say that *M* represents the time in which *A*, *B*, and *C* will first conjunct in their present position. The unit of time— say, a day or a year—which measures number *M* is the greatest time, or greatest common measure, numbering [or measuring] these velocities.

If, now, a part of *M* is taken which is denominated by the greatest of the numbers representing the velocities, namely, number *E*, we obtain the time it takes for *A* to complete one circulation; if a part [of *M*] denominated by *F* were taken, we get the time in which *B* completes one circulation; and the part [of *M*] denominated by number *G* yields the time it takes for *C* to complete one circulation. A ratio of these parts is reciprocally proportional to the ratio

inveniantur *A* / minimus: minus *AP* numeris *L*

587 qui: et *B*

588 primitus coniungentur: coniungentur primo *B* primo coniungentur *L*

589 numerum: numeri *U* / *post* tempus *hab.* *L* mensurans istum numerum *M* / vel: aut *F*

589–90 maxima mensura: maxime mensurationis *A*

590 communis earum *tr.* *B* earum *A* / vel:

sive *F* / *post* annus *add.* *B* vel hora et cetera

591–92 ?velocitatis *A*

592 fecit *A*

593 *post* si *hab.* *P* A facit / si abscindatur *om.* *U* / veniet: proveniet *B* / quo *om.* *FP*

593–94 F veniet…numero *om.* *U*

594 fecit *A* / in *om.* *AL*

595 C: G *B*

595–96 econverso: econtraria *L* econtrario *U*

so proportioni velocitatum. Et hec omnia facilius et brevius patent exemplo quam demonstratione formata.

Verbi gratia, sint numeri velocitatum 6 et 5 et 4, ergo numeri minimi qui sunt proportionales econtrario sunt 10 et 12 et 15. Cum itigur 60 sit minimus ab eis numeratus, dico quod 60 est numerus temporis totius revolutionis horum trium et numerat tempus in quo primo coniungentur ubi nunc sunt. Sint ergo 60 dies, ergo dies est maximum tempus numerans vel mensurans velocitates predictas. Ergo *A*, quod est mobile velocius, complet unam circulationem in sexta parte huius totius temporis, scilicet in 10 diebus; et *B* in quinta parte, scilicet in 12 diebus; et *C* in quarta parte, scilicet in 15 diebus. Et isti numeri, scilicet 10, 12, 15, se habent proportionaliter econverso illis tribus primis positis qui erant 6 et 5 et 4, ut patet intuenti.

Sic, igitur, apparet in quanto tempore *A* facit unam circulationem et per precedentem manifestum est quod post 6 circulationes ipsius *A* primo coniungentur ista tria in puncto *d*. Igitur inventum est tempus in quo primo redibunt ad locum in quo nunc sunt. Hoc est itaque peryodus revolutionis horum qua completa et peracta coniunctiones et omnes aspectus rursum incipient fieri penitus sicut ante. Et de quotlibet mobilibus inveniretur hoc modo.

Conclusio sextadecima. Tempus reperire in quo huiusmodi mobilia sive in puncto in quo nunc sunt sive in alio primitus coniungentur.

Invenietur per sextam conclusionem tempus in quo *A* et *B* primitus coniungentur; atque iterum per eandem reperietur tempus in quo *C* et *B* primo coniungentur, et etiam tempus in quo *A* et *C* primo coniungentur. Quibus sic habitis queratur per tricensimamsextam septimi Euclidis tem-

---

596 proportio *F* proportionum *P* / facilius: faciliter *AFP*
598 sint: sicut *P* / et[1] *om. BL* / et[2] *om. B* / *ante* numeri *hab. U* isti / minimi *om. F* / qui: que *P*
599 sunt[1]: sint *A* / econtrario: econverso *B* / et[1] *om. AB* / et[2] *om. B*
599–600 minimus: numerus *L*
600 eis: eo *L* / est *om. FP* pars *U* / numerus *om. BFPU*
601 trium: talium *U* / numerat: numerale *A*
602 Sint *om. P* / vel: sive *F*
603 velocitates predictas: dictas velocitates *B* / quod est *om. B* / velocius: velox *U*
604 huius *om. B*
605 scilicet: B *P* / in[2] *om. B* / quarta: 4 *A* /

parte *om. AB*
606 in *om. B* / scilicet *om. BU* B *P* / *post* 12 *hab. A* et
606–7 proportionaliter *om. U* proportionabiliter *A*
607 *ante* tribus *hab. U* numeris / primis *om. B* primo *U* / et[1] *om. AB* / et[2] *om. B*
609 sicut *P* / igitur *om. B* / apparet: atque *AP* patet *F* / A *om. B* / facit: fecerit *A* faciant *B*
610 precedentem: predicta *L* / quod *om. B*
612 locus *P* / quo: ergo *F* / *post* sunt *hab. B* et / itaque *om. AB* igitur *L*
613 *post* horum *hab.* B 3 / qua: que *A* / *post* completa *hab. A* est / et peracta *om. B* / et[2]: est *A* / *ante* rursum *hab.* B ?contro /

of velocities. But all these things can be shown more readily and in briefer compass by an example rather than by formal demonstration.

For example, let the numbers representing the velocities be 6, 5, and 4, so that the least numbers [forming ratios correspondingly] reciprocal to these are 10, 12, and 15. Since 60 is the lowest number numbered [or measured] by them, I say that 60 is the number representing the time of the whole revolution of the three mobiles, and hence represents the time it takes for the three mobiles to conjunct first in their present point of conjunction. Should the 60 be 60 days, the day will be the greatest [unit of] time numbering [or measuring] the aforesaid velocities. Therefore A, the quickest mobile, completes one circulation in $\frac{1}{6}$ of this total time, namely, in 10 days; B [accomplishes this] in $\frac{1}{5}$ of the total time, namely, in 12 days; and C does it in $\frac{1}{4}$ of the total time, or in 15 days. It is obvious by inspection that 10, 12, and 15 [form ratios that] are reciprocal to [those formed by] the first three numbers posited, namely, 6, 5, and 4.

Thus, the time it takes for A to make one circulation is known, and by what has preceded it is clear that after 6 circulations of A the three mobiles will again be in conjunction in point d, their first place of conjunction. The time in which they will first return to their present position has now been found. Moreover, this is the period of their revolution, and when it has been completed and finished, the conjunctions and all the aspects begin again exactly as before. The period of revolution can be determined in this way for any number of mobiles.[9]

Proposition 16. [How] to find the time in which such mobiles will first conjunct either in the point in which they are now, or in another point.

By the sixth proposition, the time can be found when A and B will first be in conjunction; and, again, by the same proposition, the time can be found when C and B, as well as A and C, will first be in conjunction. After finding these things, we [then] use the thirty-sixth [proposition] of the seventh [book]

---

rursus *FU*

614 penitus sicut ante: sicut ante penitus *B* / mobilibus: mobilium *A* modis *F* / *post* mobilibus *hab. B* similiter / inveniretur: invenientur *U* / hoc: isto *F*

615 *post* modo *hab. P* et cetera

616 Conclusio sextadecima *om. FP* Propositio sextadecima *U et ante* Tempus *mg. hab. B* sextadecima *et post* huiusmodi *mg. hab. A* 16 / reperire *om. B* invenire *LU* / in quo: quando *ALU* quo *B* / ante

sive *hab. B* primo coniungentur

617 in quo: quo *BU* ubi *L* / primitus coniungentur: reperire *B*

618 primitus: primo *B*

619 atque: et *B* / eundem *FP* / reperietur: invenietur *L*

619–20 et…coniungentur² *om. L*

620 et² *om. B*

621 Quibus: Istis *L* / sic *om. B* similiter *F* / queritur *U* / septimi: aut *P*

pus minimum ab hiis tribus temporibus numeratum et in huius temporis termino hec tria mobilia primitus coniungentur quod statim patet exemplo.

625 Verbi gratia, in casu iam posito, patebit per sextam quod *A* et *B* primo coniungentur sexagensima die et sic consequenter de 60 in 60 diebus; et per eandem scietur quod *B* et *C* coniungentur sexagensima die; et *A* et *C* tricensima die, et sic semper de 30 in 30. Et ergo cum 60 sit minimus numerus numeratus ab istis qui sunt 60, 60, 30, sequitur quod sexagen-
630 sima die primitus coniungentur, et hoc erit in puncto *d* in quo nunc sunt quia in isto casu non est nisi unus punctus in quo hec tria possunt coniungi.

Sed gratia exercitii ponatur alio modo ita quod velocitas *A* sit sicut 7, et velocitas *B* sicut 5, et velocitas *C* ut 3. Tunc per arismeticam invenietur quod minimi numeri qui sunt econverso proportionales sunt 15, 21, 35, et
635 minimus numeratus ab eis est 105. Ergo in fine 105 dierum primo coniungentur ubi nunc sunt. Sed tamen interim alia vice coniungentur quia *A* et *B* primo coniungentur transactis 52 diebus cum dimidio quod invenietur per sextam conclusionem hoc modo. Quia *A* facit unam circulationem in 15 diebus, ergo in die transit $\frac{1}{15}$ circuli; et *B* $\frac{1}{21}$. Ergo subtrahendo $\frac{1}{21}$ ab
640 $\frac{1}{15}$ restant $\frac{6}{315}$; et dividendo denominatorem, scilicet 315, per numeratorem, scilicet per 6, veniunt 52 cum dimidio, quod est propositum.

Et per eandem viam invenietur quod *A* et *C* coniungentur vicensimasexta die cum $\frac{1}{4}$, ac etiam *B* et *C*, quinquagensimasecunda die cum dimidio. Et minimum tempus ab istis numeratum est 52 [diebus] cum dimidio; ergo

622 *ante* ab *hab. A* et / huius *obs. B* huiusmodi *A* / temporis: partis *F*
623 quod: et patet per precedentem et *B*
625 *post* sextam *add. B* conclusionem
626 *post* coniungentur *hab. U* in / sexagensima *corr. ex* vicensima *ABU* secundo *F* decima *L* duo *P* / 60¹ *corr. ex* 20 *BFPU* vicensima *A* 10 *L* / *ante* in *hab. U* diebus / 60² *corr. ex* 20 *BU* 30 *FP* vicensimam *A* 10 *L* / diebus *om. U* diem *A*
627 B et C: C et B *LU* / *post* coniungentur *hab. B* primo *et hab. LU* in / sexagensima *corr. ex* duodecima *ABLU* 12 *FP* / *post* die *hab. B* et sic consequenter de 12 in 12 diebus / *post* et² *hab. AB* quod
628 tricensima *corr. ex* 10 *AFLP* decima *BU* / 30¹ *corr. ex* 10 *ABFLPU* / 30² *corr. ex* 10 *ABFLPU et post* 10 *hab. B*

diebus / Et² *om. B* / ergo cum *tr. B* / 60: sexagensima *P* / minimus *om. FLP*
629 numerus *om. AU* / *ante* ab *hab. B* et / qui: que *AP* / qui sunt *om. B* / 60, 60, 30 *corr. ex* 20, 12, 10 *ABFLPU*
630 primitus: ista 3 primo *B* / et...in¹: in ?6 *B* / erit: est *U* / *ante* puncto *hab. U* ipso / *post* sunt *hab. B* coniuncta
631 quo: quolibet *P*
632 sicut: ut *B*
633 et¹ *om. B* / *ante* sicut *hab. L* sit / sicut: ut *B* / et² *om. B* / ut: sicut *U* / invenitur *BFU*
634 minimi numeri *tr. B* / econverso: econtrario *AU* / *post* econverso *hab. B* illis / *post* 15 *hab. U* et / 35: 25 *L* / *post* et *hab. AL* quod
635 minimus: numerus *FLP* / numeratus ab eis: ab eis numeratus *AL* / *post* primo *hab. B* simul

of Euclid to determine the least time numbered [or measured] by these three times, and at the end of this time these three mobiles will be in conjunction for the first time. This is immediately evident by an example.

For example, in the case already presented, it will be obvious by the sixth proposition that A and B will conjunct for the first time on the sixtieth day, and continually thereafter every 60 days; and, by the same proposition, it will be determined that B and C will first conjunct on the sixtieth day [and thereafter every 60 days]; and that A and C will first conjunct on the thirtieth day and every 30 days thereafter. Because 60 is the least number numbered [or measured] by 60, 60, and 30, it follows that on the sixtieth day the three mobiles will first be in point d—their present position—since in this example there is only one point in which these three can conjunct.

But [now] let us put the example in another way so that the velocity of A is as 7, that of B as 5, and C as 3. Then, by means of arithmetic, it will be found that 15, 21, and 35, are the least numbers [that form ratios correspondingly reciprocal to 7, 5, and 3], and that 105 is the least number numbered [or measured] by them. At the end of 105 days, they will first be in conjunction where they are now. But, in the interim, they will conjunct in another place since A and B will conjunct first after moving for $52\frac{1}{2}$ days—and this is found by the sixth proposition in the following way. Since A makes one circulation in 15 days, it will traverse $\frac{1}{15}$ of a circle in one day; and B [makes one circulation in 21 days, so that it traverses] $\frac{1}{21}$ [of a circle in one day]. The result of subtracting $\frac{1}{21}$ from $\frac{1}{15}$ is $\frac{6}{315}$; and by dividing the denominator, i.e., 315, by the numerator, i.e., 6, we get $52\frac{1}{2}$, which is what we sought.

In the same way, it will be determined that A and C will conjunct after $26\frac{1}{4}$ days, and B and C after $52\frac{1}{2}$ days. The least time measured by these times is $52\frac{1}{2}$ days, after which these three mobiles will first conjunct [simultaneously].

---

636 interim: iterum *FP* / *ante* alia *hab. A* coniungentur / *ante* quia *hab.* B patet
637 dimidio: divisione *L* / quod invenietur: ut patet *B* quod invenitur *U*
638 conclusionem *om.* B / A: 7 B
639 transibit *A* / *post* B *hab.* B in 21 diebus ergo in die / $\frac{1}{21}$ ^{1}: 12^{am} *L* / $\frac{1}{21}$ ^{2}: 12^{am} *L* unam 22^{am} *P* / ab: de B a *L*
640 $\frac{1}{15}$: 15 *FP* 15° *L* / restant *om. FP* provenient B circuli remanent *U* / $\frac{6}{315}$: 315 *L* 315 *P* / *post* et *hab.* A di- / denomina-

torem: demoninationem *FPU* / *infra* scilicet *scr. et del.* B est / numeratoram: numerationem *FP*
641 per *om.* B / veniunt: provenient B venient *U*
642 via *P* / inveniretur *AU* invenitur *F* / coniungerentur *U*
643 ac: et *B* / *post* C *hab.* B coniungentur
643–44 Et…dimidio *om.* L
644 [diebus] *om. AFPU*

645 tunc ista tria primitus coniungentur. Et de quotlibet mobilibus per easdem
regulas conformiter diceretur.

*Conclusio septimadecima. Coniunctiones totius revolutionis seu totius
peryodi numerare videlicet quotiens coniungentur quousque sint iterum in
puncto eodem in quo nunc sunt.*

650 Propter hoc enim quod motus sunt regulares semper est equalitas de
tempore inter quaslibet coniunctiones proximas, ergo non oportet nisi
accipere tempus totius revolutionis horum trium mobilium quod tempus
invenitur per quintamdecimam conclusionem, et istud tempus dividere per
tempus quod est inter duas coniunctiones proximas quod tempus habetur
655 per conclusionem immediate precedentem et exibit numerus quesitus.

Verbi gratia, in exemplo ultimo posito tempus totius revolutionis illorum
trium erat 105 dies, et tempus inter duas coniunctiones sibi invicem proximas erat 52 dies cum dimidio per quod dividatur 105 dies et veniet 2. Ergo
tantummodo in duobus locis possunt coniungi ita quod tertia coniunctio
660 erit ubi fuit prima, et quarta ubi erat secunda, et sic in infinitum semper
redeundo sicut dicebatur de duobus mobilibus in septima conclusione. Et
omnino consimiliter agendum est de 4, aut quotlibet mobilibus nunc
coniunctis. Et nota quod si numeri locorum in quibus coniunguntur bina
et bina sunt incommunicantes numquam poterunt omnia simul coniungi
665 nisi tantum in uno loco. Et si sint communicantes in totidem punctis
possunt coniungi quotus est maximus numerus in quo communicant
numeri supradicti et illi numeri facillime habentur ex conclusione undecima.

Verbi gratia, sit velocitas ipsius *A* ut 20, et velocitas *B* ut 10, et velocitas

645 ista: illa *L* / ista tria primitus: primitus
   ista tria *A* / *post* primitus *hab. B* simul /
   post coniungentur *hab. B* scilicet in
   puncto opposito illi in quo primo nunc
   coniuncta sunt sic sunt 2 puncta verbi
   gratia possunt coniungi in hoc casu
   patebit / Et *om. U*
647 Conclusio septimadecima *mg. hab. L*
   post totius *om. FP* Propositio septimadecima *U et ante* coniunctiones *mg. hab.*
   B septimadecima *et mg. hab. A* 17 /
   totius² *om. B*
648 sint: veniant *L* / iterum: interim *U*
649 puncto eodem *tr. B* / in *om. B*
650 enim: vero *F* / equalitas: equaliter *FP*
   equalis *A*
650–51 de tempore: tempus *B*
652 accipere *om. A* capere *L* accipe *P* / quod:

quas *P*
652–53 tempus invenitur: scitur *B*
653 invenietur *FP* invenientur *U* / dividere:
   dividetur *U*
654 tempus habetur: haberetur *B*
655 conclusionem immediate precendentem:
   immediate precedentem conclusionem
   AL conclusionem precedentem immediate *B* immediate conclusionem precedentem *U* / exibit: exit *B*
656 illorum: istorum *A*
657 et: per *FPU* / coniunctiones *om. B* / sibi
   invicem *om. BP* sibi invicis *L* sibi *U*
658 erat: est *B* / dimidio: medio *B* / dividantur *U* / *post* dividatur *hab. B* tempus / dies² *om. FLPU* / venient *ALU*
   provenit *B*
659 tantummodo: tantum *B* / *ante* possunt

Applying these same rules, a similar case can be made for any number of mobiles.[10]

Proposition 17. [How] to number the conjunctions of a whole revolution or period—i.e., how many conjunctions will occur until the mobiles are again in the same point where they are now.

Because the motions are regular, there is always an equal time interval between any successive conjunctions. To determine the number of conjunctions in a revolution, one need only take the time of the whole revolution of these three mobiles, which can be found by the fifteenth proposition, and divide it by the time interval between two successive conjunctions, which can be determined by the proposition immediately preceding.

For instance, in the last example, the time of the whole revolution of these three mobiles was 105 days, and the time between successive conjunctions was $52\frac{1}{2}$ days, which when divided into 105 days produces 2. Hence there can be only two places of conjunction, so that a third conjunction occurs where the first was, and a fourth where the second occurred, and so on ad infinitum proceeding always in the manner described in the seventh proposition where two mobiles were involved. And the same argument must be applied to four mobiles, or to any number of them now in conjunction. Note, however, that if the numbers of places in which the mobiles will conjunct two at a time are incommunicant [or prime to one another], all the mobiles will never be able to conjunct simultaneously, except in one place only. But if these numbers are communicant, the number of points in which they can conjunct equals the greatest number that measures the numbers mentioned above; and these numbers can be found very easily by the eleventh proposition.

For example, let the velocity of *A* be as 20, that of *B* as 10, and *C* as 7.

---

*hab. U ?*coniungi / ita quod: itaque *F* / tertia: 3 *AB*

660 fuit: erat *L* / erat: fuit *U* / sic in infinitum *om. B*

661 *ante* dicebatur *hab. B* primo / dicebatur: declarabatur *U*

662 omnino consimiliter: omnino similiter *BU* ideo consimiliter *L* / agendum est *tr. F* arguenda est *A* / de 4 aut: ?regula de *A* / *post* 4 *hab. B* aut 5 aut 6 aut 7

663 Et nota: datis suis velocitatibus et numeris proportionibus *B* / quod: sed *B* / *ante* numeri *scr. et del. A* unum / in *om.*

B / coniungentur *B*

664 numquam: necquiquam *P* / omnia: omnino *A*

665 tantum in uno loco: in uno loco tantum *B* / tantum: tamen *P* / *ante* loco *hab. U* modo / Et: ut *FPU* / sint: sunt *AB* / totidem: tot *A*

667 facillime: faciliter *P*

667–68 conclusione undecima *tr. B*

669 ipsius *om. B* / ut[1]: sicut *F* / et[1] *om. B* / *ante* B *hab. A* ipsius / ut[2]: sicut *F* / *ante* et[2] *hab. P* C / et[2] *om. B*

670   *C* ut 7. Ergo per undecimam puncta coniunctionum *A* et *B* sunt 10, et
puncta coniunctionum *C* et *B* sunt 3, et puncta coniunctionum *A* et *C* sunt
13. Et isti tres numeri sunt incommunicantes, ergo ista tria mobilia non
possunt coniungi nisi in uno loco tantum. Sed si velocitas *A* esset ut 19, et
velocitas *B* ut 13, et velocitas *C* ut 10, tunc illi tres numeri punctorum in
675   quibus coniunguntur bina et bina essent 6, et 3, et 9; et quia maximus in
quo communicant est ternarius, ideo in tribus punctis possunt coniungi
tria simul.

    Conclusio duodevicensima. Locum prime coniunctionis sequentis assignare.
Sicut in octava conclusione dictum est de duobus mobilibus adiuvante
680   sexta conclusione ita penitus dicendum est de pluribus adiuvante sextadecima.

    Verbi gratia, in exemplo prius posito, quia velocitas *A* erat sicut 7, et *B*
5, et *C* 3, et tempus totius revolutionis est 105 dies, igitur *A* facit 7 circulationes in isto toto tempore; ergo facit unam in septima parte huius
685   temporis, videlicet in 15 diebus. Ergo qualibet die transit de isto circulo $\frac{1}{15}$.
Sed per sextamdecimam apparet quod prima coniunctio sequens erit
transactis 52 diebus cum dimidio, ergo illo tempore *A* pertransit de suo
circulo $\frac{52}{15}$ cum dimidia. Ergo punctus in quo primo coniungentur hec tria
distabit a puncto in quo nunc sunt per $\frac{7}{15}$ cum dimidia, et hoc est per medi-
690   etatem circuli et est punctus oppositus *d* puncto.

    Et idem eveniret operando per motum ipsius *B* et etiam per motum *C*.
Et ita agendum est de quibuscumque mobilibus quotquot essent.

    *Conclusio undevicensima. Numerum et seriem punctorum reperire in
quibus umquam talia plura mobilia coniungentur.*
695   Simili modo penitus quantum ad hoc sicut actum est de duobus mobili-

---

670 ut: sicut *F* / per *om.* L / *post* undecimam
add. B conclusionem / et¹ *om.* BU
670–71 10...sunt¹ *om.* F
671 et B...et C *om.* A
672 numeri: termini *A*
673 si: ibi *U* / esset: est *AP* sit *B*
674 ut²: sicut *F* / illi *om.* B / numeri: termini *A* / in *om.* B
675 et¹ *om.* B
676 communicat *P* / *post* punctis *hab.* B
tantum / possunt *om.* A possent *FP* /
post coniungi *hab.* B ista
677 *post* tria *hab.* B mobilia / *post* simul *hab.*
P et cetera
678 Conclusio duodevicensima *om.* P Propositio duodevicensima *U et post* simul
(*linea* 677) *hab.* F incipit 18 capitulum *et*

ante Locum *mg. hab.* B duodevicensima
et mag. hab. A 18
680 penitus dicendum est: dicendum est penitus *A* / *post* penitus *hab.* B hic / dicendum est *tr.* F
680–81 sextadecima: 16 *AFPU et post* 16
hab. U conclusione
682 prius: iam *B* / quia *om.* B / erat *om.* A /
sicut: ut *BF* / *post* et *add.* B velocitas
683 *ante* 5 *hab.* B ut / C 3: velocitas C ut
3 *B* / totius: totiens *A* / A *om.* U / fecit
A / 7 *rep.* L / *post* 7 *hab.* A revolutiones
vel
684 isto toto tempore: toto isto tempore *B*
isto tono tempore *P* isto tempore toto
U / septima: alia *U* / huius: istius *AB*
685 temporis *om.* L / videlicet: scilicet *B* / 15

Then, by the eleventh proposition, A and B have 10 points of conjunction, C and B have 3 points, and A and C have 13 points. But these three numbers are incommunicant, so that the three mobiles can conjunct in only one place. However, should the velocity of A be as 19, the velocity of B as 13, and that of C as 10, the three numbers representing the points in which they will conjunct two at a time are 6, 3, and 9. But since their greatest common measure is 3, they can conjunct simultaneously in three points.[11]

Proposition 18. [How] to determine the place of the first conjunction following [a present conjunction].

Just as in the eighth proposition, [where] two mobiles were discussed with the aid of the sixth proposition, so now, more than two mobiles will be considered with the aid of the sixteenth proposition.

For instance, in the example given previously, since the velocity of A was as 7, the velocity of B as 5, that of C as 3, and the time of the whole revolution is 105 days, it follows that A makes 7 circulations during this total time; therefore, it makes one circulation in $\frac{1}{7}$ of this time, namely, in 15 days. Hence, on any day whatever, A traverses $\frac{1}{15}$ of its circle. But by the sixteenth proposition, we see that the first conjunction following will have occurred in $52\frac{1}{2}$ days, so that A traverses $\frac{52}{15}$ of its circle plus $\frac{1}{2}$ [of $\frac{1}{15}$ of it]. The point in which these three mobiles will first conjunct will be separated from the point in which they are now by $\frac{7}{15}$ plus $\frac{1}{2}$ [of $\frac{1}{15}$], that is, by half of the circle; and this point is opposite point d.

The same results are obtained by operating with the motion of B, or with that of C. Whatever the number of mobiles, this procedure must be followed.[12]

Proposition 19. [How] to find the number and sequence of points in which several [three or more] such mobiles will always conjunct.

Now just as in the tenth proposition, [where] two mobiles were considered

---

obs. B 25 F / *post* Ergo *hab. A* de / isto: suo *B* / $\frac{1}{15}$: ?5^{am} *A* 25^{am} *F*

686 *post* sextamdecimam *hab. B* conclusionem / apparet: 13 *A* patet *F* atque *P* / quod *om. L*

687 52: 25 *FP* / 52 diebus cum dimidio: diebus $52\frac{1}{2}$ *U* / dimidio: medio *B* / illo: isto *A* / tempore A: et ea *P* / *A om. U* / *post* de *hab. F* illo

687–88 suo circulo *tr. FL*

688 quo: qua *F*

689 distabunt *U* / dimidio *FP* $\frac{1}{2}$ *U*

690 d puncto *tr. B*

691 operando *om. U* / motum^1: modum *F* /

etiam: similiter *B* / *post* motum^2 *hab. B* ipsius

692 est *om. AFLP* / quibuscumque: quibusdam *F* quibusque *P* / quotquot essent *om. A* / *post* essent *add. FP* et cetera

693 Conclusio undevicensima *om. FP* Propositio undevicensima *U et ante* Numerum *mg. hab. B* undevicensima *et post* plura (*linea* 694) *mg. hab. A* 19 / reperire *tr. B post* coniungentur (*linea* 694)

694 umquam: numquam *FP* / talia: plura *A* tali *U* / plura mobilia *tr. B* talia mobilia *A*

695 similis *A* simile *L* / modo: non *U* / penitus *om. B*

bus in decima conclusione cum auxilio septime, ita in proposito agendum est cum iuvamine conclusionis septimedecime. Et ex illa serie et processu per puncta coniunctionum patet quando et qualiter mutantur coniunctiones de triplicitate in triplicitatem; et alii aspectus similiter quoniam sicut in decima conclusione dictum est de duobus qualiter non semper procedunt coniunctiones de uno puncto ad proximum, sed quandoque fuerit in quibusdam motibus intervalla. Ita potest contingere de tribus aut pluribus pari forma.

Conclusio vicensima. Si circuli fuerint eccentrici erit idem numerus locorum qui esset si forent concentrici sed erunt distantie temporis et spatii inequales.

Ob hoc enim quod motus sunt commensurabiles necesse est quando mobilia sunt in uno loco coniuncta quod ibidem alias coniungantur, ut prius probatum est. Ergo loca talia sunt finita et facta revolutione omnium coniunctiones iterum incipient fieri sicut ante. Sint ergo duo mobilia *A* et *B*; et sit *c* centrum mundi et centrum motus ipsius *A*; et sit *d* centrum motus ipsius *B*. Et sint nunc *A* et *B* coniuncta in linea *cdg* que transit per augem et oppositum augis et tunc se habent *A* et *B* ac si motus et circuli sint concentrici et sunt idem locus et motus verus et medius; et similiter quando sunt in opposito augis.

---

696 septime: alie *FP*
697 iuvamine: iuvamente *AF* adiutorio *B* iuvantum *P* / illa: ista *AFP*
698 *ante* -ta *in* puncta *mg. hab. B* ?notanda vel ?nota / *post* qualiter *hab. B* et quo loco / mutentur *A*
698–99 coniunctiones: puncta coniunctionis *L* / *post* coniunctiones *hab. B* ?planetarum superiorum
699 triplicitate in triplicitatem: una tripliciter ad aliam *B* / triplicitatem: triplicitate *P* / alii: a *P* / similiter: signetur *A* / quoniam: quando *A* / sicut *om. A*
700 dictum: demonstratum *F* / est *om. L* / *ante* qualiter *hab. L* mobilibus / *infra* qualiter *mg. hab. L* figuram / procedent *P*
701 coniunctiones *om. L* / uno *om. AFPU* / proximam *A*
701–2 in quibusdam motibus *om. A* saltando per *B* in quibusdam *U*
702 potest: possunt *P*
704 Conclusio vicensima *corr. ex* Conclusio 28ᵃ *om. FP* Propositio vicensima *U et*

ante Si *mg. hab. B* vicensima *et post* erit *mg. hab. A* 20 / circuli fuerint *tr. B* / eccentricus *B* / erit idem *tr. B* idem *L*
705 *post* locorum *hab. B* punctorum / esset: enim *P* / forent: essent *L* / erunt: essent *F*
706 *ante* inequales *hab. B* eorum
707 enim: vero *F* aut *L* / *post* motus *hab. B* eorum / necesse est *tr. F*
708 uno: eodem *B* / ibidem: idem *A*
709 *post* facta *hab. U* tali
710 coniunctionum *U* / coniunctiones iterum *tr. B* / incipiunt *FLP* reincipient *B* / sicut: ut *B* / ergo *om. A*
710–11 A et B *om. A A B F*
711 sit¹ *om. B* / mundi: medii *U* / et centrum *om. U* / *ante* centrum² *hab. B* g / *post* A *hab. U* stelle / sit² *om. BU*
711–12 et³...ipsius B *om. L*
712 *post* B² *hab. F* consequenter
713 habent: habebunt *U* / et³ *om. AFLU* / *ante* sint *hab. B* sui / sint: sicut *P*
714 et²: duo *U* / verus et medius: medius et verus *AB* vero et medius *F* / et⁴ *om. B* / similiter: sunt *A*

with the help of the seventh proposition, so, similarly, in this proposition [where more than two mobiles are involved] the help of the seventeenth proposition is required. When and how conjunctions are altered from one triplicity to another, as well as how other aspects are altered, becomes obvious from this [proposition] by means of the order and passage through the points of conjunction. For just as in the tenth proposition, [where] we discussed how the conjunctions of two mobiles do not always proceed from one point to the very next point but that in some motions there is sometimes an interval [or jump], so also can three or more mobiles be considered in the same way.[13]

Proposition 20. If the circles should be eccentric, the number of places [of conjunction] would be the same as if they were concentric, but the intervals of time and the distances [between conjunctions] would be unequal.

Now, as previously demonstrated, when mobiles have been in conjunction in one place, it is necessary, because of their commensurable motions, that they have conjuncted in that same place at other times. The number of such places is finite, but after all [the mobiles] have completed a revolution, the conjunctions will repeat all over again, just as before. Let there be two mobiles, *A* and *B*; and let *c* be the center of the world and the center of *A*'s motion; and let *d* be the center of *B*'s motion [see Fig. 22]. Assume that *A* and *B* are now in conjunction on line *cdg*, which passes through the aux [i.e., the greatest distance from the earth] and the point opposite the aux [i.e., the shortest distance from the earth]. Then *A* and *B* are related as if the motions and circles were concentric, for the true and mean motions are the same at that place [i.e., at the aux]; the same may be said when they are in the point opposite the aux.

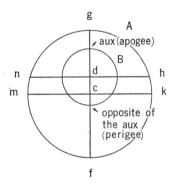

Fig. 22

That portion of the figure to the right of diameter *gf* is based on, but not identical with, figures in MSS *A* (Vat. lat. 4082, fol. 101v, in the margin below col. 2) and *L* (Florence, MS Ashburnham 210, fol. 164r, in the margin below col. 1). The remaining part of the figure has been added to cover the other cases discussed; the verbal descriptions have been inserted for convenience.

Ponatur ergo quod *B* pertranseat quartam partem sui circuli in aliquo tempore et in fine illius temporis sit *B* in linea *dh*; et in eodem tempore *A* pertranseat totum suum circulum et quartam partem eiusdem et in fine erit in linea *ck*. Et statim sequitur quod si motus essent concentrici tunc
720 essent coniuncta, et quod nunc *A* est ante *B* propter eccentricitatem [circulorum]. Sed positum est quod *A* est velocius ergo *A* et *B* ante finem illius temporis fuerint coniuncta respectu centri *c* et etiam in linea vel puncto propinquiori puncto *g*. Et consimiliter videbitur de aliis coniunctionibus nisi quod ab alia parte dyametri *gdcf*, videlicet a sinistris, contingit
725 econtrario scilicet quod tardius coniungentur quam si concentrice moverentur ita quod motus medius qui ymaginatur a *c*, si esset concentricus, ab una parte dyametri addit supra motum verum et ab alia verus motus addit supra medium. Et sicut dictum est de coniunctione duorum ita dicendum est de tribus vel pluribus.
730 Patet ergo quod distantie temporis et spatii aliter sunt quam si motus essent concentrici. Et si essent concentrici tunc forent equales distantie utrobique propter regularitatem motuum. Ergo nunc sunt huiusmodi distantie temporis et spatii inequales, et hoc est verum nisi in casu ubi non fierent coniunctiones nisi in auge et in opposito augis.
735 Omnia itaque dicta ante istam conclusionem sunt ad motus medios referenda. Et hec conclusio docet ad motus veros cuncta conformiter applicare non obstantibus eccentricis nec etiam epicyclis.
Conclusio vicensimaprima. Quecumque dicta sunt de coniunctione duorum vel plurium mobilium consimiliter intelligenda sunt de oppositione et de
740 *quocumque alio aspectu sive modo se habendi.*
Verumtamen distinguendus est aspectus trinus ante coniunctionem ab aspectu trino ipsam coniunctionem sequenti; et sic de quolibet aspectu seu

---

717 *post* sit *mg. hab. B figuram* / A *om. U sed mg. hab.* forte deest A
718 eius *B* / *post* fine *hab. B* illius
719 essent: esset *P* erunt *B*
720 A: k *U*
721 [circulorum] *om. AFLPU*
722 fuerunt *BLU* / c *om. U* / vel: sive *F* in *U*
723 consimiliter: similiter *F*
724 *ante* ab alia *scr. et del. U* linea vel p / alia: illa *F* / dyameter *A* / gdcf: gcdf *U*
725 concentrici *U*
725–26 moveretur *B*
726 *infra* qui ymaginatur *mg. hab. A figuram* / esset: est *A* enim *P*
727 addat *U* / supra: super *U* / motum: mo-

dum *P* / verum *om. L* / *post* alia *add. LU* parte / verus motus *tr. F* verus *B* verus modus *P* unius motus *L*
728 addat *U* / Et *om. P* / ita: igitur *P*
729 dicendum est *tr. AL* / tribus vel pluribus: coniunctione 3 4 vel plurium *B* / vel: sive *F* aut *L* / *post* vel *hab. ALP* de
730 distantia *L*
731 essent[2]: enim *P* / forent: essent *BL* / equales distantie *tr. B*
732 regularitatem: raritatem *AL* / huiusmodi: huius in *P*
733 et spatii *om. L*
734 fierent: fuerint *B* fuerunt *U* / in[2] *om. B* / augis: eius *B*

Let us assume that *B* traverses $\frac{1}{4}$ part of its circle in a certain time, and that it is on line *dh* at the end of that time; and during the same time, let us suppose that *A* traverses its whole circle plus $\frac{1}{4}$ part of it, so that at the end of that time it will be on line *ck*. It follows directly that if the motions were concentric, the mobiles would be in conjunction; however, because of the eccentricity of the circles, *A* now precedes *B*. But *A* was assumed to be quicker [than *B*], so that, with respect to *c*, they will have conjuncted before the end of this time interval in a line, or point, nearer to point *g*. Similar statements can be made about other conjunctions, except that when any conjunctions occur on the other side of the diameter *gdcf*—i.e., on the left side—it happens in a contrary way, i.e., the mobiles will conjunct later than if they were moved concentrically. Thus, if the mean motion, which can be imagined [as measurable] from point *c*, were concentric, it would add to the true motion on one side of the diameter, while on the other side the true motion would add to the mean motion. The same reasoning that was applied to two mobiles must also be applied to three or more mobiles.

It appears obvious, then, that the intervals of time and distance are different than those that would arise if the motions were concentric. For if they were concentric, the distances would be everywhere equal, because of the regularity of the motions. But now it is true that the intervals of time and distance are unequal, except in the case where the conjunctions occur in the aux and [the point] opposite the aux.

Everything said prior to this proposition must be taken as applying to mean motions. But this proposition shows [us] how to make a similar application of these things to true motions, despite the use of eccentrics and even epicycles.

Proposition 21. It must be understood that anything said about the conjunction of two or more mobiles applies also to opposition, and to any other relational aspect, or mode.

However, we must distinguish a trinal aspect before conjunction from a trinal aspect following this conjunction. This distinction must be made for

---

735 *ante* Omnia *mg. hab. B* notanda / itaque: etiam *B* / dicta *om. A* / motus medios *tr. B* motus medias *L*

736 *infra* referenda *mg. hab. L figuram* / referenda: inferenda *U* / Et: etiam *B* / hec *om. A* / cuncta conformiter *tr. F* confert *U*

737 nec etiam: et *BU*

738 Conclusio *om. ABFP* propositio *U* / vicensimaprima *om. P* vicensimanona *L* *et post* epicyclis (*linea 737*) *hab. F* incipit

vicesimum primum capit *et ante* Quecumque *mg. hab. AB* vicensimaprima

739 intelligenda *om. A* / *ante* et *hab. B* eorum / de *om. P*

740 quocumque: quecunque *B* quoque *P* / sive: seu *B*

742 ipsam: post *BU* / *ante* sequenti *hab. U* ipsam / sequente *L* / sic *om. U* / seu: sive *F*

742–43 sequenti…modo: sic *?*quartus et sextilis vel alius modus *B*

modo se habendi exceptis coniunctione et oppositione quoniam omnis alter aspectus est dupliciter, scilicet ante coniunctionem et post una vice a
745 dextris et alia a sinistris. Quibus sic intellectis omnia predicta possent de quolibet aspectu omnino similiter demonstrari sicut iam de coniunctionibus probata sunt.

Conclusio vicensimasecunda. Consimilia applicare ad idem mobile quod pluribus motibus moveretur.

750 Verbi gratia, posito quod sol moveatur solum duobus motibus, scilicet proprio et diurno, et sit *A* primus punctus Cancri in spera nona qui semper de die in diem describit unum eundem circulum, scilicet tropicum estivalem; et sit *B* centrum corporis solis qui uno anno describit eclipticam motu proprio. Et nunc sint *A* et *B* in puncto *d* ymaginato immobili. Cum igitur
755 isti motus sint commensurabiles, ut supponitur, erit proportio motuum rationalis. Si ergo hec proportio fuerit multiplex sequitur quod *A* et *B* numquam erunt simul nisi in *d* puncto quod patet statim exemplo. Ut si *A* faciat 100 circulationes in tempore in quo *B* facit unam tunc in fine istarum 100 circulationum *A* et *B* erunt in *d* puncto, et non ante, et sic semper de
760 100 in 100 diebus.

Si autem fuerit alia proportio quam multiplex in eius denominatione ponetur fractio. Denominator itaque fractionis ostendit quot sunt puncta in quibus *A* et *B* poterunt convenire. Ut si in tempore in quo *B* facit unam

743 quoniam: quia *B*
744 dupliciter: duplex *U* / scilicet: sive *A* / coniunctionem et post: et post coniunctionem *U*
745 dextrum *U* / a sinistris: ad sinistrum *U* / sic *om. AB* / omnia: primam *L* / predicta: dicta *B* / possent *tr. U post* aspectu (*linea 746*)
746 omnino similiter *om. B* / iam *om. B*
746–47 coniunctionibus probata sunt: coniunctione *B*
748 Conclusio *om. ABFP* propositio *U* / vicensimasecunda *om. P* tricensima *L et post* sunt (*linea 747*) *hab.* F incipit 22 capitulum *et ante* Consimilia *mg. hab. A* vicensimasecunda *et ante* ad idem *mg. hab. B* vicensimasecunda / Consimilia applicare *tr. B ante* posito (*linea 750*)
749 movetur *BU*
750 Verbi gratia *om. BU* / moveretur *A* / solum duobus motibus: duobus motibus solum *B* solum motibus duobus *U* /

ante motibus *scr. et del. U* tribus / scilicet: ut puta *ABL*
751 et sit: sitque *U* / Cancri: centri *ABL* / *post* spera *mg. hab. B* de mobilibus motis pluribus motibus / spera nona *tr. BU* / qui: que *FP*
752 scilicet: videlicet *AFU* videt *P*
752–53 estivale *A* estivalis *L*
753 qui: que *P* / *post* qui *hab. A* in
754 *post* sint *hab. B* simul / Cum: cumque *U*
755 sunt *L* / *ante* erit *hab. B* hic / *ante* motuum *hab. B* horum
756 hec: ista *F* / fuerit: erat *L*
757 erunt: essent *FP* / d puncto *tr. B* / quod: questio *P* / statim *om. B* / Ut si: sit ut *U*
758 faciat: fecerit *A* / in[1] *om. B* / facit: fecit *A*
759 100 *om. B* / d puncto *tr. AB* / *post* et[2] *hab. B A* / *post* ante *hab. B* nec post / *post* sic *hab. B* sit
761 fuerit alia *tr. A* / quam: quod *P* / *ante* multiplex *hab. U* non

every aspect, or relationship, except conjunction and opposition, since every other aspect is twofold, namely, before and after conjunction—[i.e.,] one way from the right side, and the other from the left side. If this is understood, everything said before can be demonstrated for any aspect in the same manner as was done for conjunctions.

Proposition 22. [How] propositions [that are] similar [to those demonstrated previously for two or more distinct mobiles] can be applied to one and the same mobile moving with several [simultaneous] motions.

For example, let it be assumed that the sun has only two motions, namely, proper and diurnal; and on the ninth sphere, let A be the first point of Cancer, which describes the same circle every day, namely, the summer tropic; and let B be the center of the sun describing the ecliptic in one year with a proper motion [see Fig. 23]. [Finally] assume that A and B are now in point d, which

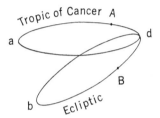

Fig. 23

Similar figures appear in MSS A (Vat. lat. 4082, fol. 102r, in the margin below col. 1) and L (Florence, MS Ashburnham 210, fol. 164r, in the margin below col. 2), but I have altered the position of the circles (a, b) and distinguished them from the points or bodies on the circles (A, B; only lowercase letters appear in the manuscripts).

is imagined as motionless. Now on the assumption that these motions are commensurable, the ratio of the motions will be rational. If this ratio were multiple, it follows that A and B will never be together simultaneously except in point d. This can be made clear at once by an example, for if A should make 100 circulations in the same time that it takes B to make one circulation, then at the termination of these 100 circulations—and not before—A and B will be in point d; and this will always occur every 100 days.

But if the ratio of the motions is not multiple, a fraction must be assumed in its denomination. Furthermore, the denominator of the fraction indicates the number of points in which A and B can meet. Thus, if A makes $100\frac{1}{2}$

761–62 poneretur A
762 *post* fractio *hab.* B et / itaque *om.* B
    atque A / quot: quod P
763 poterunt: possunt A / *ante* convenire

hab. B simul / *post* Ut si *scr. et del.* U
deest forte *?et ?2 / in² *om.* B / facit:
faceret L

circulationem ipsum *A* faceret 100 cum dimidia. Ergo cum *B* moveatur ad
765   motum ipsius *A* sequitur quod *A* et *B* se invenirent in puncto opposito
ipsi *d* qui erat punctus primus; et sequenti vice invenirent se in *d*. Et sic
non essent nisi duo loca fixa in quibus *A* et *B* possent convenire. Et si in
tempore quo *B* facit unam circulationem *A* faceret 100 circulationes cum $\frac{1}{3}$
parte unius, tunc essent tria puncta fixa precise in quibus *B* posset attin-
770   gere ad *A*; et si $\frac{1}{4}$ essent quatuor, et sic semper. Rursum quoque si in
tempore in quo *B* facit unam circulationem *A* faceret 100 et $\frac{2}{5}$ unius tunc
essent quinque loca in quibus convenirent quoniam tempore illo *A*
pertransiret circulum aliquotiens et $\frac{2}{5}$ et iterum alia vice totiens et $\frac{2}{5}$, et sic
de aliis.

775   Ex quo patet numerus et ordo punctorum celi quibus sol intraret primum
punctum Cancri, et ita de ceteris gradibus zodiaci. Unde manifestum est si
sol non moveatur nisi duobus motibus quod si annus solaris mensuraretur
precise diebus integris numquam sol intraret Cancrum nisi ipso existente
in uno meridiano; et iterum alia vice in eodem et numquam in alio. Et si
780   annus contineat precise aliquot dies et aliquam vel aliquas quartas diei
sicut 365 dies et $\frac{1}{4}$, tunc sunt tantummodo 4 puncta in ipso circulo tropico
ab invicem equaliter distantia quibus sol potest intrare primum punctum
Cancri; et ita de aliis gradibus.

    Si vero fuerint tres motus tunc per modum iam dictum combinentur
785   bini et bini et iterum iste combinationes invicem comparentur fere eodem
modo sicut prius dicebatur de coniunctione trium mobilium; et ita de 4
vel de pluribus.

    Ex quibus omnibus sequitur quod quotquot fuerint motus dummodo
sint omnes invicem commensurabiles quacumque dispositione data semper
790   sunt certa loca numerata in quibus illud mobile potest taliter se habere, et

---

764 cum: in *B*
765 motu *A* / se *om. U* / *post* se *hab. B*
  primo / invenirent: convenirent *U* sed
  mg. hab. alibi invenirent se
766 et[1] *om. A* / inveniret *AFP* invenient *L* /
  post in *hab. B* puncto / Et[2] *om. B*
767 *infra* nisi duo *mg. hab. A figuram* / *ante*
  loca *hab. A* puncta / possunt *P*
768 *ante* quo *add. FP* in / facit: fecit *A* /
  faceret: facerit *B* / circulationes *om. B*
769 parte *om. U* / essent: sunt *P* / in *om.*
  B / *post* quibus *hab. B* A et B possunt
  convenire vel / posset *om. B* possent *U*
769–70 attingere: pertingere *A*
770 *post* si[1] *hab. B* centum faceret / *post*

quatuor *hab. B* loca si cum centum
  unam quintam essent 5 loca / *post* sic
  hab. B de aliis / si[2] *om. U*
771 in quo *om. A* quo *B* / B facit: fecerit B *A*
772 essent: sunt *AP* / in *om. B* / convenirent: quereret *P* / quoniam: quandoque
  A quia *B* / tempore illo A: in isto tempore *U* / illo: isto *A* illa *B*
774 de aliis: semper *B*
775 quo: quibus *F* / et...celi: punctorum
  celi et ordo iterum *B* / intraret: intrat
  BP
776 punctum *om. U* / ceteris: ceteribus *B* /
  ante Unde *hab. P* et / *post* est *hab. BU*
  quod *et B rep.* quod / si *om. U*

circulations in the same time as B makes one circulation, it follows that they would be found in a point opposite d, the first point; and after the next sequence [of circulations], they would be found in d [itself]. [In this example] there are only two fixed places where A and B could meet. But B could meet A in exactly three fixed points, if B makes one circulation when A makes $100\frac{1}{3}$; and if [A makes] $100\frac{1}{4}$ [circulations], there would be four [points of contact], and so on. Also, if A should complete $100\frac{2}{5}$ circulations while B makes one circulation, there would be five places where they could meet, since in that time A would traverse its circle a certain number of times plus $\frac{2}{5}$ of it, and during another [equal] time [period] would again make just as many circulations plus $\frac{2}{5}$ [of one]; and this applies to all [such cases].

The number and arrangement of the celestial points on which the sun could enter the first point of Cancer—as well as other degrees of the zodiac—become apparent from all this. Thus if the sun were moved with only two motions and the solar year were measured by an exact number of integral days, it is obvious that the sun could enter [the first point of] Cancer only when it coincided with one [particular] meridian; and after another [equal interval of] time it would again [enter Cancer] on the same meridian, and never on another. But if the year contained a certain number of days exactly plus $\frac{1}{4}$ part, or parts, of a day—as, for example, $365\frac{1}{4}$ days—there would be only four equidistant points on this tropical circle on which the sun could enter the first point of Cancer; and the same applies to other degrees [of the zodiac].

If, however, three motions were involved, then, by the method already outlined, they will be taken two at a time and their combinations mutually compared in quite the same way as in our previous discussion of a conjunction of three mobiles; the same procedure would be followed for four or more motions.

From all this, it follows that whatever the number of motions involved—provided they are all mutually commensurable—there is always, for any given disposition, a certain number of fixed places—and no more—in which that

---

777 quod: et AB / mensuretur F

778 integris: numeris P / numquam sol $tr.$ BF numquam A / Cancrum: in cancro L / ipso: eo B

779 uno $om.$ U

780 contineat precise $tr.$ AB precise continet L / aliquot $om.$ B / aliquam vel $om.$ B / vel: sive F seu L / diei $om.$ B dies U

781 sicut $om.$ A / dies $om.$ U diebus B / $\frac{1}{4}$: una quarta AFP unam quartam BL /

tantummodo $om.$ U tantum B

782 invicem: invicis L initio U

784 vero: enim A / tres: duo FPU / combinentur: continentur A combineretur P

785 iste: ille A / invicem: ad invicem A / comparentur: combinentur U

786 prius $om.$ B

787 vel: sive F

789 sunt P / omnes $om.$ BL / invicem $om.$ A

790 in $om.$ B / potest: poterit A

non in pluribus. Et cum processerit per omnia illa loca taliter se habendo
rursum incipiet procedere penitus similiter sicut ante.

   Verbi gratia, si sint precise 4 puncta fixa in quibus sol potest intrare
Cancrum et sit *d* unum illorum, dico quod incipiendo a puncto *d* sol
795 procedet per 4 annos continue et cotidie describendo novam girationem
quousque in fine quarti anni redeat ad *d* punctum et tunc resumet cursum
suum sicut a principio, et sic semper. Et eodem modo si fuerint plura talia
puncta sed tunc in maiori tempore fieret huiusmodi revolutio. Et conformiter diceretur de luna et de quolibet aliorum planetarum.

800   Si igitur omnes motus quibus sol movetur sint commensurabiles ipse
sol in spatio ymaginato immobili describit centro suo unam viam seu
lineam finitam carentem punctis terminantibus ad modum linee circularis,
non tamen circularem sed ex multis girationibus seu spiris quodammodo
intextam quarum quedam incipiunt versus tropicum Cancri et procedunt
805 versus tropicum Capricorni quas velut ordiens sol peragit; et alias econtrario rediens intersecantes priores, quasi intexendo, describit. Sunt quoque
per totum totidem gyri seu spire quotis diebus revolvitur sol de puncto
celi ymaginato immobili iterum ad eundem. Opusque huiusmodi texture
inter duos tropicos simile est quadrangulis, oblongis, sive rombis vel
810 losengis, in quorum area vel spatio intrinseco numquam fuit, nec erit, in
perpetuum centrum solis. Hec etiam via semel peragratur a sole in una
peryodo et rursum incipit super eandem procedere sicut ante; et de luna
et de aliis diceretur hoc modo.

   Unde quodlibet mobile pluribus motibus per se sumptum habet certam
815 peryodum qua peracta renovatur iterum et sic infinities, et que potest

791 processerint *A* / illa: ista *A*
792 incipiet *om. P* / similiter: similis *B*
793 si *om. B* / fixa *om. L* / in *om. B*
794 unum *om. BFP* primum *U* / illorum: istorum *AP*
795 procedit *L* / *post* et *hab. F* cum / cotidie: continue *AFP* totidie *B*
796 d punctum *tr. B* / et *om. U* / tunc *om. P*
797 Et² *om. BU* / si: sicut *FPU* / fuerunt *F* / plura talia *tr. A*
798 tunc *om. BFPU* / fieret: fuerit *U sed mg. hab. U* alibi forte scitur
798–99 conformiter: confirmiter *A*
799 planetarum *om. AFPU*
800 movetur *om. P*
801 viam *om. F* via *P* / seu *om. F* sive *L*
802 lineam: linea *P* / carentem: carens *AFPU* / punctis: punctualiter *A* / circularis: circularum *AF*
803 circularem: circularis *U*
804 intextam: in dextram *P* / *post* versus *hab. B* primum / tropicus *P* / Cancri et procedunt *om. L*
805 *ante* versus *hab. B* ?girativo / capricornis *B* / velut ordiens sol: sol velud ordiens *B* / ordiens: oriens *FP* ordines *U* / alias *om. P*
805–6 econtrario: ?econverso *B*
806 intexando *F* / Sunt quoque: suntque *B*
807 per totum *om. A* / gyre *AB* gire *FL* giro *P* / seu: sive *F* / spiro *P* / quotis: quot *U* / revolvitur sol: revolvuntur sol *A* sol *B* sol revertitur *L*
808 *ante* iterum *hab. B* revolvitur / huiusmodi: huius *A et post* huius *scr. et del. A* gyre

mobile can be related in the same way. And when it has proceeded through all these places maintaining this same relationship, it will begin, once again, to run through them exactly as before.

For example, let *d* be one of only four fixed points on which the sun can enter [the first point of] Cancer. I say that if the sun begins from point *d* and moves continuously for four years describing daily a new spiral until it returns to point *d* at the end of the fourth year, it would then resume its course from the beginning; and so on, endlessly. In the same manner, this would also happen if there were more points of this kind, but then such a revolution would be completed in a greater time. The same thing could be said about the moon and any of the other planets.

Therefore, if all the motions by which the sun is moved were commensurable, the sun's center, moving within a space imagined as immobile, would describe a finite path, or line, that is like a circular line without terminating points. And yet it does not describe a circular path. Indeed, its path is composed of many circular-like or spiral lines that are [interlaced or] interwoven in a certain way. Some of these lines begin toward the tropic of Cancer and proceed in the direction of the tropic of Capricorn, which is the path the sun follows if beginning its motion there; and at other times the sun returns in the opposite direction describing lines that intersect those made previously—as if weaving a pattern. Also the number of circular-like lines, or spirals, between the two tropics equals the number of days that it takes the sun to depart from and return to the point which is imagined as fixed in the heavens. The [end] product of such an interweaving between the two tropics is similar to a quadrangle, oblong, or rhombus. But throughout all eternity, the center of the sun never was, nor will ever be, inside the area or inner space of these [figures]. Furthermore, once this path has been traversed by the sun in its period, it begins again to move over the same path as before. The same thing could be said of the moon and the other planets.

Thus, any mobile taken independently, but having several motions, has a fixed period upon the completion of which it begins again; and this goes on

---

809 duos *om. BFPU* / est *om. A* / oblongis *om. B* / sive: sicut *A* / vel: aut *F*
809–10 sive...losengis *om. L*
810 *infra* in quorum *mg. hab. A figuram* / area *obs. B* / vel: et *B* sive *F* / fuit: fuerit *L* / nec: et sic *B*
811 Hec: ista *F* hac *U* / viam *P* / semel *obs. B* solis *A* / uno *A*
812 *post* peryodo *hab. B* vel revolutione /

super *om. FP* / eadem *F* / *post* sicut *hab. F* vero / ante *obs. B* / luna: linea *L*
813 et de aliis *om. U* / de *om. B* / *post* aliis *hab. B* planetis / dicetur *U* / hoc: isto *F*
814 quotlibet *A* / mobile *om. AP*
815 qua: que *BFLP* / renovatur iterum *tr. A* / *post* sic *hab. U* in / et² *om. L* in *B* / que: qui *U*

vocari annus magnus istius mobilis, consimiliter quelibet duo mobilia celestia simul sumpta complent cursum suum certa peryodo temporis qua transacta reincipiunt ut prius; et sic de tribus aut quotlibet. Et potest dici annus magnus ipsorum sicut dicunt quidam de sole et octava spera quod

820 annus magnus istorum duorum est 36,000 anni solares. Sed annus magnus omnium planetarum et octave spere est valde multo maior. Et, breviter, si omnes motus celi sunt commensurabiles invicem necesse est quod omnium simul sit una maxima peryodus qua finita renovatur non eadem sed similis vicibus infinitis, si mundus esset eternus.

825 *Conclusio vicensimatertia. Si aliqua mobilia talia nunc sunt coniuncta semper distabunt commensurabiliter a puncto coniunctionis et inter se.*

Hoc intelligendum est de distantia adepta secundum circuli portionem et penes angulos descriptos supra centrum. Sint *A* et *B* et *C* in puncto *d*; et sit *A* velocius, et *B* medium, et *C* tardius. Cum ergo moveantur commen-

830 surabiliter, ergo in quolibet instanti post instans presens spatia ab ipsis pertransita erunt commensurabilia, scilicet arcus *dC* et arcus *dB* et arcus *dA* qui sunt distantie eorum a puncto *d*. Ergo etiam distantie eorum ab invicem sunt commensurabiles quia quecumque distant ab aliquo tertio commensurabiliter distant etiam ab invicem commensurabiliter ut faciliter

835 probaretur per nonam decimi Euclidis. Ex quo patet quod quotienscumque describunt angulos supra centrum illi anguli sunt commensurabiles.

Conclusio vicensimaquarta. Si tria nunc sunt coniuncta quandocumque

816 annus magnus istius: magnus annus illius *B* / mobilis *obs. B* / consimiliter: similiter *F*

817 celestia *om. B* / *ante* cursum *hab. U* simul / certa *om. L* / peryodo: perioda *F* / qua: que *AF*

818 reincipiunt *om. U* / *post* tribus *hab. B* aut 4 / aut: et *A* sive de *F*

819 annus magnus *tr. AB* / ipsorum: istorum *U* / dicunt quidam: dicunt *obs. B* quidem dicunt *F* / solo *P*

819–20 sicut…istorum *om. U*

820 duorum *om. BP*

821 est: esset *FL* erit *U* / valde *om. L* / multo *om. B*

822 sunt: sicut *P* / invicem: adinvicem *B*

823 simul *om. A* / *ante* una *hab. U* unus annus maximus et / maxima peryodus *tr. A* / eandem *FP*

823–24 similis vicibus: vicibus simul *U*

825 Conclusio *om. ABFP* propositio *U* /

vicensimatertia *om. FP* tricensimaprima *L et ante* Si aliqua *mg. hab. A* 23 *et mg. hab. B* vicensimatertia / mobilia *om. BFPU* / talia *U sed mg. hab. U* alibi tria / sint *AL*

827 Hoc: istud *F* hic *L* / intelligendum est *tr. FU* / adepta: accepta *L* / portiones *U*

828 descriptos supra centrum: supra centrum descriptos *F* / et² *om. BL*

829 et¹: coniuncta *B* / et² *om. BU* / et³ *om. U* / *C om. B* / moveantur: commoveantur *F*

829–30 commensurabiliter: incommensurabiliter *B*

830 instanti: tempore *U* / instans presens *tr. B* presens *U* / ipsis: hiis *A* eis *B*

831 erunt: essent *FU* / dC *tr. L* / dB *tr. L* / arcus³ *om. U*

831–32 et arcus dA *om. A*

832 qui: que *AFPU* / etiam *tr. B post* in-

an infinite number of times. This period can be called the Great Year of that mobile, just as, in a similar way, any two celestial mobiles that might be taken complete their courses simultaneously in a fixed period of time, after which they begin over again as before; and this can be said about three mobiles, or [indeed] any number of them. These mobiles can also be said to have a Great Year, just as some say that 36,000 solar years constitute a Great Year of the sun and the eighth sphere. But a Great Year of all the planets and the eighth sphere is very much longer. In brief, if all the celestial motions are mutually commensurable, it is necessary that there be one maximum [or perfect] period for all of them at the same time, [a period] which is repeated after its termination. But if the world were eternal, the perfect period [or Great Year], although it would always be similar, would not be repeated in the same place, but, rather, in infinite [different] places.

Proposition 23. If any such mobiles are now in conjunction, they will always be separated from that point of conjunction, and from one another, by distances that are commensurable.

Distance, here, must be understood as measured by a sector of the circle and the angles described around the center. Let *A*, *B* and *C* be conjuncted in point *d*; and let *A* be quickest, *B* intermediate [in speed], and *C* slowest. Since they are moved with commensurable speeds, the distances traversed by them in any instant after the present instant will be commensurable—i.e., arcs *dC*, *dB*, and *dA*, which are the distances of the mobiles from point *d*, will be commensurable. Also, the distances separating them are mutually commensurable, since any [two] mobiles that are separated from a third mobile by commensurable distances are also separated from one another by commensurable distances. This is easily proved by the ninth [proposition] of the tenth [book] of Euclid. It is apparent from all this that any number of angles described around the center by these mobiles are commensurable.

Proposition 24. If three mobiles are now in conjunction, then whenever two of

---

vicem (*linea 833*)
832–33 ab invicem: adinvicem *L*
833 sunt: erunt *L* super *P* / commensurabiles: mensurabiles *A* / quia quecumque: quecumque *FU* queque *P*
834 distant...invicem: etiam ab invicem distant *U* / etiam ab invicem *BFP* inter se *A* et inter se *L*
835 probatur *B* / decimi: ideo *P* / quotienscumque: quotiens *L*

836 describunt *om. A* / *ante* angulos *scr. et del. B* super / super *U* / sunt *om. FP* / *post* commensurabiles *hab. P* et cetera
837 Conclusio *om. ABFD* propositio *U* / vicensimaquarta *om. FP* tricensimasecunda *L* et *ante* Si tria *mg. hab. A* 24 *et mg. hab. B* vicensimaquarta / tria nunc *tr. L* / *post* tria *hab. B* talia / quandocumque: quecumque *A* quandoque *P*

duo eorum precise erunt coniuncta tertium distabit ab ipsis secundum angulum commensurabilem recto sive per portionem commensurabilem toti circulo.

840   Quoniam loca in quibus possunt coniungi duo ipsorum, sicut *A* et *B*, distant inter se equaliter, ut patet ex decima conclusione, ergo hec loca vel puncta dividunt circulum per portiones equales ad angulos equales in centro, quorum quilibet sit *G*. Ergo angulus *G* aliquotiens sumptus precise consumit totam superficiem circa centrum que videlicet 4 rectos. Ergo
845   angulus *G* est commensurabilis angulo recto et ergo quandocumque duo coniunguntur distant a puncto in quo tria coniungebantur per angulum commensurabilem recto, qui sit *K*. Ergo angulus rectus est commensurabilis angulo *K*. Sed angulus quo illa duo tunc distant a tertio mobili est commensurabilis angulo *K* per precedentem. Ergo est commensurabilis
850   angulo recto quoniam quecumque uni et eidem sunt commensurabilia inter se sunt commensurabilia, ut patet ex decima.

   Conclusio vicensimaquinta. Que proportiones motuum possunt per fractiones phisicas adequari quibus scilicet utuntur astrologi sive punctualiter tabularii et que non assignare.

855   Ex conclusionibus decima, undecima, septimadecima, undevicensima, et vicensimaprima, patet quod loca seu puncta coniunctionum duorum mobilium sive plurium sunt certo numero numerata; et similiter est de aliis aspectibus aut dispositionibus quibuscumque. Et secundum hoc totus circulus dividitur in partes totidem per eadem puncta predicta. Ergo per
860   nullam divisionem que non communicaret cum divisione facta per hec puncta possent haberi coniunctiones aut aspectus precise ergo nec tales motus possent per aliam divisionem tabulari ut patet etiam per tertiam conclusionem.

   Sed nulla divisio secundum aliquam proportionalitatem communicat

---

838 duo: duorum *U sed mg. hab. U* alibi duo / ab ipsis *om. L*

839 commensurabilem[1]: commensurabiles *B* / per *om. P* propter *A* / portionem: proportionem *AFP*

840 ipsorum: eorum *U* / sicut: si *BFP* sunt *A* scilicet *U*

841 distent *FP* / hec: ista *F* / vel *om. B* sive *F*

842 per *om. P* / portiones: proportiones *AFP* / ad angulos equales *om. U*

843 G²: ergo *P*

844 videlicet *om. B*

845 G: ergo *P* / angulo recto *tr. A* / quandocumque: quandoque *P*

846 coniunguntur: coniungentur *L* / distant *tr. A ante* duo (*linea 845*)

847 commensurabile *A* / rectus: recto *B*

848 *post* K *hab. L* per precedentem / illa: ista *A* / tunc distant *tr. F* / mobile *A* / est: et *B*

850 angulo *om. B* / quoniam: quia *B*

851 inter…patet: prout patet evidenter *F* / *post* decima *hab. F* et sic finis capituli huius *et hab. P* et cetera

852 Conclusio *om. ABFP* propositio *U* / vicensimaquinta *mg. hab. B ante* Que *om. FP* 25 *U* tricensimatertia *L et post* possunt *mg. hab. A* 25 / possint *A*

853 phisices *B* / quibus scilicet *tr. B* / utunt *L*

them are exactly in conjunction, the third will be separated from them [either]
by an angle that is commensurable to a right angle, or by a sector of the circle
commensurable to the whole circle.

Since it is obvious by the tenth proposition that an equal distance separates
the places in which two of these mobiles, say, *A* and *B*, can be in conjunction,
these places, or points, divide the circle into equal angles at the center by
[means of] the equal sectors. Let *G* be any one of these angles. Therefore, angle
G, taken a certain number of times, exhausts the whole surface around the
center, which is obviously [equal to] four right angles. It follows that angle
G is commensurable to a right angle and, consequently, whenever two mobiles
conjunct, they are separated from the point where the three were conjuncted
by an angle that is commensurable to a right angle. Let us call this angle *K*.
Therefore, a right angle is commensurable to angle *K*. But, in terms of what
has already been said, the angle which then separates these two from the third
mobile is commensurable to angle *K*, and, consequently, is commensurable to
a right angle, since, by the tenth [book of Euclid], it is clear that any [quan-
tities] commensurable to one and the same [quantity] are commensurable to
each other.

Proposition 25. How to determine ratios of motions that can be compared
through the use of physical [or sexagesimal] fractions—i.e., those [fractions]
used or precisely tabulated by astronomers—and those that cannot be so compared.

On the basis of the tenth, eleventh, seventeenth, nineteenth, and twenty-first
propositions, it is obvious that the places, or points, of conjunction of two
or more mobiles are fixed in number; the same thing may be said about any
other aspects or dispositions. The entire circle is also divided into as many
parts as there are points. Thus no exact conjunctions or aspects could occur
if [some of] the points made by another division of the circle are not shared
[in common] with [some of] the points made by this [first] division. Therefore,
such motions as there are [in the first division] cannot be calculated in terms
of the other division; and this is evident by the third proposition.

But no division of any proportionality has a common measure with the

---

855 *post* decima *add. P* et / *post* undecima
    add. FLP et / *post* septimadecima *add.*
    FLP et
856 et: 20ᵃ *U* / seu: sive *FU* vel *B* / *post*
    puncta *hab. U* duarum
857 sive: vel *B* / sunt: sub *A* super *U* / *ante*
    numero *hab. A* minimo
858 aut: et *B* sive *F* / dispositionibus *rep.*

F / *ante* secundum *hab.* L quod / secun-
    dum: similiter *F* signaret *P*
859 in: per *U* / predictam *F*
861 aut: et *B* sive *F* / precisi *FLP* predicti
    A
862 per¹ *om. U* / divisionem *om. P* / etiam
    om. B 2 *U*
864 aliquam *om. A* / communicant *P*

865   cum divisione astronomicarum tabularum nisi illius proportionalitatis
numerus immediate sequens unitatem communicet cum 60. Excepto forte
uno puncto nec est communicatio possibilis in pluribus, ut patet per tertiam
conclusionem, et quibus numeris dividitur circulus in omni proportione
motuum. Hoc habetur per conclusiones decimam, undecimam, et septi-
870   mamdecimam. Ergo ex illis conclusionibus potest haberi qui motus cir-
culares possunt per tabulas astronomicas punctualiter adequari. Si igitur
sunt precise 7 puncta fixa, aut 13, et cetera, in quibus sol potest intrare
primum punctum Cancri, tunc numquam per communes tabulas haberetur
introitus eius in Cancrum nec in aliquod aliud signum, sive gradum; nec
875   per eas posset haberi motus eius precisus, aut locus, nec vera quantitas
anni. Et similiter si essent 7 puncta precise, vel 19, aut 23, et cetera, in
quibus sol et luna possent coniungi numquam etiam per communes tabulas
posset eorum coniunctio adequari; et ita de aliis aspectibus et de aliis
motibus et planetis. Sufficit tamen astrologo quod coniunctio sit in tali
880   gradu, vel in tali minuto, vel secundo, et cetera, licet ignoret in quo puncto
illius minuti; aut sufficit quod error ipsius astrologi non deprehendatur per
visum cum aliquo instrumento.

Hec de celi motibus generaliter dicta sufficiant supposita commensurabilitate ipsorum. Hic incipit secunda pars huius tractatus.

---

865 *post* cum *hab. U* aliqua / astronomicarum tabularum: tabularum *?astrono-
mice B* / proportionalitatis: proportionabiliter *A*
866 numerus *om. P*
867 est communicatio: quo *A* / ut *om. L* /
patet *om. A* / per *om. BLU*
868 et: in *L* / numeris: *?numeretis P*
869 motuum: mobilium *U* / Hoc: illud *F* /
haberetur *F* / decimam *om. A* / undecimam *om. P* undecima *F* / et *om. U*
869–70 septimamdecimam: 17 *A*
870 ex *rep. U* / potest: *?possit A* / qui: quid
L
871 possint *A* / adequari: tabulari *U*
872 sunt *om. A* sint *L* / fixa *om. BF* / in
om. B
873 numquam: nequiquam *P* / communes
tabulas *tr. B*
874 nec in aliquod: vel *B* / aliquid *LP* / aliud
om. U / graduum *P*
875 aut: nec *A* ac *P* / *post* locus *hab. B*
eius / nec: sive *F* / vera: una *FP* *?vide-

tur U*
876 anni: simpliciter *L* / Et *om. B* / *post*
precise *hab. FP* puncta / vel: ter *A* sive
F / 19: 9 *A* / aut: vel *B* sive *F* / et cetera
om. L etiam *A* et etiam *P* / in *om. B*
877 *ante* per *scr. et del. A* per / communes
tabulas *tr. B*
878 coniunctio: *?coniuncter P* / et de aliis:
vel *B* / de² *om. AL* / et²: vel *?aliis B*
880 vel¹: aut *LU* sive *F* / in tali *om. U* tali
B / vel²: aut *FLP* / *ante* secundo *hab. U*
tali / secundo: 2 *A*
881 ipsius: eius *B* / astrologi *om. B* astro-
logium *P* / non *om. P*
883 Hec: et hoc *A* hoc *B* / celi motibus *tr.*
BL / sufficiant *om. P*
883–84 commensurabilitate ipsorum *tr. U*
884 Hic...tractatus *om. FP* explicit prima
pars istius operis incipit secunda *A* sequitur secunda pars huius *B* explicit
prima pars huius tractatus de commensurabilitate vel incommensurabilitate
motuum supercelestium *U*

division used in astronomical tables, unless the number following immediately after unity in that proportionality measures [or is a factor of] 60. With the exception of one point, there is no other possible common measure for any more [terms in these proportionalities] as is obvious by the third proposition. But the numbers [representing these divisions are the ones that] divide the circle into ratios of motions. This is shown by the tenth, eleventh, and seventeenth propositions. By utilizing these propositions, one can determine which circular motions can be compared exactly by means of the [common] astronomical tables. Thus, if there are exactly 7, or 13, fixed points, and so on, on which the sun can enter the first point of Cancer, the common tables will never furnish the [exact] information concerning its entry into Cancer, or [for that matter] into any other sign, or degree; nor could the sun's exact motion, [or] position, or the true length of the year [be determined from these tables]. In a similar way, this would also be true for conjunctions of the sun and moon, if they conjuncted in exactly 7, or 19, or 23, fixed points, and so on. Not a single conjunction of theirs would be exactly expressible in terms of the common tables. And the same reasoning applies to other aspects, motions, and planets. An astronomer is satisfied if he knows that a conjunction is in a certain degree, or minute, or second, and so forth, even though he does not know in what [precise] point of a particular minute it occurs; or, he is content when an error is visually undetectable through use of an instrument.

What has been said thus far, based on the assumption that the celestial motions are commensurable, must suffice. The second part of this treatise begins here.

[Secunda pars]

In hac secunda parte istius operis supponatur quod aliqui motuum celestium sint incommensurabiles et quid inde sequatur restat ostendere.

Conclusio prima. Si duo talia mobilia incommensurabiliter mota nunc 5 *sint coniuncta numquam alias in puncto eodem coniungentur.*

Sint *A* et *B* in puncto *d*. Igitur si aliquando post conveniant in puncto *d* sequitur quod in isto tempore quodlibet illorum fecerit aliquot circulationes precise. Sit itaque *E* numerus circulationum quas fecerit *A*, et sit *G* numerus circulationum quas fecerit *B*. Cum ergo proportio velocitatum sit 10 sicut proportio pertransitorum in eodem tempore sequitur statim quod velocitates istorum mobilium sunt sicut duo numeri *E* et *G*. Ergo velocitates iste sunt commensurabiles quod est contra positum.

Verbi gratia, si *A* et *B*, que nunc sunt in puncto *d*, iterum aliquando post coniungantur in eodem puncto, ponatur quod *A* tunc fecerit 5 circulationes 15 et quod *B* fecerit 3. Tunc patet quod velocitates ipsorum se habent sicut isti numeri 5 et 3, et quod *A* movetur velocius quam *B* in proportione superpartiente duas tertias.

Et per idem probabitur quod numquam fuerunt coniuncta [ante] in puncto *d*; eodem modo de quolibet alio aspectu. Unde si *A* sit in puncto *d*

1 [Secunda pars] *om. ABFLPU*

2 In…operis: incipit istius operis pars secunda in qua *U* / secunda *om. FP* / istius: huius *BFL* / supponatur: supponitus *U* restat ostendere quid sequatur si ?presumatur *B* / quod *om. A* / motuum: mota *F*

3 *ante* celestium *hab. F* nunc / *post* sint *hab. P* et / incommensurabile *A* / et quid…ostendere: pro quo hec prima conclusio *B* / ostendere: ostende *P*

4 Conclusio prima *om. AFP* propositio prima *U* prima conclusio *mg. hab. B ante* Si duo / talia *om. A* / talia mobilia *tr. L* / mota *om. A*

5 puncto eodem *tr. BU* / *post* eodem *mg. hab. B* de duobus mobilibus

6 *ante* in puncto[1] *hab. B* nunc iuncta *et hab. L* duo mobilia / *post om. BFPU*

7 quod: que *A* / in *om. L* / isto: nullo *L* illo *U* / quotlibet *U* / illorum: ipsorum *AL* eorum *B* / aliquos *P*

8 itaque: ergo *B* / circulationum: earum *B* / fecerit: fecit *AF* facit *L* / sit[2] *om. B*

8-9 E…numerus *om. P*

9 circulationum: earum *B* / fecerit: fecit *AF* facit *L* / ergo: B *A*

10 sicut *om. L*

11 velocitates…Ergo *om. A* / istorum mobilium *tr. FP* / sunt: sint *L*

[Part II]

In the second part of this work, it is assumed that some of the celestial motions are incommensurable. We must now show what follows from that assumption.

Proposition 1. If two mobiles have moved with incommensurable velocities, and are now in conjunction, they will never conjunct in that same point at other times.

Assume that A and B are in point d. If, at some later time, they should meet in point d, it follows that during that time each will have made a certain number of exact circulations. Let E be the number of circulations that will have been made by A, and G the number that will have been made by B. Since a ratio of velocities varies as the ratio of the distances traversed [by these velocities] during the same time, it follows directly that the speeds of these mobiles are related as the two numbers E and G. Therefore, the velocities are commensurable, which is contrary to what was assumed.

For example, if A and B are now in point d, and at some later time they are again in conjunction in the same point, assume that A will have made 5 circulations and B will have made 3. It is obvious, then, that their velocities are related as the numbers 5 and 3, and that A is moved more quickly than B by a superpartient two-thirds ratio.

And one can show in the same manner that they were never before in conjunction in point d. All this is applicable, in just the same way, to any other aspect. Thus if A is in point d, and B is in the point opposite [to d], they will

---

12 est: erit *L* / positum: suppositum *FP*

13 *post* sunt *hab. B* iuncta / aliquando post tr. *U*

14 coniungentur *U* / *post* puncto *hab. B* d / tunc *om. BU* / fecerit: fecit *A*

15 quod[1] *om. BUL* / fecerit *om. BU* / Tunc *om. U* / ipsorum: istorum *A* iste *B* eorum *U* / *post* se *hab. P* nunc / se habent: sint proportionales *U*

16 movetur velocius: velocius moveretur *U* / in *om. L*

17 superpartiente: superbipartiente *AU* / duas *om. U* / tertias *U sed mg. hab.* 10[am] vel 19[am]

18 probatur *B* patet *U* / [ante] *om. AFLPU*

19 *ante* de *hab. B* intelligitur *et post* de *hab. B* oppositione et / Unde: ut *B*

20 et *B* sit in puncto opposito numquam alias opponentur ipsis existentibus
in punctis istis ut per idem demonstraretur. Est igitur revolutio eorum
infinita et verius loquendo nulla est.

    Conclusio secunda. Si duo nunc sint coniuncta numquam alias coniungen-
tur in puncto distanti a puncto in quo sunt per partem circuli commensura-
25 *bilem suo toti.*

    Quoniam si cuilibet commensurabilium addatur aliquid quod est
utrique commensurabile tota fient commensurabilia ut statim sequitur ex
octava et nona decimi Euclidis. Cum igitur quotlibet circulationes quotlibet aliis circulationibus sint commensurabiles si utrique aggregato
30 addatur portio commensurabilis toti circulationi uni tota erunt commensurabilia. Igitur si *A* et *B* coniungerentur in puncto distante a puncto *d*
quodlibet illorum tempore intermedio complevisset precise aliquot circulationes et partem illam circulationis commensurabilem suo toti. Ergo
motus eorum essent commensurabiles eo quod pertransita eodem tempore
35 forent commensurabilia.

    Verbi gratia, si *A* et *B* coniungerentur in puncto distante a puncto *d* per
$\frac{1}{4}$ partem circuli et quod *A* tunc fecisset 5 circulationes cum $\frac{1}{4}$ parte unius
circulationis et *B* fecisset 3 circulationes cum $\frac{1}{4}$, tunc proportio motus *A* ad
motum *B* esset sicut proportio 5 et $\frac{1}{4}$ ad 3 et $\frac{1}{4}$, videlicet sicut 21 ad 13 et est
40 proportio $1\frac{8}{13}$. Consimiliter est dicendum de quolibet alio aspectu. Unde
si *A* nunc sit in puncto *d* et *B* in puncto opposito per idem demonstrabitur
quod numquam alias opponentur ipsis existentibus in punctis ab istis distantibus commensurabiliter capiendo distantiam secundum circuli portionem; et ita de qualibet alia dispositione. Et de preterito argueretur eodem
45 modo, sicut dictum est de futuro. Sic ergo distantia seu portio circuli inter

---

20 sit *om. ABU* / *post* puncto *hab. B* illi
21 *ante* ut *hab. B* nec ante ?opposita sint /
    ut: et *L* vel *U* / demonstratur *LU* de-
    monstretur *P* / eorum: ipsorum *ALU*
22 *post* infinita *scr. et del. A* versus / nulla
    est *tr. B* / *post* est *hab. L* revolutio
23 Conclusio secunda *om. FP* propositio
    secunda *U* secunda *mg. hab. AB ante*
    Si duo / nunc sint *tr. LP* sunt nunc *A*
    nunc sunt *U* / coniuncta: iuncta *A* /
    numquam: ?necquiquam *P*
24 *ante* sunt *hab. L* nunc
26 cuilibet: cuiuslibet *A*
27 utrique: cuilibet *U* / commensurabile:
    commensurabilem *L* / fient: sibi erunt
    L / ut statim sequitur: patet *L*

28 et nona decimi: nona et decima *L*
29 sint: sicut *P*
30 portio: proportio *A* / uni *om. A* / erunt:
    essent *F*
31 coniungantur *F* coniungentur *L* / distan-
    te: ?*B*
31–32 A et…tempore *om. U*
32 quolibet *A* / illorum: eorum *AL* / *post*
    intermedio *hab. A* qui / complevisset:
    complevis sed *P* / aliquot *om. U*
33 *ante* partem *hab. FP* per / illam *om. B* /
    suo toti *om. A*
34 esset *B*
35 forent: fuerint *U*
36 *post* B *hab. B* nunc / distanti *U* / a punc-
    to *om. A*

never be in opposition at other times in these very same points. This could be demonstrated in the same manner as before. Hence, the [period of] revolution of these mobiles is infinite; indeed, it might be truer to say that there is no period of revolution at all.

Proposition 2. If two mobiles are now in conjunction, they will never conjunct at other times in any point separated from their present point of conjunction by a part of the circle commensurable to the whole circle.

From the eighth and ninth [propositions] of the tenth [book] of Euclid, it follows at once that if something is added to each of two commensurable quantities, and is commensurable to each of them, the wholes will be commensurable. Therefore, if to each of two sets of circulations—there may be any number of circulations in each set—there is added a part of a circulation commensurable to a whole circulation, the wholes will be commensurable. Now if A and B are in conjunction in some point separated from point d, then, in the time that has intervened [since their conjunction in d], each will have completed an exact number of circulations plus some part of a circulation commensurable to a whole circulation. Their motions must, therefore, be commensurable, since the distances traversed in the same period of time would be commensurable.

For example, if A and B are in conjunction in some point that is separated from point d by $\frac{1}{4}$ part of the circle, and if A had by then completed $5\frac{1}{4}$ circulations and B, $3\frac{1}{4}$ circulations, the ratio of the motions of A and B would be as $5\frac{1}{4}$ to $3\frac{1}{4}$—i.e., as 21 to 13, which is a ratio of $1\frac{8}{13}$. The same may be said for any other aspect. Thus if A is presently in point d and B is in the point opposite d, then, by taking a part of the circle commensurable to the whole circle, one can demonstrate by the same method that they will never again be in opposition in points that are separated from their present points of opposition by distances that are commensurable to the whole circle. Any other disposition can be treated in this manner. And the same argument applies to the past as well as the future. One may, therefore, conclude that the distance,

---

37 $\frac{1}{4}$¹: 4 *A* / partem *om. AU* / tunc fecisset *tr. B* / ante $\frac{1}{4}$² *hab. F* quinta / $\frac{1}{4}$²: 4 *A* / parte *om. U*

38 circulationis *om. B* / $\frac{1}{4}$: 4 *A* / ante tunc *hab. A* parte unius

39 motum: modum *P* / $\frac{1}{4}$¹: unius quarte *FLP* unius 4 *A* quarte *B* / 3: tertiam *AF* tertia *P* / $\frac{1}{4}$²: unam quartam *AFLP* quartam *B* / 21: 12 *L* / et³: etiam *A*

40 $1\frac{8}{13}$ *om. L* / est dicendum *tr. F* / post de *hab. B* oppositione et / quotlibet *A*

41 nunc *om. A*

42 quod: quot *P* / opponerentur *A* opponetur *B* / ab *om. B* / istis: invicem *U*

42–43 *post* distantibus *hab. B* istis / distantibus commensurabiliter *om. P*

43 capiendo *om. P*

44 preterito: prito *P* / arguetur *PU* arguatur *L*

44–45 de²…modo: eodem modo argueretur de preterito *B*

45 dictum est *om. L* / seu: vel *B* sive *F*

duas coniunctiones proximas est incommensurabilis toti circulo; et ita de
quolibet alio aspectu; et similiter tempus inter duas tales coniunctiones
tempori circulationis unius mobilis et tempori circulationis alterius, et
similiter ista tempora circulationum sunt incommensurabilia inter se.

50   *Conclusio tertia. Numquam altero eorum existente in puncto in quo nunc*
sunt ambo ipsa distabunt per partem circulo commensurabilem.

Sint ut prius *A* et *B* in puncto *d* et quoniam moventur incommensurabiliter semper pertransita ab eis erunt incommensurabilia. Igitur quandocumque *A* fecerit aliquot circulationes precise et erit in puncto *d* tunc ipsum
55   *B* compleverit partem circulationis incommensurabilem toti aut circulationem seu circulationes cum tali parte incommensurabili toti. Ergo tunc *B*
distabit ab *A* per portionem circuli incommensurabilem suo toti. Et ita *A*
distabit ab ipso *B* per portionem circuli incommensurabilem toti circulo
quotienscumque *B* fecerit aliquot circulationes et erit in puncto *d* ut per
60   idem probaretur. Igitur tunc *A* et *B* respicient centrum ad angulum incommensurabilem angulo recto et ergo altero eorum existente in puncto *d*
numquam aspicient se aspectu sextili, nec etiam aspectu quarto, vel trino,
vel opposito, nec aliquo aspectu commensurabili istis. Et etiam econverso
in quibuscumque punctis aspiciant se per partem circuli commensurabilem
65   toti in nullo illorum punctorum possibile est ipsa in posterum coniungi
nec in preterito fuisse coniuncta.

Conclusio quarta. Nulla est circuli tam parva portio in qua talia duo
mobilia non coniungantur in posterum et in qua non fuerint [in preterito]
aliquando coniuncta.

70   Sint nunc *A* et *B* in puncto *d*, et punctus *e* sit locus prime coniunctionis

46 coniunctiones proximas *tr.* B circulatio-
    nes proximas *P* / est *om.* U et *B* / incom-
    mensurabilis: commensurabilis *BFLPU*
47 *post* aspectu *hab.* B vel dispositione
    celi / et *om.* B / *post* coniunctiones *hab.*
    B est commensurabile
48 tempori[1] *om.* U / et[1] *om.* U / et[2] *om.* U
49 ista: illa *BL* alia *U* / tempora: pars *L* /
    incommensurabilia: commensurabilia
    BFPU / *post* se *add.* LP et cetera
50 Conclusio tertia *om.* FP propositio ter-
    tia *U* tertia *mg. hab. A post* in[2] *et mg.*
    hab. B *ante* Numquam / eorum *om.* A /
    in[2] *om.* B
51 circuli *U*
53 semper...incommensurabilia *om.* A /
    semper *tr.* B *post* eis
53-54 quandoque *FP*

54 aliquos *P* / et erit *om.* A / erat *B* / ip-
    sum: ipsemet *A*
55 compleverit *om.* FP fecit *B* pertransit *U*
56 seu: vel *B* sive *F* aut *U* / Ergo *om.* A /
    B *om.* F
57 distabat *L* / portionem: positionem *P* /
    suo: sunt *P*
57-58 A distabit *tr.* B
58 B *om.* A / per: ad *U* / incommensurabi-
    lem: commensurabilem *U sed mg. hab.*
    U alibi forte incommensurabilem
59 aliquos *P*
60 tunc A et B: A et B tunc *B* / respicient:
    circumstabunt *A*
60-61 tunc...ergo *om.* U
61 et *om.* B / ergo *om.* P igitur *L* / eorum
    om. U / puncto d *tr.* B
62 aspicient se *tr.* F aspiciunt se *A* / aspectu

or sector of a circle, that separates two successive conjunctions is incommensurable to the whole circle; and this may be said about any other aspect. And similarly can the time between two such conjunctions be related to the time it takes each mobile to complete one circulation; and these times of circulation are likewise mutually incommensurable.

Proposition 3. Whenever one of the two mobiles is in the point where they are now, they will never be separated by a part [of the circle] commensurable to the [whole] circle.

As before, assume that *A* and *B* are in point *d*. Since they are moved with incommensurable velocities, the distances that they have traversed will always be incommensurable. Whenever *A* will have made an exact number of circulations and arrives in point *d*, *B* itself will have completed a part of a circulation incommensurable to a whole circulation; or, it will have made a complete circulation, or circulations, plus a part which is incommensurable to a whole circulation. [Mobile] *B* will then be separated from *A* by a part of the circle that is incommensurable to the whole circle. And, in the same manner, it can be shown that when *B* is in point *d* after making an exact number of circulations, *A* will be separated from it by a part of the circle incommensurable to the whole circle. When the center is used as a reference [point], *A* and *B* form an angle that is incommensurable to a right angle, so that when one of them is in point *d*, they will never be related in sextile, quartile, or trinal aspects; nor, indeed, will they be in opposition or in any aspect commensurable to these aspects. Conversely, it is impossible for these two mobiles to enter into future conjunctions, or to have had conjunctions in the past, in any points that are separated by an angular distance that constitutes a part of the circle which is incommensurable to the whole circle.

Proposition 4. No sector of a circle is so small that two such mobiles could not conjunct in it at some future time, and could not have conjuncted in it sometime [in the past].

Assume that *A* and *B* are now in point *d*, and that point *e* is the place of

sextili *tr. B* / aspectu² *om. B* / quarto: 4 *A*
63 vel: aut *F* / istos *A* / econverso *om. L*
64 punctis aspiciant: aspiciant *A* positis aspicientibus *U*
65 illorum: istorum *AU sed mg. hab. U* alibi illorum / illorum punctorum *tr. L* / ipsa *om. A* ipsam *L* / posterum: futurum *B*
66 preterito fuisse: puncto fieri *L* / *post* coniuncta *hab. F* et sic finis est huius presentis capituli

67 Conclusio quarta *om. P* propositio quarta *U* incipit quarta conclusio *F* quarta *mg. hab. A post* parva *et mg. hab. B ante* Nulla / *ante* est *scr. et del. A* et / qua: quo *U*
68 posterum: futurum B / fuerunt *A* / [in preterito] *om. AFLPU*
69 aliquando coniuncta: adhuc coniuncta *A* coniuncta olim *U*
70 nunc: vero *U*; *tr. B post* B / punctus: primus *A* / sit *om. P*

sequentis. Cum igitur _A_ et _B_ moveantur regulariter semper erit equalitas de tempore inter quaslibet duas coniunctiones proximas; et similiter de spatio ita quod quantus est arcus inter punctum _d_ et punctum prime coniunctionis sequentis tantus est arcus inter punctum secunde et punctum
75 tertie, et sic deinceps quasi replicando arcum _de_ infinities et applicando circulo. Et quoniam arcus _de_ est incommensurabilis circulo per tertiam conclusionem statim precedentem, ergo post aliquot replicationes alicuius arcus equalis ipsi _de_ excedetur circulus et transibit iste arcus ultra punctum _d_ secando arcum _de_ in puncto _g._ Et omnino consimiliter secabitur alter
80 arcus qui erat inter coniunctionem secundam et tertiam et sic consequenter donec totus circulus erit sic divisus quod nulla pars remanebit indivisa maior quam sit arcus _dg._

Et postea quia arcus _de_ est incommensurabilis circulo rursum per talem replicationem secabitur arcus _dg_ et fiet unus arcus minor qui sit _dk_ et
85 tandem circulus erit sic divisus quod nulla pars ipsius erit maior quam sit arcus _dk._ Et sic procedetur in infinitum semper dividendo circulum in partes minores in infinitum propter incommensurabilitatem et infinitatem coniunctionum et replicationum. Ergo nulla pars circuli restabit quin aliquando ad ymaginationem sit divisa per hunc modum sicut qui repli-
90 caret costam quadrati super dyametrum quousque excederet, et iterum abscinderet illum excessum secundum et replicaret ut prius et acciperet tertium excessum, et sic procederet in infinitum. Tunc in infinitum diminueretur ille excessus secundum cuius quantitatem semper divideretur dyameter, igitur nulla pars dyametri remanet toto tempore indivisa; et ita
95 est quodammodo in proposito.

Ergo nulla erit tam parva [circuli] portio quin aliquando coniungantur

71 equalitas: equaliter _AFLP_
71–72 erit…tempore: equale tempus _B_
72 et _om. BU_
72–73 de spatio: equale spatium _B_
74 sequentis tantus: quantus _L_ / est _om._ _U_ erit _B_ / _post_ punctum[2] _hab._ _B_ sequentis coniunctionis
75 quasi…infinities _om._ _P_ / _post_ de _scr. et del._ _A_ e / _ante_ infinities _scr. et del._ _L_ est incommensurabilis ?circulo / applicando: replicando _A_
76 circulo[1]: circulum _L_ / incommensurabilis: commensurabilis _BFPU_ / tertiam: secundam _P_
77 _infra_ conclusionem _mg. hab._ _L_ figuram / conclusionem statim: huius _B_ / post:

per _L_ / aliquot: aliquas B aliquos _P_
77–78 alicuius arcus _tr._ _U_
78 excedetur _obs._ _B_ exceditur _A_
79 omnino _obs._ _B_ / consimiliter: similiter _FL_ / sectabitur _A_
80 erit _U_ / coniunctionem secundam _tr._ _AL_
81 erit sic divisus: sit divisus sic _B_ / pars _om._ _L_ / remaneat _U_
81–82 indivisa maior _tr._ _L_
83 _infra_ Et postea _mg. hab._ _A_ figuram / de _corr. ex_ dg _ABFLPU_ / incommensurabilis: commensurabilis _BFPU_
84 sectabitur _A_ / fiet: hec _A_ / unus arcus _tr._ _U_
86 procederetur _A_ proceditur _L_ / in[1] _om._ _A_

the first conjunction following. Therefore, since *A* and *B* are moved regularly [or uniformly] the time between any two successive conjunctions will always be equal. The distances will also be equal, so that the arc separating point *d* and the point of the first conjunction following [namely, *e*] is equal to the arc separating the second and third points of conjunction, and so on by successively repeating arc *de* an infinite number of times and applying each of them to the circle. Now since arc *de* is incommensurable to the circle, by the third proposition immediately preceding, it will happen that after an arc equal to *de* is applied a certain number of times, the circle is surpassed and this arc crosses beyond point *d* by cutting arc *de* in point *g*. The arc lying between the second and third points of conjunction will be cut in exactly the same manner, and so on in succession until the whole circle will be so divided that no part greater than arc *dg* will remain undivided.

And then, since arc *de* is incommensurable to the circle, marking off another such sequence [of arcs equal to *de*] will result in the division of arc *dg*, producing a smaller arc, *dk*, until finally the circle will be divided in such a way that no part of it will be greater than arc *dk*. In virtue of the incommensurability [involved] and the infinite number of successively linked conjunctions, this process can be carried into infinity by always dividing the circle into smaller parts ad infinitum. Thus, no part of the circle will remain but that it could not, at some time, be imagined as divisible in this way, just as one might apply [or superpose] the side of a square onto its diagonal, until it exceeds the diagonal; then one could cut off this excess and apply it to [the diagonal] a second time, just as before; one could then take the excess [length again and apply it] a third time. Proceeding in this way ad infinitum, the excess by which the diagonal is divided would be diminished into infinity so that in the whole time no part of the diagonal would remain undivided. This is similar to what would occur in our discussion [of the arcs on the circle].

Therefore, no sector of the circle will be so small but that at some time in

---

87 in *om. A* / incommensurabilitatem: commensurabilitatem *BFPU sed U mg. hab.* forte incommensurabilitatem / infinitatem: infinities *AFP*

88 replicationem *FP* / quin: quando *L*

89 aliquando *om. B* / ad *rep. A* / *ante* sit *hab. B* aliquando / sit: sic *AL* / qui: igitur *L*

91 abscindet *A* abscindetur *F* excederet *U* / secundum *corr. ex* tertium *ABFLPU* /

et acciperet: accipet *A*

92 tertium *corr. ex* quartum *ABFLPU* / procedetur *F* procedet *L* / infinitum[1]: infinita *P*

93 ille: iste *A* / cuius: sui *U*

94 dyameter: dyametri *B* dyametrum *U* / remaneret *BLP* / toto tempore *tr. B* / indivisa: divisa *U*

96 erit: est *U* / [circuli] *om. AFLPU* / quin: quando *P* / coniungatur *AFU*

in futurum in aliquo puncto illius quod est propositum; et ita de preterito supposita eternitate motuum. Unde in tali circulo coniungentur *A* et *B* infinities et semper in novo puncto per primam conclusionem huius partis
100 et equaliter distabit secundus punctus a primo et tertius a secundo, et sic de aliis. Et protrahendo lineam de primo ad secundum, et de secundo ad tertium, et sic deinceps describeretur in circulo una figura infinitorum angulorum equalium se invicem secantium et quilibet talis angulus erit incommensurabilis recto ut faciliter probaretur. Igitur nulla pars circuli
105 carebit in perpetuum istis angulis sed inter quecumque puncta circumferentie fierent infiniti tales anguli ita quod per equalitatem istorum angulorum et equidistantiam ipsorum et multiplicationem eorum in infinitum posset demonstrari de quacumque parte circuli quod impossibile est ipsam vacare ab huiusmodi angulis.
110   In quorum angulorum descriptione et talium punctorum multiplicatione diligens theoricus spectare potest modum mirabilem quo ex incommensurabilitate et regularitate motuum oritur quedam ut ita dicam rationalis irrationalitas, regularis difformitas, uniformis disparitas, concors discordia. Et cum summa inequalitate que ab omni equalitate degenerat equis-
115 sima atque ratissima ordinatio perseverat.
     Item supposita eternitate motuum et in quolibet puncto coniunctionis ymaginata una divisione per significationem statim patet quod per coniunctiones preteritas nulla pars circuli remanet indivisa et quod nulla pars restabit que non dividatur etiam per futuras. Et tamen in nullo puncto
120 dividetur circulus in posterum in quo fuit divisus in preteritum et sic restant adhuc infinita puncta in quibus nulla cecidit divisio. Et adhuc infinita

98 coniungentur *om. P*
99 et *om. U* / novo: alio *U*
101 Et *om. U*
102 sic *om. B* / describetur *A* / in *om. U* / in...figura: figura una in circulo *U* / una *om. B*
103 *post* equalium *hab. A* et infinitorum angulorum inequalium et infinitorum laterum equalium *et hab. L* et infinitorum laterum equalium / se: sibi *U* / angulus *om. B*
104 ut: et *P* / probatur *B*
105 perpetuum: propositum *P* / sed: si *U* / queque *P*
106 fient *A* / per *om. L* / istorum *om. L*
107 equidistantia *A* / ipsorum: eorum *B* / *ante* eorum *scr. et del. A* ipsorum / eorum *om. U* / in *om. A*

108 quaque *P* / impossibile est *tr. U* / vocare *P*
110 angulorum descriptione *tr. U* / multiplicationem *U*
111 theoricus: theologus *F* / spectare potest: videt spectare *U* / modum mirabilem *tr. B* medium mirabile *P*
111–12 incommensurabilitate: commensuratione *LP*
112 et regularitate motuum: motuum et regularitate *B*
113 irrationabilitas *A* irregularitas *U* / *post* disparitas *hab. B* vel diversitas
113–14 concors discordia *tr. AFPU* discordationis *L*
114 summa *om. A* ?sonia *B* / inequalitate: equalitate *U* / ab omni: alteri *U sed mg. hab. U* alibi ?ab ?omnia

the future the mobiles could not conjunct in some point of it, and this is what we have proposed; and, assuming that the motions are eternal, this also applies to the past. In such a circle, *A* and *B* will conjunct an infinite number of times, but always in a new point—by the first proposition of this part. And the distance separating the second point from the first point of conjunction equals the distance separating the third from the second point of conjunction, and so on. By drawing a line from the first [point of conjunction] to the second, and from the second to the third, and so on successively, there would be described in the circle a figure with an infinite number of equal angles mutually intersecting. It could easily be proved that any such angle would be incommensurable to a right angle. Therefore, through all eternity no part of the circle will lack these angles, but an infinite number of such angles are made between any points [of conjunction] on the circumference, so that by the equality and equidistance of these angles and their infinite multiplication, it could be shown that it is impossible for any part of the circle to be without such angles.

In the description of these angles and from the multiplication of such points, a diligent mind can consider the wonderful way in which some things arise from the incommensurability and regularity of motions, so that I could utter such [expressions] as "rational irrationality," "regular non-uniformity," "uniform disparity," "harmonious discord." Thus, by means of the greatest inequality, which departs from every equality, the most just and established order is preserved.

Furthermore, assuming that the motions are eternal and that every imaginable point of conjunction has been marked off, it is immediately apparent that no part of the circle remains undivided through all past conjunctions; and no part will remain that could not be divided by future conjunctions. And yet, in the future, the circle could not be divided in any point where it was divided in the past, so that an infinite number of points remain in which no division has yet fallen. And this is yet another infinite [set of points]—namely, all those

---

115 atque: seu *A* et *BU* / rarissima *F* rotis-
   sima *P*
116 et *om. ABL*
117 divisione *U sed mg. hab. U* alibi divisione vel diminutione vel distantia / per[1]
   om. P / significationem: signationem *BL*
   figurationem *U* / quod: quia *U*

118 circuli: preteriti *U*
119 restabit: restaurabit *A* / etiam *om. U*
   vel *B* / puncto: potest ?ta *A*
120 divideretur *L* / in[2] *om. B* / fuerit *B* sint
   L / preterito *L* / et *om. B*
120–21 restant adhuc *tr. B* restat adhuc *A*
121 cecidit: accidit *U*

alia, videlicet omnia que incommensurabiliter distant ab aliquo predictorum ut patet per secundam conclusionem huius partis.

125 Quod autem quilibet de prius dictis angulis sit incommensurabilis recto probatur. Et sit *o* centrum circuli, et sit *d* punctus in circumferentia in quo *A* et *B* nunc sunt coniuncta; sitque *e* primus punctus in quo postea coniungentur, et sit *g* secundus. Protractis itaque chordis describatur angulus *deg* quem dico esse incommensurabilem recto. Ducantur enim linee ad centrum et fiat angulus *doe* qui est incommensurabilis recto per secundam
130 huius. Ergo reliqui duo anguli istius trianguli *doe* simul sumpti sunt incommensurabiles recto quia cum illo qui est circa centrum valent duos rectos. Et cum isti reliqui duo sunt equales ergo quilibet eorum est incommensurabilis recto; igitur angulus *oed* est incommensurabilis recto; ergo angulus sibi equalis, videlicet *oeg*, est incommensurabilis recto. Ergo totus
135 angulus *deg* est incommensurabilis recto et tenent iste consequentie per octavam et nonam decimi Euclidis.

Conclusio quinta. Quolibet puncto coniunctionis dato, in infinitum prope illum punctum huiusmodi mobilia coniungentur et in infinitum prope fuerunt coniuncta.

140 Detur punctus *d* et signetur alter punctus propinquus qui sit *c*. Ergo per precedentem coniungentur inter *c* et *d*. Et si rursum signetur alter punctus

122 videlicet *om. B* / omnia que *tr. B* / incommensurabiliter: commensurabiliter *A*

122–23 predictorum: punctorum *A*

123 ut *om. B* / patet *om. AB* / per *om. LP* / conclusionem *om. B* / infra conclusionem huius *mg. hab. A figuram*

124 autem: ante *P* / incommensurabilis: commensurabilis *A*

125 Et[1] *om. B* / sit[2] *om. B* / ante punctus *scr. et del. A* tempus / in[2] *om. B*

126 nunc sunt *tr. B*

126–27 postea coniungentur *tr. U*

127 sit g *tr. B* / secundus: ?sedens *A* / post secundus *hab. B* et cetera / angulus *om. U*

128 incommensurabilem: commensurabilem *A* / enim: vero *F*

128–29 *post* centrum *add. B* o

129 qui: que *FP* / incommensurabilis: commensurabilis *L*

130 *post* reliqui *scr. et del. A* anguli / istius: illius *A* huius *B* / sumptu *A*

131 illo: isto *A* / circa *om. A* / *post* centrum

add. B o

132 cum *om. F* quoniam *AL*

133 igitur...ergo *om. P* / oed *corr. ex* doe *AB* deo *FU*

133–34 angulus...Ergo *om. L*

134 videlicet: scilicet *B* / *post* recto *hab. F* et

134–35 infra totus angulus *mg. hab. B figuram*

135 tenet *F* / iste: ille *F* / consequente *AP* consequentia *B* ?consequens *F*

135–36 infra per octavam *mg. hab. B figuram*

136 octavam et nonam: nonam et octavam *L* octavam decimam *U* / *post* Euclidis *hab. P* et cetera

137 Conclusio quinta *om. FP* quinta conclusio seu propositio *U* quinta *mg. hab. A post* da- *in* dato *et mg. hab. B ante* Quolibet

138 illum punctum *om. L* / mobilia *om. L* / et: etiam *U*

140 signetur *obs. B* signeretur *U* / qui: et *B* / c: g *L*

141 inter: interim *U* / c et d: d et c *AL* ?c d *U* / rursus *U* / signetur *obs. B*

that are separated from any of the aforementioned points [of conjunction] by a distance that is incommensurable [to the whole circle], as is obvious by the second proposition of this part.

Let us now demonstrate that any one of the angles mentioned before is incommensurable to a right angle. Let *o* be the center of the circle, and *d* a point on the circumference where *A* and *B* are now in conjunction [see Fig. 24]; let *e* be the first point in which *A* and *B* will afterward enter into conjunc-

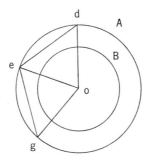

Fig. 24

This figure is very nearly identical with figures appearing in MSS *A* (Vat. lat. 4082, fol. 103r, in the margin below col. 2), *B* (Paris, MS BN lat. 7281, fol. 266v, bottom margin), and *L* (Florence, MS Ashburnham 210, fol. 166v, in the margin below col. 1).

tion, and let *g* be the second such point. After the chords are drawn [connecting these points of conjunction], let angle *deg* be described. I say it is incommensurable to a right angle. For let lines be drawn to the center making angle *doe*, which is incommensurable to a right angle by the second [proposition] of this [part]. It follows that the two remaining angles of triangle *doe* are incommensurable to a right angle, because when they are simultaneously added to the central angle [*doe*] we have the equivalent of two right angles. Since these two remaining angles are equal, each of them is incommensurable to a right angle, so that angle *doe* is incommensurable to a right angle. Therefore, the angle which is equal to *oed*—namely, *oeg*—is incommensurable to a right angle. Consequently, the whole angle *deg* is incommensurable to a right angle. All these consequences hold by virtue of the eighth and ninth [propositions] of the tenth [book] of Euclid.

Proposition 5. Such mobiles will conjunct infinitely close to any given point of conjunction, and have already [in the past] conjuncted infinitely near to it.

Let point *d* be the given point and *c* another point assigned near to it. By the preceding [proposition], the mobiles will conjunct between points *c* and *d*. And if another point, say, *f*, were assigned halfway between, it is again

duplo propinquior puncto *d* qui sit *f*, adhuc per eandem patet quod coniungentur inter *d* et *f* et sic in infinitum appropinquando.

   Conclusio sexta. Possibile est tria [mobilia] vel plura nunc coniuncta alias
145 *coniungi quorum quodlibet respectu cuiuslibet movetur incommensurabiliter.*

   Sint *A* et *B* et *C* in puncto *d*. Ergo *A* et *B* alias coniungentur propter motuum inequalitatem. Ponatur igitur quod postea coniungantur in puncto *e* qui distat a puncto *d* incommensurabiliter per secundam huius. Ergo tunc *A* perficit unam circulationem vel aliquot circulationes cum
150 aliqua portione circulationis que est arcus *de*; et similiter *B* complevit aliquot circulationes cum eadem portione. Sed spatia pertransita ab *A* et *B* sunt incommensurabilia et quotlibet circulationes quotlibet circulationibus precise sumptis sunt commensurabiles, ergo arcus *de* additus utrobique facit illa pertransita incommensurabilia. Ergo possibile est quod commen-
155 surabilibus addatur idem vel equale et tota fient incommensurabilia.

   Ponatur ergo quod *A* et *B* et *C* sint postea simul in puncto *e*. Ergo tunc quodlibet eorum descripsit aliquot circulationes et cum hoc arcum *de*. Sed arcus *de* additus cum quotlibet circulationibus inequaliter numeratis reddit tota aggregata incommensurabilia. Ergo pertransitum seu descrip-
160 tum ab *A* est incommensurabile toti pertransito ab ipso *B*, et similiter toti pertransito a *C*; et similiter pertransitum a *B* est incommensurabile pertransito a *C*. Ergo posito quod *A* et *B* et *C* fuerint coniuncta in puncto *d* et iterum sint coniuncta in puncto *e* stat quod quodlibet eorum respectu cuiuslibet incommensurabiliter movetur, ergo non repugnat illa incommen-
165 surabiliter moveri et alias coniungi quod est propositum.

   Conclusio septima. Possibile est quod sint tria aut plura [mobilia] nunc

142 *ante* duplo *hab. B* in *et hab. L* qui est in / *infra* puncto *mg. hab. L figuram* / d *om. U* / *ante* per *hab. FP* hoc / quod *rep. A*

143 inter *obs. B* / *post* appropinquando *hab. P* et cetera

144 Conclusio sexta *tr. U om. FP* sexta *mg. hab. A post* nunc *et ante* possible *mg. hab. B* / [mobilia] *om. AFLPU* / vel: sive *F*

146 Sint: sicut *P* / et¹ *om. B* / et C *om. A* / *post* puncto d *mg. hab. B* de tribus vel pluribus

147 Ponatur *obs. B* / quod postea *om. B* / coniungentur *A*

148 e *om. A* o *U*

149 perfecit *F* / vel: sive *F* / aliquos *P*

150 aliqua: ista *A* illa *L* / que: qui *A* / et *om. B*

151 aliquos *P* / spatia pertransita: spaciam pertransitam *P*

152 *post* sunt *hab. U* ?equalia / quotlibet circulationes *om. P*

153 prescise *U* / *ante* sunt *hab. U* que / commensurabiles: incommensurabiles *L*

154 illa: ista *A* / est *om. P*

154–55 commensurabilibus: incommensurabilibus *A*

156 ergo¹ *om. BL* / et¹ *om. BU* / et C *om. A* C *B* / simul *om. L* / *infra* Ergo tunc *mg. hab. P* circulationes et

156–57 tunc quodlibet eorum: quotlibet eorum tunc *B*

157 quotlibet *L* / aliquos *P* / Sed: et *U*

159 aggregata: agrega *A* / *infra* Ergo *mg. hab. A* duas figuras / *post* Ergo *hab. A* ?positum / *ante* pertransitum *hab. L*

obvious—by the same [preceding proposition]—that the mobiles will conjunct between d and f. In this manner their conjunctions will approximate infinitely close [to d].

Proposition 6. It is possible that three or more mobiles moving with mutually incommensurable motions are now in conjunction and will conjunct again at other times.

Assume that A, B, and C are in point d. Because their motions are unequal, A and B will enter into conjunction at other times. Let us suppose that they were conjuncted afterward in point e which, by the second [proposition] of this [part], is separated from point d by a distance that is incommensurable [to the whole circle]. Therefore, A completes one circulation, or several circulations, plus some part of a circulation equal to arc de; similarly, B has completed a certain number of circulations plus the same part of a circulation [namely, arc de]. Now the distances traversed by A and B are incommensurable; but any number of exact circulations is commensurable to any other exact number of circulations, so that it is the addition of arc de to each [of these exact numbers of circulations] that makes the distances traversed incommensurable. Therefore, it is possible that when the same thing or equal things are added to commensurables, the wholes become incommensurable.

Let us now assume that A, B, and C are afterward simultaneously in point e. Each of them would then have described a certain number of circulations plus arc de. But when arc de has been added to integrally, though differently, numbered sets of circulations, the total aggregates [or collections of circulations] are incommensurable. Hence the distance traversed or described by A is incommensurable to the total distance traversed by B, and also to that traversed by C; similarly, the distance traversed by B is incommensurable to that traversed by C. The assumption being made that A, B, and C conjuncted in point d, and then again in point e, one can maintain that any one of them is moved incommensurably with respect to any other. It is not inconsistent, then, that these mobiles are moved with incommensurable speeds, and [yet] have been in conjunction at other times—and this is what was proposed.

Proposition 7. It is possible that three or more mobiles whose motions are

---

totum / seu: vel B sive F

160 toti[1]: toto B / pertransito...toti *om.* F / ipso *om.* B

161 a[1]: ab ipso U / pertransitum: transitum F / est: C F / incommensurabile: commensurabile A

161–62 et similiter...a C *om.* P

162 et[1] *om.* B / et[2] *om.* B / fuerunt A / coniuncta: iuncta B

163 e: et P / quod: quot P / quotlibet BL

164 *post* cuiuslibet *hab.* B eorum / incommensurabiliter movetur *tr.* B / moveretur A / repugna P / illa: ista B

164–65 incommensurabiliter: commensurabiliter $BFLP$

166 Conclusio septima *om.* $AFPU$ septima *mg. hab.* B *ante* Possibile / aut: sive F / [mobilia] *om.* $AFLPU$

coniuncta quorum quilibet motus sunt incommensurabiles que numquam
poterunt alia vice coniungi ita quod ex quadam incommensurabilitate
sequitur ipsa infinities coniungi et ex alia non sequitur.

170   Sint ut prius *A* et *B* et *C* in puncto *d*. Resumatur itaque quoddam demonstratum ex secunda conclusione prime partis, scilicet quod si continuum
dividatur in infinitum una divisibilitate secundum proportionalitatem
rationalem, deinde idem ymaginetur dividi secundum unam aliam proportionalitatem rationalem que proportionalitates sint incommunicantes.
175   Tunc nullus punctus unius divisibilitatis est aliquis punctus alterius
divisibilitatis, nisi unus solus, si illud continuum sit circulus; quamvis talia
puncta divisionum communicarent si proportionalitates essent communicantes.

Eodem modo necesse est esse de proportionalitatibus secundum pro-
180   portiones irrationales. Si enim una sit secundum medietatem duple et alia
secundum medietatem octuple tunc sunt communicantes; sed si una fiat
secundum medietatem duple et alia secundum medietatem triple tunc sunt
incommunicantes. Ergo possibile est dividi continuum in infinitum per
puncta incommensurabiliter distantia et tamen nullus punctus unius
185   divisibilitatis erit aliquis punctus alterius divisibilitatis nisi forte unus. Si
illud continuum est circulare, ergo possibile est in circulo fieri duas tales
divisibilitates. Quod a puncto *d* per ordinem incipient due series numerorum infinitorum incommensurabiliter distantiam que due series non amplius communicabunt in aliquo puncto. Ergo possibile est quod omnia
190   puncta infinita alia a puncto *d* in quibus coniungentur *A* et *B* sunt incom-
municantia cum illis in quibus coniungentur *B* et *C* igitur numquam alias
coniungentur *B* et *C* quando nec ubi coniungentur *A* et *B*. Ergo *A* et *B* et *C*
numquam alias coniungentur.

Verbi gratia, posito quod *A* et *B*, que nunc sunt in *d*, postea coniungan-
195   tur in *e*, deinde in *f*, et cetera; et tunc *d*, *e*, *f*, et cetera, equaliter distant

---

167 coniuncta: iuncta *B* / quilibet *om. B* /
     sint *F*
169 coniungi: iungi *B* / *ante* alia *hab. L*
     quadam / sequitur² *om. B*
170 et¹ *om. B* / et² *om. B* / resumantur *AF* /
     itaque *om. B*
171 quod *om. B* / si *om. AL* h *F* b *P* / *ante*
     continuum *hab. B* totum
172 proportionalitatem: proportionem
     ABPU
173 unam aliam: aliam *BU* unam *L*
173–74 deinde...rationalem *om. A* / propor-
     tionalitatem: proportionem *BU*

174 proportionalitates: proportiones *B* /sunt
     U
175 divisibilitatis: indivisibilitatis *A* / aliquis
     om. B
176 continuum *om. AU*
177 *post* divisionum *hab. B* huiusmodi / es-
     set *B*
179 *ante* Eodem *hab. B* et / proportionalita-
     tibus: proportionibus *FL*
180 irrationales: rationales *U* / enim: vero
     F in *U* / *post* secundum *add. BFLU* unam
181 octuple: 8 *A* duple *FL* / fiat: sit *B*
183 dividi continuum *tr. A*; *om. FLPU*

mutually incommensurable are now in conjunction but can never conjunct in another place. Therefore, in one case of incommensurability it follows that the mobiles can be in conjunction an infinite number of times, but in another case this does not follow.

As before, let *A*, *B*, and *C* be in point *d*. Let us now reconsider something that was demonstrated in the second proposition of the first part—namely, that if a continuum were divided into infinity by one rational [geometric] proportionality, it could also be imagined as divisible by [yet] another rational [geometric] proportionality; and these proportionalities are incommunicant [or prime to each other]. If this continuum were a circle, then, except for one point only, no point in one division is a point in the other division. However, if the proportionalities were communicant, there would be points in common between the two divisions.

The same distinctions apply to divisions made by proportionalities consisting of irrational ratios. Thus if one proportionality is based on half of a double ratio [i.e., $(2:1)^{1/2}$], and the other is based on half of an octuple ratio [i.e., $(8:1)^{1/2}$], these are communicant; but if the proportionalities are based on half of a double ratio and half of a triple ratio [i.e., $(3:1)^{1/2}$], they are incommunicant [or prime to each other]. Hence it is possible for the continuum to be divided by points that are separated by incommensurable distances, so that no point in one division is also a point in the other division—except, perhaps, one point. If the continuum is circular, two such divisions are possible in a circle. Now since these two series of infinite numbers will begin in sequence from point *d* and they are separated from each other by incommensurable distances, the two series will not meet in [or share] any point. It is possible, then, that [beginning] from point *d*, all the other infinite number of points in which *A* and *B* will conjunct are incommunicant with those in which *B* and *C* will conjunct, so that *B* and *C* will never conjunct in the same places where *A* and *B* conjunct. Therefore, *A*, *B*, and *C* will never conjunct at other times.

For example, assume that *A* and *B*, which are now in *d*, are afterward in conjunction in *e*, then in *f*, and so on [see Fig. 25]. Then, because of the reg-

---

184 nullius *U* / unius *om. U*
185 nisi: non *P* / forte unus *tr. B* forte *U*
186 in circulo fieri: fieri in circulo *B*
188 distantium *LU*
190 a: in *A* / puncto *om. L*
190–91 d in...illis *om. AU*
191 illis: istis *P* / *ante* igitur *hab. P* et / igitur: et *U sed mg. hab. U* alibi igitur
191–92 igitur...B et C¹ *om. A*

192 quando *om. B* quam *U* / nec: nunc *U* / et³ *om. B* / et⁴ *om. B*
194 que *om. A* / nunc: non *U sed mg. scr. et del. U* alibi nunc / sint *A* / postea: posita *U*
194–95 coniungentur *A*
195 e¹: m *U sed mg. hab. U* alibi in e / *ante* deinde *hab. U* et / et cetera et *om. F* et *A* et cetera *B* / d e f et cetera *om. F*

propter regularitatem motuum. Sitque totus circulus ad arcum *de* in medietate proportionis sexqualtere. Item *B* et *C*, que sunt in puncto *d*, postea coniungantur in *g*, deinde in *h*, et cetera; et totus circulus ad arcum *dg* sit in medietate proportionis sexquitertie.

200   Cum ergo medietas sexqualtere et medietas sexquitertie sint proportiones incommensurabiles, ut patet ex libro *De proportionibus proportionum*, sequitur quod predicte divisibilitates sunt incommunicantes. Et ergo *A* et *B* et *C* numquam alias coniungentur et ita de preterito et de pluribus mobilibus.

205   Notandum tamen quod non oportet circulum dividi per talia puncta secundum proportionem velocitatum vel aliquam sibi commensurabilem proportionem quod statim patet in proportionibus rationalibus si enim *A* moveretur in quadruplo velocius quam *B* tunc per undecimam prime partis patet quod circulus dividetur in 3 partes, non in 4, per ista puncta coniunc-
210   tionum et ad quamlibet illarum partium totus circulus esset triplus.

Conclusio octava. Si fuerint tria vel plura nunc coniuncta que omnia commensurabiliter moveantur preter unum cuius motus sit aliis incommensurabilis numquam alias coniungentur nec alia vice fuerunt coniuncta.

Sint *A* et *B* et *C* in puncto *d* et ponatur quod *A* et *B* moveantur commen-
215   surabiliter. Tunc patet per corollarium decime prime partis quod quilibet punctus in quo possunt coniungi *A* et *B* distat a puncto *d* commensurabiliter; sed quilibet punctus in quo possunt coniungi *B* et *C* distat a puncto

196 propter: per *FP* / regularitatem: raritatem *A* / ad arcum: arcuum *FLP* est que *U*
197 que: qui *F* / *ante* sunt *hab. B* nunc / puncto *om. B*
198 postea: posita *U* / coniungentur *A* iungantur *B* / cetera: ergo *F*
199 sexquitertie: sesqualtere *A*
200 sexqualtere: sesquitertie *A* / sexquitertie: sesqualtere *A*
202 quod: quot *P* / predici *F* / sint *F* / Et¹ *om. B*
203 et ita: nec *B* / *post* et³ *hab. A* ita
203–4 et³...mobilibus: coniuncta sunt aliquando ita est de 4 mobilibus et quotlibet pluribus *B* / mobilibus: motibus *AU*
205 Notandum: intelligendum *B* / *ante* tamen *hab. F* est
206 vel: aut *F* / aliquem *F* / sibi: ?est *A* vel *P*
207 proportionem *om. U*

208 movetur *U* / in *om. BFL* / quadruplo *obs. B* / quam: quod *A* / undecimam: nullam *P* / prime partis: ?partis prime *U*
209 divideretur *AF* dividitur *B* / *ante* non *hab. A* et / per *obs. B*
210 illarum: istarum *A* / est *A*
211 Conclusio octava *mg. hab. L post* nunc *tr. F; om. P* propositio octava *U* octava *mg. hab. A post* nunc *et mg. hab. B ante* Si fuerint / vel: sive *F* / nunc *obs. B*
212 motus: modus *P* / *post* sit *hab. A* ab / aliis: alter *U*
213 fuerint *BFLP*
214 et¹ *om. B* / et² *om. B*
215 per *om. BL* / quelibet *P*
216 possunt *obs. B* / puncto d *tr. L*
216–17 A et...coniungi *om. U*
217 *post* possunt *mg. hab. A* figuram / distant *A*

ularity of the motions, points *d*, *e*, *f*, and so on, are separated by equal distances. Let the whole circle be related to arc *de* as half of a sesquialterate ratio. Also, assume that *B* and *C*, which are in point *d*, are afterward in conjunction in *g*, then in *h*, and so on. [Finally] let the whole circle be related to arc *dg* as half of a sesquitertian ratio [namely, as $(4:3)^{1/2}$].

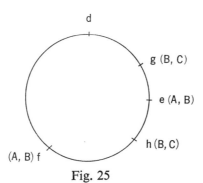

Fig. 25

Only MS *A* (Vat. lat. 4082, fol. 103v, in the margin below col. 2) contains a figure similar to the one presented here.

Now since half of a sesquialterate and half of a sesquitertian are incommensurable ratios, as is shown in the *De proportionibus proportionum*, it follows that the aforementioned divisions are incommunicant. Thus, *A*, *B*, and *C* will never conjunct at other times; this applies also to the past and to more [than three] mobiles.

It should be noted, however, that the circle ought not to be divided by such points in accordance with a ratio of velocities, or any ratio commensurable to it. This is immediately obvious in rational ratios, for if *A* were moved four times quicker than *B*, the points of conjunction would divide the circle into three, not four, parts; and the whole circle would be triple any one of these parts. This is shown by the eleventh proposition of the first part.

Proposition 8. If three or more mobiles are now in conjunction moving with commensurable motions, except for one whose motion is incommensurable to the others, they will never conjunct elsewhere [in the future], nor have they conjuncted in another place [in the past].

Let *A*, *B*, and *C* be in point *d* and assume that *A* and *B* have commensurable speeds. By a corollary of the tenth proposition of the first part, it is clear that any point in which *A* and *B* can conjunct must be separated from point *d* by a distance that is commensurable [to the whole circle]; but, by the second [proposition] of this part, any point in which *B* and *C* can conjunct is separated

d incommensurabiliter per secundam huius. Ergo nullus punctus in quo *A* et *B* de cetero coniungentur est aliquis illorum in quibus *B* et *C* de cetero coniungentur. Ergo *A* et *B* et *C* numquam in posterum coniungentur.

Verbi gratia, posito quod *A* moveatur velocius quam *B* in duplo ergo per undecimam prime huius numquam *A* et *B* coniungentur nisi semper in puncto *d*. Et quia *B* et *C* moventur incommensurabiliter per positum, sequitur per primam huius quod numquam coniungentur in puncto *d*. Ergo *A* et *B* et *C* numquam amplius simul coniungentur; et eodem modo argueretur de preterito et ita de pluribus mobilibus.

Conclusio nona. Omnia tria aut plura mobilia aut numquam simul, aut semel solum, aut infinities toto eterno tempore coniungentur.

Quod possibile sit ipsa numquam coniungi de commensurabiliter motis patet per duodecimam prime partis. Et idem potest contingere de incommensurabiliter motis quod patet etiam satis per septimam huius, quia possibile est quod puncta in quibus *A* et *B* coniungentur et puncta in quibus *B* et *C* coniungentur non communicent nec in uno nec in pluribus. Sed quod possint coniungi toto eterno tempore solum semel demonstratum est per duas conclusiones immediate precedentes. Quod autem infinities possint coniungi patet de commensurabiliter motis per quartamdecimam prime, et de motis incommensurabiliter per sextam huius.

Quod vero non possint coniungi bis vel ter precise in toto eterno aut secundum aliquem alium numerum probatur sic. Sint *A* et *B* et *C* in puncto *d* et postea coniungantur secundo in puncto *e*. Et sit tempus *K* inter istas duas coniunctiones; signeturque alter punctus *g* qui tantum distat a puncto *e* quantum *e* distat a puncto *d*. Cum ergo quodlibet istorum moveatur regulariter, sequitur quod transacto postea tanto tempore quantum est

---

218 d *om. L* / *post* huius *hab. B* partis
219 coniungantur *P*
219–20 est...coniungentur[1] *om. A*
220 et[1] *om. B* / et[2] *om. B*
221 moveatur velocius *tr. B* / quam...duplo: in quod duplo quam B *B* / quam: quod *A*
222 undecimam: 5 *A* / primi *F* / semper *om. L* sint *U*
225 et[1] *om. B* / et[2] *om. B* / simul *om. L* / et[3] *om. B*
226 *post* preterito *hab. B* quod numquam coniuncta sunt / *post* ita *hab. B* etiam argueretur
227 Conclusio nona *mg. hab. L post* mo- in mobilia *om. FP* propositio nona *U* nona

mg. hab. A post* mobilia *et mg. hab. B ante* Omnia / *post* Omnia *hab. U* aut / aut[2]: et *A* / *post* aut[2] *hab. F* etiam
228 tempore coniungentur *tr. U*
230 patet *tr. B ante* de (*linea 229*) / prime partis *tr. A* prime huius *B* / Et: cum *U*
230–31 per duodecimam...quod patet *rep. F* / incommensurabiliter: commensurabiliter *LP*
231 motis *om. L* motus *B* / quod: ut *B* / patet etiam satis: etiam satis patet *U* / etiam *om. L* / septimam: aliam *FP* 7 *A* / quia: quod *A*
232 A et *om. P* / et puncta *om. U*
234 Sed: et *B* / coniungi *tr. B post* tempore / solum semel: semel tantum *B*

from *d* by a distance that is incommensurable [to the whole circle]. It follows that no point in which *A* and *B* will conjunct is a point in which *B* and *C* will conjunct. Therefore, *A*, *B*, and *C* will never conjunct in the future.

For example, if it is assumed that *A* is moved twice as quickly as *B*, then, by the eleventh proposition of the first part, *A* and *B* will never conjunct except in point *d*. And since, by assumption, *B* and *C* have incommensurable speeds, it follows, by the first proposition of this part, that they will never conjunct in point *d*. Hence *A*, *B*, and *C* will never again be in conjunction at the same time; and the same argument is applicable to the past and to more [than three] mobiles.

Proposition 9. Through an eternal time, three or more mobiles will either never conjunct, conjunct only once, or conjunct an infinite number of times.

If these mobiles were moved with commensurable velocities, it is possible, by the twelfth proposition of the first part, that they might never conjunct. And the same thing could happen if their velocities were incommensurable. This is obvious by the seventh proposition of this part, since it is possible that the one or more points of conjunction of *A* and *B* are not common to the points of conjunction of *B* and *C*. However, the two immediately preceding propositions have demonstrated that these mobiles might conjunct only once through all eternity. And [finally, the possibility] that they might conjunct an infinite number of times is made clear for commensurable velocities in the fourteenth proposition of the first part, and for incommensurable velocities in the sixth proposition of this part.

But through all eternity such mobiles could not conjunct exactly twice, or three times, or any other number of exact times. This can be shown in the following way. Assume that *A*, *B*, and *C* are in point *d*, and that afterward they conjunct in a second point, *e*. Also let *K* be the interval of time between the occurrence of these two conjunctions; and let *g* be another point whose distance from point *e* is equal to the distance of *e* from point *d*. Since each of these mobiles is moved uniformly, it follows that after the passage of an

235 est *om.* B *?tamen* U / *post* conclusiones *hab.* U omnes / autem *om.* B
236 possunt *AU* / de commensurabiliter: incommensurabiliter *A*
237 primi *L* / *post* prime *hab.* B partis / motis incommensurabiliter *tr.* BL
238 vero *om.* B / possunt *U* / vel: aut B sive *F* / prescise *U* / in toto eterno *tr.* B *post* coniungi / aut: sive *F*
239 aliquem alium *tr. FP* alium *AB* / *infra*

alium numerum *mg. hab.* A figuram / numerum: terminum *U* / et[1] *om.* B / et[2] *om.* B
240 d *om.* P / et[1] *om.* B / coniunguntur *P* / secundo: duo *BFLPU* / tempus K *tr.* A / K *tr.* B *post* coniunctiones (*linea 241*) / istas: has B
241 signeturque: signetur B signetur quod *P* / distet *A*
242 Cum: quod *A* / istorum: eorum B

tempus *K* ipsum *A* pertransiverit totum circulum totiens quotiens fecerat in
245   tempore *K* et tantum arcum cum hoc quantus est arcus *de*, videlicet arcum
eg. Ergo *A* tunc erit in puncto *g*. Et per idem probabitur propter regularitatem motuum quod *B* tunc erit in puncto *g*; et similiter *C*. Ergo in fine
tanti temporis sicut erat tempus *K* tunc iterum tertio coniungentur et
iterum quarto in fine tanti temporis et sic infinities.

250   Et si moveantur commensurabiliter hoc fiet in punctis finitis et in quolibet eorum coniungentur infinities. Si vero moveantur incommensurabiliter hoc erit in punctis infinitis et in quolibet coniungentur solum semel, ut
patet ex conclusionibus prius positis. Sic ergo mensura coniunctionum
trium vel plurium mobilium in toto tempore quodammodo transit de
255   unitate ad infinitatem ad unitatem omnes numeros refutando sicut etiam
tempus eternum non potest dividi in tempora infinita nisi una divisione;
nec potest dividi divisionibus finitis quin partes sint inequales; sed infinitis
bene potest.

   Conclusio decima. Si tria aut plura incommensurabiliter moveantur
260   *numquam essent ita propinqua quin aliquando sint propinquiora quantum-*
libet in infinitum.

   Signato namque quovis puncto, ut gratia exempli signato *d* puncto,
tempus in quo *A* de puncto *d* redit ad eundem infinities sumptum. Cum
sint tempora incommensurabilia per tales replicationes ubique secabunt
265   se invicem secundum ymaginationem et nihil remanebit indivisum secundum divisionem sic ymaginatam, ergo aliquando quando *A* erit in *d*
parvum tempus deficiet quin etiam *B* sit in *d*, ergo erunt satis propinqua.
Et adhuc aliquando minus tempus postea deficiet quin *B* sit in *d* per istam
incommensurabilitatem, ergo adhuc erunt propinquiora et sic in infinitum.

244 pertransivit *F* / *post* fecerat *hab.* *B*
      primo
244–45 in tempore *om.* *U*
245 tantum…hoc: cum hoc tantum arcum
      B / quantus: probatur *P* / est arcus *tr.*
      L / videlicet: scilicet *B* videt *P*
246 eg *corr. ex* dg *ABFLPU* / A tunc *tr.*
      A A *B* / probatur *U*
248 sicut: quantum *B* / tertio coniungentur
      tr. *U*
249 quarto: quatuor *P*
250 *post* si *hab.* *B* hoc / hoc: illud *F* / fieret
      P / in: ita *P*
251 vero: autem *B*
252 *post* quolibet *add.* *B* eorum / solum
      semel ut: semel tantum *B*

253 Sic: si *A*
254 vel: aut *B* sive *F* / in *om.* *A* / quodam-
      modo: quoque *P* / quodammodo tran-
      sit *tr.* *B*
255 ad unitatem *om.* *A* unitate *F* et *post*
      unitate *scr. et del.* *F* necessitate / refu-
      tando: ?reservando *L* / sicut: sint *A* sic
      B
256 tempus *om.* *U* / tempora infinita *tr.* *L*
257 quin: quando *P* / sed: licet *B* / infinitis:
      infinitum *U sed mg. hab.* *U* alter si in-
      finitis bene potest
258 potest: possit *B*
259 Conclusio decima *om.* *FP* propositio
      decima *U* decima *mg. hab.* *AB ante* Si
      tria / aut: sive *F* / *post* plura *add.* *F*

interval of time equal to K, mobile A will have traversed the whole circle a certain number of times plus an arc equal to *de*, namely, arc *eg*—i.e., it will have traversed the same distance that it regularly traverses in time K. [Mobile] A will then be in point g. By the same procedure, and because of the uniformity of the motions, one can prove that B, as well as C, will then be in point g. At the end of the next time interval equal to K, they will again, for the third time, be in conjunction; at the end of another such interval, [they will conjunct] a fourth time, and they will do this an infinite number of times.

Now if their velocities are commensurable, the points of conjunction are finite in number, but they will conjunct an infinite number of times in any such point. However, should their velocities be incommensurable, they will conjunct an infinite number of times, but only once in any particular point, as is evident by earlier propositions. Thus the measure of [the number of] conjunctions for three or more mobiles through an eternity of time moves from unity to infinity to unity by rejecting all numbers, just as an eternity of time cannot be divided into infinite times except by one division; nor can it be divided into finite divisions unless the parts are unequal; but it can be divided into infinite divisions.

Proposition 10. If three or more mobiles are moved with incommensurable velocities, they could never be so close but that at some other time, they could not be even closer; and so into infinity.

After having assigned some point, say, d, it takes a certain time for A to return to d; and [A] does this an infinite number of times. Since wherever we might imagine the mutual separation [of these mobiles], the times of their circulations are incommensurable, and nothing will be left undivided by a division imagined in this way, then at some time when A will be in point d, B will also be in d a short time later, and they will be quite close. And yet, at some other time, in virtue of the incommensurability [of their motions], B will arrive at d in an even shorter interval of time after [A has been there], and they will be even closer together, and so on into infinity. The relationship

mobilia / moveantur *om. P*
260 essent: erunt *BLU* / aliquando *om. L* / sint: erunt *L*
260–61 quantumlibet: quamlibet *P*
261 in *om. P*
262 quovis puncto: coniunctionis puncto *A* puncto coniunctionis *B* conclusionis puncto *P* / ut *om. BL* / signato d puncto: d *B*
263 sumptis *U*
264 sit *F* / tempora incommensurabilia *tr. F*

265 invicem *om. A* / nihil: ipse *U*
266 sicut *U* / quando: quia *A*
267 quin: quando *AP* / *post* d hab. *L* per istam incommensurabilitatem / erunt: essent *FP*
268 ad *U* / aliquando: quandoque *F* / minus tempus postea: postea minus tempus *B* in postea minus tempus *U* / per: propter *A*
269 incommensurabilitatem: incommensurationem *P* / erunt: essent *F*

270   Et eodem modo diceretur de *C* mobili respectu utriusque istorum sigillatim
et ita de quotlibet mobilibus. Ergo non erunt ita propinqua quin adhuc
sint propinquiora in futurum. Et ubique per totum circulum erit talis
approximatio eo modo quo dictum est de coniunctione duorum mobilium
in quarta conclusione huius partis. Sic itaque omnia commensurabiliter
275   mota habent certam distantiam sue approximationis et sepe sunt ita
propinqua quod numquam possunt plus appropinquari; adhuc loquendo
de illis que numquam coniungentur omnia simul ut patuit per tertiamdecimam prime huius. Sed de motis incommensurabiliter, ut nunc dictum est,
non oportet.

280   Posito ergo quod omnium aut plurium planetarum motus sint incommensurabiles, scilicet quilibet motus cuilibet aut saltem quod nulli tres
motus sint invicem commensurabiles quamvis essent commensurabiles
bini et bini, dico ergo quod necesse est si ita sit illos planetas in eodem
gradu aliquando convenire, et aliquando in eodem minuto, et quandoque
285   in eodem secundo, et tertio, et quarto, et sic in infinitum approximando et
tamen numquam punctualiter coniungentur. Et adhuc hoc sequitur de
quolibet gradu celi, minuto, secundo, tertio, quarto, et cetera, ita quod
aliquando erunt in primo minuto Cancri et aliquando in primo secundo
Capricorni; et sic de aliis.

290   *Conclusio undecima. Que de coniunctionibus duorum aut plurium mobili-*
um dicta sunt pari ratione intelligenda sunt de omni alio aspectu seu modo
se habendi.

      Quod docet vicensimaprima prime de motibus commensurabilibus
idem proponit de incommensurabilibus presens conclusio. Si enim duo
295   mobilia nunc sunt coniuncta numquam alias in puncto eodem coniungentur, ut dicit prima conclusio huius. Ita per idem demonstrabitur quod si

---

270 Et *om. B* / dicetur *U* / sigillatum *A*
271 et *om. B* / *post* ita¹ *hab. B* diceretur /
   quolibet *A* / Ergo: et ideo *U*
272 in futurum Et *om. U* / per totum *rep.*
   L / erat *B*
273 eo: eodem *P* et *U* / quo: quod *AP* / dictum est *tr. F* tantum est *A*
274 conclusione *om. B* / *ante* commensurabiliter *scr. et del. B* in
275 approximationis: appropinquationis *U*
276 numquam: non *B* / appropinquare *U* /
   ad *U*
277 illis: istis *P* / omnia *om. L* / patet *ALU* /
   per *om. L*
277–78 tertiamdecimam: 13 *A*

278 primi *FLP*
278–79 ut…oportet: non oportet ut dictum
   est nunc *B*
280 Posito: pono *U* / omnium aut plurium:
   plurium aut omnium *A* / sint: sicut *P*
281 quelibet *P* / cuiuslibet *AP*
282 motus *om. A*
283 *ante* bini¹ *hab. U* invicem / planetas *obs.*
   B
284 et¹ *om. L* / aliquando²: quandoque *F* /
   quandoque: aliquando *L*
285 secundo *om. A* / et¹ *om. AB* / tertio: 3
   A
286 hoc: illud *F*
287 secundo: 2 *A* / tertio: 3 *A* / *ante* quarto

of mobile *C* to each of the others singly can be treated in the same way; and the same may be said for any number of mobiles. They will never be so close, but that they could not approach even closer in the future. The kind of approximation described for the conjunction of two mobiles in the fourth proposition of this part will occur everywhere on the circle. And so all mobiles moving with commensurable velocities have a fixed distance of proximity and are often so close that they can come no closer. Furthermore, in speaking of commensurable motions, the mobiles will never all conjunct simultaneously as was made clear in the thirteenth proposition of the first part. But this does not hold for mobiles moving with incommensurable velocities, which is what we are considering here.

Let us assume that the motions of all, or most, of the planets are incommensurable—i.e., any motion is incommensurable to any other motion, or, at the very least, no three motions are mutually commensurable, although they could be commensurable if taken two at a time. I say, then, that these planets would, at some time, arrive in the same minute, and at some time in the same second, and at some time in the same third, or fourth, and so on, approaching closer and closer into infinity, and yet they will never conjunct exactly. And this applies to any degree, minute, second, third, fourth, etc., of the sky, so that at some time they will be in the first minute of Cancer, and at some other time in the first second of Capricorn; and the same may be said of the other subdivisions.

Proposition 11. By a parity of reasoning, it must be understood that what has been said about the conjunctions of two or more mobiles applies to every aspect or relationship.

What the twenty-first proposition of the first part shows about commensurable motions, the present proposition proposes to show for incommensurable motions. The first proposition of this part shows that if two mobiles are now in conjunction, they will never conjunct in that same point at other times. The same proposition can be utilized to demonstrate that if they were in opposition,

---

add. FLPU et

288 aliquando[1]: quandoque *F* / erunt: essent *F* / minuto: puncto *B* / aliquando[2]: quandoque *F* / secundo: 2 *A*

289 capricornis *B* / *post* aliis *hab. B* punctis

290 Conclusio undecima *om. AFP* propositio undecima *U* undecima *mg. hab. B ante* Que / aut: sive *F*

291 alio *om. A* / seu: sive *F* / *post* seu *hab. B*

dispositione vel

293 *ante* vicensimaprima *hab. L* undecima / *post* prime *hab. B* partis / motibus: mobilibus *A*

294 *ante* idem *hab. U* et / *post* presens *hab. B* hec / enim: vero *F* tamen *U*

295 *ante* nunc *hab. B* huiusmodi / alia *P*

296 conclusio *om. ABU*

296–97 quod...opposita *om. U*

sint opposita vel quomodolibet aliter numquam consimiliter sc habebunt ipsis existentibus in eisdem punctis in quibus sunt. Et conformiter dicendum est de conclusionibus quarta, quinta, sexta, septima, et etiam de
300 octava. Unde talia tria vel plura in quolibet instanti sic se habebunt quod numquam ita se habuerunt alias nec umquam poterunt alias taliter se habere.

Et de nona similiter quoniam duo esse coniuncta et reliquum eisdem opponi. Aut accidit solum semel in toto eterno tempore, scilicet si motus
305 sint incommensurabiles; aut infinities, scilicet si sint commensurabiles; aut numquam. Et ita dicetur de aliis modis se habendi.

Supposita namque incommensurabilitate motuum et eternitate, pulchrum est considerare qualiter talis constellatio sicut esset una coniunctio punctualis evenit semel solum in toto tempore infinito et quomodo ab
310 eterno futura erat necessario pro hoc instanti nulla simili precedente aut sequente. Nec est querenda ratio quare magis evenit tunc quam alias nisi quia tales sunt velocitates motuum et immutabiles voluntates moventium.

Et si constellationes sint cause inferiorum effectuum continue erit talis dispositio quod numquam erit similis in hoc mundo cumque notabiles
315 aspectus respiciant totam unam speciem. Non videtur inopinabile, loquendo naturaliter, quod una magna coniunctio planetarum cui numquam fuit similis producat aliquod individuum cui non fuerit simile in specie et quod incipiat esse nova et invisa prius species aut substantie aut accidentis sicut dicit Plinius de egritudine libro vicensimosexto quod "sensit facies homi-
320 num novos et omni evo…incognitos…morbos." Et forte possible est quod talis species incepta numquam desineret si mundus perpetuaretur aut

---

297 vel: aut *F* / quomodolibet: quolibet *AFU* / alter *AP*

298 ipsis: ?in primis *P*

298–99 dicendum est: dicemus *L*

299 *ante* conclusionibus *scr. et del. U* punctus / quarta…septima: 4 5 6 7 *A* / sexta: 6 *F* / et: aut *U* / de² *om. BU*

300 octava: 8 *A* / vel: sive *F*

301 numquam…alias¹: numquam alias sic se habuerunt *A* numquam alias ita se habuerunt *B* ita numquam se habuerunt alias *L* numquam ita se habebunt alias *P* numquam se habuerunt ita alias *U* / *ante* alias² *hab. U* se / taliter se *om. U* sic se *A*

302 habere *U sed mg. hab. U* vel in duo etiam consimiliter / *post* habere *scr. et del. U* de ceteris

303 Et de nona *om. U* et ita nona *F* / *post* nona *add. B* conclusione / similiter: consimiliter *U* / coniuncta: puncta *A* / eiusdem *P*

304 Aut: ut *F* / solum semel *om. B sed hab.* solum semel *post* tempore / in *om. A*

305 infinities…commensurabiles *om. L* / scilicet *AFPU*; *om. B* / *ante* commensurabiles *scr. et del. B* in

306 aut numquam *om. A* / Et *om. B* / diceretur *AFU* / *post* aliis *hab. B* dispositionibus vel / modum *B*

307 eternitatem *P* / *post* eternitate *scr. et del. B* motuum *et add.* temporum

309 eveniet *A* et erit *P* / solum: tantum *B* / toto *om. U*

310 erat: eveniat *F* / pro hoc instanti: ad hoc instans *A* / *post* precedente *hab. B* ali-

or in any other relationship, they will never enter into the same relationship in the very same points where they now are. And this reasoning can be extended to the fourth, fifth, sixth, seventh, and also the eighth propositions [of this part]. Thus three or more such mobiles will, in any instant, be related as they never were related before, and as they never will be related [in the future].

A similar extension can be made for the ninth [proposition of this part] since two of the mobiles could be in conjunction with the remaining one in opposition to them. Then, either this [opposition] occurs only once through all eternity— i.e., if the motions are incommensurable—or, if the motions are commensurable, it occurs an infinite number of times, or it never occurs. The same reasoning could be applied to the other kinds of relationships.

On the assumption of the incommensurability and eternity of motions, it is truly beautiful to contemplate how such a configuration as an exact conjunction occurs only once through all of infinite time, and how it was necessary through an eternal future that it occur in this [very] instant with no conjunction like it preceding or following. One cannot even find a reason as to why it happens at that time [rather] than at another time, unless it be because the velocities of motions and the unalterable inclinations of moving bodies are [simply that way].

And if [celestial] configurations are the continuous causes of effects in the lower regions, then, whenever extraordinary aspects concern a whole species, there would occur such a disposition that never again will there be one like it in this world. Speaking naturally, it does not seem inconceivable that a great conjunction of the planets, different from anything that happened before, could produce some individual unlike any other which would begin as a new and previously unseen species in either substance or accident, just as Pliny, in the twenty-sixth book [of his *Natural History*], says that with regard to sickness "the face of man has been afflicted with new and unknown diseases in every age."[1] And, if the world were eternal, perhaps it is even possible that once such a species has come into being, it would never cease to exist; or it might

---

quando / aut: vel *B* nec *L*
311 *ante* Nec *mg. hab. B* notanda
312 *post* tales *hab. F* res / immutabiles: in-commensurabiles *A* / voluntates: velo-citates *A*
313 sunt *A* / *ante* effectuum *add. B* actium vel / erit talis *tr. B*
314 quod: que *P* / similis *om. U* simul *P* / cumque: quandoque *A*
315 aspectus: respectus *P* / respicient *A* aspi-ciant *LU* / inoppinabilia *P*

316 una *om. A* / coniunctio: ?*P* / fuit: erit *L* / *post* fuit *add. B* aut erit
317 *post* individuum *hab. U* in specie / non: numquam *AU* / fuerat *A*
318 *post* esse *hab. B* et / invisa: non visa *B* / prius species *tr. U* presens species *FLP* species *B* / substantia *U*
319 vicensimosexto: 26 *AF* / facies: sciens *F* / facies *U sed mg. hab. U* alibi sensus
320 novas *A*
321 *ante* incepta *hab. U* semel / aut *om. U*

quod aliquando desineret virtute alterius constellationis. Et sic de similibus corollariis que ex dictis possunt elici.

    Conclusio duodecima. De eodem mobili quod pluribus motibus movetur
325   *enunciare consimilia prius dictis.*

    Ita quod sicut dictum est de motibus commensurabilibus in vicensima-
secunda conclusione prime huius sic in proposito intelligendum est de motibus incommensurabilibus. Verbi gratia, posito quod sol moveretur solum
duobus motibus incommensurabilibus et perpetuis ut motu proprio et
330   diurno, sit itaque *A* primus punctus Cancri qui unum eundemque circulum
cotidie describit, videlicet tropicum estivum; et sit *B* centrum corporis
solis quod uno anno describit eclipticam. Sintque *A* et *B* in puncto *d* ymaginato in spatio immobili, et hoc est *A* et *B* esse coniuncta.

    Dico igitur quod *A* et *B* numquam erunt simul alias in puncto *d* quod
335   probabitur omnino similiter sicut probata est prima conclusio. Si ergo
sol intret primum punctum Cancri in aliquo meridiano numquam alias
intrabit, nec intravit, hoc signum ipso existente in isto eodem meridiano
nec etiam ipso existente in meridiano distanti ab isto commensurabiliter ut
probatur per secundam huius. Sint quoque infinita puncta in isto circulo
340   *A* ubique dispersa in quorum quolibet *B* existens intravit Cancrum et alia
infinita in quorum quolibet *B* existens intrabit in posterum idem signum
ita quod nulla est huius circuli tam parva portio in qua non sit aliquis
meridianus talis quod *B* existente in eo ipsum *B* erat in primo puncto
Cancri et in qua non sit aliquis meridianus talis quod *B* aliquando existens
345   in eo erit in primo puncto Cancri sicut de coniunctione dictum est in quarta

322 aliquando: quandoque *F* / alterius: ali-
    cuius *B* / sic *om. FP*
322–23 similibus: consimilibus *L*
323 corollariis: ?corrogatis *P* / que: quem
    P / ex…elici: possunt elici ex dictis *B*
324 Conclusio duodecima *om. AP*; *tr. F*
    propositio duodecima *U* duodecima *mg.*
    hab. B ante De / motibus movetur *tr. U*
325 enunciare…dictis: consimilia prius dic-
    tis enunciare *B* enunciare similia prius
    dictum est *U et post* enunciare *mg. hab.*
    B de motis pluribus motibus *et post*
    similia *mg. hab. U* 12
326 Ita quod *tr. U* / dictum est *tr. F*
326–27 vicensimasecunda: 22 *A*
327 conclusione *om. ABPU* / huius: partis
    B / intelligendum est *tr. F*
328 sol: so *P* / moveretur solum *tr. U* mo-
    veatur solum *B*

329 ut motu: aut motu *A* scilicet *B*
330 sit itaque A: sitque A *A* sit *B* / primus
    punctus: in primo puncto *A* / *ante* qui
    hab. B *A* / eundemque: et eundem *AU*
331 cotidie *om. U* / videlicet: scilicet *U* /
    estivum: estivalem *U* / B *tr. B post* solis
    (*linea 332*)
332 eclipticam: eclipsicam *P* / *post* eclipti-
    cam *hab. A* motu proprio / Sintque:
    sint itaque *A* sint etiam *B*
333 immobili *U sed mg. hab. U* alibi in-
    commensurabili
334 igitur *om. A* / erunt: essent *F* / simul
    alias: alias in simul *A* / *ante* quod *hab. F*
    meridiano distantia *et ante* meridiano
    scr. et del. F nec ipso existente in
335 probatur *BF* / omnino: animo *P* / *post*
    conclusio *add. B* huius
336 intraret *P* / cancrum *P* / in *om. P*

at some time cease to exist because of the power of another configuration. Similar things may be said about similar corollaries that are deducible from what has been said.

Proposition 12. Propositions enunciated previously [for two or more distinct mobiles] will now be applied to [one and] the same mobile moving with several [simultaneous] motions.

Now just as in the twenty-second proposition of the first part of this work, the motions discussed were commensurable, so in this proposition they are to be taken as incommensurable. For example, assuming that the sun is moved with only two incommensurable and perpetual motions—namely, proper and diurnal—let *A* be the first point of Cancer which daily describes one and the same circle, namely, the summer tropic; and let *B* be the center of the sun's body which describes the ecliptic in one year. And [finally] let *A* and *B* be in point *d* imagined as motionless in space—i.e., *A* and *B* are in conjunction there.

I say, then, that *A* and *B* will never be simultaneously in point *d* at other times, and this can be demonstrated exactly as was the first proposition [of this part]. Thus, if the sun should enter the first point of Cancer on some [particular] meridian, it will never enter, nor has it ever entered, this sign at other times when on this same meridian; nor, indeed, as shown in the second proposition of this part, will it enter when it is on any meridian separated from this one by a distance that is commensurable [to the whole circle of the summer tropic]. There is also an infinite set of points scattered everywhere on circle *a* in each of which *B* has entered Cancer; and there is [also] another infinite set of points in each of which *B* will enter this same sign in the future. Then, as was the case for conjunctions in the fourth proposition of this part, no part [or arc] of this circle is so small that it does not contain some meridian on which *B* appeared [in the past] when it was in the first point of Cancer; nor can it be so small that it does not contain some meridian on which *B* will appear in the first point of Cancer sometime [in the future]. One should

---

337 nec intravit *om. F* / *post* ipso *hab. B* sole sed *mg. hab.* vel signo / in isto eodem *om. A* in eodem *L* in isto eadem *P* in isto *U*

338 *ante* existente *scr. et del. B* distante / in *om. P* / distante *AB* / commensurabiliter: commensurabile *A*

339 probabitur *A* probabatur *U* / sunt *FLU* / puncto *U* / isto *om. A*

340 *infra* existens *hab. B figuram*

341 posterum: futurum *B* / *post* signum *add. B* cancri

342 portionis *A* / aliquis *om. U*

343 meridianus talis *tr. U*

343–44 existente...quod B *om. U*

344 *ante* aliquando *hab. L* existente

344–45 et in qua...Cancri *rep. F sed hab.* quandoque *in loco* aliquando

345 puncto *om. P* / dictum est *tr. F* dictum *BU*

conclusione huius et sicut dictum est de primo puncto Cancri ita intelligendum est de quolibet puncto zodiaci.

Ex quo sequitur quod omni die *B* describit unam novam spiram in spatio ymaginato immobili quam numquam alias descripsit et viam percurrit quam numquam alias peragravit. Et sic suo vestigio seu ymaginato fluxu prolongare videtur lineam girativam iam infinitam ex infinitis spiris in preterito descriptis confectam quandoque de tropico ad tropicum quasdam spiras describit et iterum alias revertendo que priores intersecant et econverso. Et ergo in quolibet puncto harum intersectionum bis existit *B* in toto eterno et in quolibet aliorum aut semel solum aut numquam.

Secundum hanc igitur ymaginationem totum celi spatium inter duos tropicos exaratur ab ipso *B* derelinquendo ex istis girationibus figuram velut opus texture aut rethis per totum istud spatium expanse. Et huiusmodi textura iam tempore preterito perpetuo fuit in infinitum inspissata et tamen adhuc continue inspissatur eo quod fit cotidie nova spira. In toto vero huiusmodi spatio non possunt signari aliqua duo puncta quin inter illa infinities fuerit *B* et nihilominus infinita sunt in eodem spatio in quibus numquam fuit nec erit; ac etiam alia infinita in quorum quolibet in tota eternitate habet esse solum semel et alia infinita in quorum quolibet habet esse bis in toto tempore infinito et nullum est in quo possit esse pluries.

Rursum statimque sequitur ex predictis si *B* nunc sit in primo puncto Cancri et in isto meridiano quod *B* numquam fuit ita prope zenith istius horizontis, nec erit in perpetuum sicut nunc est; et ita de remotione ipsius in primo puncto Capricorni.

Item si moveatur secundum circulum eccentricum vel epiciclum ratione huius describit spiras suas appropinquando ad centrum mundi et aliquan-

---

346 conclusione *om.* *B* / huius *om.* *U* / dictum *om.* *A* / est *obs.* *B*

348 *infra* sequitur *mg.* *hab.* *L* *figuram* / omni…describit: B describit omni die *B* / speram *U* *sed* *mg.* *hab.* *U* alibi spiram

349 spatio: puncto *L* / ymaginato: quanto *U*

350 percurrit: decurrit *U* / numquam alias *tr.* *B* / sicut *A* / seu: sive *F*

350–51 *ante* ymaginato *hab.* *U* suo / ymaginato fluxu *tr.* *BF*

351 lineam: licitam *P*

352 spiris: spheris *U* *sed* *mg.* *hab.* *U* alibi spiris / confectam: condectam *P*

353 quasdem *L* / speras *U* / et *om.* *LU* / iterum: interim *U*

354 punctorum *A* / harum: istarum *F* / *ante* intersectionum *hab.* *U* ?coniunctionum

354–55 bis existit B: B bis existit *A*

355 toto *om.* *B* / aliorum: alio *F* / solum: tantum *B*

356 igitur *om.* *P*

357 derelinquendo: ?relinquendo *B* delinquendo *FP* / istis: ipsis *A*

358 istud: illud *AF*

358–59 rethis…huiusmodi *om.* *P* / huiusmodi: huius *F*

359 texturam *BF* / inspissata: inspirata *B* inspissatum *U*

360 continue: cotidie *B* / inspissatur: inspiratur *B* / fit: sit *FLU* / cotidie: continue *AU* / nova spira: in nona spera *U*

understand that what has been said about the first point of Cancer applies also to any point of the zodiac.

It follows from all this that *B* describes daily a new spiral—imagined as motionless in space—which it has never before described; and it traverses a path which it has never before traversed. And so, by its track, or imagined flow, *B* seems to extend a spiral line that has already become infinite [in length] from the infinite spirals that were described in the past. This infinite spiral line is constituted from the spirals that were described between the two tropics, and from the intersections between the earlier and later spirals. Consequently, through an eternal time, *B* appears twice in any point of intersection [between these spiral lines]—and only once, or never, in any of the other points.

In accordance with what has been imagined here, the whole celestial space between the two tropics is traced by *B*, leaving behind a web- or net-like figure expanded through the whole of this space. The structure of this figure was already infinitely dense through the course of an infinite past time, and yet, nonetheless, it will be made continually more dense, since it produces a new spiral every day. Throughout the extent of such a space, *B* has appeared between any two assignable points and yet, despite this, in the very same space [between any two such points] there are an infinite number of points in which *B* never was, nor ever will be. There is also another infinite [set of points] in each point of which *B* enters only once through all eternity; and there is yet another infinite set of points through each of which *B* passes twice during the course of an infinite time; but there is no point through which it could pass more than twice.

And, again, it follows directly from what has been said that if *B* were now in the first point of Cancer, and on this meridian, it was never so near the zenith of this horizon; nor will it ever again be as it is now. The same may be said about its removal into the first point of Capricorn.

Now if *B* were moved on an eccentric circle, or on an epicycle, it would, in virtue of this [circumstance], describe its spirals by sometimes approaching,

---

361 huiusmodi: huius *A*
362 illa: ita *A* ista *FP* / infinities *om. U* / *ante* infinita *hab. B* qui
363 nec: aut *B* / ac: et *B* nec *?L et U sed mg. hab. U* alibi ut / etiam *om. L* et *P* / alia: talia *P* / toto *B*
364 eternitate: eterno *B* eterni *P*
364–65 habet esse bis: bis habet esse *B*
365 posset *U*

366 rursus *U* / statim *ABP* / nunc sit *tr. U*
367 zenith: cenith *ABFP* zeniht *U*
368 horizontis: orizontis *ABL* orisontis *FP* / nunc: inter *P* / ipsius: eius *A* cuius *B*
369 primo *om. A* / capricornis *B*
370 circulum *om. B* / epysiclum *B* episciclum *F* epyciclum *L*
371 speras *U*
371–72 aliquando *om. B* quandoque *F*

do recedendo propter quod in quocumque meridiano *B* existat numquam
alias erat precise tantum distans a centro mundi ipso existente in eodem
meridiano sed semper plus aut minus. Similiter si semel fuerit in longitu-
375   dine propiori ipso existente in aliquo meridiano, semper alias in perpetuum
erit remotius a centro mundi ipso stante in eodem meridiano.

Adhuc autem cum ipsum *B* describat continue novam spiram necesse
est quod punctus oppositus, qui dicitur nadir solis, describat cotidie conformiter novam spiram igitur et conus umbre terre peragrat continue
380   novam viam. Et ex hoc fit aliqua pars celi tenebrosa que numquam alias
secundum se totam fuit lumine solis privata. Quod etiam sequeretur ex
eccentricitate quoniam si *B* est aliquando ita prope centrum mundi quod
nulla alia vice poterit esse tam prope nec potuit ipso existente in eodem
meridiano, ut statim dictum est, oportet quod tunc conus umbre terre
385   protendatur in celo ita longe quod numquam alias in parte illa longius
protendetur. Et hoc est possibile non obstante quod orbis lune aliter
moveatur quam orbis solis, sive commensurabiliter sive incommensurabiliter.

Adhuc autem ex predicta incommensurabilitate contingeret quod annus
390   solaris medius contineret aliquot dies et portionem diei incommensurabilem suo toti. Quo posito, impossibile est precisam anni quantitatem
numeris deprehendere aut perpetuum almanach condere seu verum kalendarium invenire.

Consequenter transeundum est ad illud mobile quod pluribus motibus
395   moveretur. Et exempli causa ponatur quod luna moveatur tantum tribus
motibus, scilicet motu diurno et motu in eccentrico et motu in epiciclo et
quod isti duo sint incommensurabiles. Tunc iuxta octavam et undecimam
conclusiones huius partis diceretur quod luna semper taliter se habet
respectu centri terre et taliter illud respicit quod impossibile est alias

372 recedendo: retrocedendo *A* / quod:
    quot *P* / quocumque: quoque *P*
373 precise *om. U*
375 propiori: propinquiori *A* longiori *B*
376 stante: existente *A*
377 cum ipsum: ipsum *A* cum *B* / describat
    continue *tr. L*
378 est *om. F* / *post* punctus *hab. B* ei / op-
    posito *A* / qui dicitur *om. FLP* qui dice-
    tur *A* / *post* nadir *hab. F* vadaxit / sol *A*
378–79 cotidie conformiter *tr. A* confor-
    miter *FLPU*
379 igitur *om. U* / *post* et *hab. U* cotidie
380 celi *om. A* circuli *U*

381 sequitur *BU*
382 est aliquando *tr. F* / mundi *om. U*
383 tam: ita *U* / nec: ut *U* / *ante* potuit *hab.*
    B aliquando / patuit *U*
384 *infra* meridiano *mg. hab. F* opportet /
    statim: iam *B* / dictum *om. A* / quod
    tunc *tr. B* / terre: tempore *A*
385 protendatur: procendatur *L* / *ante* longe
    scr. et del. A quod / illa: ista *A*
386 protenderetur *AL* protendatur *F* pro-
    tendentur *P* / obstante: ?obstans *U*
387–88 sive incommensurabiliter *om. P*
389 autem: aliter *A* / incommensurabilitate:
    commensurabilitate *U*

and sometimes receding from, the center of the world. For this reason, on whatever meridian *B* might appear, it was never at any other time exactly that far away from the center of the world when on the same meridian, but always more or less remote from the center. Similarly, if it were once in perigee on some meridian, it would always, and forever, be more distant from the center of the world when it lies on that same meridian.

And yet, when *B* has continually described a new spiral, the opposite point, called the nadir of the sun, must necessarily describe daily a new spiral, so that the cone of the earth's shadow would continually traverse a new path. And because of this, some part of the heavens would become darkened which was never at any other time wholly deprived of the sun's light. This same result would also follow from the eccentricity [of the circle] because if *B* is once so near the center of the world that it can get no closer at any other place, not even on the same meridian, as was just said, the cone of the earth's shadow would then be extended so far into the heavens that it could never be extended farther at any other time. And this is possible notwithstanding the fact that the moon's orb is moved—whether commensurably or incommensurably—otherwise than that of the sun.

Furthermore, from the incommensurability mentioned above, it might be [the case] that the mean solar year contained a certain number of days and a part of a day incommensurable to a whole day. If this were assumed [to be true], the precise length of the year could not possibly be expressed in numbers, nor could a perpetual almanac be established, or a true calendar found.

We must now move on to consider a mobile that is moved with more [than two] motions. By way of example, let us suppose that the moon is moved with only three motions—namely, a diurnal motion, motion on its eccentric circle, and motion on its epicycle; and let the [latter] two motions be incommensurable. Then, by the eighth and eleventh propositions of this part, it could be said that the moon is always related to the center of the earth in a way that

---

390 solaris: solis *B* / aliquos *P* / portioni *P*
390–91 incommensurabilem: incommensurabilitate *P*
391 precise *B*
392 perpetuum almanach *tr. B* / condere: quidem *P*
392–93 kalendario *P*
394 illud *om. B*
395 moveretur: movetur *BU* / moveatur

tantum *tr. U*
395–96 moveatur...motibus: movetur 3 motibus tantum *B*
396 motu¹ *om. B* / et motu² *om. B* / eccentrice *P* / motu³ *om. B* / in *om. FL* / epyciclo *BL* episciclo *FP*
397 sunt *AB* / octavam: 9 *A*
398 conclusiones *om. B* / partis *om. B* / dicetur *A* / habeat *A* haberet *U*

400 ipsam consimiliter se habere. Item sit *A* centrum corporis lune, et *B*
centrum epicicli, et *c* mundi, et *d* sit punctus in celo superpositus longitu-
dini propiori; fiatque linea *cd* circumscribatur que motus diurnus; sint
igitur *A* et *B* nunc in linea *cd*. Ergo eodem modo quo probata est prima
conclusio huius probabitur quod numquam *A* et *B* convenient simul in
405 linea *cd* quoniam quotienscumque *A* erit in linea *cd* a parte ipsius *c*, tunc
ipsum *A* fecerit aliquot circulationes precise circa suum parvum circulum;
et similiter quotienscumque *B* erit in linea *cd* fecerit aliquot circulationes
precise circa suum magnum circulum. Ergo motus essent commensurabiles
si aliquando post in eodem instanti essent in linea *cd*. Ergo si nunc luna
410 sit propinquissima centro mundi, scilicet in inferiori puncto epicicli in
longitudine propiori eccentrici, impossibile est ipsam alia vice fore aut
fuisse ita propinqua ipsi centro nec in eodem meridiano nec in alio. Igitur
si luna nunc sit opposita soli dyametraliter ipsa existente in inferiori
puncto epicicli et in longitudine propiori eccentrici et sole existente in
415 longitudine propiori eccentrici sui, eclipsis lune tanta est quod impossibile
est maiorem fuisse aut aliquando futuram esse in perhenni seculorum
tempore. Hec ergo erit eclipsis maxima que nunc prima visa est similem
nec habere sequentem.

Et conformiter dicendum est de aliis aspectibus solis et lune et si luna
420 moveatur quatuor motibus aut pluribus; et sic de aliis planetis aut mobilibus quibuscumque. Et sicut dictum est de quantitate anni solaris et de
kalendario solis ita intelligendum est de mense et kalendario lune et aliorum.

---

400 ipsam consimiliter *tr. U* / sit: si *P* / et
  om. U
401 epycicli *BL* episcicli *F* / *ante* et[1] *hab. B*
  cuius / superpositus: suppositus *B*
401–2 longitudinum *P*
402 que *om. A* quod *F* / motus: modus *P*
403 nunc *tr. AB ante* A et B / eodem: eo
  FL / quo: conclusio *P* sicut *U*
404 huius probabitur: hec probatur *U* / A et
  B *om. A* / convenient ?*BF* ?communi-
  cant *A*
405 tunc: cum *F*
406 ipsius *A* / fecit *BU* / aliquos *P* / *infra*
  suum parvum *hab. B figuram*
407 et *om. B* / *post* cd *hab. B* tunc B / fecit
  B / aliquos *P*
408 precise *om. U* / *post* circulum *hab. U*
  precise / esset *P*
409 post *om. P* / esset *A* / luna: linea *AF*

410 sit *om. A* / in[1] *om. AP* / puncto: parte *B*
411 fore: esse *B* fieri *U*
412 propinquam *FLU* / nec[1] *om. B*
413 luna nunc *tr. B* / dyametraliter: dyameter *P* / in *om. P*
414 puncto *om. B* / longitudine *om. B* /
  propiori *om. U* / *post* eccentrici *scr. et
  del. U* ?longiore
415 longitudine *om. B* / *ante* propriori *hab.*
  A longiori / tanta est *tr. F*
416 *post* maiorem *hab. B* aliquando / aliquando *tr. U post* esse; quandoque *F*
416–17 in perhenni...que *om. A*
417 erit eclipsis maxima: eclipsis est maxima
  B eclipsis maxima erit *U* / nunc: nec
  ALU / prima...similem: primam visa
  est similem *FL* primam similem visa est
  A prima similem visa est *B* priorem
  similem visa est *U*

cannot possibly be repeated at any other time. Let *A* be the center of the body of the moon, *B* the center of the epicycle, *c* the center of the world, and *d* a point in the sky superposed at the point of perigee; [now] draw line *cd*, which describes a daily motion; and assume that *A* and *B* are now on line *cd* [see Fig. 8 on p. 65; also chap. 2, n. 102 on p. 65]. In the same way as we proved the first proposition of this part, we shall now demonstrate that *A* and *B* will never meet simultaneously on line *cd*, for the frequency with which *A* will be on *c*'s side of line *cd* corresponds to a certain number of exact circulations around its small circle [or epicycle]. And, similarly, the frequency with which *B* is found on line *cd* corresponds to a certain number of exact circulations around its great circle [or deferent]. Now these motions would be commensurable if, at some later instant, they should be on line *cd*. Therefore, if the moon were now closest to the center of the world—i.e., on the lowest point of the epicycle in the shortest distance of the eccentric [i.e., at perigee]—it is impossible for it to be as close [in the future], or to have been as close [in the past], whether on this very same meridian, or on another. Thus, if the moon were now opposed diametrically to the sun, and located on the lowest point of the epicycle and in the shortest distance of the eccentric [i.e., at perigee] with the sun [also] appearing at the shortest distance of its eccentric [i.e., at perigee], then the moon would be eclipsed to such an extent that no greater eclipse could have been possible [in the past], nor could there be a greater one in the unending ages to come. This eclipse, which would have been seen now for the first time, will be a total [or perfect] eclipse, and none like it will follow.

Similar things could be said about other aspects of the sun and moon, even if the moon were moved with four, or more, motions; and the same may be said about any other planets or mobiles. And what has been said about the length of the solar year and the solar calendar also applies to the month and to the lunar calendar, as well as to other planets or mobiles.

---

418 nec habere *tr. B*
419 et lune *om. U* / et³: etiam *B*
420 et sic: et *AU* similiter *B* / *post* planetis *hab. B* dicendum est
420-21 mobilibus: motibus *A*

421 est *om. B*
422 *post* mense *hab. B* lunari / *post* et² *hab. B* etiam
422-23 *post* aliorum *hab. B* planetarum

Unde universaliter certum est quod nulli motus incommensurabiles
425   possunt per numeros adequari, nec est possible coniunctiones, oppositiones, et aspectus huiusmodi motuum tabulare. Hic incipit tertia pars huius
operis.

424 *ante* certum *hab. B* verum et / certum:
verum *L*

425 possent *U* / adequari: equari *U* / *ante*
nec *hab. B* ergo nec motus planetarum
equari / est possibile *tr. B* / *post* coniunctiones *hab. A* et *et hab. B* lunarum
aut aliorum planetarum aut

426 *ante* aspectus *hab. B* ceteros / tabularii
U / *post* tabulare *hab. B* hec et alia
similia sequuntur supposita incommensurabilitate omnium motuum celestis
aut aliquorum ex eis *et hab. L* utrum
motus celestes sint commensurabiles *et*
hab. P et cetera *et hab. U* explicit secunda pars huius tractatus

426–27 Hic…operis *om. FP* incipit tertia
pars huius operis *A* sequitur tertia pars
B incipit ultima ac tertia pars tractatus
de motuum commensurabilitate vel incommensurabilitate *U*

It is universally true, then, that no incommensurable motions could be equated [or compared] by the use of numbers; nor, with such motions, is it possible to make tables of conjunctions, oppositions, and other such aspects. The third part of this work begins here.

[Tertia pars]

Cum de tribus propositis duo utrumque pertransissem ex duabus ypothesibus contradictoriis utrumque conditionaliter concludendo quid sequitur si omnes motus celi sint invicem commensurabiles, et quid si aliqui sint incommensurabiles; restat tertium quod plus appetit intellectus, non plene quietatus, donec cathegorice sit conclusum. Et quousque suppositum fiat notum, scilicet an sint commensurabiles an non. Sed dum suspenso animo hoc expedire proponerem.

Ecce mihi, quasi sompniatori, visus est Apollo, musis et scientiis comitatus, meque talibus increpat verbis pessima inquit: "Tua est occupatio afflictio spiritus est et labor interminabilis, an nescis quod rerum mundi proportiones. Nosce precisas humanum transcendit ingenium. Quod cum de sensibilibus queris a sensibus debes incipere quibus nequit deprehendi precisio punctualis. Si enim excessus imperceptibilis, ymo minor pars quam eius millesima, equalitatem tollit et proportionem mutat de rationali ad irrationalem, quomodo motuum aut magnitudinum celestium punctualem proportionem poteris agnoscere? Hoc te docuerunt si eos legisti astrologi precipui. Ait enim Albategni auctoritate Ptholomei quod 'in tanti magisterii excellentia, tam nobili tam celesti, veritatem ad

1 [Tertia pars] *om. ABFLPU* Pulchra disputacio si omnes motus celi sint invicem commensurabiles an non *R*

2 tribus propositis: propositio tribus *U* / utrumque: utcumque *AR* utique *U* / duobus *A*

3 utrumque: utriusque *F* utrique *P* / conditionaliter concludendo *tr. L*

4 sequatur *FLP* sequetur *U* / omnis *ABFL* / celi: celestes *U* / sunt *ABFP* essent *U* / invicem commensurabiles *tr. L* commensurabiles adinvicem *U* / quid: quod *RU*

5 aliqui *om. U* / sunt *R* / restabat *A* restabit *R* / plus appetit *tr. U* / intellecto *U*

6 donec: quousque *L* / Et *om. AP*

7 notu *ABU*

8 *ante* suspenso *mg. hab. P* notandum per optimum processum si motus celestes sint commensurabiles vel non / hec *FLP*

9 sompniatori: sompnianti *LU* sompniatus *R* / est *om. A* / scientiis: sonis *R* factis *U sed mg. hab. U* scientiis

10 inquit Tua est: inquit est tua *AR* est inquit tua *B*

10–11 *post* occupatio *mg. hab. B* Apollo

11 est *om. B* / et *om. P* / interminalis *R* / *ante* quod *hab. L* inquit / rerum: corporum *U*

12 Nosce: noscere *FLP* nosse *U*

[Part III]

Since two of the three things that I proposed have [now] been completed by concluding each of them conditionally from two contradictory hypotheses —namely, [1] what follows if all the celestial motions are mutually commensurable, and [2] what follows if some are incommensurable—the third thing remains to be considered because the understanding, not yet fully satisfied, seeks more [discussion] until the issue can be terminated categorically. Until now it was assumed that one could determine whether or not the celestial motions are commensurable. But while our judgment is suspended, I propose to explain this.

As one having a dream, I saw Apollo accompanied by the Muses and Sciences. And he rebuked me with these words:

"Because you are ignorant of the ratios relating the things of this world, your occupation is an affliction of the spirit and an unending labor. You should understand that exactness transcends the human mind. But, since you inquire about sensible things, you ought to begin with the senses through which [however] exact precision is undetectable. For if an imperceptible excess— even a part smaller than a thousandth—could destroy an equality and alter a ratio from rational to irrational, how will you be able to know a punctual [or exact] ratio of motions or celestial magnitudes? The distinguished astronomers have shown this to you—if you have read them. Thus, with reference to the authority of Ptolemy, al-Battani says that '[even] on the excellence of such instruction, so noble [and] so heavenly, it is not possible for anyone to

---

13 Quod: ex *A* et *B* que *U sed mg. hab. U alibi quia cum vel ?quod ?tamen ?disce et alia verba illegibilia* / de: disce *U* / sensibilibus: sensibus *AFLPR* / queris *om. U* querit *AR* / sensibus: sensibilibus *B* sensu *R* / debet *AR* / incipere: incipe *B*

14 punctuali *A* / *ante* Si *hab. B* quia / enim *om. B* vero *F* tu *U sed mg. hab. U* alibi

sit enim

15 quam eius *tr. B* / proportionem mutat *tr. R*

16 *ante* aut *hab. B* ?etiam

17 punctualem: punctualiter *R*

18 eos *om. B* / enim *om. B* vero *F*

19 in: michi *P* / *post* excellentia *hab. B* et ordinatione

20   unguem comprehendere non est cuiquam possibile.' Cuius causam assig-
     navit postea dicens quod 'forte est aliquis motus celi qui nondum homini-
     bus innotuit' aut propter tarditatem aut propter ipsius ad aliquem alium
     motum vicinitatem. Sic ergo eorum proportio ignota est nec ad eius noti-
     tiam te perducet arismetica nec geometria, que cum ad sensibilia applican-
25   tur sensibilibus principiis innituntur. Talem quoque temeritatem incre-
     pabat Plinius dicens 'miror,' inquit, 'quo tendit improbitas cordis humani
     parvuli...ausi divinare solis ad terram spatia...ac si mundi...mensura
     veniat in digitos...et tamquam plane a perpendiculo mensura celi constet.'
     Unde presumptuosi sunt qui se iactant almanach perpetuum et verum
30   kalendarium tradidisse, cum tale quid forte invenire non sit possibile. Et
     pro certo impossibile est cognoscere se aliquid huiusmodi reperisse.''

     Tunc ego, bone pater mi, intelligo quod ad talia non acceditur humanis
     viribus nisi presupposito aliquo principio presentato. Et nunc scio quod
     iudicium sensus non attingit precisionem. Et si attingat adhuc ignorat se
35   iudicasse recte pro eo quod insensibilis additio vel subtractio mutaret
     proportionem et non variaret iudicium. Non ergo vane presumam mathe-
     matica demonstratione terminare predictum problema.

     Sed o dii immortales, qui noscis omnia, cur fecistis quod homines natura
     scire desiderant et fraudato desiderio vel frustrato nobis absconditis
40   optimas veritates? Nonne mentitus est Esyodus qui propter hoc ausu
     blasphemo vos invidos appellavit? Et nonne sibi alter poeta recte obviat
     dicens "haud hoc equidem invidia nec enim livescere fas est aut nocuisse
     deos?" Sic etiam divinus Plato asserit "ab optimo" inquit, scilicet deo,

20 Cuius causam: causam huius *B et post*
   causam *mg. hab. B* Albategni Ptholomei
20–21 assignat *B*
21 dicens *om. B* / quod: quia *BU* / celi:
   cause *P* / nundum *AB*
22 innotuit: ignotuit *F* / *infra* innotuit *mg.*
   hab. L figuram / *post* tarditatem *hab. B*
   eius et vice hominum brevitatem
23 motum: modum *P* / *post* motum *hab. B*
   aliquem / vicinitatem: venientem ?cir-
   cumferentiam *A* / ignota est: est nobis
   ignota *B*
24 perducit *P* deducet *R* adducet *U* / nec:
   aut *B*
24–25 applicatur *AB*
25 quoque: vero *F*
26 inquit *om. B*
27 parvuli *om. U* / *post* ausi *hab. B* sunt
   et post sunt *mg. hab. B* Plinius secundo /

divinare: durare *AP* / *ante* ac *hab. B* et
   cetera / ac: et *A*
28 a perpendiculo: appendibile *P* / mensura celi *tr. BL* mensatur celi *P*
29 se iactant *tr. L* se vocant *R* iactitant se
   U / *ante* almanach *mg. hab. B* notanda
30 *ante* tradidisse *hab. B* reperisse et *et*
   hab. R eternumque / quid: huiusmodi
   B / forte invenire *tr. ABU* forte immediate *L* / *ante* possibile *hab. B* numquam
31 cognoscere se *tr. U* cognoscere *B*
32 Tunc: tamen *ARU sed ante* tamen *hab.*
   R ergo ?sisce gradum nec ultra proce-
   de / *ante* bone *hab. U* bene / mi *om. A*
   iam *R* / accedit *A*
33 presupposito: suppositio *R* presentato
   U / principio *om. R* / presentato *om. U*
34 attingit *B*
35 iudicasse recte *tr. U* / recte: rome *P*

understand the truth exactly.'[1] He [i.e., al-Battani] later gave the reason for this, saying that 'perhaps there is some celestial motion that is not yet known to men,'[2] either because of its slowness, or because of its proximity to some other motion. And so the ratio of these motions is unknown, and neither arithmetic nor geometry will lead you to a knowledge of it. For when these [i.e., arithmetic and geometry] are applied to sensible things, they are based on sensible principles. Pliny also rebuked such rashness saying: 'I marvel at how the meager human spirit...has dared to divine the distance between the sun and the earth... as if the measure of the world could be reduced to inches..., and as if the measure of the heavens could be determined from a plumb line.'[3] Indeed, there are presumptuous men who boast that they have produced a perpetual almanac and a true calendar when it is not even possible to find such a thing. And, surely, it is impossible for him [even] to know that he had found something of this kind."

I understand, then, my dear father, that it is not given to human powers to discover such things unless furnished with some principle assumed beforehand. And now I know that the judgment of the senses cannot attain exactness. But [even] if the senses could attain such exact knowledge, one could not know whether he had judged rightly, since an insensible [or undetectable] addition or subtraction could alter a ratio but would not change the judgment. Therefore, I do not vainly presume that the aforesaid problem is solvable by mathematical demonstration.

But, oh immortal gods who know all things, why did you make the very nature of men such that they desire to know, and then deceive or frustrate this desire by concealing from us the most important truths? Did not Hesiod speak falsely when, because of this and with blasphemous daring, he called you hateful?[4] And did not another poet rightly oppose him, saying "Surely there is no hate—for it is not possible that gods are envious or do harm?"[5] Even the divine Plato declared that "there is no jealousy in the good," namely,

36 vane *om.* *AP* sane *F*

37 terminare: determinare *AR*

38 noscis: noscitis *F* / *ante* homines *hab.* *R* omnes

39 scire *om.* *LR* / desiderant: desidantur *R* / et: in *A* / *post* fraudato *hab.* *B* huiusmodi / vel frustrato *om.* *B*

40 *ante* veritates *hab.* *B* et nobiles / Nonne: numquam *R* / Esyodus: esydorus *A* / *post* ausu *hab.* *B* ?libro *et hab.* *U* temerario ac

41 Et *om.* *B* / sibi *om.* *U* / alter poeta recte: alter poeta *B* alter porta recte *P* recte alter poeta *R* / obviat: obviavit *R*

42 haut *LPR* hanc *AF* / *post* haud *mg. hab.* *B* esyodus / hoc *om.* *U* hec *R* / equidem: quidem *ABFLPR*

43 deos *om.* *F* eos *P* / Sic etiam: sic enim *FL* sicut enim *P* et *A* / divinus: divus *B* divinas *U* / *post* Plato *mg. rep.* *B* Plato / inquit scilicet *tr.* *P* inquit *FL* / *ante* deo *hab.* *U* a / deo *om.* *FLP*

"longe invidia relegata est…cunctaque sui similia prout cuiusque natura
45   capax beatitudinis esse poterat effici voluit"; de quo Boetius "forma boni
livore carens," et cetera.

Si itaque per inventionem nostram multa scire non possumus, oro ego
ut hoc unum dubium per vestram doctrinam mihi de benigna gratia reseretur. Tunc Apollo, subridens, musas et scientias circumstantes aspexit
50   precipitque eis dicens "docete hunc quod ipse petit." Et mox Arismetica
dixit: "omnes motus celi sunt commensurabiles." Surgens, Geometria
contradixit eidem et ait: "ymo aliqui sunt incommensurabiles." Factaque
quasi litis contestatione, iussit Apollo utramque partem rationibus defendere suam causam. Meque admirante et auscultante, Arismetica prior
55   inquit:

[Oratio Arismetice]

O veritatis actor eterne et defensor invicte te decet erroris efflare nebulas
et obscuras ignorantie mentis tenebras effugare ut studiosis ingeniis veri
splendor irradiet ipsius que sinceritas equitate tui iudicii patefiat divina
60   auctoritate firmata ne quisquam ulterius de mundi mensura male sapiat
seu audeat affirmare de dignis rebus indigna. Profecto nostre sororis,
Geometrie, contraria nobis conclusio divine detrahit bonitati mundi
perfectionem diminuit celi tollit decorem infert nobis iniuriam hominibus
affert ignorantiam et totius universitatis entium pulchritudini derogare
65   videtur. Omnis namque incommensurabilium proportio seclusa est a
ratione mensurabili numerorum et inde a proprietate rei irrationalis
nuncupatur et surda. Indignum ergo videtur et irrationabile ut divina ratio

44 longa *R* / religata *RU* / sui: sibi *BR* /
   natura: notatur *P*
45 poterat: potest ac *A* / voluit: venit *P* /
   de quo: et *B*
46 livoris *R* / *post* livore *mg. hab. B* Boetius / cares *A* / et cetera *om. ALR*
47 itaque: ergo *B* / nostram: vestram *A* /
   multa *om. P* / *post* multa *hab. B* talia /
   ego *?B* ergo *A* igitur *F*
48 hoc unum *tr. U* hoc *B* / mihi *om. P*
48–49 reseraretur *?B P* reseratur F
49 Tunc: *?*ut *F*
50 Et *om. B*
51 Surgens Geometria *tr. U*
52 eidem: ei *B* / ait: dixit *FLP* / *post* ymo

   hab. *A* et / sunt incommensurabiles *tr.*
   A / Factaque: ffacta itaque *R*
53 quasi *om. FLP* / utrique *B* utrumque
   L / *post* partem *hab. U* cum
53–54 rationibus defendere *tr. B*
54 causam: tamen *A* / admirante et auscultante: ascultantem et admirantem
   A / *post* admirante *hab. R* ante
55 inquit: dixit *R* / *post* inquit *hab. P* et
   cetera
56 [Oratio Arismetice] *ante* veritatis (*linea
   57*) *mg. hab. F et ante* dixit (*vide lectiones
   in linea 55*) *mg. hab. R et post* decet (*linea
   57*) *hab. U; om. ABP*
57 *ante* O veritatis *mg. hab. B* priores aris-

God, "and he desired that all things, insofar as they were capable of such blessedness, should be as much like himself as possible."[6] And concerning this, Boethius wished that "the form of the good be without envy."[7] And one could continue in this vein.

If, then, we are unable to know many things by discovering them for ourselves, I beseech you to disclose to me, with good grace, your teachings concerning this one doubt. Then Apollo, smiling, saw the Muses and Sciences standing around and ordered them to do this, saying "Teach him what he asks." And immediately Arithmetic said: "All the celestial motions are commensurable." Geometry, rising, contradicted her and said: "On the contrary, some celestial motions are incommensurable." When this was over, Apollo ordered each side to defend their cause with reasoned arguments, as if they were litigants in a law suit. Filled with wonder, I sat listening as Arithmetic spoke first.

[The Oration of Arithmetic]

Oh speaker of the eternal truth and defender of the unvanquished, it is proper for you to blow away the fogs of error and put to flight the dark shadows of the ignorant mind, so that the brilliance of truth should radiate with its very own zealous nature and its sincerity should be revealed by the equity of your judgment, supported by divine authority. For otherwise, one would make erroneous judgments about the measure of the world, or hear unworthy things said about worthy things. Indeed, the opposite conclusion of our sister, Geometry, seems to deprive us of divine goodness, diminish the perfection of the world, destroy the beauty of the heavens, bring harm to mankind, cause ignorance, and detract from the beauty of the whole universe of beings. For every ratio of incommensurables is quite distinct from a measurable ratio of numbers, and, because of its very nature, is called irrational and absurd. It seems unworthy and unreasonable that the divine mind should connect the

---

metice / O *om. FP* / veritas *R* / auctor *A* aptor *U* / errorum *PU* / efflare: assimulare *A*

58 ignorate *P* / mentis tenebras *tr. B*
59 ipsius: ipsi *B* / tui: sui *B*
60 auctoritate firmata *tr. U* / mundi *om. U* / *ante* male *hab. B* et proportione
61 audeat affirmare *tr. B* / rebus: verbis *U* / *post* indigna *hab. B* vel falsa
62 contrarie *A* / bonitate *B*
62–63 mundi perfectionem *tr. B*

63 *post* celi *hab. B* que / tollit decorem *tr. B*
63–64 hominibus affert ignorantiam *om. B* / hominibus affert *tr. F*
65 namque: tamque *F* / est: re *A* et *P*
66 mensurabili *om. B* / inde: in *U* / proprietate: appellate *U*
67 nuncupatur: vocatur *B* nuncupantur *R* / ergo *om. U* / et² *om. B* / irrationabile: irrationale *BFP* / ratio: gratia *R*

motus celestes tali indiscreta habitudine connexisset per quos ceteri motus corporei debite ordinantur ac rationabiliter et regulariter fuerit.

70   Quamvis etiam quidam dixerit quod in mathematicis non est bonum aut melius quia conclusiones theorice non sunt de electione aut praxi, quedam tamen figure et aliqui numeri digniores habentur aut quadam secreta ratione nature aut quia rebus perfectioribus insunt. Sic enim ternarius est nobilior numerorum, ut probat Aristoteles primo celi. Unde quidem:
75   "Omnes res sunt tres, numerus ternarius in re qualibet existit nec enim invenimus istum a nobis numerum, sed eum natura docet nos"; de quo Virgilius: "numero deus impare gaudet." Sic etiam circulus apud Aristotelem dicitur perfectissima figurarum. Noster quoque Pictagoras quedam istorum a parte boni posuit, alia contra a parte mali statuit in sua co-
80   ordinatione bipartita.

Quod si contingat in numeris et figuris ita de proportionibus oportet dicere. Teste etiam Averrois super tertium celi dicente quod antiqui multum commendaverunt proportionem duplam; unde quidam: "infima supremis proportio dupla ligavit." Sunt igitur proportiones quedam aliis
85   meliores. Quo posito, quisque statim, quasi naturali instinctu, concederet proportiones rationales fore aliis, scilicet irrationalibus, digniores. Sicut ergo celestibus orbibus convenit figura perfectior ita eorum motibus congruit proportio nobilior ut corporibus que propter dignitatis precellentiam Aristoteles gloriosa vocavit nullus desit decor corporeus.
90   Ipsa si quidem irrationalis proportio aut incommensurabilitas potius

---

68 tali indiscreta *tr. B* tali discreta *P* / convexisset *ABL* connexuisset *U* / quos: quod *FP*

69 *ante* regulariter *hab. B* et / regulariter *?L* regulantur *P* / fuerint *A*

70 etiam *om. B* ?utinam *U* / dixit *AU* / *post* in *hab. U* praxi / aut: sive *F* nec *L*

71 melius: finis *A* / *post* theorice *hab. A* mathematice

72 digniores: dignorentur *A* / habentur *om. A* / quadam: quedam *F*

73 enim *om. B*

73–74 est nobilior *tr. B*

74 probat *om. B* probatur *P* / Unde quidem *om. A* / quidam *BLRU*

75 tres: tria *U* / numerus: et *B* naturas *U* / re qualibet *tr. ABR* se qualibet *F* / *ante* -istit *in* existit *mg. hab. R* nec enim invenimus / enim *om. BU*

76 a nobis numerum: numerum quasi a

nobis *B* / docuit *B* / nos: nobis *B* ut *U*

77 *ante* Virgilius *mg. hab. B* ?Virgilius / sicut *P* / etiam *om. B*

77–78 Aristotelis *A*

78 dicitur perfectissima figurarum: figurarum dicitur perfectissima *B* dicitur perfectissimus figurarum *F* / Noster: nam *A* / quoque: que *B*

79 istarum *A* / posuit: composuit *FLP* / *post* alia *hab. U* autem / contra: econtra *L* econtraria *U*

79–80 coordinatione: ordinatione *A*

80 bipartitam *A*

81 numeris et figuris: numeris *A* figuris et numeris *B* numeris et sigillas *F* / opportebit *A*

82 *ante* Teste *mg. hab. B* Averrois / etiam *om. B* / super *om. B* / tertio *BU* / dicente: dictum *B*

83 commandaverunt proportionem du-

celestial motions, which organize and regulate the other corporeal motions, in such a haphazard relationship, when, indeed, it ought to arrange them rationally and according to a rule.

Although a certain person has said that in mathematics nothing is "good" or "better"—since theoretical propositions are not a matter of selection or desire[8]—certain figures and certain numbers are, nevertheless, said to be more worthy [than others], either because of some secret order in nature, or because they exist in more perfect things. For example, among numbers, three is the most noble, as Aristotle shows in the first book of the *De caelo.** Thus: "All things are three. The number three exists in everything; nor, indeed, do we discover this number ourselves, but nature reveals it to us."[9] And Virgil could say about this that "God delights in an odd number."[10] Among figures, the circle is said by Aristotle to be the most perfect.[11] In his twofold series, our Pythagoras also placed some of these on the side of good, and others opposite, on the side of evil.[12]

But if it should happen [that] in numbers and figures [some are more worthy than others], one must say the same for ratios. [In commenting] on the third book of the *De caelo,* Averroes bears witness to this when he says that the ancients greatly commended the double ratio;[13] furthermore, someone said that "a double ratio unites the lowest things with the highest."† Thus, [we see that] some ratios are better than others. But this being asserted, everyone should concede immediately, as if by a natural instinct, that rational ratios are more worthy than the others, namely, irrationals. Therefore, just as the more perfect figure is appropriate for the celestial orbs, so [also] is the more noble ratio best suited for their motions, so that there is no lack of physical beauty to these bodies, which, in virtue of their unsurpassed dignity, Aristotle has called glorious.[14]

If, however, this [more noble] ratio is irrational, or an incommensurability,

---

* Aristotle, *De caelo* 1. 1. 268a11–19.   † Unidentified.

plam: commandaverint proportionem duplam *F* proportionem duplam commandaverunt *R* / unde quidam *om. FLP*
84 sint *A* / quedam: que *R*
85 Quo: que *A* / quasi naturali instinctu: quasi instinctu naturali *B* instinctu quasi naturali *U* / *ante* concederet *hab. U* naturaliter
86 fore *om. B* / scilicet *om. ABFPRU* / ir-

rationalibus: rationalibus *B*
87 perfectior: perfectione *F*
87–88 congruit: convenit *L*
88 proportio: perfectio *R* / nobilior *om. A* / ut: sicut *F* / que *om. U* / dignitatem *RU* / precellentiam: excellentiam *L*
89 Aristoteles *om. U* / gloriosa: glosa *F* eorum *U* / decor: de *A* / corporeus: corpore ipsa *A* corporis *U*

disproportio dicenda est atque privatio que etiam designatur nomine
privatio. Cum igitur habitus sit aliquid melius, pulchrius, et magis perfectum non videtur quod ipsa privatio imperfectionem carentiam et
turpitudinem denotans in illis ponenda sit que nulla perfectione sibi
95 privata sunt. Omnis enim pulchritudo et omne quod delectat intuitum
consistit in proportione rationali secundum perspectivos, sed etiam armonie que mulcent auditum omnia quoque pigmentaria saporum, odorumque
sunt confectiones circa commensurationem miscentur. Nec adhuc omnes
proportiones rationales inducunt sensibus voluptatem sed determinate de
100 quibus ait Aristoteles "symphonie vero pauce." Et iste proportiones
reputate sunt inter ceteras digniores; et econverso, omnis autem irrationalis proportio offendit in sonis auditum in saporibus gustum; et sic de
aliis, ut vult Aristoteles ipse. Namque quodammodo decipiunt quod non
videtur nisi propter earum disproportionem cum sensu et quamdam, ut est
105 dicere, indecentiam quod non decet imponere eternis motibus ne tanta
turpitudine celi spatiositas maculetur nec solum ledit sensum talis proportio sed etiam intellectum. Cum iuxta secundam decimi Euclidis ex
subtractione unius incommensurabilium ab altero quedam infinitas oriatur
in cuius consideratione mens hebetatur, ratio retunditur, dum illud chaos
110 infinitatis incomprehensibile animos fastidiosos efficit et contristat. Sic
ergo intellectibus non convenit nec est eis proportionalis irrationalis
proportio propter quod antiqui dixerunt animam constare quadam
numerali et armonica ratione.

Et si irrationalis proportio nostro disconvenit et displicet ingenio quo-
115 modo ponemus intelligentias motrices vitam ducentes optimam tam ina-

91 dicenda...privatio: atque privatio dicenda *U* / *ante* atque *hab. A* quam
92 privativo *U* / aliquid: quid *L*
92–93 magis perfectum: perfectius *B*
93 et: ut *A*
94 turpitudine *B* / illis: istis *P* / que nulla:
que ulla *A* si cuiuscumque *R* / perfectione: perfecit *B*
95 sint *R* / delectatur *P* / intuitum *om. FLP*
intuentem *A* / *post* intuitum *hab. B* et
auditum
96 secundum: sed *U* / sed: sic *B* / etiam
corr. ex et *BFLPRU*; *om. A*
97 mulcent: multum *A* demulcent *B* mul-
cret *P* / quoque: que *B* / odorumque:
deorumque *U*
98 sunt *om. ABRU* / commensurationem:
commensurabilitate *LU* commensura-

bilitatem *B* commensuratio *F* mensura-
tionem *R* / adhuc: ad hoc *R*
99 proportiones *om. A*
99–100 de quibus *om. R*
100 vero: ?libent *U*
101 reputate...ceteras: inter ceteras reputate sunt *B* / et *om. B* / econverso: econtraria *BR* / autem *om. ABR*
101–2 irrationalis proportio *tr. B*
102 sonoris *B* / *post* gustum *hab. B* in odo-
ribus olfactum
103 decipiunt: despiciuntur *BU* despiciunt
P desipiunt *R*
104 eorum *U* / cum: autem *B* / et: ut *U* /
ut: nec *U*
104–5 ut est dicere *om. B*
105 quod: quam *ABR* / *post* tanta *hab. B* et
tali

it ought rather to be called a "disproportion" and privation. That it is a privation is indicated by the very name which designates it. But since something better, more beautiful, and more perfect must be characteristic [of the celestial motions], it seems that this privation—denoting imperfection, deprivation, and deformity—ought not to be assumed in their motions, which should not be deprived of any perfection. For according to writers on optics, every beautiful thing, and everything that delights the sight, is based on a rational ratio;[15] but even the harmonies that delight the ear, and all colors, as well as the production of tastes and odors, are united on the basis of commensurability. And yet not all rational ratios produce pleasure for the senses, but only certain ones. For this reason, Aristotle says: "Harmonies are few, indeed."[16] And those ratios [producing the harmonies] are reputed to be more worthy than the others. But, on the contrary, every irrational ratio offends the ear by the sounds it produces, the taste by the flavors it produces, and so on, as Aristotle himself maintains.[17] To a certain degree, however, these [irrational ratios] are deceptive, for if it were not for their lack of harmony with the senses, and a certain impropriety, so to speak, it would not seem that such a ratio is improper [and unsuitable] to impose upon eternal motions. But were they imposed, the vast extent of the heavens would be defiled by such baseness. Indeed, not only would the sense involved be defiled, but also the understanding, since, by the second proposition of the tenth book of Euclid, a certain infinity arises from the subtraction of one incommensurable [magnitude] from another.[18] In considering this [infinity], the mind is rendered dull, the reason enfeebled, until that incomprehensible chaos of infinity produces loathesome and afflicted minds. Thus, an irrational ratio is neither suitable or relatable to the understanding, for which reason the ancients said that the mind conforms to a certain numerical and harmonic plan.

But if an irrational ratio is inconsistent with and displeasing to our nature, how could we assume that the motor intelligences, who lead the very loftiest

---

106 *ante* celi *hab. B* et indecentia / spatiositas: speciositas *LRU* / *post* sensum *hab. B* hec

107 sed: ymo *BF* / intellectum: ?intellectualitatem *A* / *ante* secundam *hab. F* ?veram / decimi: secundi *B*

108 altero: alto *L* / quedam: quod *P* / infinitas: infirmitas *L*

109 hebetatur: ebetatur *LRU* abetatur *P* / rotunditur *B* / dum: unde *U*

110 infinitatis: infitis *A* infinitis *F* / incom-

prehensibile: comprehensibile *A* / Sic: dic *P*

111 eis *om. R*

111–12 irrationalis proportio *tr. L* ?irrationalis *B* irregularis proportio *P*

112 quod: quid *B* / *ante* quadam *hab. U* ex

114 *post* si *hab. B* hec / irrationalis proportio *tr. B* / nostro *om. FP*

115 ponemus: ponerem *U* / vitam *obs. B* / ducentes optimam *tr. B*

mena et tristabili disparitate movere que tamen in agitatione et plausu orbium summo gaudio delectantur? Nam et si quis faceret horologium materiale nonne efficeret omnes motus rotasque commensurabiles iuxta posse? Quanto magis hoc opinandum est de architectore illo qui omnia

120 fecisse dicitur numero, pondere, et mensura? Nulla autem incommensurabilia sunt numeris mensurata. Rursumque asserit Boetius quod "omnia que a primeva rerum origine processerunt ratione numerorum formata sunt"; et dixerunt antiqui quod forme rerum sunt similes numeris. Plato quoque ait mundi opilicem quatuor corporalia elementa connexisse

125 in proportionalitate continua complectente duos numeros cubicos et duos eorum proportionaliter medios; unde Boetius: "Tu numeris elementa ligas," et cetera. Ergo potissime prima mundi incorruptibilia elementa eorumque motus considerat analogia numeralis ut si quedam iuxta nos aliquando propter variationem continuam proportionentur incommen-

130 surabiliter hoc fit, ut verbis Aristotelis utar, "quia hec inferoria plena sunt turbinis propter longe stare a commodo divino," scilicet a deo, qui talem inordinationem prope se in celestibus non permittit.

Hactenus monstravi quibus modis positio nostre contraria pulchritudinem interimit universi deorumque derogat bonitati. Nunc referre volo

135 quantitas nobis affert iniurias. Ego inter mathematicas primogenita sum teneoque primatum ita ut ob hoc Macrobius per argumenta multa concludit "antiquiorem esse numerum superficie et lineis." Si ergo celestibus expellor sedibus qua parte mundi fugiam an extra mundi limites exulabor?

O mi Jordane frustra me tam subtiliter exquisisti. Quis me dignabitur

---

116 tristabili: contristabili *U*

116–17 et plausu orbium: orbium plausu *U*

117 faceret: componeret *B*

118 materiale: male *P* / motus *om. U* / rotasque: rotas *U*

119 hoc *om. ABF* / opinandum: operandum *R* / architectorie *A* architecte *F* architectate *P* / qui: que *P*

120 *ante* numero *hab. U* in / et *om. U*

121 mensurata: commensurata *L* / Rursumque: rursum *B* / asserit Boetius *tr. B* / quod *om. U*

123 et *om. P* ideo *BLU* immo *F* / dixerunt antiqui *tr. B* dixerunt *L* / rerum: fierint *U*

124 quoque: etiam *B* / corporalia elementa *tr. L* corruptibilia elementa *R* / connexisse: convexisse *LU*

125 duos^1: dies *R*

126 eorum: eos *A* / unde Boetius *om. R* et Boecius *B* / Tu: qui *BFLPRU*

127 incorruptibilia: incommutabilia *U*

128 consederat *BL* confederat *FPR* / numeralis: naturalis *RU*

129 aliquando *om. U* quandoque *F* / continuam *om. R* continuationem *U*

130 hoc: illud *F* / Aristotelis: eorum *P* / quia: scilicet *U* / hec: ista *F* / plena sunt *tr. A*

131 turbinis *corr. ex* turbine *ABFLPRU* / stare: distare *A* / a^2 *om. B* / qui: quia *R*

133 *ante* monstravi *hab. U* quidem / positioni *LU* / contrariam *U*

134 deorumque: deorum *B* deique *U*

134–35 Nunc...iniurias *om. R*

135 nobit *A* / afferat *A* infert *U* / primogenita: prior genita *P* / sum: sui *A* sic *F*

136 teneorque *P* / primatum: principatum

[kind of] life, move with such unpleasant and shameful irregularity and yet [also believe] that they are filled with the greatest delight in the excitement and applause of the orbs? For if someone should construct a material clock would he not make all the motions and wheels as nearly commensurable as possible? How much more [then] ought we to think [in this way] about that architect who, it is said, has made all things in number, weight, and measure?[19] However, no incommensurable things are measured by numbers. And, again, Boethius asserts that "everything that proceeded from the very origin of things was formed with reference to numbers;"[20] and the ancients said that the forms of things are like numbers.[21] Plato also said that the maker of the world joined four corporeal elements in continuous proportionality embracing two cube numbers and their two mean proportionals.[22] For this reason, Boethius could say: "You bind the elements with numbers,"* and so on. Thus, the most potent primary, incorruptible elements of the world, and their motions, reflect a numerical relationship, so that if certain things near to us are sometimes proportioned incommensurably by virtue of a continuous variation, this happens—and I use the words of Aristotle—"because things in these lower regions have twisting [and confusing] motions as a result of their remoteness from a proper divinity"—that is, from God, who would not permit such disorder near himself in the heavens.[23]

Up to this point, I have shown the ways in which a position contrary to ours would destroy the beauty of the universe and detract from the goodness of the gods. Now [however] I wish to set out the way in which quantity [or magnitude] brings harm to us. Among the mathematical disciplines, I [Arithmetic] am the firstborn and hold the first rank, for which reason Macrobius concludes, by means of numerous arguments, "that numbers precede surfaces and lines."[24] If, therefore, I am expelled from the celestial regions, to what part of the world shall I flee; or will I be banished beyond the boundaries of the world?

Oh my dear Jordanus, in vain did you investigate me so subtly![25] Who will

---

* Boethius, *De consolatione philosophiae*, bk. 3. 9; *Consolation of Philosophy*, trans. Stewart, p. 264, line 10.

F / ob hoc *om. F* et *P* / argumenta multa *tr. L* argumenta *R*

137 superficie et lineis: superficie et linea *B* lineis et superficie *L* scilicet lineis superficie et corporibus *U* / *post* celestibus *hab. B* a

138 expellor sedibus *tr. U* / qua parte: que propter *A* / *post* mundi[1] *hab. B* stabo aut / fugiam *om. A* / extra: circa *R* / exulabo *LRU*

139 subtiliter: subtitur *P* / exquesisti *L* exquesiti *P* / *post* me[2] *hab. R* de

140   respicere si mei numeri ad celestes motus nequeant applicari? Et si musica
       ad sonos contrahit numeros, quare non poterit eos astrologia suis motibus
       coaptare cumque per me numerentur? Et astra cur motus eorum non
       poterunt meis numeris mensurari? Quid me vultis? a stellato solio degra-
       dare? a domo mea hereditaria destituere? Certe non potestis. Cum ipso
145   enim summo omnium principe uno et trino ubique etiam supra sidera
       habito qui "mundi regna triformis" cuncta regit et attingens a fine usque
       ad finem fortiter disponit omnia suaviter, scilicet armonice.

           Ipsa irrationali asperitate a cunctis rebus exclusa maxime vero irratio-
       nalis proportio propulsa creditur a celi motibus concentus agentibus
150   mellitos. Omnis namque talis proportio est in armonica dissonans
       extranea ideoque ab omni consonantia aliena magis enim apta est luctibus
       horridis tristis inferni quam celi motibus miscentibus melodias musicas
       mulcentes magnum miro moderamine mundum.

           Itaque celesti privaretur honore nostra iocunda filia musica dulcisona
155   que tamen multis phisicis testibus partem habet in regno celorum. Unde
       noster Pictagoras supremam continentiam se profitebatur audisse qua
       secundum ipsum celi quodammodo Dei gloriam enarrare videntur cum
       in omnem terram exeat sonus eorum totumque repleat orbem et per cuius
       dulcedinem mundi opifex nexum moderat universe fabrice mundi.

160   Narrat etiam Tullius in sexto *De republica* sompnium Scipionis dicentis:
       "quid...est," inquit Scipio, "qui complet aures meas tantus ac tam dulcis
       sonus?" Responditque quem videbat in sompnio avus eius Scipio: "Hic
       est," inquit, "ille qui intervallis distinctus imparibus sed tamen pro rata
       partium ratione distinctus pulsu et motu ipsorum orbium efficitur et acuta

140 respicere: aspicere *B* inspiretur *U* / si
    om. L / ad celestes motus *om. A* ad
    motus celestes *B* / nequeunt *R* / appli-
    cari: explicari *A*
141 poterat *R* / *ante* suis *hab. A* a
142 coaptare: applicare *U* / cumque: cum
    vero *F* / *post* numerentur *hab. B* celi /
    Et *om. F* / cur: cui *P*
143 potuerunt *R* / mensurare *P* / Quid: si
    quidem *U*
143–44 degradarie *A*
144 *ante* a *hab. U* et
145 enim *om. B* vero *F* / summo *om. L* /
    omnium *om. A* ?omni *P* / etiam: et *F*
146 qui: que *P* / regna *om. BFLP* / regit:
    regnat *U* / et *om. B* / usque *om. B*
147 disponit: disponeres *U* / scilicet: secun-
    dum *P*

148 irrationali: rationali *F* irrationabili *L* /
    asperitate: aspitate *A* / a *om. U* / cunctis
    rebus exclusa: contactis exclusa *A* cunc-
    tis exclusa rebus *L* rebus cunctis ex-
    clusa *U* / vero: ergo *B*
148–49 irrationalis: irrationabilis *L*
149 agentibus: mellices *P*
150 talis *om. F* / dissonans: dispositione *U*
150–51 dissonans...consonantia *om. A*
151 *ante* extranea *hab. L* et
152 horridis: cordis *U* / inferni *om. A*
153 moderamine: modulamine *ARU* mode-
    ratione *P*
154 honore *om. R* / filia *om. F* / musica dul-
    cisona: dulcissima musica *U*
155 tamen *om. F* / phisicis testibus *tr. BL*
156 noster: non *F* / se profitebatur *tr. R* /
    qua: quam *A*

deem me worthy of respect if my numbers are incapable of application to celestial motions? And if music can reduce numbers to sounds, why cannot astronomy join these numbers to its motions whenever they are numbered by me? And why are the motions of the stars not measurable by my numbers? What do you want from me? Do you wish to oust me from my starry throne, to force me from my hereditary home? Surely, you will not achieve this. For the greatest prince of all, himself one and three everywhere—even beyond the stars—who rules all the "kingdoms of the triformed universe,"[26] extends powerfully from one end to the other and arranges all things pleasantly [and agreeably], that is, harmonically.

If the difficulties resulting from irrationality were eliminated from all things —and especially if the irrational ratio were driven out—it is believed that lovely harmonious music would issue forth from the celestial motions. For every such ratio is discordant and strange in [its] harmony and, consequently, foreign to every consonance, so that it seems more appropriate to the wild lamentations of miserable hell than to celestial motions that unite, with marvelous control, the musical melodies soothing a great world.[27]

Furthermore, our delightful daughter, sweetly sounding Music, would be deprived of celestial honor, even though she participates in ruling the heavens, as many physical occurrences bear witness. Thus, our Pythagoras, acknowledged to possess the greatest moderation, heard how the musical harmonies seem to explain [or account for] the glory of the heavens—i.e., the glory of God. When their sounds issue forth over the entire earth and fill the whole orb, the creator of the world regulates the bonds of the whole fabric of the world by the sweetness of these sounds.

In the sixth book of the *Republic*, Tully recounts the *Dream of Scipio* where Scipio says: "What is this great and pleasing sound that fills my ears?"[28] And his grandfather, whom Scipio had seen in his dream, answered: "That," he replied, "is a concord of tones separated by unequal but nevertheless carefully proportioned intervals, caused by the rapid motion of the spheres themselves.

---

157 secundum ipsum *om. B* se ipsum *R* / Dei gloriam: Dei gratiam *A* gratiam Dei *R* / enarrare: narrare *R*

158 et *om. F* ut *R* quia *U*

159 mundi opifex *tr. B* / moderatur *R*

160 Narrat etiam Tullius: Tullius etiam narrat *B* narrat enim Tullius *R* narrat Tullius *U* / in *om. B* / sexta *U* / dicentis *om. U*

161 inquid *P* / qui: quod *U* / complet: re-plet *BU*

162 sonus *om. F* / Responditque: respondit *B* / quem *om. A* / videbat: audiebat *A* videt *L* videbit *U* / avus: aurum *A* / Scipio *om. U*

163 ille: iste *P* / distinctus *corr. ex* coniunctus *AFLRU* iunctus *B* coniunctis *P* / imparibus: temporibus *FLP* / tamen: tantum *A* / rata: rapta *A*

164 ipsorum orbium *tr. U*

165 cum gravibus temperans varios equabiliter concentus efficit;...et ille
summus stellifer celi cursus eius conversio est concitatior, acuto et excitato
movetur sono, gravissimo autem hic lunaris atque infimus," et cetera.

Quod exponens Macrobius sic ait huic: "Plato in *Republica* sua cum de
sperarum celestium volubilitate tractaret singulas ait Sirenas singulis
170 orbibus insidere significans sperarum motu cantum numinibus exhiberi.
Nam Siren deo canens Greco intellectu valet. Theologi quoque novem
Musas octo sperarum musicos cantus et unam maximam continentiam
que conficitur ex omnibus esse voluerunt. Unde Esyodus in *Theogonia* sua
octavam Musam Uraniam vocat," id est celum intelligens octavam orbem
175 stelliferum.

Et subdit: "Ideo canere celum etiam theologi comprobantes sonos
musicos sacrificiis adhibuerunt.... In ipsis quoque hymnis deorum per
stropham et antistropham metra canoris versibus adhibebantur ut per
stropham rectus orbis stelliferi motus per antistropham diversus vagarum
180 regressus predicaretur; ex quibus duobus motibus primus in natura
hymnus dicandus deo sumpsit exordium. Mortuos quoque ad sepulturam
prosequi oportere cum cantu plurimarum gentium instituta sanxerunt
persuasione hac qua post corpus anime ad originem dulcedinis musice,
id est ad celum, redire credantur.

185 "Nam ideo in hac vita omnis anima musicis sonis capitur ut non solum
qui habitu cultiores verum universe quoque barbare nationes cantus quibus
vel ad amorem virtutis animentur vel ad molliciem voluptatis resolvantur
exerceant quia in corpus defert memoriam musice cuius in celo fuit conscia
et ita delenimentis canticis occupatur ut nullum sit tam immite tamque
190 asperum pectus quod non oblectamentorum talium teneatur affectu." Aves
quoque, et cetera.

---

165 equabiliter: equaliter *FR* equalitatis *U* /
et *om. U* / ille: iste *A*
166 cursus *om. A* / conversione *U* / est *om.*
RU / concitator *AFLP* concitatur *U* /
excitator *A*
167 gravissime *P* / autem: consequenter *A*
ante *F*
168 huic *om. A* hinc *U* ?hic *F* / sua *om. B*
169 volubilitate: voluntate *F* / tractavit *R*
170 orbibus: pedibus *F* / significans: sig-
nans *AFPR* significanter *B* factas *U* /
cantum *om. R* / numinibus: universe *U*
171 Siren: syrenam *U* / intellectu: intellectibus *P* / valet *om. P*
172 octo: scilicet *FLPU* / musicas *B* musices

F / *ante* cantus *hab. L* et
173 conficiatur *U* / ex *obs. B* / voluerant *U* /
Esyodus: oxiodorus *U* / Theogonia:
theologia *LPR* ?theologica *A* / sua *om.*
L
174 octavam[1]: sextam *F* / Uraniam vocat
tr. B Urania vocat *A* / id est: per *A*
176 etiam *om. U* et *A* unde *R*
177 musices *P* / ipsis: his *U*
178 antistropham: antiphonam *R* / versibus
corr. ex vocibus *ABFLPRU* / adhiben-
tur *A*
179 vagarum *corr. ex* vagorum *ABFLPRU*
180 regressus: gressus *B* / duobus: duorum
R / in natura *om. U*

The high and low tones blended together produce different harmonies...and the outermost sphere, the star-bearer, with its swifter motion gives forth a high-pitched tone, whereas the lunar sphere, the lowest, has the deepest tone,"[29] and so on.

In explaining this, Macrobius also says: "In a discussion in the *Republic* about the whirling motion of the heavenly spheres, Plato says that a Siren sits upon each of the spheres, thus indicating that by the motions of the spheres divinities were provided with song; for a singing Siren is equivalent to a god in the Greek acceptance of the word. Moreover, cosmogonists have chosen to consider the nine Muses as the tuneful song of the eight spheres and the one predominant harmony that comes from all of them. In the *Theogony*, Hesiod calls the eighth Muse Urania,"[30] that is, "the sky," understanding by this the star-bearing eighth sphere.

And he [Macrobius] adds: "That the priests acknowledged that the heavens sing is indicated by their use of music at sacrificial ceremonies.... In the hymns to the gods, too, the verses of the strophe and antistrophe used to be set to music, so that the strophe might represent the forward motion of the celestial sphere and the antistrophe the reverse motion of the planetary spheres; these two motions produced nature's first hymn in honor of the Supreme God. In funeral processions, too, the practices of diverse peoples have ordained that it was proper to have musical accompaniment, owing to the belief that souls after death return to the source of sweet music, that is, to the sky.

"Every soul in this world is allured by musical sounds so that not only those who are more refined in their habits, but all the barbarous peoples as well, have adopted songs by which they are inflamed with courage or wooed to pleasure; for the soul carries with it into the body a memory of the music which it knew in the sky, and is so captivated by its charm that there is no breast so cruel or savage as not to be gripped by the spell of such an appeal."[31] Indeed, even birds are affected by it.[32]

---

181 hymnus: ymnis *A* hipnus *F* himnis *R* / dicendus *U*

182 cum cantu *om. U* cum cantibus *F* cum cantum *L* / plurimarum: plurimo *A* / gentium: per agentium *A* / *ante* sanxerunt *hab. R* esse

183 persuasione hac *tr. B* persuasiones hac *L* pro suavitate hac *R* / qua post: quia post *B* potest *U* / corpus anime *tr. U*

184 id est *om. A* scilicet *B*

185 anima *om. P* / solum: sonum *P*

186 qui: que *BFLPU* / cultores *A* ?alcorreres *F* / quoque *om. A*

187 vel[1] *om. U* sive *F* / amorem: morem *A* / virtutis: veritatis *U*

188 exerceant: excrescant *B* / *ante* in[1] *hab. L* qui / conscia: ?cencia *BFP* tripendia *U*

189 et ita *om. A* / delenimentibus *U* / occupantur *L* / *ante* tamque *hab. B* poetus

190 pectus *om. A*

Unde in *Policratico:* "Equissimum est animam cognati generis gratia mansuescere et omnem dediscere alienationem cum ei proprie concentus originis et nature melioris archana resultant."

195 Nunc revertor ad Macrobium qui postea dixit: "Ipsa autem mundi anima viventibus omnibus vitam ministrat: 'Hinc hominum pecudumque genus viteque volantum et que marmoreo fert monstra sub equore pontus.' Iure igitur capitur musica omne quod vivit quia celestis anima, qua animatur universitas, originem sumpsit ex musica. Hec dum ad speralem
200 motum corpus mundi impellit sonum efficit 'qui intervallis est distinctus imparibus sed tamen pro rata partium ratione distinctus,'" hec ille.

Sed et Boetius affirmat "quod non frustra a Platone dictum sit mundi animam musica convenientia fuisse compactam"; et rursum inquit: "non potest dubitari quin anime nostre et corporis status eisdem quodam-
205 modo proportionibus videatur esse compositus quibus armonice modulationes copulantur."

Et idem Boetius trinam posuit musicam. Primam vocavit "mundanam" de qua dixit: "Mundana, in his maxime inspicienda est, que in celo vel compage elementorum vel temporum varietate visuntur." Et subdit:
210 "Unde non potest ab hac celesti vertigine ratus ordo modulationis absistere. Iam vero quatuor elementorum diversitates contrariasque potentias nisi quedam armonia coniungeret quomodo fieri posset ut in unum corpus ac machinam convenirent?"

Alter etiam in *Policratico* de musica dicit quod: "virtutis sue potentia
215 specierum⟨que⟩ varietate et sibi famulantibus numeris universa complectitur, omnium que sunt et que dicuntur, dissidentem et dissonam multi-

---

192 *post* Unde *hab. B* quidam / *post* animam *mg. hab. R* equissimum est / *post* generis *hab. L* animam / gratia *om. A*
193 mansuescere: mansuefacere *A* / et omnem: omnem *ABFP* omnemque *U* / dediscere: desistere *L* / cum ei: tunc eis *A* / *post* proprie *hab. A* virtutis
194 resultat *U*
195 revertar *R* / qui *om. U* que *P* / *ante* dixit *mg. hab. R* ipsa autem mundi anima / postea dicit *BP* dicentem postea *U*
195–196 mundi anima *tr. U*
197 viteque: vite quod *B* / et: est *L* / marmores *AP* / fert: fiet *B* sunt *FP* / monstra: monstratur *A* monstro *F* monstre *P*
198–99 animatur: riantur *A*
199 speralem: specialem *L* specialiter *U*
200 corpus: corporum *B* / *ante* sonum *hab.*

U et / est distinctus *tr. B*
201 imparibus: temporibus *L* / pro: per *PU* / ille: iste *P*
202 Sed et *om. U* sed *L* / *post* Boetius *hab. A* asserit
202–61 affirmat....marcescit: quoque in musica sua de mundana musica aut habere Platone *?*plura scripsit. Hermes etiam pater philosophorum atque Cassiodorus eloquentissimus *?*plurima de mundana atque celesti musica sive armonia scripserunt scitu iocundissima videantur eorum scripta desuper *U*
203 fuisse *om. L*
203–4 inquit non potest: non potest inquit *R*
204 quin: quoniam *B* / statius *R*
205 compositus: compotus *F*
207 *infra* musicam Primam *mg. hab. A* dixit

In the *Policraticus* we read: "It is most fitting that the soul, thanks to a kindred element, calm itself and forget all resentment when harmonies of like origin with itself and mysteries of nature in her kindlier aspects are revealed in sound."[33]

Now I return to Macrobius, who says later: "The World-Soul, moreover, provides all creatures with life: 'Thence the race of man and beast, the life of winged things, and the strange shapes ocean bears beneath his glassy floor.'* Consequently it is natural for everything that breathes to be captivated by music since the heavenly Soul that animates the universe sprang from music. Thus, while it quickens the body of the world to spherical motion, it produces tones 'separated by unequal but nevertheless carefully proportioned intervals.'"[34]

But Boethius also supports the very same views saying "it was not without reason Plato remarked that the World-Soul is united in harmony with music."[35] And again he says that "it cannot be doubted but that the relationship between our soul and bodies appears, in some measure, to be composed of the same proportions as those which link melodies harmonically."[36]

The same Boethius holds that there are three types of music. He calls the first "wordly" and says of it: "Wordly music should be looked for especially in phenomena that occur in the sky, either in a union [or meeting] of elements, or in the diversity of the times [in which the phenomena occur]."[37] And he adds: "Thus the established order of harmony cannot be omitted from any consideration of the celestial revolutions. For unless a certain harmony should unite the diversities and contrary powers of the four elements, how could they join to form one body and structure?"[38]

Another similar opinion concerning music is found in the *Policraticus* where the author says: "Because of the great power exercised by it, its many forms, and the harmonies that serve it, it embraces the universe; that is to say, it

---

* Virgil, *Aeneid* 6. 728; *Virgil*, trans. Fairclough, Loeb Library, vol. 1, p. 557.

---

mundana in hiis / *post* Primam *hab.* F que *et hab.* P quam

208 inspicienda: insipientia R

209 compage elementorum *tr.* L / subdit: subait A

209–10 Et...ordo *om.* R

210 Unde: unum P / non *om.* P nec F / hic P / *ante* vertigine *hab.* L virgine

211 Iam: in illam P / *post* diversitates *hab.* P que / *ante* potentias *hab.* B impressiones

212 nisi quedam: non quidam P / armonica

A armoniam F / coniungerent F / posset *om.* R

213 ac: et B aut L / conveniret AB

214 de musica dicit: de musica dico A dixit de musica R / quod: que P / virtutis sue potentia: virtutes sive potentie F

215 famulantibus: famulis P / *ante* numeris *hab.* P ?auribus

216 que *om.* L / dissidentem: dissidere A / dissona A

tudinem proportionum suarum, id est, inequali quadam equitatis lege
concilians. Hac etenim celestia temperantur et humana reguntur.... Unde
et archano quodam meatu nature secretis⟨que⟩ cuniculis vivacitatis sue
220   vigore per universa discurrit, et cuivis nature substantieque conformis
rationis sensus vite in singulis providentie dispensationis decreto efficaciam
modulatur."

Hermes etiam, pater philosophorum, sic inquit: "Nec immerito in hominum cetum Musarum chorus est a summa divinitate demissus, scilicet
225   ne terrenus mundus videretur incultior, si modorum dulcedine caruisset,
sed potius ut musicatis hominum cantilenis concelebraretur laudibus qui
solus est omnia aut pater est omnium, atque ita celestibus laudibus nec in
terris armonie suavitas defuisset."

Hec etiam rationalis proportio celestem musicam faciens in sacris elo-
230   quiis tangitur ubi dicitur: "Quis enarrabit celorum rationem ac concentum
celi quis dormire faciet?"

Verumtamen, Plato voluit motum celi rationalem fieri sine sono; nec
placet Aristoteli quod motus orbium faciat strepitum sive sonum audibilem. Quid ergo erit ne symphonia sine sono certe sic non enim intellexe-
235   runt philosophi mundi musicam fore auribus sensibilem sed animo
intelligibilem mente capabilem.

Et hoc secundum Hermetem a paucissimis qui "pura mente prediti, sor-
titi sunt celi suspiciendi venerabilem curam." Unde rursum idem: "Musicam nosse nihil aliud est nisi cunctarum rerum ordines scire queque ratio
240   divina sortita est: ordo enim rerum singularum in unum omnium artifici
ratione collatus concentum quemdam melo divino dulcissimum verissi-
mumque conficit."

Quod autem hec mundana musica non sit perceptibilis auribus corporeis

---

217 id est *om.* A et P / quadam: quedam L
218 etenim: enim P
219 arcoano F arcoane P / cuniculis: cumu-
     lis A / suo A
220 per universa discurrit: discurrit per uni-
     versa L / cuivis *om.* A / *ante* substantie-
     que *hab.* A tunc / substantieque: sub
     est que BFP
221 providentie: providentur A / decreto:
     decetero A
223 etiam: et L; *tr.* B *post* philosophorum /
     in *om.* R
223–24 in...chorus: musarum chorus in
     hominum cetum L
224 a: de R / dimissus ABFLR

225 terrenis A / modorum: mundus A / ca-
     ruisse L
226 ut: in A / *post* ut *hab.* L in / homini
     BFLR / cantilenis: cantelnas F / con-
     celebraretur: conservaretur A celebrare-
     tur FLP
226–27 qui...laudibus *om.* F
227 est[2] *om.* A / itaque A / *post* ita *hab.* L in
228 *post* defuisset *hab.* B et cetera
230 dicitur: dici R / ac: aut BLPR sive F
231 quis *om.* R
232 motum celi *tr.* FLP
233 *infra* strepitum *mg. hab.* B audibilem
234 erit *om.* R / ne: et nec B / sicut P
235 mundi musicam *tr.* B

reconciles the clashing and dissonant relations of all that exists and of all that is thought and expressed in words by a sort of ever-varying but still harmonious law [derived] from its own symmetry. By it the phenomena of the heavens are ruled and men are governed.... Hence, by a kind of course through concealed passages, it pervades the whole universe with its own vital force. Sense har monizing with reason regulates and renders efficient the life of each nature and substance by decree of providential dispensation."[39]

Also Hermes, the father of philosophers, says the following: "And not without good reason has the supreme deity sent down the choir of the Muses to dwell among mankind. The earthly part of the universe would have seemed but rude and savage, if it had been wanting in sweet melody; and lest this should be, God sent the Muses down, to the intent that men might adore with hymns of praise Him who is all things in one, the Father of all, and that thus sweet music might not be lacking upon earth, to sound in concord with the singing of his praise in heaven."[40]

Furthermore, the making of celestial music by rational ratios is touched upon in sacred texts where it is said: "Who can declare the order of the heavens, or who can make the harmony of heaven to sleep?"[41]

However, Plato proposed that the celestial motions are made rational without sound;[42] and Aristotle was not pleased with the idea that the motion of the orbs could make noise or audible sound.[43] What else, then, but that the [celestial] harmony must be without [audible] sound, for truly the philosophers judge that worldly music would not be sensible to the ears, but rather that it would be intelligible to the soul and comprehensible to the mind.

And according to Hermes, to very few men, who "are endowed with mind uncontaminate, has fallen the high task of raising reverent eyes to heaven." Again, the same [Hermes] says: "To know the science of music is nothing else than this—to know how all things are ordered, and how God's design has assigned to each its place; for the ordered system in which each and all by the supreme Artist's skill are wrought together into a single whole yields a divinely musical harmony, sweet and true beyond all melodious sounds."[44]

That this mundane music is imperceptible to corporeal ears is asserted by

---

236 menti *F* ?intellectu *P* / capabilem: capa- cibilem *F*
237 hoc: illud *F* / a paucissimis *om. L* a paucis *A* a paucitissimis *F* / qui: que *AFLP* / predicti FP
238 suscipiendi *AFLP* / rursum idem *tr. F* / *post* idem *hab. A* in
239 nosce *AB* noscere *FLP* / aliud *om. P* /

ordinem *BLR* ordine *P* / queque: qua- cumque *A* quecumque *BFLR*
240 enim: vero *F* / ?omni *ABFLP* / artifice *B* artificie *F* artificem *L* artificum *R*
241 ratione *om. A* / quemdam: quidem *A* / divino: dei *F*
242 conficit: desistit *A* consistit *P*

ille eloquentissimus Cassiodorus asserit dicens: "Armonia celi humano
245 sermone non potest explicari quam ratio tantum animo dedit sed auribus
natura non prodidit."

Quod vero in celo sit musica idem manifestius ostendit quando de lyra
ait: "Hanc igitur," scilicet lyram, "ad imitationem varie testudinis Mer-
curius dicitur invenisse quam…astrologi inter stellas requirendam esse
250 putaverunt persuadentes celestem esse musicam quando lyre formam
comprehendere potuerunt inter sidera collocatam."

Unde quidam poeta: "Solis opus cithara studium lyra mercuriale."

Rursum, Cassiodorus recommendans musicam ait: "Quid enim illa pres-
tantius, que celi machinam sonora dulcedine modulatur et nature conve-
255 nientiam ubique dispersam virtutis sue gratia comprehendit?" hec ille.

Inde antiqui ymaginati sunt 7 planetarum tetracordum duplex sim-
plexque supremi orbis stelliferi monocordum quibus celi armonicum
tripudium amena taciturnitate conficitur. Nec solum in celo sed supra
posuerunt musicam. Unde Cassiodorus: "Dicunt enim debere credi, ut
260 beatitudo supercelestis illis musicis oblectationibus perfruatur que nec fine
deficit nec aliqua intermissione marcescit."

Nunc mihi libet ostendere quantam affert hominibus cecitatem qui celi
velocitates negat proportionari numeris. Nam si est ita nullus poterit
umquam aspectus precognoscere, coniunctiones predicere, previdere
265 effectus. Ymo latebit astrologia omni evo incognita ac etiam inscibilis ut
probatum est ante; non erit igitur inter mathematicas numeranda. Et si
velocitates celi sunt incommensurabiles et inscibiles cur mundi opifex "os
homini sublime dedit celumque videre iussit et erectos ad sidera tollere
vultus?" "Quid mentem traxisse polo, quid profuit altum erexisse caput?"

---

244 ille: iste *A* / asserit *om. F*
245 quem *A* / ratio tantum: ideo non *P*
246 prodidit: perdidit *AR*
247 Quod: que *P* quid *R* / vero: autem *B* ergo *R* / idem *om. R*
248 ait *om. A* / ad imitationem *corr. ex* admiratione *AFLR* admirationem *B* divinatione *P*
248–49 Mercurii *B*
249 quam: quos *A* / requirendam: inquirendam *L*
250 liram forme *R*
251 poterunt *P* / collocatam: collocandam *B* collocata *R*
252 quid *P* / poeta *om. L*
253 *post* recommendans *mg. hab. A* Cassiodorus / enim *om. L* / illa: ista *A*

256 Inde: item *A* / 7 *om. R* a *FP* / planetorum *P*
258 Nec: non *R* / celo: ?celeste *B*
259 *ante* Unde *hab. B* melodiam
260 illis *om. F* / oblectationibus: oblecta rationibus *R*
261 defficis *F* / *ante* marcescit *hab. B* et / marcessit *BL* / *post* marcescit *hab. BP* et cetera *et hab. L* inconvenientia si non sint commensurabiles motus
262 liberet *AU* / quanta *P* / cecitate *P* / qui: que *A* / *ante* celi *hab. B* motuum
263 proportionare *R* / ita: ista *P*
263–64 poterit umquam *tr. ABU*
264 aspectus precognoscere coniunctiones: coniunctiones et aspectus planetarum precognoscere *B* aspectus cognoscere

the most eloquent Cassiodorus, who says: "The harmony of the heavens is not explicable in human terms, for reason furnishes [the understanding of it] to the mind only, but nature does not make it known to the ears."

That music does, however, exist in the heavens is made quite clear by the same [Cassiodorus] when he says. "Mercury is said to have invented this— i.e., the lyre—"in imitation of a changeable stringed instrument which... astronomers believe can be perceived among the stars, since they visualize the shape of a lyre formed from a collection of stars. These stars, shaped as instruments, produce the music of the heavens."[45]

Indeed, a certain poet says: "The cither is the work of the sun, the lyre of Mercury."[46]

Again, in recommending music, Cassiodorus says this about it: "For what could be more excellent than that which regulates the fabric of the heavens by sweet sound and comprises the harmony of nature in virtue of its ubiquitous power?"[47]

For this reason, the ancients imagined that a double tetrachord pertains to the seven planets, and a simple monochord to the outermost star-bearing sphere; and with a wonderful silence, there is produced from these an harmonic dance. Not only did they assume music in the heavens, but even beyond it. Thus, Cassiodorus says: "They say, indeed, that one ought to believe that supercelestial blessedness is made complete and fulfilled through musical delights that never end and never cease for a moment because of any weakness."[48]

I will now show how great a blindness is produced in men who deny that the celestial velocities are proportioned by numbers. For if this were true, no one could ever foresee aspects, or predict conjunctions, or learn of effects beforehand. Indeed, astronomy would lie hidden [from us] in every age, unknown and even unknowable, as was shown before; it would no longer be counted among the mathematical disciplines. But if the celestial velocities are incommensurable, why did the maker of the world "give to man an uplifted face and bade him stand erect and turn his eyes to heaven?" "Of what avail that man derived his intelligence from above?"[49]

---

coniunctiones *L* aspectus vel coniunctiones eorum ?vel effectus precognoscere *U* / *ante* predicere *hab. U* aut

264–65 previdere effectus *om. U; tr. B* providere effectus *R*

265 *post* Ymo *hab. R* lata / *post* latebit *hab. B* omnino / *post* incognita *mg. hab. B* astronomia / ac: et *B*

266 probatum est ante: ante probatum est *A* / erit igitur: ergo erit *B* / mathemati-

cos *FL* mathematices *P* / si: celi *A*

266–70 non erit...ignorantia *om. U*

267 celi: motum celestium *B* / sint *F* / *post* et *hab. B* prorsus / *post* os *mg. hab. B* Ovidius

268 hominis *A* / erectes *P*

269 *ante* Quid[1] *hab. B* Ovidius / Quid[1]: quod *P* / caput *om. FLP* / *post* caput *hab. B* et cetera Claudius

270   Nec solum ex hoc sequitur hominum perpetua ignorantia, sed quod cecus philosophorum venerandus olim graviter erravit cum de re ignota mentiri presumpsit. "Inquit Aristoteles, Grecorum philosophorum princeps et dominus verique perhemnis amicus," quod "tempus est numerus motus celi," quod non foret si motus celi non posset numero mensurari.

275   Omnes quoque pariter tabule astrologice essent false. Numquam etiam redirent corpora celestia ad statum similem illi dispositioni in qua nunc sunt completo illo magno anno omnium circulationum quem multi philosophi esse affirmaverunt.

    Unde Plato in *Timeo* loquens de tempore dixit: "In quo," scilicet tem-
280 pore, "sit admiranda varietas proventuum celestes tramites undique stellis variantibus. Est autem intellectu facile, quod perfectus temporis numerus perfectum annum compleat tum demum cum omnium octo circumactio-num cursus peracti velut ad originem atque exordium alterius circumactio-nis revertentur, quem semper idem uniformisque motus dimetietur."

285   Et rursum: "stellarum vero choreas et alterius applicationes ad alteram variosque giros, quos edunt admirabili venustate iuxta ambitus circulorum reditusque et anfractus ad eas sedes ex quibus progresse sunt...."

    Reverti necesse est quod non minus pulchre Apuleius explicat ita dicens: "Esse autem stellarum necesse est certos ambitus legitimis circulis perpetuo
290 servatos quos vix hominum solertia comprehendit. Unde fit ut et magnus ille vocitatus annus facile noscatur cuius tempus implebitur cum vagan-tium stellarum comitatus ad eundem perveniret finem novumque sibi exordium et itinera per vias mundi reparaverit."

270 Nec: ut *A* / ex hoc sequitur: sequitur ex hoc *L* / homini *A* hominem *P* / ho-minum perpetua: perpetua homini *R* / sed quod: ?qui etiam *U*
271 *post* olim *hab. B* tam / erraverit *L*
272 *ante* mentiri *hab. B* sibi / mentiri: gra-viter *F sed post* presumpsit *hab. F* men-tiri / Inquit *om. U* inquid *L* / *post* In-quit *hab. B* autem
273 et dominus *om. B* / virique *L* ubique *P* veri *U* / quod: inquit *U* / *post* numerus *hab. B* et mensura
274 non[1]: ?nonne *U* / foret; est *A* / motus celi *om. R* / numero *om. U* / mensurari: commensurari *AU*
275 pariter *om. ABL* / *ante* false *hab. B* ?pro-tinus / *post* false *hab. B* et impossibiles
276 statum similem: consimilem statum *A* / illi dispositioni *tr. U* dispositioni *B* /

in: michi *F*
277 *post* sunt *hab. B* aut aliquando fuerunt / omnium circulationum *om. L* omnium coniunctionum *R* circulationum *U*
279 Unde: uti *U* / dixit: et cetera *U* / *post* dixit *mg. hab. B* Plato / scilicet: qui-dem *A*
279–93 In quo....reparaverit: atque Apollo-nius platones explicans ?ita *U*
280 sit: fit *ABR* / undique: ?omnino *L*
280–81 stellis variantibus *tr. FLP*
281 intellectum *BLP* / perfectus: profectus *AB* / perfectus temporis numerus: nu-merus temporum perfectus *L* / tempo-rum *ABFP*
282 tum: causam *A* / demum: tamen *A* / cum *om. B* / omni *AFP*
282–83 circumactionum: circulationum *P*
283 atque: vel *A*

Not only does it follow from all this that men would be in perpetual ignorance, but that even the philosophers, once venerated, were blind and gravely erred, having invented things about what is unknown. "Aristotle, the prince and master of Greek philosophers and eternal friend of truth, says"[50] that "time is the number [or measure] of celestial motion,"[51] which would be impossible if the celestial motions were not measurable by number. Furthermore, all astronomical tables would be equally false. The celestial bodies would never return to the same state, namely, to their present disposition, after the completion of a Great [or Perfect] Year—its existence has been affirmed by many philosophers—embracing the circulations of all [celestial bodies].

Thus, speaking of time in the *Timaeus*, Plato says: "This," namely, time, "must produce wonder because of the variety in the number [of heavenly bodies] and the divergencies in their celestial paths. It is, however, easy to comprehend that the perfect number of time terminates the perfect year at the very moment when the paths traversed by the eight revolutions return to the origin and commencement of another revolution, which is always measured by a same and uniform motion."

And once again [he says that time produces] "the dance of the stars, their relationships to one another, and the varied courses which they produce with wonderful elegance as a consequence of the revolutions of their circles, and the turning round and return to the places from which they started...."[52]

That it is necessary for them to return is not a demeaning of their beauty, as Apuleius explains in the following way: "It is necessary, moreover, that there be fixed revolutions of the heavenly bodies preserved perpetually in appropriate circles that are barely comprehended by the human mind. And so it happens that this is called a great year, whose time [or period] is easily known and will be completed when the band of wandering stars reaches the same destination and makes a fresh beginning renewing its journeys through the paths of the world."[53]

---

284 uniformisque: conformisque *L*
285 Et *om. B* / vero: velut *A*
286 quos *om. A* / edunt: educit *A* agunt *B* adunt *F* odunt *P* / venustate: venus date *P* / circulorum: circulorumque *AFLP*
287 reditusque: reditus *AFLP* / et anfractus *om. B* / progressi *A*
288 quod: quid *P* / Apuleius *corr. ex* Apollonius *ABFLPR* / ita *om. B* / *post* dicens *mg. hab. B* Appollonius

289 ambitos *P*
290 servates *P* / quos: quod *BFLP* / comprehendit: deprehendit *B* / ut et: ut *B* etiam *R*
291 noscatur: cognoscatur *A* / tempus: temporis *P*
291–92 vagantia *AFLP*
292 novumque: novum *A*
293 et itinera *om. A* / reparaverint *AFLPR*

Sic quoque de divinis oraculis scriptum est: "Oritur sol, et occidit, et ad
295 locum suum revertitur: ibique renascens gyrat per meridiem, et flectitur ad
Aquilonem: lustrans universa in circuitu pergit spiritus, et in circulos suos
revertitur."

Et tamen demonstratum est prius quod si motus quibus sol movetur
sint incommensurabiles impossibile est ut in circulos suos revertantur.
300 Item Macrobius ait: "Cum ad idem unde semel profecta sunt astra redie-
rint, eandemque totius celi descriptionem longis intervallis retulerint, tum
ille vertens annus appellari potest in quo vix dicere audeo quam multa
hominum secula teneantur."

Ob hoc etiam voluerunt isti philosophi effectus hic inferius reverti
305 similes vel eosdem sicut alibi scribitur: "Quid est quod fuit ipsum quod
futurum est. Quid est quod factum est? ipsum quod faciendum est. Nihil
sub sole novum nec valet quisquam dicere: Ecce hoc recens est: iam enim
precessit in seculis, que fuerunt ante nos," et cetera.

Hoc autem non contingeret si aliqua foret incommensurabilitas in moti-
310 bus celi sed semper provenirent constellationes et effectus quales non fuerunt in perpetuum. Et tamen Platonici dixerunt eosdem homines iterum
redituros illa magna revolutione peracta. Unde etiam Claudius: "…certis-
que ambagibus evi/rursum corporeos anime mittuntur in artus." Item Virgilius: "Has omnis, ubi mille rotam volvere per annos,…/rursumque
315 incipiunt in corpora velle reverti." Et intelligit per mille annos totum illud
maius tempus. Hoc etiam predicavit Pictagoras de quo ait Ovidius in
Fastis quod "…posse renasci nos putat," et cetera.

Hiis rationibus, hiis testimoniis, contenta sum. Quibus credo sufficienter

294 Sic…divinis: in divis quoque *U* / de:
aut *B* / *post* de *hab. AR* sole in / oraculis
scriptum est: scribitur oraculis *U* / Oritur sol *tr. B* / et² *om. A*
294–95 ad locum: cetera *U*
295 *post* suum *hab. P* et / revertetur *A* / ibique renascens gyrat *om. P*
295–96 suum…spiritus *om. U*
296 peragit *R* / spiritus *?F; om. B*
298 est *om. A* / quod: et *U* / quibus *om. U* /
movetur *om. U*
298–99 Et tamen…revertantur *om. L*
299 sint: sicut *P* / circulos suos *tr. AR* / revertatur *AFPU*
300 Item: unde *A* etiam *B* / profecta: pro-
specta *A* perfecta *P* perfectam *R*
300–17 Macrobius…et cetera *om. U*
302 vertens: virtutes *P* / quam: quod *A* qua

BFR
303 homines *P*
304 voluerunt isti philosophi: isti philosophi
voluerunt *B*
305 vel: licet *A* aut *F* / alibi: in eloquiis
sacris *B* / scribitur: describitur *A* dicitur *R* / est: omne *A* / fuerit *A* sint *FP*
306 Quid: quod *ABR* / ipsum: id *B* / faciendum: sciendum *A*
307 est *om. AP*
308 precessit: processit *BFLPR* / et cetera:
etiam *A* et etiam *P*
309 Hoc: illud *F* / aliqua: qua *AFPR* / foret:
esset *B* / si…incommensurabilitas: incommensurabilitas si qua fieret *L*
310 semper: secundum hoc *B* / provenient *A*
proveniret *FP* / ante et *hab. B* dispositiones in celestis / *post* effectus *hab. B*

This has also been written among divine pronouncements: "The sun riseth, and goeth down, and returneth to his place: and there rising again, / Maketh his round by the south, and turneth again to the north: the spirit goeth forward surveying all places round about, and returneth to his circuits."*

And yet it was demonstrated earlier that if the motions are incommensurable, it is impossible that they all return [to the same place] on their circles. Thus Macrobius says: "When all the stars have returned to the same places from which they started out and have restored the same configurations over the great distances of the whole sky, then alone can the returning cycle be called a year; how many generations are contained in a great year I scarcely dare say."[54]

For this reason, also, the philosophers wished [or thought] that similar, or identical, effects should be repeated here below [on earth] just as it is written elsewhere: "What is it that hath been? the same thing that shall be. What is it that hath been done? the same that shall be done. / Nothing under the sun is new, neither is any man able to say: Behold this is new: for it hath already gone before in the ages that were before us,"† and so forth.

However, this would not happen if there were any incommensurability in the celestial motions; but, rather, constellations and effects would occur that were not of an enduring or perpetual kind. But the Platonists, however, have said that the [very] same men would return again after the completion of this great revolution.[55] Claudius also writes: "...and after a fixed cycle of years thou sendest souls once more into mortal bodies."[56] And Virgil says: "All these, when they have rolled time's wheel through a thousand years....conceive desire to return again to the body."[57] By "a thousand years" Virgil understands a much greater period of time. Pythagoras also proclaimed this, and Ovid said of him in the *Fasti* that he "thought that we could be born again,"[58] and so forth.

I rest content now with these arguments and witnesses. From them, I believe

---

* Ecclesiastes 1:5–6; Douay-Rheims Version (Oresme's quotation agrees with Hetzenauer's edition of the Latin Vulgate).

† Ecclesiastes 1:9–10; Douay-Rheims Version (Oresme's quotation agrees with Hetzenauer's edition of the Latin Vulgate).

---

in ?terris / non: numero L
311 tamen *om. A* / *post* eosdem *hab. F* in
312 redituros: reverturos A / peracta: facta L pacta P transacta R / Unde *om. B* / Claudius: Claudianus A
312–13 certisque: ceterisque A
313 evi: vel R / Item: et B
314 omnes B / rursumque: rursum B
315 corpora: incorporare P / intelligit: ?in-

cepit L
316 maius: longum A / predicit B / *ante* Pictagoras *hab. B* ille
317 Fasti L / quod: que P / posce A
318 *post* testimoniis *hab. U* atque aliorum philosophorum et poetarum Macrobi Virgilii Claudii et cetera superius / sum: super U

probasse opinionem meam quam peto divine maiestatis auctoritate con-
320   firmari.

Ac contra Geometria cepit partem oppositam tali oratione defendere et
fulcire.

[Oratio Geometrie]

Soror nostra, pater, verborum prodiga, sententie parca implens divinas
325   longis ambagibus aures nihil efficaciter demonstravit. Et si apparenter
locuta est nihil tamen prohibet quedam falsa esse probabiliora quibusdam
veris. Suam nihilominus partem minus probabilem opinor quod ostendam
fortioribus licet paucioribus persuasionibus; argumentaque eius reiciam
non respondendo per ordinem ad singula, sed tangendo breviter unde
330   possit nostrum propositum concludi et rationes ipsius annullari.

Dicit enim quod in suis proportionibus rationalibus consistat quedam
pulchritudo atque perfectio quod non nego. Verumtamen celestia multo
ampliori fulgent decore si corpora sint commensurabilia et motus incommensurabiles; aut si aliqui motus sint commensurabiles et alii incommen-
335   surabiles qui omnes sunt regulares quam si cuncta essent commensurabilia,
ut scilicet irrationalitate et regularitate commixtis, regularitas irrationalitate varietur, et irrationalitas regularitate debita non fraudetur
propter quod etiam quilibet motus simplex spericus est secundum partes
subiectivas difformis et secundum partes temporis regularis.

340   Nam sive irrationalis proportio sit nobilior sive non, earum tamen

319–20 auctoritate confirmari: oraculo confirmari *AR* confirmari *FP* confirmari
auxilio *L* sententia confirmari *U*
321 Ac: et *BFLU* / contra: vero *U* / Geometria cepit *tr. FLP* / tali oratione *tr.*
B / defendere: offendere *P*
322 *post* fulcire *hab. B* dicens
323 [Oratio Geometrie] *hab. L post* confirmari auxilio (*see variants for lines*
319–20) *et ante* et contra *mg. hab. F* (*see*
variants for line 321); *om. APRU* respondit geometria *mg. hab. B ante* et contra
(*see variants for line 321*)
324 Soror nostra pater: pater soror nostra
B / *post* nostra *hab. U* o / pater: pacis
FL
325 *post* nihil *hab. U* locuta est quod
326 est: sit *B* / tamen prohibet *tr. R* tamen

pluribus *A* prohibet *B* / esse probabiliora *tr. A*
326–27 esse…veris: probabiliora veris quibusdam esse *B*
327 quod: et hoc *B*
328 *post* paucioribus *hab. L* rationibus / eis *F*
329 *post* non *hab. P* reiciam / *post* ordinem
hab. FLP non respondendo / ad: per
FLP et per *B* / singula: ?singularia *P* /
breviter: brevis *P*
330 posset *RU* / nostrum *om. L* / et rationes: rationesque *U* / ipsius: illius *ABU*
331 rationalibus: rationabiliter *A*
332 atque: et *B* / quod: quedam *F* / *ante* celestia *hab. L* ?an / multo *om. L* / *post*
multo *hab. B* et
333 ampliori: applicari *FP* / fulgent: multo
fulgeret *F* fulget *A* / sunt *B*

that I have adequately proved my point of view, which I pray may be upheld by the authority of divine majesty.

And against this, Geometry took the opposite side [and sought] to defend and support it with this oration.

[The Oration of Geometry]

Oh father, our sister, lavish with words and spare in [good] judgment, filling your divine ears with long digressions, has demonstrated nothing effectually. But if she has spoken clearly, nothing prevents certain false propositions from being more probable than certain true ones.[59] Nevertheless, I believe her side is less probable and will show this with stronger arguments and far less special pleading. [However] I shall not repudiate her arguments by replying to them separately and in order, but will destroy her reasons by discussing them briefly as they bear upon the inferences to be drawn from our proposals.

Now she maintains that there is a certain beauty and perfection in her rational ratios, and I do not deny this. However, the heavens would glitter with even greater splendor if the bodies were commensurable and their motions incommensurable, or if some motions were commensurable and others incommensurable, where all are regular [and uniform], than if all were commensurable. By mixing together irrationality and regularity, the regularity would be varied by the irrationality, and the irrationality, with regularity bound to it, would not be deprived. It is because of this that any simple spherical motion is non-uniform with respect to its subjective parts and regular as regards temporal parts.

Now whether or not an irrational ratio is more noble [than a rational ratio],

---

334 aut: ac *A* / motus sint *tr. FLP* motus *A* motus sunt *R* / commensurabiles: incommensurabiles *L* / alii: aliquando *P* aliqui *U*

334–35 incommensurabiles: commensurabiles *L*

335 *post* qui *hab. B* tamen / quam: quod *P* / cuncta: omnia *L* / essent: sint *B* / commensurabilia: incommensurabilia *P*

336 irrationalitate: rationabilitate *A* / irrationalitate et regularitate: irregularitate et irrationalitate *U* / commixta *F*

336–37 irrationalitate: irrationabilitate *A* ir-

regularitate *R*

337 *ante* varietur *hab. BU* et / irrationalitas: irrationabilitas *A* / fraudetur: fundetur *P*

338 quod: hoc *R* / quilibet: quibus *RL* quelibet *P* / *ante* spericus *hab. U* motus / est *om. B*

339 *ante* difformis *hab. B* est

340 *post* Nam *hab. AFPR* et / *post* sive[1] *hab. U* regularis / irrationalis: rationalis *LR* / *post* proportio *hab. R* irrationali / sit nobilior *tr. L* / eorum *AB* ipsarum *L*

congrua commixtio pulchrior est singularitate uniformi; sic enim videmus in aliis. Unde mixtum ex elementis melius est optimo elemento; et celum insignius quam si essent stelle ubique per totum; ymo universum est perfectius propter corruptibilia et etiam propter monstra; cantusque consonantiis variatus dulcior quam si fieret continue optima consonantia, scilicet dyapason; et pictura variis distincta coloribus speciosior colore pulcherrimo in tota superficie uniformiter diffuso. Sic etiam celorum machina nullo carens decore tali varietate componitur ut corpora numero singulumque eorum pondere, id est magnitudine, motusque mensura constent. Que mensura si esset numeralis frustra videretur dictum numero et mensura. Hec ergo mensura ad continuitatem illam refertur que non potest per numeros dimetiri. Et dum eam comprehendere non possumus ipsam irrationalem et incommensurabilem appellamus. Solet si quidem sepe contingere ut homo subtilis in multa variatione pulchritudinem percipiat cuius diversitatis ordinem homo rudis non advertens totum estimat fore confusum, sicut irrationalem proportionem vocamus quam nostra ratio capere nequit. Et ipsam tamen distincte cognoscit dei ratio infinita et divino conspectui loco suo posita placet celestesque circuitus efficit pulchriores.

Ad hoc autem quod arguit nos sibi facere iniuriam et de primogenitura qua se dicit precellere, respondemus quod nullam mensuram nullam proportionem habet in numeris quam non habeamus in nostris magnitudinibus et cum hoc infinite alie reperiuntur in continuis quarum nulla invenitur in numeris. Habemus igitur quicquid habet et multo plus que ergo est primogenitura sua. Nec etiam proportiones numerales privamus a celo. Sed si cum eis in celo ubi omnia relucent sunt alie nullum inde patitur Arismetica detrimentum.

341 congrua commixtio *tr.* R congrua mixtio BF / est *om.* AL / *post* est *hab.* B et ?decertior / enim *om.* B

342 Unde: quia B / *post* mixtum *hab.* A quod est / *post* melius *hab.* B et nobilius / et *om.* A

343 essent: sunt A / ubique: undique B

343–44 est perfectius *tr.* B

344 cantusque: et cantus B cantus L

344–45 consonantiis: disonantiis R

345 *ante* quam *hab.* F est / fieret *om.* A esset B / continue *tr.* F *post* consonantia / *post* consonantia *hab.* B factus

346 et *om.* F / varia U / distinctis U / *ante* colore *hab.* F est

346–47 pulcherrimo: pulcrior A

347 in *om.* B / uniformiter *om.* L / diffuso: superfuso LPU superfusio A / sicut A

348 decore: docere A / componitur *om.* U

349 *ante* mensura *hab.* B eorum / constent: consistant B constant L constet P

350 Que: quia U / *ante* si *hab.* U ea / est A

351 illam: istam A ?atque F

352 possimus R / ipsam: eam L

353 et *om.* AP

354 *ante* percipiat *hab.* B et decorem

355 ordinem: ordinationem L

356 sic R

357 nequit *obs.* B non potest R / Et[1] *om.* ABR

an harmonious union of them is more excellent than a separate [and inde-
pendent] uniformity. Indeed, we see this in other things. Thus, a mixture of
elements is better than the best element; the sky is more wonderful than [it
would otherwise be] if the stars were distributed everywhere; the whole universe
is more perfect because there are corruptible things—and even monsters—in
it; a song with its consonances varied is sweeter than if it were constituted
continually from the best consonance [that was unvaried], namely, a diapason;
and a picture decorated with different colors is more beautiful than one in
which the most beautiful color is spread uniformly over the entire surface.
Even the structure of the sky, which lacks no adornment, is constituted out
of such variety that the bodies are determined by number, and each body by
weight—that is, magnitude—, and the motions by measure. If "measure" were
numerical, there would be no good reason to express celestial motions by both
number and measure. Therefore, this "measure" is relevant to that continuity
which numbers cannot measure. When the measure is indeterminable, we call
it irrational and incommensurable. It happens frequently that a subtle man
perceives the beauty in much diversity, while an ignorant man, who fails to
consider the whole, thinks that the sequence [of events] in this diversity is
confused, just as he does not realize that what we call an irrational ratio is
part of our order and plan. And yet the infinite plan of God distinctly recognizes this diversity which, put in its proper place, is pleasing to the divine
sight and makes the celestial revolutions more beautiful.

Let us reply now to her charge that we do her harm and [also] consider
the matter of the "firstborn," a distinction she claims for herself. We say that
in numbers there is no measure and no ratio that is not included within our
magnitudes; but along with these there can be discovered in continuous
quantities an infinite number of other [ratios and measures], none of which
is found among numbers. Therefore, we have what it has, and much more,
so that we are the firstborn [of Mathematics]. Nor do we deprive the heavens
of numerical ratios. But if, in addition to these numerical ratios, there are other
ratios in the heavens, where everything shines forth, Arithmetic suffers no loss.

---

358 conspectui: spectui *U* / loco suo *tr. B* /
    celestusque *A* que celestes *R* / circuitus:
    motus *B*
360 Ad hoc: adhuc *A* / nos sibi: nobis *R* /
    primogenituram *A*
361 dicit: ?dic *AFP* / precellere: precedere *L*
362 quam: quem *A* / *ante* non *hab. AL* nos /
    nostris *om. A*

363 quarum: quorum *B*
364 *post* in *hab. B* discretis et / *post* habet
    hab. A ipsa / multo *om. U* cum hoc *B* /
    est *om. A* / *post* est *hab. B* de qua se
    iactat
365 etiam *om. U*
366 Sed…celo om. *F* / si *om. AL* / sint *U*
367 detrimentum: decrementum *P*

De sermone vero quo celi musicam predicat, sic dicimus quod credi non debet testibus invicem discordibus. Unus enim dicit quod illa armonia fit

370 cum sono audibili, alter negat; unus asserit quod orbis supremus resonat acutius, alter dicit quod non sed infimus. De quibus ait Plinius in libro secundo *Naturalis historie:* "Dicunt Saturnum moveri Dorio Mercurium pthongo, Iovem Phrygio, et in reliquis similia, iocunda magis quam necessaria subtilitate." Quasi diceret quod voluntarie ficta sunt et sine

375 ratione. Et concesso quod facerent consonantiam tacitam. Tamen non est proportio sonorum sicut proportio velocitatum sive enim chorda sive tympanum percutiatur fortiter aut debiliter, tarde vel velociter nihil inde mutatur. Soni gravitas numquam ob hoc variatur, acuties. Nec sonus fistule propter sibili velocitatem continuam acuitur continue licet quandoque

380 transeat quasi subito ad duplam acutiem, et ita de aliis musicis instrumentis. Unde Pictagoras non malleantium motum non lacertorum mensuravit vigorem sed malleorum proportionem quesivit quantitatem que eorum per pondera novit.

Igitur acuties soni non sequitur velocitatem sed potius corporum sonan-

385 tium quantitatem aut figuram; aut rigorem, ut in chordis; aut asperitatem maiorem aut minorem, ut in rota plaustri; aut quantitatem propulsi aeris ut vult Aristoteles in *De probleumatibus;* aut plura istorum vel similium. Et ergo nec proportio sonorum est sicut proportio velocitatum. Sed quod plus est si consonantia esset secundum velocitates corporum sonan-

390 tium, adhuc proportiones velocitatum celi non sunt symphonice. Dicunt enim quod solis ad Venerem sit dyesis que consistit in proportione istorum numerorum 256 et 243. Et tamen nullus ponit motus solis et Veneris taliter proportionari et universaliter nullorum motuum celi proportio est

---

368 sermone: secundo *A* / vero *om. U* / celi musicam predicat: predicat celi musicam *A* / sic *om. BU* sicut *U*

369 testibus invicem discordibus: discordibus invicem testibus *B* testibus suis invicem discordibus *R* invicem discordibus *U* / unum *A* / Unus enim: quia unus *B* / enim: vero *F* / illa: isti *A*

370 audibili: audibiliter *BFLP* / negat: nominat *P* / resonet *ARU*

371 alter dicit *tr. U* / quibus: quo *F* / ait: dicit *U* / *ante* Plinius *hab. B* ipse / in *om. BFLPR* / libro *om. U*

372 secundo *om. A*

372–73 Dicunt…similia: vide ibidem ubi Plinius dicit talia *U*

373 pthongo: prehongo *A* / Phrygio *corr. ex*

frigio *ABFLPR*

374 subtilitate *om. U* / *post* voluntarie *scr. et del. F* fixa / ficta sunt *tr. F* ficta sint *AU*

375 facerent: faciant *U* / *post* facerent *hab. R* aliquam

376 porportio[1]: consonantia *L* / sonorum: suorum *R* / enim: eius *P* / sive[2]: aut *BL* etiam *FP*

377 fortiter *om. L* / tarde: tardelis *P* tardius *U* / vel *om. P* sive *F* / velociter: velocius *U* / inde: michi *P*

378 ob: ab *R* / variabitur *U* / sonus: sompnus *A* / fistuli *A*

379 sibili: fistuli *A* / licet: sed *F*

380 transeat: pertranseat *U* / acutiem: acuitatem *B* / et ita: sic est *B* est ita *F*

As to the discussion where she lauds celestial music, we say that, in virtue of such mutually conflicting witnesses, she ought not to be believed. For one witness insists that the harmony occurs with audible sound, while another denies it; one asserts that the outermost sphere sends forth the sharpest [or loudest] sound, while another denies it claiming this for the lowest sphere. Concerning all this, Pliny, in the second book of his *Natural History*, says: "They [i.e., the Pythagoreans] say that Saturn and Mercury move in the Dorian mode, Jupiter in the Phrygian, and similarly with the other planets—a refinement more entertaining than convincing."[60] Pliny as much as says that these are assigned arbitrarily and without reason. But I concede that they might produce a silent harmony. However, a ratio of tones does not vary as a ratio of velocities, for whether a string or a drum is struck strongly or weakly, slowly or quickly, nothing is changed by this. A bass [or deep tone] is consequently never altered; nor is a high [or sharp] tone. [Thus] the tone of a pipe is not made continuously sharper by the continuous velocity of the whistling, although sometimes it doubles the sharpness almost instantaneously. And this applies to other musical instruments.[61] This explains why Pythagoras did not measure the motion of the hammering or the force of the blows, but [instead] sought the ratio of the hammers, a quantity which he knew by their weights.[62]

Hence the sharpness of a sound [or tone] does not vary as a velocity but rather as the quantity, or figure, of the sounding bodies; or as the tension of the sounding bodies, as in strings; or as their greater or lesser roughness, as in the wheel of a wagon; or, as Aristotle would have it in the *Problems*,[63] the velocity varies as the amount of air that is propelled; or it might vary as several of these [factors], or as ones similar to them. And so, a ratio of sounds [or tones] is not related as a ratio of velocities. But if a consonance [or harmony] did vary as the velocities of the bodies emitting the sounds, even then the ratios of the celestial velocities would not be harmonic. For they say that the relationship obtaining between the sun and Venus is a *diesis*, a ratio consisting of the numbers 256 and 243. But no one [really] supposes that the motions of the sun and Venus are related in this way, and generally [no one really believes]

---

381 maleatium *A* / motum: modum *L* / la-
    certorum: ?brachorum *U* / mensuravit:
    mensurant *R*
382 quesivit: que sunt *P* / earum *L* / *ante*
    per pondera *mg. hab. R* per pondera
385 quantitatum *A*
387 vult *om. B* volt *R* / in *om. U* / *post* pro-
    bleumatibus *mg. hab. B* Aristoteles
388 Et *om. B* / nec *om. FP* / proportio²

om. A
388–90 Sed…sonantium: corporum nam si
    sit *U*
389 est *A*
390 *post* adhuc *hab. U* tamen
391 sit: est *L*
392 et¹ *om. A* / 243: 5243 *A* / nullus *om.*
    P / *post* solis *hab. A* scilicet
393 taliter: sic *AB* / celi *om. L* / est: sit *R*

secundum aliquam consonantiam principalem. Si igitur spere celi faciant
395 aliquam consonantiam in movendo, illa non est attendenda penes velocitates motuum sed penes capacitates sperarum, seu penes orbium quantitates. Negata itaque commensuratione motuum dummodo commensuratio
corporum celestium concedatur, in nullo preiudicatur musice quin ipsa in
suo celesti choro sensibiliter intonet aut intelligibiliter plaudat. Ymo placet
400 nobis quod in deorum palatio cum suis Musis tripudiet spericamque pulsans citharam appareat in conspectu Apollonis gratissima ioculatrix.

Adhuc autem alia via procedemus. Que est ista cantilena que placeret
sepe aut multotiens repetita? Nonne talis uniformitas gignit fastidium?
Ymo certe, et novitas plus delectat. Nec esset reputatus cantor optimus sed
405 cuculus, qui non posset modulos musicos variare qui sunt variabiles in
infinitum. Nunc vero si omnes motus celi sunt commensurabiles necesse
est eosdem, vel similes, motus et effectus infinities iterari, si mundus semper
duraret. Et similiter necesse est esse illum magnum annum qui ante oculos
deitatis non est nisi sicut dies hesterna que preteriit ymo minus. Qua
410 propter iocundius atque perfectius videtur et etiam divinitati convenientius
quod non totiens repetatur idem sed quod novas et dissimiles prioribus
constellationes effectusque varios semper producat ut illa seculorum pro-
lixa series, quam Pictagoras per cathenam auream intellexit, non redeat
in circulum sed recte procedat sine fine semper in longum quod tamen
415 non fieret absque aliqua incommensurabilitate motuum celestium.

Nec est credendum in hoc illis philosophis poeticis quoniam sicut de
aliis dictum est ipsi invicem non concordant in quantitate illius anni
magni; nec dicta eorum possunt stare cum experientiis ab astrologis hactenus observatis.

394 celi: celestes *U*
394–95 faciant…movendo: aliquam in mundo faciant consonantiam *U*
395 movendo *AR* modo *FLP* motu *B* / illa:
ista *AFP* / non est attendenda: attendenda est *U*
395–96 velocitates…penes[1] *om. U*
396 capacitatem *U*
396–97 orbium quantitates *tr. B*
397 itaque: ergo *B* igitur *U* / commensuratione: commensurabilitate *U* / post
motuum *hab. B* celi
398 corporum celestium concedatur: concedatur corporum *B* / celestium *FLPRU*
supercelestium *A* / quin: quoniam *FP* /
ipsa *om. AR*
400 suis Musis *tr. U* / tripudiat *R*

400–401 pulsans: propulsans *A*
401 appareat: apariat *A* / *ante* ioculatrix
hab. B iocundissimaque
402 autem *om. BF* / via procedemus: vice
procedemus *R* procederemus via *U* /
ista: illa *PR* / placet *B* pulset *U*
403 sepe aut multotiens: multotiens et sepe
A / aut: sive *F* ac *R* / Nonne: nam *R*
404 *ante* certe *mg. hab. R* notanda / certe
om. P / et novitas *om. B* novitas *F* / esset
om. A est *P* / reputus *LP* reputatur *A*
405 *post* posset *hab. B* sciret / modulos:
modos *AFLPRU* / qui[2]: que *P*
406 motus: modus *P* / celi sunt: celestes
sint *U*
407 vel *om. B* / sive *F* / iterari: reiterari *A*
408 Et *om. B* / similiter: simpliciter *U sed*

that any celestial motions are related as any one of the principal concords [or harmonies]. However, should the celestial spheres produce some concord while moving, this ought not to be measured in terms of the velocities of the motions, but rather by the volumes of the spheres, or the quantities of the orbs. Furthermore, although the commensurability of the celestial motions was denied, while the commensurability of the celestial bodies was granted, there is yet no way to judge whether, in its celestial setting, the music issues forth sensibly or intelligibly. It is, indeed, very pleasing to us that, in the palace of the gods, the most delightful jester should dance with the Muses and, while strumming a spherical cither, appear in full view of Apollo.

Now let us proceed along another path. What song would please that is frequently or oft repeated? Would not such uniformity [and repetition] produce disgust? It surely would, for novelty is more delightful. A singer who is unable to vary musical sounds,[64] which are infinitely variable, would no longer be thought best, but [would be taken for] a cuckoo. Now if all the celestial motions are commensurable, and if the world were eternal, the same, or similar, motions and effects would necessarily be repeated. Also, of necessity, there would be a Great Year, which, in the eyes of God, has no existence— indeed, it has even less existence than a day that has passed into yesterday. For this reason it seems more delightful and perfect—and also more appropriate to the deity—that the same event should not be repeated so often, but that [on the contrary] new and dissimilar configurations should emerge from previous ones and always produce different effects. In this way, the far-stretching sequence of ages, which Pythagoras knew as the golden chain, would not return in a circle, but would always proceed endlessly in a straight line. This could not happen, however, without some incommensurability [obtaining] in the celestial motions.

On this matter one should not give credence to those poetic philosophers, who, as in other matters already mentioned, fail to agree with one another about the length of the Great Year. Nor, indeed, are their statements compatible with the phenomena observed thus far by astronomers.

mg. hab. U alibi similiter / illum: istum *A* / qui: que *L*

409 nisi *om. R* / sicut *om. U* / hesterna: externa *F* / que *om. B* aut *U*

410 atque: et *B* / videtur *om. L* / et *om. A* / etiam *om. B* / convenientius *U sed mg. hab. U* alibi convenire

411 quod[1]: que *F* / repetat *A* / novas: nonas *BU* / et *obs. B; om. R*

415 absque: sine *R* / aliqua *om. B*

416 est credendum *tr. B* / illis *om. U* / philosophis poeticis *tr. A*

417 ipsi: ipsis *AB* / invicem: adinvicem *L* / invicem non concordant: non concordant invicem *B* / illius: istius *P*

417–18 anni magni *tr. B*

418 dicta eorum *tr. B*

420   Hanc autem incommensurabilitatem aliquis forte magis credet ex eo
quod nunc dicam. Nam prius demonstratus est quod si omnes motus celi
sint commensurabiles impossibile est solem et lunam coniungi vel opponi
toto eterno preter quam in paucis punctis celi; et ita de aliis aspectibus et
de reliquis planetis. Unde infiniti essent meridiani nec in quibus nec prope
425   quos sol existens umquam intraret Arietem; et sic de ceteris signis. Similiter essent multi gradus celi in quibus aliqui planete non possent convenire
nec etiam tres vel plures possent quamlibet approximari; et sic de consimilibus consequentiis iam probatis. Quarum consequentia non contingeret
incommensurabilitate motuum posita et que non sunt verisimillia vel
430   congrua pulchritudini ipsius universi. Ut quid enim privarentur alique
partes ecliptice a coniunctione solis et lune, vel ab aliqua notabili con-
stellatione? Magis igitur dicendum est quod ipsius ecliptice nulla est tam
parva portio in qua non sint vel fuerint aliquando sol et luna coniuncti;
et sic de similibus que incommensurationem motuum consequuntur.
435   Illud autem quod obicitur de hominum ignorantia non concludit, sufficit enim prescire coniunctionem vel eclipsim futuram esse infra aliquem
gradum, aut minutum, vel secundum, vel tertium, ipsius mobilis; sive
temporis nec oportet predicere in quo puncto aut in quo instanti, quoniam
ut ait Plinius "celi mensura non venit in digitos." Et secundum Ptholo-
440   meum non possumus in talibus comprehendere veritatem ad unguem.
Qui igitur in talibus sic pronuntiat ut non appareat error notabilis sufficienter et pulchre iudicasse videtur. Et si omnes motus precise ab hominibus
noscerentur iam non oporteret amplius observationes facere nec vigili

---

420 autem: ergo *A*
421 prius: primo *A* / *post* prius *hab.* *U* dic-
    tum est et / est *om.* *AU* / motus celi *tr.*
    U
422 sunt *ABFP* essent *U* / *ante* commensura-
    biles *hab.* *U* aut sint / solem et lunam:
    lunam et solem *F*
423 *ante* toto *hab.* *U* in / preter: propter *FP* /
    paucis punctis celi: punctis celi paucis
    B / *post* aspectibus *hab.* *B* similiter
424 de *om.* *A* / Unde: similiter *B* / nec[1] *om.*
    B / nec[2] *om.* *B*
425 quos *om.* *B* / umquam intraret *tr.* *U*
    numquam intraret *BR* / *post* Arietem
    hab. *B* vel Cancrum
426 multi: aliqui *L* / grada *B* / celi *om.* *P*
427 etiam: et *R* / pluris *P* / possunt *R* / et
    sic: nec *A*
427–28 consimilibus: similibus *FR*

428 *ante* probatis *hab.* *B* ante / consequen-
    tia: consequentie *U* / contingerent *FLPU*
429 et *om.* *B* / que *om.* *A* / vel: nec *ABR*
    nihil *P*
430 ipsius universi *tr.* *U* universi *FLP* / pri-
    vareretur *P* / aliquas *A* aliqui *P*
431 solis et lune: lunarum vel aliorum plane-
    tarum *B* / ab *om.* *B* / notabili: nobili
    AFP
432 est[1] *om.* *P* / ipsius *om.* *BU* / ecliptice *tr.*
    B post portio (*linea 433*) / nullam *A* /
    post nulla *hab.* *U* ?nec / est[2]: esse *A*
433 parvam portionem *A* / sunt *R* / vel: sive
    F nec *R* aut *U* / fuerint aliquando *tr.* *A* /
    aliquando: quandoque *F* / *post* luna *hab.*
    B et cetera.
434 sic: ita *BU* sicut *P* / similibus: consimili-
    bus *L* aliis *B* / incommensurationem: in-
    commensurabilitatem *B* incommensura-

But perhaps from what I shall say now someone may believe this incommensurability more readily. For it was demonstrated previously that if all celestial motions were commensurable, it would be impossible for the sun and moon to be in conjunction or opposition through all eternity, except in a few points in the sky. This applies also to other aspects and the remaining planets. There would exist, then, an infinite number of meridians in which, and near to which, the sun could never enter Aries. The same may be said about other signs [of the zodiac]. In a similar way, there would be many degrees in the sky where some planets could never meet—not even three, or more—however much they might approximate to one another. And the same applies to consequences of a similar kind that have already been demonstrated. [However], none of these consequences, which are improbable as well as inappropriate for the beauty of this world, would occur if the incommensurability of these motions were assumed. For why should some parts of the ecliptic be deprived of a conjunction between sun and moon, or of some extraordinary configuration? Rather, one should be able to say that there is no part of the ecliptic so small that the sun and moon would not conjunct there sometime, or have not already conjuncted there. And the same may be said for similar events that might follow from the incommensurability of these motions.

Furthermore, those who object that man would be ignorant [of celestial events if these motions were incommensurable] are unconvincing. For it is enough to know beforehand that a future conjunction, or eclipse, of this mobile falls below a certain degree, minute, second, or third; nor is it necessary to predict the exact point or instant of time [in which these will occur], since, as Pliny says, "the measure of the heavens is not reducible to inches."[65] And according to Ptolemy, we are unable to determine the exact truth in such matters.* In these matters, then, anyone who announces results that are free of noticeable error would seem to have determined things adequately and beautifully. But if men knew all motions exactly, it would not be necessary to make

---

* See above, III.18–20.

---

bilem *U sed mg. hab. U* alibi incommensurabilitatem motuum ?consequitur / motum *U* / *post* motuum *hab. B* etiam / consequitur *A*
435 homini *R* / concludit: conclusit *U*
436 enim *om. L* / prescire: scire *L* / eclipsim futuram *tr. B* / infra: intra *P*
436–37 aliquem gradum *tr. U*
437 *post* tertium *hab. B* et cetera
438 temporis: tempus *R* / in quo...in^2: punctu aut *U*
439 ut *om. P* / celi mensura *tr. B*
440 ad unguem: ?adiungiam *F*
441 appareat error notabilis: error notabilis appareat *B*
442 pulchre: plane *U* / iudicasse: iudicatur *R* / Et: ut *L* / hominibus: omnibus *U*
443 oportet *U*

cura celi circuitus annotare. Melius ergo fuit ut de rebus tam excellentibus
445 aliquid esset scitum et semper restaret aliquid ignotum et ulterius inquirendum quod quadam pregustata dulcedine generosos animos a terrenis
abduceret et excitato desiderio perhemniter detineret in tam alti negotii
venerabili exercitio occupatos. Si etiam isti motus punctualiter essent noti
et ille maximus annus foret possibilis, iam omnia ventura totusque ordo
450 proventuum futurorum possent ab hominibus previderi possentque sibi
fabricare almanach perpetuum de cunctis effectibus mundi. Et sic forent
similes diis immortalibus quorum et non hominum est prenoscere tempora
et momenta futura que soli divine subiacent potestati. Ymo forte repugnat
de quibusdam futuris quod ab homine prenoscantur. Et videtur quedam
455 superbia credere posse pertingere ad prescientiam contingentium futurorum quorum tantum quedam subsunt aliqualiter celesti virtuti.

Magis igitur ponenda est incommensuratio motuum celestium ex qua
hec inconvenientia non sequuntur. Que quidem incommensurabilitas
adhuc aliter ostenditur quoniam sicut alibi probatum est quibuslibet
460 ignotis magnitudinibus demonstratis verisimillius est illas esse incommensurabiles quam commensurabiles sicut quacumque ignota multitudine
proposita magis verisimile est quod sit non perfectus numerus quam perfectus. Ergo de proportione quorumlibet duorum motuum nobis ignota
verisimillius et probabilius est ipsam esse irrationalem quam rationalem,
465 si non obstet alia ratio quod non videtur in proposito consideratis predictis.

Nondum finierat Geometria propositum et ecce Apollo iubet silentium
satisque se reputat informatum. Ac ego, non immerito, admiratus tanta-

444 cura: curare *P* / celi *om. P* / *post* celi
hab. U motus sive / de: ab *U* / rebus
om. U
445 esset: est *A* fuit *R* / restaret aliquid *tr. B*
aliquid *P* / et² *om. BL*
446 *ante* quod *hab. FLP* quam / quadam:
quidem *U*
447 *ante* abduceret *hab. A* valde / et *om. L* /
excitato: ex tanto *A* excitatoque *L*
448 *ante* Si *hab. U* sed et / Si etiam *tr. B*
sed etiam *FP* / essent: est *A*
449 annus foret: motus foret annus *U* /
foret: est *A* / *post* possibilis *hab. B* et
scitus
450 proventuum futurorum *om. U* / possent
tr. U post hominibus; posset *ABFLPR* /
ante ab *hab. U* et / *post* hominibus *hab.*

B precognosci et / possentque: et possent *B* possetque *F*
451 *post* de *hab. B* motibus et / mundi:
huiusmodi *R* / Et *om. B* / forent: fore-
mus *A* esset *B*
452 similes diis immortalibus: diis immortalibus similes *B* / homini *U* / est *om. R* /
prenosce *APR* prenosse *L*
452-53 tempora…futura: momenta et tempora futura *U*
453 momenta futura *tr. AR* / divina *L* / *post*
divine *hab. B* maiestatis / potestati: aut
sunt reposita *B* / forte *om. B* / repugnat
om. U et tr. B post futuris (*linea 454*)
455 pertingere: attingere *L* pertangere *P* /
prescientiam: scientiam *U*
456 tantum quedam *tr. F* tamen quedam

further observations, or to record the celestial revolutions with attentive care. As far as such excellent things are concerned, it would be better that something should always be known about them, while, at the same time, something should always remain unknown, so that it may be investigated further. For such an inquiry, acquired with a sweet taste, would divert noble minds from terrestrial things and, with their desire continually aroused, totally engage and engross those [already] occupied in so respectable an exercise of high-minded endeavor. If, however, these motions were known punctually, and this Great [or Perfect] Year were possible, then, surely, all things to come, and the whole order of future events, could be foreseen by men, who could then construct a perpetual almanac based on all the effects of the world. They would become like the immortal gods. But it is the gods, not men, who know future times and moments, which are subject to the divine power alone. Indeed, it would be very repugnant that men should come to know about future events beforehand. It seems arrogant of them to believe that they can acquire a foreknowledge of future contingents, only some of which are subject to celestial powers.

It seems better, therefore, to assume the incommensurability of the celestial motions, since these difficulties do not follow from that [supposition]. Indeed, incommensurability is shown in yet another way, for, as demonstrated elsewhere,[66] when any two unknown magnitudes have been designated, it is more probable that they are incommensurable than commensurable, just as it is more probable that any unknown [number] proposed from a multitude of numbers would be non-perfect rather than perfect. Consequently, with regard to any two motions whose ratio is unknown to us, it is more probable that that ratio is irrational than rational—provided that no other consideration intervenes that was not taken into account in what has already been discussed.

Scarcely had Geometry finished what she had proposed, when, behold, Apollo, believing himself adequately informed, ordered silence. And I, not

---

AB causam quedam R / celestium B
457 est om. U / incommensurationem U
458 hec inconvenientia: huiusmodi convenientia U / non sequuntur: non sequitur AB sequuntur F consequantur LP consequuntur U
458–59 Que...ostenditur om. B
459 alibet P / quibuslibet: quibusdam A
460 est om. FP / ante illas hab. U dicere / illas: istas AF
461 quam commensurabiles om. P / quacumque: quantumcumque AU quan-

tumque B quoque P
462 magis om. L / sit non tr. BU
463 nobis ignota: incognita U
464 probabilius: ?probabimus P
465 quod: que B / consideratis: considerare A
467 Nondum: nunc F / finierat: fierat A / Apollo: Appollonibus A / iubet om. A
468 se om. AB / ?reputa FP / non om. A / admiratus: admiratione B
468–69 ante tantaque hab. B ?crebatus

que rerum novitate attonitus hec intra me cogitabam. Cum cuicumque
470 vero consonet omne verum, cur sunt discordes iste veritatis parentes? Et
quid est quod loquuntur rethoricis persuasionibus aut topicis probationi-
bus que solent uti solum demonstrationibus omnem aliam argumentatio-
nem aspernantes? Cur incertioris scientie modum eis insolitum accepe-
runt? Vidensque pater Apollo cordis mei cogitatum "ne" inquit "estimes
475 veram esse discordiam inter istas evidentis veritatis clarissimas genitrices
seriose, Enim ludunt et inferioris scientie stilum deludunt. Nos quoque
cum illis fabulando iocose formam dubii iudicis induemus. Processusque
et causas earum visitabimus, inde statim pronuntiabimus in figura iudicii
veritatem." Cumque summo desiderio sententiam expectarem, et ecce
480 sompnus abiit, dubia conclusio restat et ipse nescio quid super hoc iudex
decrevit Apollo.

469 *post* Cum *hab. B* omne verum / cuicumque *obs. B* cuique *PU*
470 vero *om. A* / omne verum *om. B* verum
esse *A* esse verum *P* / sint *A* / veritates
AU / parentes: partes *U*
471 loquitur *ABU* / rethoricis persuasionibus *tr. FP* disputationibus rethoricis *L*
472 solum: solis *AR* / *post* solum *hab. B* rec-
tissimis / demonstrationibus: probatio-
nibus *U*
472–73 argumentationibus *F*
473 aspernantes: spernantes *F* / insolitum:
solitum *F*
474 Vidensque: videns autem *B* / pater
Apollo *tr. R* / estimas *R*
475 evidentis: evidenter *R sed mg. hab. R*
evidentis / genitrices: ?animatrices *B*
476 inferiorum *P* / quoque *obs. B*
477 illis: eis *B* istis *P* / iudicis: ?iudicimus
A / induemus: ?inducimus *A* inducias
U / processumque *B*
478 pronuntiabimus *tr. B post* veritatem

(*linea 479*)
478–79 in…veritatem: iudicii veritatem in
figura *L*
479 *post* Cumque *hab. B* ego
480 *post* ipse *hab. B* prorsus
481 decreverit *AU* / *post* Apollo *hab. A* Ecce
finem sine fine. Explicit tractatus de incommensurabilitate motuum celestium
editus per magistrum Nicholaum Orem
et per me Petrum Defita Padue undecima novembris 1401 deo gracias amen:
scriptum *et hab. B* Explicit huius tractatus de commensuratione motus *et hab.*
L Explicit de commensurabilitate motuum celestium *et hab. P* Explicit no-
bilis tractatus magistri Jordani de Nemore de motibus celestibus et cetera. Si
motus celestes sint commensurabiles vel
non *et hab. U* Explicit tractatus de
commensurabilitate et incommensurabilitate motuum celestium magistri Nicolai Ab horeym

without reason, was astonished and confounded at the novelty of so many things, so that these thoughts occurred to me. Since every truth seemed consonant with each side, in what sense are these disagreements the parents of truth? And what do they really contribute with their rhetorical persuasions or sophistical proofs, seeing how contemptuous they are of every other kind of argument usually employed solely in demonstrations? Why do they adopt the method of more uncertain knowledge—a method to which they are unaccustomed? Perceiving the thoughts in my mind, father Apollo said: "Do not seriously believe that there is a genuine disagreement between these most illustrious mothers of evident truth. For they amuse themselves and mock the stylistic mode of an inferior science. In conversing with them, we also jestingly adopt the manner of doubt in our judgments. But we shall see the advances and causes of these things, for straightaway we shall announce the truth in the form of a judgment." With the most ardent desire did I await his determination, but, alas, the dream vanishes, the conclusion is left in doubt, and I am ignorant of what Apollo, the judge, has decreed on this matter.

Reference Matter

Commentary

Prologue

1 ''Zenonem...tulerunt'' (Prol.4–6). Seneca, *Dialogorum liber VIII: Ad serenum de otio*, chap. 6. Oresme's quotation from Seneca's dialogues is identical with the text given in the edition of Emil Hermes, p. 239.

2 "quo...pulchrius" (Prol.18–19). Cicero, *De natura deorum* 2. 40. 104; trans. Rackham, p. 223. In Rackham's Latin text we find "nihil potest admirabilius esse" (p. 222) where Oresme has only "nihil admirabilius." Since the two versions are substantially alike, I have followed Rackham's translation.

3 "nulla...prestantior" (Prol.19–20). Ibid. 2. 62. 155. Oresme's quotation varies sufficiently from Rackham's text to warrant a full quotation from Rackham (p. 272): "nulla est enim insatiabilior species, nulla pulchrior et ad rationem sollertiamque praestantior;...." The divergences between the two versions have necessitated a somewhat different translation than that given by Rackham. This quotation from Cicero is not identified in Zoubov's Russian translation of the *De commensurabilitate*. The same Ciceronian lines were also cited by Oresme in his *De visione stellarum* (see fol. 31r of MS Florence, Biblioteca Nazionale, Conv. Soppr., San Marco, J.X.19) and *Tractatus contra astrologos* (in *Studien zu den astrologischen Schriften des Heinrich von Langenstein*, ed. Pruckner, p. 236). See also G. F. Vescovini, *Studi sulla prospettiva medievale* (Turin, 1965), p. 199, n. 10; on pp. 195–204, Vescovini discusses and partially summarizes the *De visione stellarum*.

4 "primum...possent" (Prol.21–22). Cicero, *De natura deorum* 2. 56. 140; trans. Rackham. In Oresme's quotation, the word "homines" replaces "eos humo" in Rackham's text (p. 256). I have modified Rackham's translation (p. 257) accordingly.

5 "ut scias...imposui" (Prol.22–27). Seneca, *Dialogorum liber VIII: Ad serenum de otio*, chap. 5. Oresme's version of this quotation from Seneca is nearly identical with what appears in Hermes' edition of the *Dialogues*, p. 236. The few trivial differences are as follows (Oresme's version is recorded first): naturam: illam; locum nobis: nobis locum; omni: omnium; vultum: voltum; flexibili: flexili.

6 "cetera...cupientes" (Prol.36–38). Cicero, *De natura deorum* 2. 41. 104–5; trans. Rackham. Except for a minor difference in word order (Oresme has "animus potest expleri" where Rackham has "expleri potest animus"), Oresme's quotation is identical with Rackham's text (p. 222).

7 "hunc...reperi" (Prol.41–44). The words "postquam scripseram alibi" (Prol. 44) probably refer to the *Ad pauca respicientes*, which, I have argued (Grant, *Oresme PPAP*, p. 80), was an earlier version of the *De commensurabilitate*. Incidentally, I am indebted to Professor Marshall Clagett for calling my attention to the fact that the opening line of the *Ad pauca respicientes* (Ad pauca respicientes de facili enunciant ut dicit Aristoteles [see Grant, *Oresme PPAP*, p. 382]) is a

direct quotation from the medieval Latin version of Aristotle's *De generatione et corruptione* 1. 316a. 9–10 rather than a general sentiment attributed to Aristotle. In the context of Aristotle's discussion, the opening line of the treatise might better have been rendered as "'Examining [only] a few things,' as Aristotle says, 'there are people who speak out much too readily,'" rather than "Concerning some matters, as Aristotle says, there are people who speak out much too readily" (ibid., p. 383). The sense of the first paragraph of the treatise is not really affected by Clagett's discovery since the thought that "there are people who speak out much too readily" implies a tendency to utter pronouncements hastily and without all of the evidence.

8 "Sed....benigne" (Prol.44–48).   For the relevance of this passage in dating the *De commensurabilitate*, see above, p. 4. Although Oresme's reason for submitting his *De commensurabilitate* to the Fellows and Masters of the University of Paris is a significant and revealing one—namely, that the content of the work was novel (see above, pp. 101–3)—he sought similar approval for a number of other treatises. For example, in a prologue to the *Algorismus proportionum*, he explains that he has submitted his work for correction to Philippe de Vitry, Bishop of Meaux, in order to forestall and thwart potential critics (see my translation in "Part I of Nicole Oresme's *Algorismus proportionum*," *Isis*, vol. 56 [1965], p. 328). For similar instances and for a general summary and evaluation of the nature of other prefaces by Oresme, see Clagett, *Nicole Oresme and Medieval Geometry*, chap. 3, sec. E.

Part I

1 "Numerus....reperitur" (I.2–23).   Except for the definitions of incommunicant and communicant proportionalities (I.10–17), all of these definitions are found in Campanus of Novara's thirteenth-century edition of Euclid's *Elements*, the edition that Oresme and most medieval scholastics used. The definitions of prime and composite numbers (I.2–3), mutually prime or incommunicant numbers (I.4–6), mutually non-prime or communicant numbers (I.6–9), are given in bk. 7, defs. 5–8 of *Euc.-Campanus* (p. 168) substantially as Oresme has them and in the same order. The definition of proportionality (I.10) is identical with Euclid, bk. 5, def. 4 (ibid., p. 204). Oresme's definition of commensurable magnitudes (I.18–20) is a combination of Euclid, bk. 10, def. 1, and bk. 10, prop. 5, while his definition of incommensurable magnitudes (I.20–23) draws upon bk. 10, def. 2, and the negation of bk. 10, prop. 6. For commensurable magnitudes Campanus states:

bk. 10. def. 1:   "Magnitudes are said to be communicant [or commensurable] which are numbered [or measured] by a common magnitude." ("Quantitates quibus fuerit una quantitas communis eas numerans dicentur communicantes."—*Euc.-Campanus*, p. 243.)

bk. 10, prop. 5:   "Between any two communicant [or commensurable] magnitudes, there is a ratio as between a number to a number."

("Omnium duarum quantitatum communicantium est proportio tanquam numeri ad numerum."—Ibid., p. 247.);
and for incommensurable magnitudes:

bk. 10, def. 2:   "Magnitudes are said to be incommensurable which are not
numbered [or measured] by a common magnitude." ("Quibus
vero non fuerit una communis quantitas eas numerans dicentur incommensurabiles."—Ibid., p. 243).

and the negation of

bk. 10, prop. 6:   "If two magnitudes have a ratio as one number to another,
it is necessary that they be communicant [or commensurable]."
("Si fuerint due quantitates quarum sit proportio unius ad
alterum tanquam numeri ad numerum, eas duas communicantes esse necesse est."—Ibid., p. 247.)

Incommunicant and communicant proportionalities (I.10–17) are nowhere
defined or discussed in Euclid's *Elements*. Two geometric proportionalities are
incommunicant, or incommensurable, when none of the terms of one series is a
term of the other series, as, for example, $(2:1)^n$ and $(3:1)^n$, where $n = 1, 2,$
3, 4,...; they are communicant, or commensurable, when the terms of the one
series are parts of the other—i.e., one is a subset of the other—as $(4:1)^n$ is part
of the series $(2:1)^n$. The same relationships are expressed in Chapter 3, Proposition VI of the *De proportionibus proportionum* (Grant, *Oresme PPAP*, pp. 228–32,
lines 117–55), where, however, Oresme uses the terms "incommensurabiles" and
"commensurabiles" in place of "incommunicantes" and "communicantes."

2 "Commensurabilitatem....centrum" (I.24–31). That the *De commensurabilitate*
is a later and much-improved version of the *Ad pauca respicientes* is indicated by
the differences in Oresme's definitions and discussions of commensurability and
incommensurability of circular motions. First, in the earlier *Ad pauca*, Oresme did
not specifically define commensurable circular motions (he does in the *De commensurabilitate*, I.26–29). His sole definition of incommensurable circular motions
is as follows (pt. 1, prop. IV): "...mobiles are said to be moved incommensurably
with respect to the center when they describe incommensurable angles in equal
times" (Grant, *Oresme PPAP*, p. 395). Obviously, this agrees perfectly with the
second part of his definition in *De commensurabilitate*, I.30–31. But Oresme then
enumerates four ways in which such incommensurability can occur (see Grant,
Oresme PPAP, p. 395 and pp. 432–33):

(1) when the circumferences of the circles are incommensurable and the curvilinear velocities are equal;

(2) when the circumferences of the circles are equal or commensurable, but two
mobiles move with incommensurable curvilinear velocities;

(3) when the circumferences are equal but the distances traversed are incommensurable;

(4) and, finally, when the circles are incommensurable and the curvilinear
velocities are incommensurable.

This superfluous fourfold enumeration is simply omitted from the *De commensurabilitate* and in place of it we find the definition in I.30–31 actually completed

when Oresme adds (I.29–30) that incommensurable velocities also arise when equal distances (expressed either as total circulations or by equal angles) are traversed in incommensurable times.

3 "Dicetur...differt" (I.60–62).   Differences arising from eccentric motions are considered later. But of the thirty-seven propositions in the treatise, only Part I, Proposition 20 and Part II, Proposition 12 consider motions on eccentric circles.

4 "Si....primo" (I.68–79).   Why did Oresme formulate the enunciation of Part I, Proposition 1 (I.68–71) as the converse of 9. 11 of Euclid's *Elements*, rather than give 9. 11 itself, which he quotes verbatim in I.72–75? Although Euclid's proof is applicable to any infinite geometric progression, his enunciation does mention a last term, whereas in Oresme's enunciation of Proposition 1 any prime number that measures the first term after unity is ipso facto a measure of any term in that progression ad infinitum. Therefore no last term need be mentioned. Since infinite geometric progressions play a significant role in the *De commensurabilitate*, Oresme probably thought it more relevant and appropriate to offer a converse version of Euclid 9. 11 and so avoid mention of a last term.

5 "Sed...Euclidis" (I.105–8).   Euclid 7.38 reads: "If any number has any part whatever, the number assigned to that part will number it." ("Si numerus aliquis partem quotamcunque habeat, numerabit ipsum numerus ad illam partem dictus."—*Euc.-Campanus*, p. 197.) In Campanus's example, "...every number having a third [part] is numbered [or measured] by three, and every number having a fifth [part] is numbered [or measured] by five, and so on." ("...omnis numerus habens tertiam numeratur a ternario, et habens quintam a quinario, sicque de caeteris."—Ibid.)

6 "Sed...Euclidis" (I.279–81).   Heath's translation of Euclid 7.34 (*Euclid's Elements*, vol. 2, p. 336) reads: "given two numbers, to find the least number which they measure." This is achieved, as Oresme tells us, by multiplying the two given numbers. In the printed version of Campanus's edition of Euclid, 7.34 does not correspond to the proposition intended by Oresme. However, 7.35 seems to qualify as a considerably altered, and somewhat obscure, version of it which includes the beginning of the proof as given in Heath's translation of the proof of 7.34. Here is my rather uncertain translation of what Campanus has in the enunciation of 7.35: "Any two numbers whatever produce a least number numbered [or measured] by them when the lowest numbers of their ratio [i.e., after it is reduced to mutually prime numbers] are multiplied, the greater multiplying the lesser [number] and the lesser multiplying the greater." ("Quilibet duo numeri minimos numeros suae proportionis maior minorem et minor maiorem multiplicantes minimum ab ipsis numeratum producunt."—*Euc.-Campanus*, p. 194.)

7 "Recolligendo....inveni" (I.423–62).   Five propositions, or parts of propositions, support the assertions made in I.424–33 and therefore underlie the five-step procedure from which Oresme says a general rule or law can be generated (I.423–24). The line numbers of these five statements and the propositions to which they refer are as follows:

(1) I.424–26 refers to Proposition 11.
(2) I.426–27 refers to Proposition 10, I.368–71.

(3) I.427–29 refers to Proposition 5, I.254–57.

(4) I.429–31 can be derived from the propositions mentioned in 1–3, or by use of Proposition 6.

(5) I.431–33 refers to Proposition 8.

Using these five propositions, Oresme says that the sequence of conjunctions can be determined. For example (I.434–48), if $V'_A . V'_B = 8:3$, then, by Proposition 11, since $V_A - V_B = 5$ there exist five points of conjunction, which, as shown in Proposition 10, divide the circle into five equal parts. Multiplying $V_A \cdot V_B = 8 \cdot 3$, we find that 24 days is the period of revolution of bodies A and B. Using these data, divide the period of revolution by the number of points of conjunction and obtain $4\frac{4}{5}$ days (using Proposition 6, we could derive this in another way; see I.440), the interval between any two immediately successive conjunctions. And now, by invoking Proposition 8, we find that $\frac{3}{5}$ of a circle will separate any two immediately successive conjunctions (i.e., for mobile A, $4\frac{4}{5} \cdot \frac{1}{3} = 1\frac{3}{5} - 1 = \frac{3}{5}$; and for B, $4\frac{4}{5} \cdot \frac{1}{8} = \frac{3}{5}$).

The sequence of conjunctions is now determinable (I.449–62). Let e, g, k, f, and h be five equidistant points arranged in that order (Fig. 26). Assume further

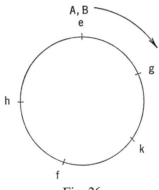

Fig. 26

This figure is found in MSS A (Vat. lat. 4082, fol. 100r, in the margin below col. 2) and L (Florence, MS Ashburnham 210, fol. 162v, in the margin below col. 1), where, however, the points are joined to form a star-shaped figure.

that A and B are now in conjunction at e but moving clockwise toward g. Since we have already determined that successive conjunctions will be separated from each other by $\frac{3}{5}$ of a circle, the next conjunction after e will occur in f, then in g, then in h, and so on. Thus two points of conjunction, namely, g and k, are by-passed as A and B move from conjunction in e to their next conjunction in f; and, similarly, points of conjunction h and e are bypassed as they move from conjunction in f to conjunction in g; and so on ad infinitum. However, if the ratio of velocities were $V_A : V_B = 9:4$ (I.453–56), there would again be five equidistant points of conjunction (by Proposition 11), but now three points of conjunction would always be bypassed between any two immediately successive conjunctions. This is explained by the fact that now the distance that separates any two successive conjunctions will be $\frac{4}{5}$ of a circle since the period of revolution is now $9 \cdot 4$, or 36 days, which when divided by 5 gives $7\frac{1}{5}$ days as the time between successive conjunctions; and when $7\frac{1}{5}$ is multiplied by $\frac{1}{9}$ we get $\frac{4}{5}$, and when multiplied by $\frac{1}{4}$ we have $1\frac{4}{5}$, or just $\frac{4}{5}$ when the whole circle is subtracted from it. Hence the first conjunction following conjunction in e will be in h (omitting or bypassing points g, k, and f), the next in f, and so on. Oresme then seems to say that three points of conjunction would also be bypassed if $V_A : V_B = 11:6$ (I.456–57). However, by computing, as before, the distance separating two successive conjunctions we find that only $\frac{1}{5}$ of a circle will separate them and, consequently, no points will be omitted. That is, after conjunction in e, A and B will conjunct next in g, then in k, then in f, and so on (see Fig. 26), following the same pattern. Finally, if the ratio of velocities is $V_A : V_B = 12:7$ (see I.458–59), the distance separating any two successive conjunctions is $\frac{2}{5}$ of a circle, and it follows that the immediate neighbor, in a clockwise direction, of any present point of conjunction will always be bypassed. Thus, after conjunction in e, the very next conjunction will occur in k, then in h, g, f, and so on.

In all these cases, the difference between the velocities, namely, $V_A - V_B$, was constant, although the ratio of velocities, $V_A : V_B$, differed. Oresme could, therefore, easily determine the sequence of conjunctions and the number of points of conjunction that would be bypassed. But if the differences between the velocities should vary from example to example. Oresme admits that he has no rule for determining the manner in which points of conjunction would be bypassed or omitted. If the differences between velocities varied from example to example, the total number of points of conjunction would also vary and, consequently, the equidistant divisions of the circle. Thus, a different set of points would have to be considered for each case and no rule could be given.

8 "Si....ratione" (I.564–78).   Let A, B, and C be in conjunction in point d. Since their velocities are assumed to be commensurable, it follows that after some interval of time each mobile will have completed an integral number of circulations that is directly proportional to its velocity. Consequently, mobiles A, B, and C will once again be in conjunction in d. For example (I.573–77), let $V_A = 6$, $V_B = 5$, and $V_C = 4$. Therefore, when A completes 6 circulations, B 5, and C 4, they will again conjunct in point d. The same procedure can, of course, be applied to more than three mobiles.

9 "[Tempus]....modo" (I.579–615).   In Proposition 15 (for which there is no counterpart in the *Ad pauca respicientes*), Oresme determines the time interval between two successive conjunctions of three mobiles in point d. If E, F, and G represent the velocities of mobiles A, B, and C, respectively, then, because in equal times distance traversed is proportional to velocity, the three mobiles will always conjunct in point d when A will have completed E circulations; B, F circulations; and C, G circulations. Furthermore, since time is inversely proportional to velocity (I.581–83), let the times of circulation for A, B, and C be represented by H, K, and L, respectively. Thus when any two of these three times are related, they will form a ratio that is reciprocal to the ratio formed by the corresponding velocities

among E, F, and G (e.g., $H:K$ and $E:F$ are reciprocal ratios since $H:K = F:E$).[1]

By Euclid 7.36,[2] some number, say, M, can be found that will be a least common multiple of H, K, and L, and therefore will represent the time in which A, B, and C will first conjunct again in d, their present point of conjunction. The time in which each mobile completes one circulation (I.591–96) is given for A by $1/E \cdot M$; for B by $1/F \cdot M$; and for C by $1/G \cdot M$. Since we have seen that any ratio of times is reciprocal to the corresponding ratio of velocities, it follows that a ratio of parts of M is produced by a ratio of velocities that is reciprocal to it. Thus a ratio of parts of M representing the times of single circulations, such as $(M:E)/(M:F) = F:E$, is produced by the ratio of velocities $H:K$, where $H:K = E:F$.

For example (I.598–99), let the velocities of A, B, and C be related as 6, 5, and 4, respectively; and let 10, 12, and 15 be the numbers that represent the times that will produce reciprocal ratios when taken in corresponding order. Since 60 is the least common multiple of 10, 12, and 15, and since the day can be assumed to be the unit of time, it follows that A, B, and C will conjunct simultaneously in their present point of conjunction in 60 days, their period of revolution. Single circulations will be completed by A in 10 days (i.e., $\frac{1}{6} \cdot 60$), by B in 12 days (i.e., $\frac{1}{5} \cdot 60$), and by C in 15 days (i.e., $\frac{1}{4} \cdot 60$). It also follows that when they conjunct again in 60 days, A will have made 6 circulations; B, 5 circulations; and C, 4.

10 "Tempus....diceretur" (I.616–46). From some given initial point of conjunction of three mobiles, Proposition 16 shows how to determine the time interval before the next conjunction, wherever that may occur. If, once again, the three mobiles are A, B, and C, Proposition 6 will enable us to find the times of the next conjunction for two mobiles only—i.e., for A and B, B and C, and A and C. By Euclid 7.36 (see below, the Commentary, fn. 2, p. 333), we can find the least common multiple of the times for these three pairs of mobiles, and this will give us the time of their next conjunction.

Two examples are presented. The first (I.625–31) uses the data of the previous proposition (see above, the Commentary, n. 9 for I.579–615) where the initial conditions are such that conjunction of the three mobiles is possible in only one point, namely, d. After departing from conjunction in d, A and B conjunct first on the sixtieth day (by Proposition 6, $\frac{1}{10} - \frac{1}{12} = \frac{1}{60}$, so that every day A gains $\frac{1}{60}$ of a circle over B) as will B and C (as with A and B, the difference in the daily distances is $\frac{1}{60}$ of a circle [i.e., $\frac{1}{12} - \frac{1}{15}$]); but A and C will conjunct next on the thirtieth

---

1. When associated with some form of the term *proportio* ("ratio"), the terms *econtrario* (I.582–83 and 599) and *econverso* (I.595–96, 607, and 634) have been translated as "reciprocal" rather than "inverse" or "converse," since Oresme has used reciprocal in the modern sense.

2. In *Euc.-Campanus*, Book 7, Proposition 36 (p. 194) reads: "Given any numbers you please, to find the least number which they

number [or measure]." ("Propositis quotlibet numeris minimum ab eis numeratum reperire.") In the modern edition, Book 7, Proposition 36 is concerned with finding the least common multiple for a maximum of three terms, but as Heath says (*Euclid's Elements*, vol. 2, p. 341), "The process can be continued *ad libitum*, so that we can find the L.C.M., not only of three, but of as many numbers as we please."

day ($\frac{1}{10} - \frac{1}{15} = \frac{1}{30}$, so that A will gain daily $\frac{1}{30}$ of a circle over C). Since 60 is the least common multiple of 60 and 30, the three mobiles will enter into their next conjunction on the sixtieth day, which will occur in point d.

In the second example (I.632–46), the velocities of A, B, and C are as 7, 5, and 3, respectively, and 15, 21, and 35 are, in that order, the least numbers that will form correspondingly reciprocal ratios. Since 105 is the least common multiple for these terms, the three mobiles will conjunct again in their present point of conjunction, d, in 105 days. However, they will also conjunct in yet another place in only $52\frac{1}{2}$ days. This results from the fact that every day A and B move $\frac{1}{15}$ and $\frac{1}{21}$ parts of a circle, respectively, so that $\frac{1}{15} - \frac{1}{21} = \frac{6}{315}$, and $\frac{315}{6} - 52\frac{1}{2}$, the time of the first conjunction of A and B after departure from d; and also from the motions of B and C, which daily move $\frac{1}{15}$ and $\frac{1}{35}$ parts of a circle, respectively, so that $\frac{1}{15} - \frac{1}{35} = \frac{4}{105}$, and $\frac{105}{4} = 26\frac{1}{4}$, signifying that B and C will conjunct $26\frac{1}{4}$ days after departure from d. Since $52\frac{1}{2}$ is the least common multiple of $52\frac{1}{2}$ and $26\frac{1}{4}$, it follows that A, B, and C will conjunct again in $52\frac{1}{2}$ days in a point diametrically opposite point d; and $52\frac{1}{2}$ days later, or 105 days after their initial departure from d, these three mobiles will once again conjunct in d. This pattern will be repeated ad infinitum.

No counterpart to this proposition is found in the earlier *Ad pauca respicientes*.

11 "Coniunctiones....simul" (I.647–77). Extending Proposition 7 to cover the case of three mobiles, Oresme, in Proposition 17, shows how to find the number of conjunctions in a period of revolution when three or more mobiles are in motion. This is achieved directly by dividing the period of revolution (found by Proposition 15) by the time between two successive conjunctions (found by Proposition 16). Using the data of the second example of Proposition 16 (in I.656 Oresme refers to I.632–46), we obtain two points, or places, of conjunction after dividing 105 days by $52\frac{1}{2}$ days.

Referring to Proposition 11, Oresme observes (I.663–77) that if the numbers representing the difference between any two velocities are communicant, that is, have a common measure, then that number which is their common measure represents the number of points in which the three mobiles can conjunct simultaneously. Thus if $V_A = 20$, $V_B = 10$, and $V_C = 7$, then, by Proposition 11, A and B have 10 points of conjunction (i.e., $V_A - V_B = 10$), B and C have 3 points of conjunction (i.e., $V_B - V_C = 3$), and A and C have 13 points (i.e., $V_A - V_C = 13$). But since 10, 3, and 13 have no common measure, mobiles A, B, and C cannot simultaneously conjunct in any point. This is obvious from the fact that when each of these three sets of points divides the circle into equal parts, no point of one set, or division, is a point in either of the other two sets. However, should the velocities of A, B, and C be 19, 13, and 10, respectively, then $V_A - V_B = 6$, $V_B - V_C = 3$, and $V_A - V_C = 9$, so that the three mobiles will conjunct in 3 points, since 3 is the common measure of 6, 3, and 9.

Although no proposition of the *Ad pauca respicientes* corresponds to Part I, Proposition 17 of the *De commensurabilitate*, Part 1, Proposition VI of the *Ad pauca* mentions that three mobiles moving with commensurable speeds will conjunct in a finite number of places and that these conjunctions will repeat ad infinitum (Grant, *Oresme PPAP*, p. 398, lines 160–65 and p. 94, n. 19). All this is not made explicit in Proposition 17, but it is certainly implied.

12 "Locum....essent" (I.678–92). The objective of Proposition 18 is to find the location of the first point of conjunction that immediately follows a present conjunction of three mobiles. Once again Oresme employs the data from the second example in Proposition 16 (I.632–41). The velocities of A, B, and C are 7, 5, and 3, in that order, and the period of revolution is 105 days. Since A makes 7 circulations in 105 days (although not mentioned, this follows from Proposition 14) and one circulation in 15 days, it will traverse $\frac{1}{15}$ of its circle in 1 day, and in $52\frac{1}{2}$ days it will have traversed $3\frac{1}{2}$ circulations (i.e., $52\frac{1}{2} \cdot \frac{1}{15} = 3\frac{1}{2}$). From Proposition 16, it is known that A, B, and C will first conjunct in $52\frac{1}{2}$ days, from which it follows that the three mobiles will conjunct in a point halfway around the circle in a point diametrically opposite d. This place could also have been determined by working with the specific data for mobile B or C.

13 "Numerum....forma" (I.693–703). In Proposition 19, Oresme does little more than assert (I.693–97) that with the aid of Proposition 17 we can find the number and sequence, or order, of points of conjunction for three or more mobiles, just as was done earlier for two mobiles in Proposition 10 (with the aid of Proposition 7). Since Proposition 17 enables us to determine the total number of conjunctions of three mobiles in a period of revolution, and since these points of conjunction divide the circle into equal parts, it is only necessary to find the distance separating any two successive conjunctions of three or more mobiles. This can be achieved by the application of Propositions 18, which, curiously enough, Oresme fails to cite. Its significance should have been apparent, if only by analogy with Proposition 10.

After the order and sequence of the conjunctions have been determined, we can discover when and how conjunctions are affected or altered as they occur in the different signs of the zodiac, or, as Oresme puts it, as they occur in different triplicities (I.697–99).[3]

---

3. A triplicity ("triplicitas") was a systematic grouping of three signs of the zodiac. We should recall that the order of the signs beginning with Aries is: (1) Aries, (2) Taurus, (3) Gemini, (4) Cancer, (5) Leo, (6) Virgo, (7) Libra, (8) Scorpio, (9) Sagittarius, (10) Capricornus, (11) Aquarius, and (12) Pisces. Since it was a common astrological belief that the four sublunar elements (earth, water, air, and fire) were governed and organized by the twelve signs of the zodiac, the signs were grouped in threes.

Each group was called a triplicity, and was believed to have the nature of the particular element which it controlled. Thus Aries, Leo, and Sagittarius constituted the fiery triplicity; Capricornus, Taurus, and Virgo formed the earthy triplicity; Cancer, Scorpio, and Pisces made up the watery triplicity; and Gemini, Libra, and Aquarius the airy triplicity. Between any two successive signs of any triplicity, three successive signs are omitted. Thus for any one sign of a triplicity, the remaining two signs will be the fourth and eighth signs from it.

Oresme is probably repeating a current astrological notion that the potency and influence of a triplicity becomes efficacious, or perhaps even ascendant, when it serves as the location of a conjunction. By using Proposition 19 and assuming commensurable planetary motions, we could discover the triplici-

Part II

1 "sensit...morbos" (II.319–20). Pliny *Natural History* 26. 1. 1; 10 vols., Loeb
Classical Library (London and Cambridge, Mass., 1938–62), vol. 7, trans. W. H. S.
Jones, p. 265. In the text accompanying Jones's translation, the Latin (p. 264),
including the words omitted in Oresme's quotation, reads: "Sensit facies hominum
et novos omnique aevo priore incognitos non Italiae modo verum etiam universae
prope Europae morbos,...." ("The face of man has also been afflicted with new
diseases, unknown in past years not only to Italy but also to almost the whole of
Europe,..." [p. 265].) In his *Le Livre du ciel et du monde*, bk. 1, chap. 34, 57c,
Oresme quotes this passage again, substantially as Jones has it (see Menut, *Oresme
du ciel*, p. 243). The same quotation is alluded to in his *Questiones super de celo*,
bk. 1, quest. 24 (see above, chap. 2, n. 90).

Part III

1 "Ait...possibile" (III.18–20). Al-Battani, *De scientia stellarum* (in Alfragani,
Rudimenta astronomica [Nuremberg, 1537]), chap. 1, fol. 4r. Cf. below, III.438–39.
In the 1537 edition of al-Battani's *De scientia stellarum*, this quote differs in the
following ways from what is given by Oresme (Oresme's version is presented
first): tanti: tanta enim; tam celesti veritatem: tamque coelesti veritate. In Chapter 10 of his *Tractatus contra astronomos* (in *Joannis Gersonii opera omnia*), Pierre
d'Ailly, who included substantial portions of material plagiarized almost verbatim
from Part III of Oresme's *De commensurabilitate*, repeats the brief quote from
al-Battani (it is taken from III.10–23 of the *De commensurabilitate*; see above,
chap. 3, n. 121) as follows: "in tanti magesterii excellentia veritatem ad unguem

---

(Note 3 continued)
ties in which conjunctions will occur in ever
recurrent patterns. For example, if the ratios
of commensurable velocities are such as to
produce a sequence of conjunctions where
three signs are omitted between any two
successive conjunctions, every conjunction
would occur in the same triplicity (provided
count is made in the order of the signs as
given above). If, however, two signs are
omitted, one conjunction will occur in every
triplicity before the pattern repeats. For
example, if conjunction occurs first in Aries
(fiery triplicity), it will occur next in Cancer
(watery triplicity), then Libra (airy triplicity),
and finally in Capricornus (earthy triplicity)
before this same pattern repeats ad infinitum. Knowledge of the order and time of

these conjunctions would permit us to find
the triplicities in which they occur and,
knowing the nature of each triplicity, to predict the nature of the terrestrial effects. All
this seems implied in Oresme's remark that
"When and how conjunctions are altered
from one triplicity to another, as well as how
other aspects are altered, becomes obvious
from this [proposition] by means of the
order and passage through the points of
conjunction" (I.697–99).
    The triplicities are enumerated by both
Robertus Anglicus and Cecco d'Ascoli in
their commentaries on the *Sphere of Sacrobosco*. See Thorndike, *Sphere of Sacrobosco*,
pp. 167 (Latin), 218 (English) for Robertus
Anglicus, and 382 for Cecco d'Ascoli.

comprehendere non est cuiquam possibile" (col. 799). See also the Commentary, n. 2 for III.21–22.

2 "forte...innotuit" (III.21–22). In Chapter 52 of al-Battani's *De scientia stellarum*, the following brief section seems to be the source of Oresme's quotation, which he altered somewhat to fit the sense of Apollo's discourse: "...at si propter aliquem coelestem motum, qui nec nobis nec illis innotuit..." (fol. 81r).

Although these two successive quotations in III.19–20, 21–22 are from widely separated chapters in al-Battani's *De scientia stellarum*, they are probably based on a single passage in bk. 1, chap. 2, of Ptolemy's *Tetrabiblos* (or *Quadripartitum* as it was known in the Middle Ages), which has been quoted above on pp. 108–9.

3 "miror...constet" (III.26–28). Pliny, *Natural History* 2. 21. 87; 10 vols., Loeb Library, vol. 1, trans. Rackham, p. 230. Oresme omitted parts of this quotation in order to adapt the passage for Apollo's speech. The entire passage follows: "Mirum quo precedat inprobitas cordis humani parvolo aliquo invitata successu, sicut in supra dictis impudentiae ratio largitur. Ausique divinare solis ad terram spatia eadem ad caelum agunt, quoniam sit medius sol, ut protinus mundi quoque ipsius mensura veniat ad digitos. Quantas enim dimetiens habeat septimas, tantas habere circulum duo et vicesimas, tamquam plane a perpendiculo mensura caeli constet." (Rackham's translation [p. 231]: "It is marvellous to what length the depravity of man's intellect will go when lured on by some trifling success, in the way in which reason furnishes impudence with its opportunity in the case of the calculations above stated. And when they have dared to guess the distances of the sun from the earth they apply the same figures to the sky, on the ground that the sun is at its centre with the consequence that they have at their finger's ends the dimensions of the world also. For they argue that the circumference of a circle is $\frac{2.2}{7}$ times its diameter, as though the measure of the heavens were merely regulated from a plumb line!") The numerous omissions as well as differences between the texts have necessitated a different translation for Oresme's version.

4 "Nonne...appellavit" (III.40–41). In *Orem O soizmerimosti*, p. 367, n. 5, Zoubov cites as the source of this reference Hesiod's *Theogony*, 278, which is unrelated to the action attributed by Oresme to Hesiod. Oresme did not possess Hesiod's *Theogony*, and the source for his remark is at present unknown. Using Aristotle's mention of Hesiod (*De caelo* 3. 1. 298b28–29) as a point of departure, Oresme, in his *Le Livre du ciel et du monde*, bk. 3, chap. 2, 159a (Menut, *Oresme du ciel*, p. 586), cites Ovid's "grant livre"—i.e., the *Metamorphoses*—for Hesiod's view that all things are constituted from chaos (Hesiod's opinion is actually quoted by Aristotle in *Metaphysics* 1. 4. 984b27–28). However, Ovid's *Metamorphoses* does not contain anything that could have provided the reference to Hesiod made in the *De commensurabilitate*.

5 "haud...deos" (III.42–43). Claudian, *De raptu Proserpinae*, bk. 3. 27–28. Oresme's quotation differs only trivially from Platnauer's text (*Claudian with an English Translation*, trans. Maurice Platnauer, 2 vols., Loeb Classical Library [London and New York, 1922], vol. 2, p. 346). Oresme has "invidia" for "invideo," "nec" for "neque," and "aut" for "vel."

6 "ab...voluit" (III.43–45). Plato, *Timaeus* 29D, in *Timaeus a Calcidio translatus* (ed. Waszink), p. 22. Oresme's version differs from Waszink's edition by omitting "porro" after "optimo," transposing "invidia longe," omitting the words "itaque consequenter" (at ellipsis), and replacing "cuncta" with "cunctaque." In *Phaedrus* 247a, a similar assertion about the absence of jealousy in the gods could not have served directly as Oresme's source (as Zoubov has it in *Orem O soizmerimosti*, p. 368, n. 2), since no medieval Latin translation of *Phaedrus* is known, and none was probably made.

7 "forma...carens" (III.45–46). Boethius, *De consolatione philosophiae*, bk. 3. 9. The few words quoted here are identical with the text in Stewart's edition (*The Consolation of Philosophy*, trans. H. F. Stewart, in *The Theological Tractates* [trans. H. F. Stewart and E. K. Rand], Loeb Classical Library [New York and London, 1918], p. 264, line 6).

8 "Quamvis...praxi" (III.70–71). Perhaps this statement is based upon Averroes' commentary on Aristotle's *Physics*, bk. 4, text 74 (Junctas ed., vol. 4, fol. 75v, col. 1) where he says: "...non enim dicimus quod lineae exeuntes a centro circuli ad peripheriam sunt equales ut circulus sit nobilior, sed hoc contingit naturae circuli. Et ideo nobilius non invenitur in formis mathematicis; hoc enim, scilicet nobilitas, non accidit formis, nisi secundum quod est finis."

9 "Omnes...nos" (III.75–76). *De vetula*, bk. 3, verse 42. The *De vetula*, falsely ascribed to Ovid, was probably written by Richard de Fournival (ca. 1201–ca. 1260; see D. M. Robathan, "Introduction to the Pseudo-Ovidian *De Vetula*," *Transactions of the American Philological Association*, vol. 58 [1957], p. 206). In the printed edition that I have examined (the second, separately paginated treatise in *Brunellus Vigelli et Vetula Ovidii...*, ed. S. Closius [Wolfenbüttel, 1662], p. 80), we find "Res omnes" instead of Oresme's "Omnes res" and "nos extraximus" where Oresme has "enim invenimus." In the *Le Livre du ciel et du monde*, bk. 1, chap. 1, 4c, Oresme actually cites the *De vetula* by title and quotes the first two lines of verse 42 (these two lines immediately precede his quotation in *De commensurabilitate*, III.75–76), followed directly by everything from "Omnes" to "existit," just as in III.75 of our treatise. Approximately 23 lines below, in 4d, he repeats everything from "numerus" to "nos" exactly as it appears in *De commensurabilitate* III.75–76. See Menut, *Oresme du ciel*, pp. 48–50.

10 "numero...gaudet" (III.77). Virgil, *Eclogues* 8. 75. This four-word quotation agrees with Fairclough's Latin text (*Virgil with an English Translation*, trans. H. Rushton Fairclough, 2 vols., Loeb Classical Library [London and New York], vol. 1, p. 60). However, Fairclough's translation (vol. 1, p. 61), "In an uneven number heaven delights," has been altered to read "God delights in an odd number." That "deus" must be rendered as "God" (and not "heaven") is clear from the context in which Oresme repeats this quotation in his *Le Livre du ciel et du monde*, bk. 1, chap. 1, 5a (Menut, *Oresme du ciel*, p. 50), where we read: "Et dit Virgile que Dieu se esjoist de nombre nomper: Numero deus impare gaudet; ...."

11 "Sic...figurarum" (III.77–78). Aristotle, *De caelo* 2. 4. 286b22–24. The Moerbeke translation of *De caelo*, which Oresme often used, renders the relevant pas-

sage as: "Itaque, si perfectum prius imperfecto, et propter hoc utique prior erit figura circulus."—*Simplicii philosophi acutissimi commentaria in quatuor libros De celo Aristotelis*, trans. Wm. Moerbeke (Venice, 1540), fol. 64, col. 2.

12 "Noster...bipartita" (III.78–80). Aristotle, *Metaphysics* 1. 5. 986a22–27. Aristotle lists ten pairs of opposites, or contraries, employed by some members of the Pythagorean school. Among these are "good-bad," "odd-even," and "square-oblong." Because they correspond to the side of good, odd numbers are superior to even, and square figures to oblong. In this way, Pythagoras is used in support of Arithmetic's claim (III.70–73) that some numbers and mathematical figures are better than others.

13 "Teste...duplam" (III.82–83). Averroes, text 14 of his commentary in bk. 3 of Aristotle's *De caelo*, Junctas ed., vol. 5, fol. 182v, col. 1. In his remarks on *De caelo* 3. 299b32–300a13, Averroes says that if the number of planes in a body determines the weight of the body, "it will be possible that the ratio of the planes of a heavier to a lighter body could be in a double ratio, or, depending on the [number of] planes involved, in some other ratio. But all the ancients value this ratio [i.e., the double ratio] highly and say it is the cause of many generated things." ("...possible erit ut proportio superficierum gravioris ad superficies levioris sit in proportione duplici, aut in alia proportione secundum superfices. Et omnes antiqui magnificant hanc proportionem et dicunt ipsam esse causam plurium generatorum.")

14 "Sicut...corporeus" (III.86–89). Zoubov (*Orem O soizmerimosti*, p. 369, n. 6) cites *De caelo* 2. 287a as a general reference underlying Oresme's mention of Aristotle in III.89. But nothing in 287a seems properly relevant to the context in which Oresme has invoked the name of Aristotle. On numerous occasions, however, Aristotle praised the beauty, order, and regularity of celestial bodies and perhaps Oresme intended here only a general allusion to such remarks.

15 "Omnis...perspectivos" (III.95–96). Perhaps Oresme had in mind the kind of statement that appears in the *Optics* of Vitello, where number is included in a long list of things that contribute to our comprehension of beauty (some others are magnitude, location, figure, corporeity, etc.) and about which Vitello says: "Number also makes things beautiful to the sight and because of it the places of many separate stars in the sky are more beautiful than the places of few stars; and [the flames of] many candles are more beautiful than few." ("Numerus etiam facit pulchritudinem in visu et propter hoc loca coeli multarum stellarum distinctarum sunt pulchriora locis paucarum stellarum, et plures candelae sunt pulchriores paucis."—*Opticae libri decem*, bk. 4, sec. 148, p. 184, in Alhazen, *Opticae thesaurus* ...[Basel, 1572].) The relevance of this to Arithmetic's remark in III.95–96 is simply that if numbers contribute to our perception of beauty, then only rational ratios can produce such effects since irrational ratios are not formed by integers.

16 "symphonie...pauce" (III.100). Aristotle, *De sensu* 3. 440a2. This three-word quotation from Aristotle appears as two words—"symphoniae paucae"—in the text published with the commentary of Thomas Aquinas (edited by R. M. Spiazzi [Turin and Rome, 1949], p. 31, col. 2, sec. 50). For a quotation of the Aristotelian

passage that concludes with the words discussed here, see the Commentary, n. 17 for III.101–3, below.

17 "omnis...ipse" (III.101–3). Probably Aristotle, *De sensu* 3. 439b25–440a2; 4. 442a12–17. Although Arithmetic claims it to be an Aristotelian belief that "every irrational ratio offends the ear by the sounds it produces, the taste by the flavors it produces, and so on," this is not made explicit by Aristotle, although he did believe that the most agreeable colors, sounds, and tastes arise from simple numerical or rational ratios. Thus, in discussing the existence of a plurality of colors other than black and white, Aristotle says:

> ...we may suppose that [of this plurality] many are the result of a [numerical] ratio; for the blacks and whites may be juxtaposed in the ratio of 3 to 2, or of 3 to 4, or in ratios expressible by other numbers; while some may be juxtaposed according to no numerically expressible ratio, but according to some relation of excess or defect in which the blacks and whites involved would be incommensurable quantities; and accordingly, we may regard all these colours [viz., all those based on numerical ratios] as analogous to the sounds that enter into music, and suppose that those involving simple numerical ratios, like the concords in music, may be those generally regarded as most agreeable; as, for example, purple, crimson, and some few such colours, their fewness being due to the same causes which render the concords few.—*De sensu* 3. 439b25–440a2.

The same reasoning is extended to tastes when Aristotle says:

> As the intermediate colours arise from the mixture of white and black, so the intermediate savours arise from the Sweet and Bitter; and these savours, too, severally involve either a definite ratio, or else an indefinite relation of degree, between their components, either having certain integral numbers at the basis of their mixture, and, consequently, of their stimulative effect, or else being mixed in proportions not arithmetically expressible. The tastes which give pleasure in their combination are those which have their components joined in a definite ratio.—*Ibid.* 4. 442a12–17.

18 "Cum...oriatur" (III.107–8). Heath's translation of Euclid 10. 2 is as follows (*Euclid's Elements*, vol. 3, p. 17): "If, when the less of two unequal magnitudes is continually subtracted in turn from the greater, that which is left never measures the one before it, the magnitudes will be incommensurable." (The corresponding proposition in Campanus's edition is substantially the same; see *Euc. Campanus*, p. 245.) Since no common measure is attainable, the process of subtraction between two incommensurable magnitudes is unending and Arithmetic stresses the infinite, and therefore incomprehensible, process involved. If one thought two magnitudes commensurable that were really incommensurable, he would continue this fruitless subtractive process in vain. In Arithmetic's view, the mind is so repelled by such an infinite process that it is not really conceivable for the world to be so constructed as to permit incommensurable relationships between the most fundamental quantities—and above all not between celestial motions.

19 "de...mensura" (III.119–20). The reference in Book of Wisdom 11:21 reads: "but thou hast ordered all things in measure, and number, and weight."—*The Holy Bible*, Douay Version of the *Old Testament* (New York, 1914). ("Sed omnia in mensura, et numero, et pondere disposuisti."—*Liber sapientiae* 11:21,

in *Biblia sacra vulgatae editionis*, ed. P. M. Hetzenauer [Oeniponte, 1906].)

20 "omnia...sunt" (III.122–23).   Boethius, *De institutione arithmetica* 1. 2. 14–15. Oresme's quotation differs considerably from the corresponding passage in Friedlein's edition of this work (p. 12), where we read: "Omnia quaecunque a primaeva rerum natura constructa sunt, numerorum videntur ratione formata." No variant readings for this sentence appear in Friedlein's edition.

21 "et...numeris" (III.123).   Very likely the Pythagoreans are intended when Arithmetic speaks of ancients who said that the forms of things are like, or similar to, numbers. Aristotle's *Metaphysics* 1. 5. 985b25–986a3 and 1. 6. 987b10–12, 23–30 would have provided a popular and readily available source for this general remark.

22 "Plato...medios" (III.124–26).   Based on Plato, *Timaeus* 31B–32B in *Timaeus a Calcidio translatus* (ed. Waszink), pp. 24–25. For a translation of Plato, *Timaeus* 31B–32B, see Cornford, trans., *Plato's Cosmology*, pp. 43–44. The substance of Arithmetic's reference to Plato is contained in 32B of Cornford's translation:

> Now if it had been required that the body of the universe should be a plane surface with no depth, a single mean would have been enough to connect its companions [i.e., fire and earth] and itself; but in fact the world was to be solid in form, and solids are always conjoined, not by one mean, but by two. Accordingly, the god set water and air between fire and earth, and made them, so far as was possible, proportional to one another, so that as fire is to air, so is air to water, and as air is to water, so is water to earth, and thus he bound together the frame of a world visible and tangible. (I have added the words in brackets.)

In Thomas L. Heath's representation (*A History of Greek Mathematics*, 2 vols. [Oxford, 1921], vol. 1, p. 89), "if p^3, q^3 are two cube numbers, $p^3 : p^2q = p^2q : pq^2 = pq^2 : q^3$, the means being of course in continued geometric proportion."

23 "Ergo...permittit" (III.127–32).   Pseudo-Aristotelian *De mundo* 6. 397b29–35; ed. W. L. Lorimer in *Aristoteles latinus*, vol. 11 (Rome, 1951), p. 69. The substance, and indeed some of the very words, in these lines are drawn directly from the *De mundo* (only the first few lines, from "Wherefore" to "confusion," render the quotation in III.130–31):

> Wherefore the earth and the things upon the earth, being farthest removed from the benefit which proceeds from God, seem feeble and incoherent and full of much confusion; nevertheless, inasmuch as it is the nature of the divine to penetrate to all things, the things also of our earth receive their share of it, and the things above us according to their nearness to or distance from God receive more or less of divine benefit.—Trans. E. S. Forster, *The Works of Aristotle* (Oxford, 1931), vol. 3.

Of the two medieval translations of the *De mundo*, Oresme's form of the quotation points to Nicholas of Sicily's version as his probable source, for there we read: "Propter quod terra et que super terram videntur, quia nimium distant a commodo divino, infirma esse et plena multi turbinis."—*De mundo*, ed. Lorimer, p. 69. Although Oresme says he is using the words of Aristotle, this is obviously neither a direct nor an accurate citation. The words from "Propter" to "videntur" seem to have been deliberately conflated to "quia hec inferiora" in III.130 and the remainder of the quotation has been somewhat altered or inaccurately transcribed, or perhaps even drawn from some quite different manuscript tradition. That Oresme did not use the second and anonymous translation is apparent from its wording of this same passage: "Ideoque terra et que in terra videntur, distantia plurima existentia utilitate dei, infirma et mutabilia esse et multa plena perturbatione."—*De mundo*, ed. Lorimer, p. 40.

It should be apparent that the author of the *De mundo* did not use this passage, as Arithmetic would have it, to show that terrestrial disorder and confusion produce incommensurable relationships. This consequence is drawn by Oresme on behalf of Arithmetic.

In his translation, Zoubov does not cite this as a direct quotation but directs the reader (*Orem O soizmerimosti*, p. 371, n. 5) to compare Aristotle's *De generatione animalium*, 731b, a section which seems to me wholly irrelevant to the lines under discussion.

24 "antiquiorem...lineis" (III.137). Macrobius, *Commentary on the Dream of Scipio* 1. 5. 12; trans. Stahl, p. 97. These few words from Macrobius agree perfectly with the edition of Eyssenhardt, p. 483 (*Macrobius*, ed. Franciscus Eyssenhardt [Leipzig, 1868]), and with the most recent edition by Willis, p. 17 (*Ambrosii Theodosii Macrobii Commentarii in somnium Scipionis*, ed. Jacob Willis [Leipzig: Teubner, 1963]). Since I have included some passages of Macrobius from the translation of Stahl, who utilized the Eyssenhardt but not the later Willis edition, I shall cite, where necessary and helpful, page references from both Latin editions. The book, chapter, and paragraph numeration is identical in both editions.

Macrobius believed that geometrical lines and figures were ultimately derivable from numbers, and concluded therefrom that numbers were absolutely prior to, and more fundamental than, geometrical entities. Thus, in 1. 5. 11 he says (trans. Stahl, p. 97):

Hence it is apparent that the number eight both is and is considered a solid body, if indeed one is represented by a point, two by the drawing of a line (which, as we said above, is limited by two points), and four by points arranged at right angles to each other, with lines extending between the points to form a square. When these four are duplicated and made eight, forming two equal squares, and one is superimposed upon the other, giving the figure altitude, the result is a cubical figure, which is a solid body.[1]

1. As Stahl explains (p. 97, n. 10), Macrobius's interpretation of figurate numbers differs from the usual arrangement where "the number two or any larger number may be arranged as linear; three becomes the first plane number, if the points are arranged in triangular position, and any number above three may be arranged as a plane; four becomes the first solid number in a pyramid with a triangular base, and any number above four may be solid," since "Macrobius chooses to regard four as a plane number (quadrate), the first solid with which he deals being the cube, or eight." Macrobius's choice seems to have been dictated by a desire to derive the geometric figures from even numbers, and for this purpose only the geometric progression $(2)^n$, where $n = 1, 2, 3$, would serve his purpose. This becomes clear from his explanation in 2. 2. 8–11: "The monad represents the point because, like the point, which is not a body but which produces bodies from itself, the monad is said to be not a number but the source of

At this juncture, Oresme's quotation in III.137 appears and Macrobius goes on to assert the priority of numbers (trans. Stahl, pp. 97–98):

> Thus it becomes clear that numbers precede surfaces and lines (of which surfaces consist), and in fact come before all physical objects. From lines we progress to numbers, to something more essential, as it were, so that from the various numbers of lines we understand what geometrical figures are being represented. But we have already remarked that surfaces with their lines are the first incorporeality after the corporeality of bodies and that they are nevertheless not to be separated from bodies on account of their indissoluble union with them. Therefore, whatever precedes surface is purely incorporeal; but we have shown that number is prior to surface and to lines; hence the first perfection of incorporeality is in numbers, and this is, as we previously stated, the common perfection and fullness of all numbers.

25 "O...exquisisti" (III.139).   A reference to the thirteenth-century arithmetic treatises of Jordanus de Nemore, especially the *Arithmetica*, a lengthy treatise in ten books (as yet unpublished). It was cited frequently by Oresme and many others, and since it was far superior to the *De institutione arithmetica* of Boethius, Oresme has Arithmetic invoke the name of Jordanus—inclusion of the title of his treatise would have been superfluous, for it was very well known—as the greatest expositor and interpreter of her discipline. For the little that is known about this important mathematician and mechanician, see Ernest A. Moody and Marshall Clagett, eds. and trans., *The Medieval Science of Weights* (Madison, 1952), pp. 121–23, and Clagett, *Science of Mechanics*, pp. 72–73.

26 "mundi...triformis" (III.146).   Ovid, *Metamorphoses*, bk. 15. 859; 2 vols., Loeb Classical Library (Cambridge, Mass., and London, 1946), trans. F. J. Miller, vol. 2, p. 426 (Latin), p. 427 (English). According to C. T. Lewis and Charles Short (*Harpers' Latin Dictionary* [New York, 1907]), the words "triformis mundus" (listed under "triformis"), in Ovid's phrase "mundi regna triformis," refer to a world composed of three elements—air, earth, and water. Also see C. S. Lewis, *The Discarded Image* (Cambridge, 1964), p. 39.

27 "motibus...mundum" (III.152–53).   In "Nicole Oresme et la musique" (*Med. and Ren. Studies*, vol. 5 (1961), p. 98, n. 1), Zoubov has drawn attention to the alliteration employed by Oresme in these nine words, all of which begin with the letter *m*.

28 "Narrat...sonus" (III.160–62).   Macrobius, *Commentary on the Dream of Scipio* 2. 1. 2; trans. Stahl, p. 185. Cicero's *Dream of Scipio* "was, with the exception of a

---

numbers. The first number, therefore, is two, which is like the line protracted from the point by giving it two termini. The number two doubled gives four, representing the mathematical body which is limited by four points, having length and breadth. Four doubled gives eight, the number representing a solid body, to repeat what was previously stated, that two lines placed above two others and limited by eight points produced the solid body; this explains why geometricians speak of two times two times two as a solid body. Thus with even numbers progression up to eight represents a solid body, and on this account Cicero attributed a fullness to this number in particular."—Trans. Stahl, pp. 190–91.

few brief fragments, all that had been preserved from the *De re publica* through the greater part of the Middle Ages and up until the last century. The text of *Scipio's Dream* was appended to various manuscripts of Macrobius' Commentary and was thus preserved."—Stahl, trans., p. 10. Oresme's quotations from the *Dream of Scipio*—in contrast with his citations from the *Commentary on the Dream of Scipio* by Macrobius—may, therefore, have been derived either from the text preserved and scattered through the Macrobian *Commentary* itself, or from some independent, consecutive, and integral version that could have been appended to some manuscript of the *Commentary* or circulated independently. In light of Oresme's quotations from both the *Commentary* and the *Dream* itself and based on Oresme's mistaken attribution in III.300–303 (see below for the Commentary to these lines), it seems more plausible to suppose that he drew upon a version of the *Dream* interspersed within the Macrobian *Commentary*. I shall compare Oresme's citations from Cicero's *Dream* with two editions of Macrobius's *Commentary* (those of Eyssenhardt and Willis) as well as with the independent version that forms the concluding portion of Cicero's *Republic* (at the end of bk. 6) in the edition accompanying the translation of C. W. Keyes (*De re publica, De legibus,* Loeb Classical Library [London and New York, 1928]).

    In the quotation in III.161–62, Oresme omits "hic inquam quis" after "quid" (these words appear in all three editions). Furthermore, where Oresme and Willis (p. 95) have "complet," Eyssenhardt (p. 571) and Keyes (p. 270) have "conplet." All three editions, however, have "et" where Oresme has "ac."

29 "Hic...infimus" (III.162–67).   Macrobius, *Commentary on the Dream of Scipio* 2. 1. 2–3; trans. Stahl, p. 185. Since Oresme's quotation varies from the versions found in the editions of Eyssenhardt (p. 571), Willis (p. 19), and Keyes (pp. 270–72), which are in almost complete agreement, the discrepancies recorded below give Oresme's reading first followed by the reading that appears in the three editions just mentioned: distinctus^1: disiunctus; partium: parte; distinctus^2: distinctis; pulsu: impulsu; et ille summus stellifer celi: summus ille caeli stellifer; eius: cuius; acuto et: acute (Keyes has "acuto ex"). It should be noted that I have substituted "distinctus" (III.163) where the Oresme manuscripts had "coniunctus," "iunctus," or "coniunctis." This seemed advisable not only to render the proper sense of the passage, but also because in III.200–201, where Oresme repeats the quotation from III.163–64 ("qui...distinctus"), he has "est distinctus" rather than any form of "coniunctus" or "iunctus." Because all the manuscripts of the *De commensurabilitate* were in agreement on this, it seemed proper and fitting to substitute "distinctus" rather than adopt "disiunctus," the reading of the modern editions, since the translation would be the same for either reading and the two quotations would then be almost identical.

    Despite the variations above, I have used Stahl's translation (a few minor alterations were made necessary by Oresme's introduction of a transitional sentence leading to the reply by Scipio's grandfather and because Oresme chose to omit a sentence from the quotation). Wherever possible, quotations from Macrobius's *Commentary on the Dream of Scipio* are cited in Stahl's translation because, to use the latter's own words, "this translation is based upon the interpretation that

Macrobius gave to Cicero's words. It will consequently deviate on a few occasions from Cicero's intended meaning."—Stahl, trans., p. 69, n. 1.

In *Le Livre du ciel et du monde*, bk. 2, chap. 18, 126a, Oresme, without once mentioning Macrobius, sets forth the two major opposing views concerning the production of celestial music (they are briefly mentioned in *De commensurabilitate*, III.370–71).

Some of the ancient thinkers who theorized about the music of the spheres used to say that the lowest of the heavens, the lunar sphere, has the sharpest or highest pitched sound, the topmost sphere has the loudest and lowest sound, and the other spheres intermediate between these two, with the lowest sphere always having a sound sharper and higher pitched than the one above it. Other thinkers held the contrary opinion and it had greater probability; for the amplitude of the sound depends upon the quantity or magnitude of the bodies which produce the sound and upon their shapes, but the loudness of the sound depends upon the speed of spherical motion, other things being equal. Therefore, the highest sphere should produce the heaviest and lowest sound, but it would sound louder than that of the sphere just beneath it."—Menut, *Oresme du ciel*, p. 481.

To what extent Oresme's opinion concurs with that of Macrobius is difficult to determine, since Macrobius relies solely on the respective speeds of the celestial spheres, whereas Oresme uses not only speed (indeed in *De commensurabilitate*, III.368–88, speed is eliminated as a factor) but also shape, and quantity of matter, with the last mentioned apparently the most fundamental. They seem to agree that the outermost sphere produces the loudest sound. But for Oresme it also makes the lowest and deepest tone; Macrobius, in contrast, would have it make the highest or sharpest tone. Also see below, the Commentary, n. 61 for III.375–81 on pp. 354–55 and above, pp. 68–69, n. 107.

30 "Plato....vocat" (III.168–74).   Macrobius, *Commentary on the Dream of Scipio* 2. 3. 1–2; trans. Stahl, pp. 193–94. This quotation is very nearly identical with what appears in the editions of Eyssenhardt (p. 581) and Willis (pp. 103–4) save that where Oresme has "conficitur," both editions have "confit." In one place, however, Willis disagrees with both Oresme and Eyssenhardt when he substitutes "dea" for "deo." Macrobius's references to Plato and Hesiod are to *Republic* 10. 617B and *Theogony*, line 78, respectively.

Oresme was to repeat a part of this quotation in his later *Le Livre du ciel et du monde*, bk. 2, chap. 18, 126d (Menut, *Oresme du ciel*, p. 482): "Plato—singulas ait Syrenas singulis orbibus [*editio hab.* oribus] insidere...nam Syren deo canens Greco intellectu valet."

31 "Ideo....affectu" (III.176–90).   Macrobius, *Commentary on the Dream of Scipio* 2. 3. 4–8; trans. Stahl, pp. 194–95. A comparison of Oresme's quotation from Macrobius with the editions of Eyssenhardt (p. 582) and Willis (pp. 104–5) discloses the following differences (Oresme's reading is listed first preceded by the line number in which the word or words occur): 182 after "gentium" Eyssenhardt and Willis have "vel regionum" (I have not indicated an omission in Oresme's quotation since these two words seem superfluous in the context; Stahl does not translate them); 185 solum: soli; 186 after "qui" Oresme omits "sunt"; 187 amorem: ardorem; 189 tamque: tam.

32 "Aves...cetera" (III.190–91). Oresme probably had in mind this particular remark of Macrobius (*Commentary on the Dream of Scipio* 2. 3. 10; trans. Stahl, p. 195): "Is it at all strange if music has such power over men when birds like the nightingale and swan and others of that species practice song as if it were an art with them." This likely reference to Macrobius is not mentioned by Zoubov in *Orem O soizmerimosti*.

33 "Equissimum...resultant" (III.192–94). John of Salisbury, *Policraticus*, bk. 1. 6; trans. Joseph B. Pike, *Frivolities of Courtiers and Footprints of Philosophers* (Minneapolis, 1938), p. 30. Oresme's quotation from John of Salisbury's *Policraticus* agrees verbatim with its counterpart in Webb's edition (*Ioannis Saresberiensis Episcopi Carnotensis Policratici...*, ed. C. C. I. Webb, 2 vols. [Oxford, 1909]), vol. 1, p. 40, 401b–c. In citing the edition of the *Policraticus* in J. P. Migne's *Patrologiae cursus completus*, series latina (221 vols. [Paris, 1844–64]), Zoubov has cited Volume 99, col. 401, rather than Volume 199 (*Orem O soizmerimosti*, p. 373, n. 3), an error that is repeated later on (see below, the Commentary, n. 39 for III.214–22).

34 "Ipsa....distinctus" (III.195–201). Macrobius, *Commentary on the Dream of Scipio* 2. 3. 11–12; Stahl's translation (slightly emended), pp. 195–96. Omitting transpositions of a few pairs of words, I now give the differences between the editions of Eyssenhardt (pp. 583–84) and Willis (p. 106) (in both the passage in question is identical) and Oresme's quotation (the latter's readings are presented first preceded by the line number of the particular word: 200 distinctus: disiunctus; 201 partium: parte; distinctus: distinctis.

Since Stahl (pp. 195–96) does not translate the words "speralem" and "corpus mundi"—the latter must be taken as contrasting with "celestis anima"—I have altered his translation of III.199–201 ("Hec...distinctus"). Oresme had previously included III.200–201—a direct quote by Macrobius from Cicero's *Dream of Scipio* —in an earlier quotation from Macrobius's *Commentary* in III.163–64.

35 "quod...compactam" (III.202–3). Boethius, *De institutione musica*, bk. 1. 1. Friedlein, in his edition of this work (p. 180, lines 4–5), has "coniunctam" where Oresme has "compactam."

36 "non...copulantur" (III.204–6). Boethius, *De institutione musica*, bk. 1. 1. If we ignore a transposition of two words, Oresme's version of these few lines from Boethius differs from Friedlein's edition (p. 186, lines 9–13) as follows (Oresme's reading appears first): armonice: armonicas; copulantur: posterior disputatio coniungi copularique monstrabit.

37 "Mundana...visuntur" (III.208–9). Boethius, *De institutione musica*, bk. 1. 2. Where Oresme has "inspicienda," Friedlein (p. 187, lines 24–26) has "perspicienda"; before "celo" Friedlein adds "ipso."

In addition to "worldly," or "mundane," music, Boethius also distinguishes "human" and "instrumental" types. "Et prima quidem mundana est, secunda vero humana, tertia quae in quibusdam constituta est instrumentis, ut in cithara vel tibiis ceterisque, quae cantilenae famulantur."—*De institutione musica*, bk. 1. 2 (in Friedlein's edition, p. 187, lines 20–23). Although Oresme enumerates human and worldly, or mundane, music in his *Le Livre du ciel et du monde*, bk. 2, chap.

18, 126c, his third type is "divine"—not "instrumental"—music heard only by the angels and souls of the blessed. See Menut, *Oresme du ciel*, p. 482.

38 "Unde....convenirent" (III.210–13). Boethius, *De institutione musica*, bk. 1. 2. Other than substituting "qui" for "quomodo" (III.212), Friedlein's text of the *De institutione musica* (p. 188, lines 6–10) agrees with Oresme's quotation.

39 "virtutis....modulatur" (III.214–22). John of Salisbury, *Policraticus*, bk. 1. 6; trans. Pike, *Frivolities of Courtiers*, pp. 30–31 (slightly altered). The omission indicated in III.218 corresponds to approximately 26 lines in Webb's edition of the *Policraticus*. As usual, Oresme fails to indicate the omission. His quotation differs from Webb's text (vol. 1, pp. 39–40, 401a–401d) in the following ways (Oresme's readings are given first preceded by the line numbers in which these words appear): 215 specierum⟨que⟩: specierumque; 218 et: mundana sive; 219 et: omitted in Webb's edition; secretis⟨que⟩: secretisque; 221 providentie: pro divinae. The last-mentioned variation made it desirable to alter Pike's translation of the *Policraticus* from "divine disposition" to "providential dispensation."

In citing the edition of the *Policraticus* printed in Migne's *Patrologia latina*, Zoubov (*Orem O soizmerimosti*, p. 375, n. 1) has again cited col. 401 of Volume 99 instead of Volume 199 (see above, the Commentary, n. 33 for III.192–94).

40 "Nec...defuisset" (III.223–28). *Asclepius* 1. 9; trans. Walter Scott, *Hermetica, the Ancient Greek and Latin Writings...Ascribed to Hermes Trismegistus*, 4 vols. (Oxford, 1924–36), vol. 1, p. 303. Other than spelling variations and the omission of the first "est" in III.227, the editions of Scott (vol. 1, p. 302) and A. D. Nock (*Corpus hermeticum* [Paris, 1945], vol. 2, p. 307) are in full agreement with Oresme's version.

41 "Quis...faciet" (III.230–31). Job 38:37; Douay-Rheims Version. Except for Oresme's substitution of "ac" for "et," the quotation from Job is identical with the Latin Vulgate in Hetzenauer's edition.

42 Plato...sono (III.232). Plato, *Timaeus* 37B in *Timaeus a Calcidio translatus* (ed. Waszink), p. 29; this passage reappears in Chalcidius's commentary, 103 (ed. Waszink), p. 153. In saying that Plato "proposed that the celestial motions are made rational without sound" (III.232), perhaps Oresme had in mind *Timaeus* 37B, where, in speaking of the self-moved Heaven, Plato says: "Now, whenever discourse that is alike true, whether it takes place concerning that which is different or that which is the same, being carried on without speech or sound within the thing that is self-moved, is about that which is sensible, and the circle of the Different, moving aright, carries its message throughout all its soul—then there arise judgments and beliefs that are sure and true."—Cornford, trans., *Plato's Cosmology*, p. 95. In a note, Cornford explains that the "self-moved thing" is identified with the "Heaven as a whole." See also Cornford's remarks on p. 72.

43 nec...audibilem (III.232–34). *De caelo* 2. 9. 290b12–291a7, part of the passage relevant to Oresme's reference to Aristotle, is quoted above in Chapter 2, n. 28.

44 "pura....conficit" (III. 237–42). *Asclepius* 1. 9, 13–14a; trans. Scott, *Hermetica*, vol. 1, pp. 303, 311. The first of Oresme's two successive quotations (III.237–38) from *Asclepius* 1 of the Hermetic *Corpus* is in full accord with the editions of Scott (p. 302) and Nock (p. 307). In the second (III.238–42), however, not only does

Oresme's version differ from the two modern editions, but the latter also disagree with one another. In what follows, I give Oresme's version first (preceded by the line number of the word) and then the reading from the modern editions; where the latter disagree, Nock's version (p. 312 of his edition) is placed in parentheses immediately after Scott's reading (p. 310 of the Scott edition): 238–39 musicam: musicen vero; 239 after "cunctarum" Scott adds "[omnium]" and Nock adds "omnium" without brackets; ordines: ordinem; after "queque" Scott has "sit ⟨omnes res⟩" while Nock adds only "sit"; 239–40 ratio divina sortita est: divina ratio sortita; 241 collatus: conlata⟨rum⟩ (conlatus); 242 conficit: confici[e]t (conficiet).

Despite the differences between the versions of Oresme and Scott, I have used Scott's translation, since it does adequately render Oresme's text.

45 "Armonia....collocatam" (III.244–51). Cassiodorus, *Variae*, bk. 2, letter 40. These two quotations from the *Variae* of Cassiodorus agree quite well with Theodore Mommsen's text in the *Monumenta Germaniae historica* (Berlin, 1894), vol. 12, p. 72. However, in the first quote (III.244–46), Mommsen has "harmonia vero" for Oresme's "armonia," and adds 'idonee" after "sermone." In the second quote (III.248–51), I have adopted "ad imitationem" from Mommsen (see also the variant readings). In III.249, Mommsen has "astronomi" where Oresme has "astrologi." These two terms were used interchangeably in the Middle Ages. Finally, the ellipsis in III.249 signifies that Oresme's version lacks the words "tanta utillima procurantem," which appear in Mommsen's edition.

46 "Solis...mercuriale" (III.252). Although he did not identify this quotation, Zoubov ("Nicole Oresme et la musique," *Med. and Ren. Studies*, vol. 5 (1961), p. 97) sees it as an allusion to Sidonius Appollinaris, *Carmen* 1, lines 7–8: "Arcas et Arcitenens fidibus strepuere sonoris, doctior hic citharae pulsibus, ille lyrae." (This is quoted by Zoubov from Luetjohn's edition of Sidonius's *Epistulae et Carmina* in the *Monumenta Germaniae historica*, vol. 8 [Berlin, 1887; new edition, 1961], p. 173.) While it is true that Arcas is Mercury, said to be the inventor of the lyre (see III.248–51), and that Arcitenens is a poetic epithet for Apollo, god of the sun, who was often represented with a cither, Zoubouv's interpretation is questionable, since it is evident that we have here a direct quotation from "a certain poet" (quidam poeta [III.252]), rather than a mere allusion to certain poetic lines.

47 "Quid...comprehendit" (III.253–55). Cassiodorus, *Variae*, bk. 2, letter 40. Oresme's quotation from Cassiodorus's *Variae* is in agreement (except for spelling) with that in Mommsen's edition (p. 70).

48 "Dicunt...marcescit" (III.259–61). Cassiodorus, *Variae*, bk. 2, letter 40. In Mommsen's edition of the *Variae*, "caelestis" replaces Oresme's "supercelestis" and "musicis" is omitted.

49 "os...caput" (III.267–69). Ovid, *Metamorphoses*, bk. 1. 85–86 ("os...vultus"); 2 vols., trans. Miller, Loeb Library, vol. 1, p. 9. Claudian, *De raptu Proserpinae*, bk. 3. 41–42 ("Quid... caput"); *Claudian*, 2 vols., trans. Platnauer, Loeb Library, vol. 2, p. 349. These brief quotations from Ovid and Claudian agree completely with the editions of these authors by Miller (vol. 1, p. 8) and Platnauer (vol. 2, p. 348). So that Ovid's words are more readily integrated into the flow of Oresme's discussion, I have altered the first word of Miller's translation from "gave" to

"give." Citing Ovid explicitly, Isidore of Seville includes this very quotation in his *Etymologiae*, bk. 11. 5[2] (it is employed by Isidore as evidence for believing that man's superiority over the animals stems from his erect posture). Although he puts it to a different use, Oresme may have derived the quotation directly from Isidore.

50 "Inquit...amicus" (III. 272–73). This quotation agrees precisely with the edition of the *De vetula* that I have consulted (see bibliography, under Richard de Fournival). These two lines immediately precede Oresme's earlier quotation from the *De vetula* in III.75–76 (see the discussion of these lines above, the Commentary, n. 9, p. 338). Although Zoubov identified Oresme's earlier quotation (*Orem O soizmerimosti*, p. 391, n. 23), he did not recognize and isolate as a quotation the lines under discussion here.

51 "tempus...celi" (III.273–74). See Aristotle, *Physics* 4. 12. 221a1, 221b7, 221b25; *De caelo* 1. 9. 297a15 and 2. 4. 287a24. None of these five references represents precisely the words ascribed to Aristotle, although they are fully in accord with Aristotle's views. In the Aristotelian text of the *Physics* that accompanies Thomas Aquinas's *Commentary on the Physics*, the three references to the *Physics* present some version of "tempus mensura motus est" (all three references may be found preceding Book 4, text 20 of Aquinas's *In octo libros De physico auditu sive physicorum Aristotelis commentaria*, ed. A. M. Pirotta [Naples, 1953], p. 253), but fail to include any form of the word *celum*, which is also true of the two translations (one of which is definitely medieval in origin) of Aristotle's *Physics* that accompany Averroes' commentary (see bk. 4, text 114, in the Junctas ed., vol. 4, fol. 189r, col. 2). The Aristotelian text of the *De caelo* that appears with the modern edition of Aquinas's commentary has "tempus numerus motus" (at 1. 9. 279a15) and "motuum mensura quae caeli latio" (at 2. 4. 287a24; for these two references see bk. 1, text 21, p. 101 and bk. 2, text 6, p. 171 of Aquinas's *In Aristotelis libros De caelo et mundo...*, ed. R. M. Spiazzi [Turin and Rome, 1952]). Two translations of Aristotle's *De caelo* (one of which originates in the Middle Ages) appear with the commentary of Averroes in Volume 5 of the Junctas edition and are similarly worded (see bk. 1, text 99, fol. 66v, col. 2 and bk. 2, text 28, fol. 113r, col. 2).

For this very brief quotation, Zoubov (*Orem O soizmerimosti*, p. 377, n. 2) offers without comment only the two references to the *De caelo*.

52 "In quo....sunt" (III. 279–87). Plato, *Timaeus* 39D ("In quo...dimetietur"), 40C ("stellarum...sunt"); in *Timaeus a Calcidio translatus* (ed. Waszink). In these two quotations from Chalcidius's translation of Plato's *Timaeus*, the second (III.285–87) agrees perfectly with Waszink's edition (pp. 33–34), while the first (III.279–84; p. 32 of Waszink's edition) varies as follows (Oresme's version appears first preceded by the line number in which the word appears): 281 autem: tamen; 282 tum: tunc ("tum" appears as a variant); 283–84 after "circumactionis" Waszink has "alterius"; 284 quem: quam; uniformisque: atque uniformis.

53 "Esse....reparaverit" (III.289–93). Apuleius Madaurensis, *De Platone et eius*

---

2. See the edition of W. M. Lindsay, *Isidori Hispalensis Episcopi Etymologiarum sive* *originum libri*, 2 vols. (Oxford, 1911), vol. 2.

dogmate, bk. 1, chap. 10; in *Apulei Madaurensis Opera quae supersunt*, vol. 3, *De philosophia libri*, ed. Paul Thomas (Leipzig: Teubner, 1907). Other than having "necesse est" (III.289) for "nihilo minus," Oresme's quotation from Apuleius Madaurensis (fl. 2d century A.D.) accords fully with the Teubner edition by Paul Thomas (p. 94).

Since Zoubov did not identify this quotation and was therefore unaware of Apuleius's authorship, he followed the manuscripts and retained the name "Appollonius" (*Orem O soizmerimosti*, p. 377; see variant readings for "Apuleius" in III.288).

54 "Cum...teneantur" (III.300–303).   Macrobius, *Commentary on the Dream of Scipio* 2. 11. 2; trans. Stahl, pp. 219–20. Oresme mistakenly ascribes this quotation to Macrobius, when in fact the words are those of Scipio Africanus The Elder, quoted by Macrobius directly from Cicero's *Dream of Scipio* in order to comment on it. This offers reasonable evidence for believing that in the *De commensurabilitate*, Oresme's quotations from Cicero's *Dream of Scipio* were drawn not from a separate manuscript of Book 6 of Cicero's *De republica*, which contains the *Dream of Scipio*, but rather from the text of the *Dream* as it appears in manuscripts of Macrobius's *Commentary on the Dream of Scipio*. If Oresme had used a separate version of the *Dream of Scipio*, he would not have attributed words from Cicero's text to Macrobius (see above, the Commentary, n. 28 for III.160–62, p. 344).

Comparing the texts of Eyssenhardt (p. 609), Willis (p. 128), and Keyes (p. 276) with Oresme's version, we note that after "Cum" (III.300) Keyes alone has "autem." However, all three modern editors agree in adding "cuncta" before "astra" (III.300) and "vere" after "ille" (III.302). Since Oresme omits "vere," I have altered Stahl's translation by eliminating the word "truly" after "cycle."

55 Et...peracta (III.311–12).   Cf. Origen, *Contra Celsum* 5. 20, 21 and Nemesius, *De natura hominis*, chap. 38. The return of the same individuals after the completion of every Great Year is not a Platonist doctrine, as Oresme would have it, but a Stoic one. In the *De natura hominis*, chap. 38, Nemesius (ca. 400 A.D.), Bishop of Emesa, wrote:

> The Stoics say that the planets, returning to the same point of longitude and latitude which each occupied when first the universe arose, at fixed periods of time bring about a conflagration and destruction of things, and that the universe again reverts anew to the same condition, and that as the stars again move in the same way everything that took place in the former period is exactly reproduced. Socrates, they say, and Plato will again exist, and every single man, with the same friends and countrymen; the same things will happen to them, they will meet with the same fortune, and deal with the same things.—Trans. Robbins in Ptolemy, *Tetrabiblos*, Loeb Library, p. 15, n. 3.

Although Oresme calls them Platonists rather than Stoics, this passage from Nemesius must be considered as a possible source for his statement, since the *De natura hominis*, wrongly attributed to Gregory of Nyssa, was translated into Latin in the thirteenth century.[3]

3. This translation was made by Burgundio of Pisa, who, in a dedicatory preface to   Frederick II, Holy Roman Emperor, ascribes the work to Gregory of Nyssa (fl. 2d half 4th

There are, however, at least two other possible sources where Plato or Platonists are linked directly with the Great Year and the doctrine of individual return. The first of these would have been available to Oresme in Paris. In Book 4, Distinction 43, Question 3 of the Parisian *Reportata* of his Sentence Commentary, John Duns Scotus considers whether there might be an active cause in nature that could reproduce exactly a body that had already been corrupted.[4] Concerning this he tells us that "there is a certain opinion, namely of Plato, which Augustine reports in *The City of God*, Book XII, Chapter 13. This [opinion] holds that in generable and corruptible things nature can repeat all things [exactly and] numerically the same, so that, consequently, all generable and corruptible things, which are only corporeal, can return numerically the same."[5] Although Augustine did indeed discuss this opinion, he did not attribute it to Plato, but only to "philosophers" in general,[6] a fact which Scotus himself apparently came to realize for, in his revised and ex-

---

century A.D). See *Gregorii Nysseni* (*Nemesii Emeseni*) *Peri Phuseos Anthropou liber a Burgundione in latinum translatus*, ed. Karl Burkhard in *Programme des Carl-Ludwig-Gymnasiums* (Vienna, 1891, 1892, 1896, 1901, and 1902). The attribution to Gregory of Nyssa occurs on p. 11 of the 1891 *Programm*, and a table of contents includes Chapter 38, titled "De eo quod est in nobis, quod est de libero arbitrio" (p. 13). In contrast is an earlier eleventh-century translation by Alfanus (1058–85), Bishop of Salerno. Although correctly ascribed to Nemesius, it lacks Chapter 38 and six other chapters (see *Nemesii Episcopi Premnon Physicon...a N. Alfano Archiepiscopo Salerni in latinum translatus*, ed. Karl Burkhard (Leipzig, 1917], p. v). Since I have seen only the *Programm* of 1891, which includes only the dedication and the first of 42 chapters, I do not know whether the relevant part of Chapter 38, quoted in translation above, is actually included and what its textual status might be. The later translation by Burgundio was not only more complete than its predecessor, but also came to be more widely known. See Etienne Gilson, *History of Christian Philosophy in the Middle Ages* (New York, 1955), p. 584, n. 84.

4. "An scilicet reparatio corporum habeat causam activam in natura."—Vivès edition of Scotus's *Opera omnia* (see above, chap. 3, n. 85, for full citation of this edition), vol. 24,

p. 509, col. 1. This is a subquestion in the broader context of Questio 3 which bears the title "Whether nature could be the efficient cause of a resurrection?" ("Utrum natura possit esse causa efficiens resurrectionis?"—Ibid., p. 508, col. 1.)

5. "De primo est quaedam opinio, scilicet Platonis, quam recitat Augustinus 12. *de civitate Dei cap*. 13. qui dicit quod natura potest reparare omnia eadem numero in generabilibus et corruptibilibus, et per consequens omnia generabilia et corruptibilia, quae tantum sunt corporalia, possunt redire eadem numero."—Ibid., p. 509, col. 1.

6. In responding to, and seemingly trying to avoid, the problem of what might have existed prior to the creation of our world in time, "philosophers," Augustine tells us, "have seen no other approved means of solving than by introducing cycles of time, in which there should be a constant renewal and repetition of the order of nature; and they have therefore asserted that these cycles will ceaselessly recur, one passing away and another coming, though they are not agreed as to whether one permanent world shall pass through all these cycles, or whether the world shall at fixed intervals die out, and be renewed so as to exhibit a recurrence of the same phenomena—the things which have been, and those which are to be, coinciding."
—*The City of God by Saint Augustine*, bk. 12, chap. 13; trans. Marcus Dods, 2 vols. (New

panded Oxford version of Book 4 of his Sentence Commentary, this error was corrected.[7] Oresme, who could have been aware of this section in the Parisian version, may have generalized it to cover all Platonists.

The second possible source appears in the *Contra Celsum* of Origen (185–ca. 254 A.D.), who much earlier than Nemesius gave substantially the same description of this Stoic position (*Contra Celsum* 5. 20):

> The Stoics maintain that the universe periodically undergoes a conflagration and after that a restoration of order in which everything is indistinguishable from what happened in the previous restoration of the world. All those of them who have respected the doctrine have said that there is a slight and very minute difference between one period and the events in the period before it. Now these men say that in the succeeding period it will be the same again: Socrates will again be son of Sophroniscus and be an Athenian, and Phaenarete will again marry Sophroniscus and give birth to him.[8]

(Note 6 continued)

York, 1948; reprint of 1872 ed.), vol. 1, p. 498. (In at least two editions of the Latin text this passage is in Book 12, Chapter 14: see the edition of Bernhard Dombart, *Sancti Aurelii Augustini episcopi De civitate Dei libri XXII*, 2 vols. [Leipzig: Teubner, 1877], vol. 1, p. 531; and the edition of Emanuel Hoffmann with the same title, in *Corpus scriptorum ecclesiasticorum latinorum*, vol. 40 [Vienna, 1899], pt. 1, p. 587.)

The source of Scotus's mistake in assuming that Augustine attributed the doctrine of individual return to Plato is perhaps traceable to a later passage in the same chapter, where Augustine emphatically explains what Solomon, in Ecclesiastes 1:9, 10, did not mean when he said "there is no new thing under the sun": "At all events, far be it from any true believer to suppose that by these words of Solomon those cycles are meant, in which, according to those philosophers, the same periods and events of time are repeated; as if, for example, the philosopher Plato, having taught in the school at Athens which is called the Academy, so, numberless ages before, at long but certain intervals, this same Plato, and the same school, and the same disciples existed, and so also are to be repeated during the countless cycles that are yet to be,—far be it, I say, from us to believe this."—*City of God*, trans. Dods, vol. 1, p. 499. (Although not affecting the substance of Augustine's thoughts, the translation by

Dods omits, here and there, a word or phrase that appears in the Latin text of the edition by Dombart, p. 532, and that by Hoffmann, pp. 588–89; the same may be said of the passage quoted earlier in this footnote.) It is obvious that Plato's name, and school, and disciples, are here used merely by way of example, an example which Scotus may have mistakenly interpreted as an explicit attribution to Plato. Of course, it is possible that Oresme made the same mistake independently, without knowledge of this Scotian passage.

7. In the Oxford version, usually known as the *Opus oxoniense*, Scotus says: "De primo, Augustinus 12. *Civit.* cap. 13 recitat opinionem Philosophorum aliquorum dicentium per circuitus temporum, eadem redire, quibus imponitur, quod post magnum annum, id est, post triginta sex millia annorum, omnia redirent eadem."—Dist. 43, Quest. 3 in the Vivès ed., vol. 20, p. 67, col. 1. Here it is no longer the opinion of Plato, but the "opinion of certain philosophers." Indeed, later in the same discussion (ibid., p. 68, col. 2), Scotus even quotes Augustine's example involving Plato, but it is described as an example reflecting the views of "philosophers." The relationship between the Parisian and Oxford versions of Scotus's Sentence Commentary is discussed above, pp. 116–17.

8. Origen, *Contra Celsum* 5. 20; trans. Henry Chadwick (Cambridge University

Turning next to the Pythagoreans and Platonists, Origen tells us that

> though the Pythagoreans and Platonists maintain that the whole is indestructible, yet they fall into similar absurdities. For when in certain fixed cycles the stars adopt the same configurations and relationships to each other, they say that everything on earth is in the same position as it was at the last time when the relationship of the stars in the universe to one another was the same. According, then, to this doctrine it is inevitable that when after a long period the stars come into the same relationship to one another which they had in the time of Socrates, Socrates will again be born of the same parents and suffer the same attacks, and will be accused by Anytus and Meletus, and be condemned by the council of the Areopagus."—Origen, *Contra Celsum* 5. 21; trans. Chadwick, p. 280.

Origen quite properly attributes the doctrine of the Great, or Perfect, Year to the Platonists, but then draws from it the consequence that the same individuals will return in each successive period. In this way he attributes the Stoic view to the Platonists (some Platonists may have accepted such a doctrine, but Origen is here deriving it as a consequence of their belief in a Great Year, and then unqualifiedly ascribing it to them). Now although the *Contra Celsum* does not seem to have been translated into Latin during the Middle Ages, Origen was so widely read and so influential that the substance of this passage, perhaps altered to make the Platonists appear as outright proponents of the doctrine of individual return, may have reached Oresme through some Latin source.

Aristotle was a vigorous opponent of the doctrine of individual return (see *De generatione et corruptione* 2. 11. 338b6–20), a doctrine that antedates the Stoic school. In both his middle commentary and epitome of the *De generatione*, Averroes[9] sided with Aristotle. Although the epitome of the *De generatione* was not available to Oresme in Latin,[10] it contains the more relevant comment. There,

---

Press, 1953), pp. 279–80. Later, after denying that the body will return to its original nature after corruption (see 5. 23, ibid., p. 281), Origen says: "But it is the Stoics who say that *after the body has been entirely corrupted it will return to its original nature*, because they believe in the doctrine that each world-period is indistinguishable; and it is they who say that it will again be composed *in that same first condition which it had before it was dissolved*, proving this, as they imagine, on the ground of logical necessities." The italics are the translator's. Although in 5. 23, the world-periods are said to be indistinguishable, in 5. 20 Origen has noted that the Stoics allowed for very slight, but seemingly undetectable, differences in successive Great Years (the two passages are, of course, entirely compatible). Since Oresme does not

elaborate, either of these passages could have led him to mention the return of the same men after every Great Year. For a discussion of identically recurring Great Years and those that differ minutely or undetectably, see Duhem, *Le Système du monde*, vol. 1, pp. 278–82 (the passages from Origen and Nemesius are also included).

9. *Aristotelis opera cum Averrois commentariis*, Junctas ed., vol. 5, contains both treatises.

10. It may have appeared first in Latin in the 1550 Juntine edition of the works and commentaries of Averroes in a translation made from the Hebrew by Vital Nissus, of whom nothing is known (see Kurland, trans., *Averroes on Aristotle's "De generatione,"* p. xiv).

without mentioning Stoics or Platonists, Averroes concluded that those who "argue that when the same arrangement in all of the parts of the sphere that prevailed at the time of Zaid's existence recurs, Zaid himself will return upon himself. And that is impossible as we have demonstrated."[11]

56 "certisque...artus" (III.312–13). Claudian, *De raptu Proserpinae*, bk. 1. 61–62; *Claudian*, 2 vols., trans. Platnauer, Loeb Library, vol. 2, pp. 297–99. Where Platnauer's edition (vol. 2, pp. 296–98) has "rursus," Oresme has "rursum"; otherwise the quotations are identical.

57 "Has...reverti" (III.314–15). Virgil, *Aeneid* 6. 748, 751; *Virgil*, trans. Fairclough, Loeb Library, vol. 1, p. 559. Oresme's quotation from Virgil differs only trivially from that in Fairclough's edition (vol. 1, p. 558): where Oresme has "rursumque incipiunt," Fairclough has "rursus et incipiant."

58 "posse...putat" (III.317). Ovid, *Fasti*, bk. 3. 163–64; trans. Sir James George Frazer, Loeb Classical Library (London and New York, 1931), p. 131. These few words from Ovid agree with those in Frazer's modern edition (p. 130).

59 "Et...veris" (III.325–27). Here Oresme expresses a nominalist viewpoint, namely, that some false propositions are more probable than their opposite and true propositions. Pierre de Ceffons, for example, also made this claim in the fourteenth century (see Julius R. Weinberg, *Nicolaus of Autrecourt* [Princeton, 1948], pp. 116–17). And in a quite similar vein, John of Mirecourt said: "If it is said that something is not probable because the faith is in opposition, I say: this implication is not valid, for although it follows that a proposition is not true when faith is in opposition, yet it does not follow that the proposition is not probable. Indeed, the opposites of articles of faith are more probable to us than are the articles themselves."—Ibid., pp. 120–21.

60 "Dicunt...subtilitate" (III.372–74). Pliny, *Natural History* 2. 20. 84; 10 vols., Loeb Library, vol. 1, trans. Rackham, p. 229. Oresme's quotation from Pliny varies as follows from Rackham's text (vol. 1, p. 228; Oresme's reading appears first): "Dicunt" is omitted by Rackham; moveri Dorio: Dorio moveri; "Mercurium" is also omitted by Rackham; Oresme's "frigio" has been altered to "Phrygio" (see variant readings) in order to conform with Pliny's spelling; iocunda: iucunda. Rackham's translation has been changed slightly to allow for the additions in Oresme's quotation.

61 "Tamen....instrumentis" (III.375–81). We see here that Oresme definitely rejects the view that a ratio of sounds varies as a ratio of velocities (III.375–76), a move that led him to repudiate the traditional acceptance of celestial harmony as arising from ratios of planetary velocities (III.394–97). Instead, he suggests that

11. Trans. Kurland, *Averroes on Aristotle's "De generatione,"* p. 137. I have also quoted this passage in a wider context on p. 106, above. Although the name Zaid was changed to Plato in Vital Nissus's Latin translation from the Hebrew, the substance of this passage is the same as in Kurland's translation from the Arabic. For the Latin text see *Aristotelis opera cum Averrois commentariis*, Junctas ed., vol. 5, fol. 396v, col. 1. My citation is to the volume published in 1562, not 1550, the date given by Kurland and mentioned in the immediately preceding note.

they might arise from relationships between the different quantities of matter in the planetary spheres. Thus Oresme has rejected two conflicting views mentioned in III.370–71: the one claiming that the highest, or outermost, sphere produces the sharpest, or most acute, sound because its velocity is greatest (this was held, for example, by Macrobius; see III.165–67); the other maintaining that the most acute sound is produced by the moon, the lowest, or innermost, celestial sphere (this was the view of Nicomachus of Gerasa and Boethius; see Zoubov, "Nicole Oresme et la musique," *Med. and Ren. Studies*, vol. 5 (1961), p. 100).

As Oresme explains it, there is no relationship between sharpness of sound and velocity, because "whether a string or a drum is struck strongly or weakly, slowly or quickly, nothing is changed by this" (III.376–78). He concludes that only variations in weight or quantity of matter can produce changes in sharpness, although he does not say whether a greater quantity of matter produces the more acute sound or the deeper sound (indeed, he fails to explain the sense in which "weight" or "quantity of matter" is to be understood). This is, however, made clear in the later *Le Livre du ciel et du monde*, bk. 2, chap. 18, 126a (the English translation of this passage is given in the Commentary, n. 29 for III.162–67 on p. 345, above), where Oresme specifies that it is the outermost sphere, which presumably possesses the greatest quantity of matter, that produces the lowest or deepest sound. Here, surprisingly, he also allows that velocity plays a role, explaining that the greater the celestial velocity, the greater will be the intensity or loudness of the sound. And so he concludes that the outermost sphere causes not only the lowest or deepest sound, but also the loudest. It is for quite different reasons, then, that Oresme agrees with those who insisted that the outermost sphere produces the deepest or lowest sound. In quoting this passage from *Le Livre du ciel*, Zoubov acknowledges that Oresme agreed with those who associated the deepest sound with the outermost sphere, but then, by reversing the chronology of *Le Livre du ciel* and *De commensurabilitate*, errs when he says "Mais en fin de compte Oresme rejeta l'une et l'autre hypothèses. Voici qu'on lit dans sons traité *De commensurabilitate vel incommensurabilitate motuum celi*. C'est la Géométrie qui parle:…" The quotation which Zoubov then reproduces is essentially that which is found in the present edition in III.368–71 (see Zoubov, "Nicole Oresme et la musique," *Med. and Ren. Studies*, vol. 5 [1961], p. 100 and n. 3). But the *De commensurabilitate*, frequently cited in *Le Livre du ciel*, was written long before the latter treatise, so that, contrary to Zoubov, it is Oresme's position in *Le Livre du ciel* which must be taken as final.

Oresme's separation of acute and heavy sounds on the one hand, and loud and soft on the other, represents a distinction between the pitch of a sound and its volume or intensity, respectively. This was made clear by Oresme in the *De configurationibus qualitatum*, pt. 2, chap. 15 and was detected and quoted by Zoubov in *Med. and Ren. Studies*, vol. 5 (1961), p. 101 and n. 3 (for the edited text see Clagett's edition, *Nicole Oresme and Medieval Geometry*, pp. 306–9.

62 "Unde…novit" (III.381–83). See Macrobius, *Commentary on the Dream of Scipio* 2. 1. 9–12 (ed. Eyssenhardt, p. 573, and trans. Stahl, pp. 186–87) and Boethius, *De institutione musica*, bk. 1. 10–11 (ed. Friedlein, pp. 196–98). This

account is repeated by a number of ancient authors. Here is what Macrobius says about the discovery by Pythagoras:

> He happened to pass the open shop of some blacksmiths who were beating a hot iron with hammers. The sound of the hammers striking in alternate and regular succession fell upon his ears with the higher note so attuned to the lower that each time the same musical interval returned, and always striking a concord. Here Pythagoras, seeing that his opportunity had been presented to him, ascertained with his eyes and hands what he had been searching for in his mind. He approached the smiths and stood over their work, carefully heeding the sounds that came forth from the blows of each. Thinking that the difference might be ascribed to the strength of the smiths he requested them to change hammers. Hereupon the difference in tones did not stay with the men but followed the hammers. Then he turned his whole attention to the study of their weights, and when he had recorded the difference in the weight of each, he had other hammers heavier or lighter than these made. Blows from these produced sounds that were not at all like those of the original hammers, and besides they did not harmonize. He then concluded that harmony of tones was produced according to a proportion of weights, and made a record of all the numerical relations of the various weights producing harmony.—Macrobius, *Commentary on the Dream of Scipio*, trans. Stahl, pp. 186–87; also see above, chap. 2, n. 107.

63 "*aut²...De probleumatibus*" (III.386–87).   The view attributed to Aristotle in the *Problems*[12] is formulated in response to the question, Why do "eunuchs and old men have shrill voices?" He replies (*Problems* 11. 62. 906a1–7): "Is it because the movement of air which creates a shriller sound is quicker? Now it is more difficult to move a greater amount of the same thing, and so those who are in the prime of life draw in the air in greater quantities, and therefore this air, since it travels more slowly, makes the voice deeper. In boys and eunuchs the contrary occurs, because they contain less air."

64 "*modulos*" (III.405).   In *St. Augustine's De musica, A Synopsis* (The Orthological Institute [London, 1949]), p. 12, W. F. Jackson Knight interprets St. Augustine's conception of "modulatio" in *De musica*, 1. 2. 2 and 3 as "controlled variation of sound." It seemed best, therefore, to translate "modulos" as "sounds."

65 "*celi...digitos*" (III.439).   Pliny, *Natural History* 2. 21. 87. These few words from Pliny are presented by Oresme as a direct quotation, although they are actually an alteration of part of an earlier quotation from Pliny in III.27–28 where Oresme has "mundi...mensura veniat in digitos." Since, in the context of the full quotation in III.26–28, Pliny clearly implies that men will never succeed in determining precisely the measure and dimensions of the world (see also above, the Commentary, n. 3 for III.26–28, p. 337), Oresme saw fit to change "veniat" (III.28) to "non venit" (III.439), so that he might avoid repetition of the full quotation and yet convey its substance. His alteration of "mundi" to "celi" may be explained by the fact that in III.439 the discussion is restricted to celestial aspects rather

---

12. Aristotle did not write the *Problems*, which is believed to be the work of several anonymous authors. See E. S. Forster's introduction to his translation of the *Proble-* *mata* (quoted in the Commentary, above) in *The Works of Aristotle* (Oxford, 1927), vol. 7, p. vii.

than the world at large (however, the two terms were sometimes used synonymously).

66 "sicut...probatum" (III.459).   This is beyond doubt a reference to his *De proportionibus proportionum*. See above, chap. 2, n. 113, for a summary of the steps that
led Oresme to this general claim in the *De proportionibus*; and chap. 2, n. 112, for
an explanation of why Oresme deliberately suppressed the title of the *De proportionibus* in this obvious reference to it, while citing it by title in II.201.

Appendix 1

Relevant Chapters from the
Quadripartitum numerorum by Johannes de Muris

The sigla of the three manuscripts employed in this edition of selected but relevant chapters in Johannes de Muris's *Quadripartitum numerorum*, Book I, Tract 1, are as follows:

Pa = Paris, Biblothèque Nationale, fonds latin, 7190, fols. 75r–v, 77r–79v

V = Vienna, Österreichische Nationalbibliothek, 4770, fols. 309v–311r, 311v–315v

W = Vienna, Österreichische Nationalbibliothek, 10954, fols. 119r–v, 121v–124r, 125r–v.

Since *Pa* is not only the earliest of the three manuscripts (it is of the fourteenth century, where *V* is dated in the fifteenth century and *W* in the sixteenth; see Clagett, *Science of Mechanics*, p. 126n), but also the most complete, I have tended to rely on it in preference to the Vienna manuscripts.

Johannes de Muris

Quadripartitum numerorum

Liber quartus [tractatus primus]

Capitulum 12

5   Sit ergo unum mobile quod in 3 diebus totum celum valeat peragrare; aliud vero mobile in 4 diebus, et incipiant simul moveri motibus contrariis incedentes. Queritur quando, ubi, et quotiens coniungentur antequam totius diversitatis revolutio compleatur.

    Ex datis patet quod primum mobile velocius aut fortius est secundo et move-
10 tur ad secundum in proportione sesquitertia, sicut 4 ad 3. Sit ergo totum celum partes 7 de quibus dum primum mobile transibit 4, secundum in eodem tempore ibit 3. Sit et totum tempus 7 dies. Ergo primum mobile $\frac{4}{7}$ spatii, scilicet celi, transibit in $\frac{4}{7}$ trium dierum; secundum mobile reliquas $\frac{3}{7}$ in $\frac{3}{7}$ 4 dierum. Nunc autem ex doctrina fractionum iam scisti quod $\frac{4}{7}$ trium dierum et $\frac{3}{7}$ 4 dierum
15 sunt equales et valent diem $1\frac{5}{7}$, quod est tempus prime coniunctionis duorum mobilium predictorum. In aliis coniunctionibus pari modo et septies antequam compleatur revolutio coniungentur. Tunc redibunt ambo ad initium circuli sicut prius et alie coniunctiones similes prioribus. Et per hunc modum scire posses coniunctionem lune cum capite draconis et caude; similiter de ceteris
20 planetis cum suis draconibus et de omnibus moventibus ad contrarias positiones.

Capitulum 12
  4 Capitulum 12: 12ª *Pa*
  5 3: tribus *W*
  7 *post* quando *hab. W* et
10 in *om. V*
11 dum *?V; om. Pa*/transit *V*
12 Sit: sic *W*/celi: *?W*
13 *post* secundum *hab. W* vero
14 *ante* trium *hab. W* 3

15 prima *W*
16 septies: quinquies *W sed mg. hab. W* septies
17 compleatur: ampleatur *V*/ambo *om. Pa*/ ad: ab *V*
18 ?alies ?coniunctiones *W sed mg. hab. W* aliae coniunctiones
19 possem *W*/similiter: super *W sed mg. hab. W* similiter

Johannes de Muris

The Four-Parted Book on Numbers

Fourth Book [Tract 1]

Chapter 12

Let there be, then, one mobile that can traverse the whole sky in 3 days, and another mobile that can do it in 4 days; and let them be moved simultaneously with opposite motions. Let it be sought when, where, and how often they will conjunct before a revolution of total diversity is completed.

From what has been given, it is clear that the first mobile is quicker or more powerful than the second and is moved in a sesquitertian ratio to the second, [i.e.] as 4 to 3. And let the whole sky be [divided into] 7 parts, so that while the first mobile will traverse 4 of these, the second will move through 3 in the same time. Therefore the first mobile will traverse $\frac{4}{7}$ of the distance, namely, of the sky, in $\frac{4}{7}$ of 3 days; and the second mobile will traverse the remaining $\frac{3}{7}$ [of the distance] in $\frac{3}{7}$ of 4 days. Now from the teaching of fractions, you already know that $\frac{4}{7}$ of 3 days and $\frac{3}{7}$ of 4 days are equal and make $1\frac{5}{7}$ days, which is the time of the first conjunction of the aforementioned two mobiles. Using the same method in [determining] other conjunctions, [it will be found that] they will conjunct seven times before a revolution is completed. Then both will return to the starting point of the circle as before, and other conjunctions [will occur] similar to the previous [ones]. And by this method you could know [or determine] a conjunction of the moon with the head and tail of the dragon; and similarly for the other planets with their draconic points and for all things moving toward opposite positions.

Capitulum 13

Sit circulus divisus in 24 equa super quem sunt duo mobilia *A* et *B*. Transeat autem *A* diebus singulis 5 partes, *B* vero 9; incipiant simul moveri. Queritur quando coniungentur et in qua parte circuli supradicti.

5   Responde: Sume differentiam motus unius in una die ad motum alterius, hoc est, deme minorem de maiori, scilicet 5 de 9, remanent 4 pro differentiam per quam divide totum circulum, qui est 24, exibunt 6, quod est tempus coniunctionis. Deinde duc motum *A* unius diei in 6 et sunt 30 dempto circulo remanent 6. Tunc etiam duc motum *B* unius diei in 6 et sunt 54 a quibus dempto circulo
10 quotiens potest remanent 6. Dico ergo quod *A* et *B* in 6 diebus perfectis coniunguntur 6 partibus pertransitis circuli supradicti. Et quater erunt in coniunctione antequam revolutio compleatur quod scitur dividendo circulum, qui est 24, per tempus unius coniunctionis, quod est numerus 6 dierum. Et si vis scire quot partibus movetur, scilicet in die, ista scito tempore coniunctionis, duc
15 motum ipsius *A* in tempore coniunctionis exit 30 quibus adde circulum exit 54 quos divide per tempus coniunctionis exit 9, quod est ?propositum.

Per hunc modum potes omnes planetas adinvicem coniungere in orbe signorum cognito motu diurno ciuislibet planete assignata certa signiferi quantitate; et per hunc modum etiam concludi motus lune medius et diurnus.

Capitulum 14
[De incommensuratione motuum]

Si autem sunt duo circuli incommensurabiles ut sunt dyameter et costa super idem centrum, et a puncto communi incipient duo mobilia illos duos circulos
5   pertransire numquam in parte numerali in eternum coniungerentur factis revolutionibus sempiternis. Quia posito contrario essent adinvicem commensurabiles. Et licet moverentur proportionaliter non tamen ille motus amborum habere mensuram communem, sicut gradus aut minuta vel quelibet alia fractio quantumcumque modica poneretur. Et si esset ita in celo et circulis planetarum
10 non esset mirum neminem posse precise reperire motus planetarum, sed recti-

Capitulum 13
  1 Capitulum 13: 13ᵃ *Pa*
  2 divisus in 24: in 24 divisus *W*/ante 24
    hab. *V* 4
  4 circuli supradicti *tr. V*
  5 una *V* prima *PaW sed mg. hab. W* una
  6 scilicet *om. W*/remanent *Pa*
  8 remanent *Pa* et ?*W*
  9 54: 24 *W*
  10 remanent *PaW*/ergo *om. W* igitur *V*
  13 unius coniunctionis *tr. W*

13–16 Et si . . . ?propositum *om. VW*
17 adinvicem coniungere *tr. W*
18 certa: cuncta *W sed mg. hab. W* certa
19 et per . . . diurnus *om. VW*

Capitulum 14
  1 Capitulum 14: 14ᵃ *Pa*
  2 [De incommensuratione motuum] *om.*
    VW
  4 mobilia: inaequalia *W sed mg. hab. W*
    mobilia

Chapter 13

Let there be a circle divided into 24 equal parts on which there are two mobiles, *A* and *B*. Furthermore, *A* will traverse 5 parts every day and *B* 9 parts [after] they begin to be moved simultaneously. Let it be sought when they will conjunct, and in what part of the aforementioned circle.

Answer [in this way]: Take the difference of the daily motion of one to the other, that is, subtract the lesser from the greater, namely, 5 from 9, and 4 remains as the difference by which you divide the whole circle, which is 24, and 6 will result, which is the time of conjunction. Next, multiply the daily motion of *A* by 6 and get 30, from which 6 will remain after the circle has been subtracted [from it]. Then also multiply the daily motion of *B* by 6 and get 54, from which there will remain 6 after the circle has been subtracted as many times as possible. I say, therefore, that *A* and *B* will conjunct in 6 complete days after having traversed 6 parts of the aforementioned circle. And they will be in conjunction four times before a revolution is completed, which is known by dividing the circle, namely, 24, by the time of one conjunction, namely, 6 days. And with this time of conjunction known, should you wish to know how many parts [*B*] is moved in a day, multiply the [daily] motion of *A* by the time of conjunction and obtain 30, to which you add the circle and get 54; [this number] you divide by the time of conjunction to get 9, which has been proposed.

By this method, you can conjunct all planets mutually in the orb of the signs [i.e., zodiac] upon knowing the daily motion of any planet designated by a fixed quantity [or portion] of the sky; and in this way the mean and daily motion of the moon can also be determined.

Chapter 14
[On the Incommensurability of Motions]

If two incommensurable circles with a common center are related as the diagonal and side [of a square], and [if] two mobiles begin to traverse those two circles [starting] from a common point, they will never in all eternity conjunct in the same part [or point,* even] with those revolutions that are made eternally. For with the opposite assumed, they would be mutually commensurable. And even though these motions were moved proportionally, they could not both have a common measure, as [for example] a degree, or minute, or any other fraction whatever, however small it be assumed. And if this be true in the sky and circles of the planets, it is no surprise that no one is able to find precisely the motions of the planets, but is repeatedly

---

* See above, chap. 3, n. 27.

---

5 numerali: naturali *W*/in eternum *tr. VW*
post numquam/coniungentur *W*
8 grada *W*

9 quantumcumque: quantumlibet *V*/*infra* ita *scr. et del.* Pa ?terra

ficationibus indigeret alternatim. Et si non hoc, adhuc, continuitas motus non capitur precise ab immobili instrumento. De hiis alias sermo fiet.

Capitulum 21
[De coniunctione mobilium in circulari spacio]

Sint 3 mobilia super circulum 60 quorum primum per 9, secundum per 7, tertium per 5 diebus singulis moveatur; simul incipiantque moveri et ab eodem
5   puncto circuli supradicti. Queritur quando hec tria in eodem puncto circuli coniungentur.

Responde: Tolle semper tardum a veloci sicut secundum a primo et restant 2 per que circulum divide exit 30 dies quod est tempus coniunctionis amborum. Item deme tertium a primo restant 4 per que divide circulum exit 15 dies quod
10   est tempus coniunctionis eorum. Similiter arte pari secundum coniungitur tertio in 30 diebus. Ergo nunc habes hic tria tempora, scilicet 15, 30, 30; et quoniam unum, scilicet 30, est multiplex ad aliud, scilicet 15, ipsum est, scilicet 30 dies, tempus in quo hec tria mobilia pariter sunt coniuncta 30 partibus circuli pertransitis demptis revolutionibus peragratis.

Capitulum 22

Si autem primum per 11, secundum per 8, tertium per 4 diebus singulis moveatur, dic primum cum secundo in 20 diebus et cum tertio in $8\frac{4}{7}$, secundum quoque cum tertio in 15 corporaliter copulari per modum quem anterius ex-
5   plicavi. Quod si 8 dies et $\frac{4}{7}$ septies assumpseris exibunt 60 dies quod est tempus commune in quo hec tria mobilia pariter unientur. Nam 60 multiplex est ad 20 et 15 que sunt tempora coniunctionis aliorum. Nec umquam hec tria citius coniungentur, deinde in eius multiplicibus quotiens volueris iterare. Igitur tene pro regulari generali quod inventis primis temporibus coniunctionis duorum
10   et duorum necessarium est invenire unum tempus primum numeratum ab om-

---

11 Et . . . hoc *om. VW*
12 alias: alibi *W*

Capitulum 21
1 Capitulum 21: 21ᵃ *Pa* 21 *V*
2 [De . . . spacio] *om. VW*
4 simul incipiantque *tr. V*
5 Queritur *om. VW*
8 que: quem *Pa*/dies *om VW*
9 que: quem *Pa* / exeunt *W* / dies *om. VW*
10 Similiter: super *W sed mg. hab. W* similiter

10–11 coniungitur tertio *tr. W*
11 hec *W*/scilicet *om. VW*/supra 15 *hab. Pa* diebus/*post* 15 *hab. V* et/30²: 20 *W*/supra 30² *hab. Pa* diebus / *post* 30² *hab. V* et/ supra 30³ *hab. Pa* diebus
12 est² *om. V*
13 dies *om. W* est *V*/in: 15 *V*

Capitulum 22
1 Capitulum 22: 22ᵃ *Pa*
4 cum tertio *om. W sed mg. hab.*
5 et $\frac{4}{7}$ *om. Pa sed hab.* cum $\frac{4}{7}$ *post* assump-

required [to make] corrections. But even if this were not so, as yet the continuity of motion cannot be had exactly from a fixed instrument. A discussion of these matters will be presented at another time.

Chapter 21
[On the Conjunction of Mobiles in a Circular Space]

Let there be three mobiles on a circle of 60 [parts]. In each day, let the first of these [mobiles] be moved through 9 [parts], the second through 7, and the third through 5; and let them begin to be moved simultaneously from the same point of the circle mentioned above. Let it be sought when these three will be in conjunction in the same point.

Answer [in this way]: Always subtract the slow [mobile] from the quick [mobile], as the second from the first, and 2 remains, which you divide into the circle and get 30, the time of conjunction for both of them. Likewise, subtract the third from the first and 4 remains, which you divide into the circle and get 15 days, the time of their conjunction. In a similar manner, the second is conjuncted with the third in 30 days. Now you have here three times, namely, 15, 30, and 30; and since one, namely, 30, is multiple to another, namely, 15, it—i.e., 30 days—is the time in which these three mobiles are conjuncted together when they have traversed 30 parts of the circle after having subtracted the [integral] revolutions that were traversed.

Chapter 22

Furthermore, if in each day the first [mobile] were moved through 11 [parts of the circle], the second through 8, the third through 4, then, in the manner which I explained earlier, say that the first [mobile] is united physically with the second in 20 days and with the third in $8\frac{4}{7}$ [days], and the second with the third [mobile] in 15 [days]. Now if you will take $8\frac{4}{7}$ days seven times, 60 days will result, which is the common time in which these three mobiles will be united together. For 60 is multiple to 20 and 15, which are the times of conjunction of the others. Nor will these three mobiles ever be conjuncted sooner, [so that] thereafter conjunctions can repeat in multiples of 60 as often as you wish. Therefore, take it as a general rule that after the first times of conjunction have been found [for the mobiles taken] two by two, it is necessary to find the time first numbered by all the times that were found first.

---

seris
6 communes *W*
7 citius *om. W*

9 quod *om. V*
10 et duorum *om. Pa* / necessarium: neces-
    sim *W*

nibus temporibus primis inventis, quod invenitur per ea que iam scripsimus libro primo et de hoc in semilibro fecimus iterato sermonem 48° capitulo.

Capitulum 23

Si autem per 20 dies fuerint separata moveaturque primum per 9, secundum per 7, tertium per 5, supposita predicti circuli quantitate, queritur quando adinvicem coniungentur.

5 Responde: Primum erit motum post integras revolutiones per 0, secundum per 20, tertium per 40. Ergo distantiam primi ad secundum, que est 20, divide per superlationem primi ad secundum, que est 2, exit 10 dies; item distantiam secundi ad tertium, que est 20, divide per superlationem secundi ad tertium, que est 2, exit 10 dies; similiter distantiam primi ad tertium que est 40, divide
10 per superlationem primi ad tertium, que est 4 quoniam primum movetur per 9 tertium per 5, et exit 10 dies. Dic ergo quod post tempus separationis eorum, qui est 20, coniungentur post 10 dies integros vel completos.

Aut sic et in idem redit. Sint separata per 12 dies primum erit motum per 48, secundum erit per 24, tertium per 0, abiectis revolutionibus completis. Igitur
15 deme tardum a veloci sicut secundum a primo remanent 24 que divide per superlationem primi ad secundum exit 12 dies; item tertium a primo subtrahe remanent 48 quos divide per superlationem primi ad tertium exit 12; etiam tertium deme a secundo remanent 24 quos divide per superlationem secundi ad tertium exit 12. Reseca ergo 12 de 30 quod est tempus coniunctionis eorum
20 trium et restant 18 dies. Quibus completis hec tria mobilia iterum coniungentur. Intellige me primum vocasse velociorem, secundum vero intelligo tardiorem, et adhuc tertium tardius appellatur, et ita continue si plura sint mobilia super circulum assignatum.

---

11 primis inventis: prae inventis *W* scilicet inventis *Pa*
12 hoc: his *W*/sermone *W*/48° capitulo *om.* *VW*

Capitulum 23
1 Capitulum 23: 23ᵃ *Pa* 23 *V*
2 dies *om.* *VW*
3 predicti circuli quantitate: circuli quantitate predicta *VW*
3–4 adinvicem *om.* *W*
5 erit: est *W*/post: per *W*
6 distantia *W*
7 exeunt *W*
9 similiter: super *W* *sed mg. hab.* *W*

similiter
11 exeunt *W*/post: per *Pa*
13 erit: est *W*
14 erit *om.* *W*/adiectis *W*
15 secundum a primo: a primo secundum *W*/remanet *PaW*/que: quem *Pa*
16 exeunt *W*
17 remanet *PaV*/ante etiam *hab.* *W* ergo
18 remanet *Pa*
19 de: a *W*
20 trium: ?tertium *W* *sed mg. hab.* *W* trium
21 *ante* me *scr. et del.* *V* p/intelligo: intelligo *Pa*
22 ita: prima *W* *sed mg. hab.* *W* ita/sint *om.* *W*/super *om.* *V*

This is found by those methods already described in the first book and repeated in a semi-book [where] we had a discussion in the 48th chapter.

Chapter 23

Furthermore, if they had been separated for 20 days and the first [mobile] were moved through 9 parts, the second through 7, the third through 5—[and] assuming that the quantity of the circle is as previously stated—let it be sought when they will mutually conjunct.

Answer [in this way]: After an integral number of revolutions [in 20 days], the first [mobile] will have moved through 0 [parts], the second through 20, and the third through 40. Therefore, divide the distance separating the first and second [mobiles], which is 20, by the [daily] excess of [parts which] the first [traverses] over the second, which is 2, and 10 days results; likewise, divide the distance between the second and third, which is 20, by the [daily] excess of [parts which] the second [traverses] over the third, which is 2, and 10 days results; similarly, divide the distance between the first and third, which is 40, by the [daily] excess of [parts which] the first [traverses] over the third, which is 4 since the first is moved through 9 [parts daily and] the third through 5, and 10 days results. Therefore, say that after the given time of their separation, which is 20 days, they will conjunct after 10 whole or complete days.

Or you can proceed this way and obtain the same results. [If] they have been separated [from their initial point of conjunction] for 12 days and after the integral revolutions have been subtracted, the first [mobile] will have been moved through 48, the second through 24, and the third through 0 [parts]. Therefore, subtract a slow mobile from a quick one, as the second from the first, and 24 remains, which you divide by the [daily] excess of [parts which] the first [traverses] over the second and obtain 12 days; likewise, subtract the third from the first and 48 remains, which you divide by the [daily] excess of [parts which] the first [traverses] over the third and obtain 12 days; also subtract the third from the second and 24 remains, which you divide by the [daily] excess of [parts which] the second [traverses] over the third and obtain 12. Therefore subtract 12 from 30, which is the time of conjunction for all three mobiles, and 18 days remain. The three mobiles will conjunct again after these [18 days] have passed. Understand that I have called the first quicker, the second I understand as slower, and the third as even slower, and so continuously if more mobiles should be assigned on the circle.

Capitulum 24
[De eodem aliter]

Sit idem circulus qui prius et moveantur 3 mobilia sicut ante capitulo vicen-
simo. Verumtamen de facto primum sit motum per 8, secundum per 13, tertium
per 17. Queritur quando hec tria coniungentur.

Responde: Coniunctionem binorum sic invenies. Distantiam primi in secundum, que est 5, divide per superlationem primi ad secundum, que est 2, exeunt
dies $2\frac{1}{2}$ qui sunt tempus in quo hec duo mobilia data ypothesi coniungentur.

Sed in qua parte circuli ecce? Motum diurnum primi, qui est 9, duc in coniunctionis inventum tempus et adde 8, per quos motum erat primum, exibunt $30\frac{1}{2}$,
locus primi mobilis. Rursum motum diurnum secundi mobilis, qui est 7, duc in
idem tempus coniunctionis et iunge 13, per quos motum erat secundum, exibunt
partes $30\frac{1}{2}$, locus secundi mobilis.

Si autem primum de facto sit motum per 13, secundum vero per 8, procul dubio
iam coniunctio transacta est per 2 dies et $\frac{1}{2}$ quos duc in 9 et productum deme de
13 circulo mutuato exit $50\frac{1}{2}$, locus primi mobilis. Similiter duc 2 dies $\frac{1}{2}$ in 7
productumque auffere de 8 addito circulo exit similiter $50\frac{1}{2}$, locus secundi
mobilis. Ergo in eodem puncto circuli primum et secundum mobile sociantur.
Eadem norma coniunctionem primi et tertii itemque secundi et tertii potes
concludere et erunt sic: tempus coniunctionis primi et secundi dies $2\frac{1}{2}$, locus
partes $30\frac{1}{2}$; tempus coniunctionis primi et tertii dies $2\frac{1}{4}$, locus partes $28\frac{1}{4}$; tempus
coniunctionis secundi et tertii dies 2, locus partes 27. Considera ergo te habere
binas coniunctiones iuxta ypotesim prius datam.

Sed queritur quando hec tria pariter adiungentur ubi quod tempus commune
coniunctionis illorum trium mobilium est primum et numeratum 30 dies in 30
partibus circuli. Propter quod unum volo te habere quod primum tempus consimilis dispositionis sicut data fuit ab initio taliter invenitur: sive hec tria mobilia
fuerint in coniunctione sive non, deme motum diurnum tardioris a motu diurno
velocioris sicut prius dicebatur et per residuum circulum divide exibit tempus
coniunctionis amborum. Et si sic egeris invenies tempus coniunctionis singularis
primi mobilis et secundi dies 30 et tempus coniunctionis secundi et tertii totidem

---

Capitulum 24
1 Capitulum 24: 24^a *Pa* 24 *V*
2 [De eodem aliter] *om. VW*
3–4 capitulo . . . facto: c 20 unum punctum
    de secundo *V*
4 *post* per^2 *hab. V* dies *et mg. hab. W* ?dies
8 ypothes *V*
9 diurnum *om. W*
10 quos: quod *VW*
11 diurnum *om. W*
12 *ante* motum *scr. et del. V* divide
15 2 dies et $\frac{1}{2}$: $2\frac{1}{2}$ dies *V* 2 dies $\frac{1}{2}$ *W*

16 duc *om. W sed mg. hab. W* duc / 2 dies
    $\frac{1}{2}$: $2\frac{1}{2}$ dies *V*
17 *post* productumque *hab. W* quia / auf-
    ferre: aufer *VW*/exeunt *W*
18 mobile: mobilem *Pa*
19 itemque: item *Pa*
24–26 ubi . . . circuli *om. VW*
25 30^1 *corr. ex* 330
26 volo te *om. VW*
30 sic: ergo *W sed mg. hab. W* sic
31 et^1: ?*W*

Chapter 24
[On the Same Thing in Another Way]

Let there be the same circle as before, and let three mobiles be moved as before in the twentieth chapter. However, assume that the first mobile has [already] been moved through 8 [parts of the circle], the second through 13, and the third through 17. Let it be sought when these three will be conjuncted.

Answer [in this way]: You can find the conjunction of two [mobiles] as follows. Divide the distance [separating] the first from the second, which is 5, by the [daily] excess of [parts which] the first [traverses] over the second, which is 2, and $2\frac{1}{2}$ days will result, which, from the data of the hypothesis, is the time in which these two mobiles will conjunct.

But in what part of the circle [will this occur]? Multiply the daily motion of the first [mobile], which is 9, by the time of the first conjunction [i.e., $2\frac{1}{2}$ days] and add 8, [the distance] through which the first mobile had been moved, and there will result $30\frac{1}{2}$, the place of the first mobile [when it conjuncts]. Again, multiply the daily motion of the second mobile, which is 7, by the same time of conjunction and add 13, [the distance] through which the second mobile had been moved, and there will result $30\frac{1}{2}$ parts, the place of the second mobile.

Moreover, if the first mobile were already moved through 13 [parts] and the second through 8, [then] undoubtedly the conjunction has already occurred $2\frac{1}{2}$ days ago, which [time] you multiply by 9 and subtract the product from the circle altered by 13 [i.e., $60+13 = 73$] and this gives $50\frac{1}{2}$, the place of the first mobile. Similarly, multiply $2\frac{1}{2}$ days by 7 and subtract the product from the circle, to which 8 has been added, and similarly this gives $50\frac{1}{2}$, the place of the second mobile. The first and second mobiles are therefore joined on the same point of the circle. In the same manner, you can determine a conjunction of the first and third and also of the second and third mobiles, and these will be as follows: the time of conjunction of the first and second mobiles will be $2\frac{1}{2}$ days, and they will be at $30\frac{1}{2}$ parts [of the circle]; the time of conjunction of the first and third mobiles will be $2\frac{1}{4}$ days, and they will be at $28\frac{1}{4}$ parts [of the circle]; and the time of conjunction of the second and third will be 2 days, and they will be at 27 parts [of the circle]. Consider, therefore, that you have their conjunctions two by two according to the hypothesis given previously.

But you seek when these three will be joined together wherever the common time of conjunction of these three mobiles occurs first and is numbered by 30 days in 30 parts of the circle. Because I want you to have one [time], that first time of like disposition, as given [or calculated] from the beginning, is found in this way: whether these three mobiles were in conjunction or not, subtract the daily motion of the slower [mobile] from the daily motion of the quicker, as was stated earlier, and divide the circle by the remainder [and] this will give the time of conjunction of both. And if you shall have done this, you will find the time of a particular conjunction of the first and second mobiles to be 30 days, and the time of conjunction of the second

et tempus coniunctionis primi et tertii 15. Et quoniam unum tempus numerat reliquum, scilicet 15 numerat 30, ipsum est tempus reducens dispositiones consimiles qualitercumque potuerunt variari. Et si umquam hec tria fuerint in
35 coniunctione illud tempus reducet ea ad iteratam coniunctionem; et si in illa revolutione temporis reducentis hec tria non fuerint coniuncta, nec in eternum a modo coniungentur. Et loquor michi quod nemo vixit, nec vivit, nec vivet, qui 7 planetas, ymo 3 tantum, possit in eodem puncto circuli coniungere, supposito adhuc quod latitudines non haberent sed sub ecliptica inseparabiliter
40 moverentur addito etiam quod motus eorum precissime, sint conclusi. Quod tamen numquam erit. Tamen indubitanter affero quod si umquam coniuncti fuerint 7 planete in eodem puncto signiferi, aut quavis alia dispositione ad situm consimilem, necessario reducentur rerum naturis currentibus sicut solent dum tamen sicut ante dixi circuli vel motus eorum incommensurabiles non ponantur.
45 Sed utrum hoc an illud umquam fuerit non sentio me securum. Sufficit tamen astrologo quod coniunctiones planetarum et siderum sic predicat quod sensus ostendere contrarium non valeret. Est enim bene doctrinati requiescere dum intellectus sensui satisfacit quod qui querit ultra non est de gremio sensatorum. Nolo tamen quin scientiam sensus superet intellectus in plerisque enim potest vis
50 anime rationalis ad que nequit virtus attingere sensitiva. Numquam enim sensus me docuit quod continuum in semper divisibilia dividatur aut quod spera posita super planum tangat ipsum in puncto. Nec tamen in hoc sciendo organo sensus privor.

Capitulum 25

Ex predictis infer quod duobus mobilibus in diversis partibus circuli situatis ipsa quandoque necessarium est coniungi et hanc coniunctionem redire infinities nichil obest posito quod inequaliter moveantur; si tamen equaliter sine con-
5 iunctione erunt perpetuo equedistantes.

Verumtamen tribus mobilibus aut pluribus in diversis partibus circuli collocatis stat possibile ea insimul in eodem puncto circuli numquam iungi. Quod

32 *ante* 15 *hab. VW* dies
33 scilicet: sicut *W* / dispositiones: disponens *W*
35 illud: istud *W*/coniunctionem: coniunctiones *W*
36 tempus *V*
37 michi quod: in *VW*/nec vivit *om. W*
38 ymo: primo *W sed mg. hab. W* immo/ *post* ymo 3 *mg. hab. Pa* attende/3: moties *V*/posset *W*
40 prescissime *V*
41 numquam: nec quam *W sed mg. hab. W* nunquam

42 *ante* aut *scr. et del. Pa* ?ax
43 consimilem: consuetim *W sed mg. hab. W* "alibi consimilem"/*supra* necessario *hab. Pa* non/rerum naturis currentibus: rectis naturis concurrentibus *W sed mg. hab. W* "alibi rerum naturis currentibus"
44 incommensurabiles: commensurabiles *Pa*
45 fuit *V*/securum: secuturum *W*
46 *supra* predicat *hab. Pa* ?previderet
47 valent *V*
49 superet: super *Pa* superat *W*/enim *om. VW*
50 nequid *Pa*/Numquam: non quid *W sed*

and third to be the same; and the time of conjunction of the first and third to be 15 [days]. Now since one time numbers [or measures] the other, namely, 15 numbers 30, this is the time which would bring back the like dispositions, however much they might be varied. And if these three mobiles were ever in conjunction, this time would bring them back to repeated conjunction; but if during the same time of revolution, these three had not been conjuncted, they will never from this time on be conjunct ed through all eternity. And I say that no one has lived, lives, or will live, who has demonstrated that 7 planets—indeed even 3 planets—could conjunct in the same point of the circle, even assuming that they have no [differences in] latitudes but are moved inseparably on the ecliptic, and even assuming that their motions are very exact. Indeed, this will never happen. Nevertheless, it is without any doubt that I assert that if 7 planets had been conjuncted in the same point of the sky, or were in any other disposition or similar situation, they would be returned necessarily as would be expected in the quickly moving natures of [these] things—provided that, as I said before, their circles or motions are not assumed to be incommensurable. But I am unconcerned whether it is ever this [way] or that [i.e., whether the motions are commensurable or incommensurable]. For an astronomer, however, it suffices that he can predict conjunctions of planets and stars in such a way that the senses are incapable of showing it otherwise. Indeed he has been successfully taught who rests when his understanding satisfies [or is compatible with] his senses, because one who seeks beyond does so not on the basis of things that have been perceived. Nevertheless, I do not wish to maintain that the knowledge of sense exceeds the intellect, for in most things the power of the rational souls attains to things that the sensitive power does not know how to attain. For sense can never teach me that a continuum could always be divided into divisible things, or that a sphere assumed to be on a plane could touch it in [only] one point. By [relying only on] the organ of sense, however, I am deprived of this knowledge.

Chapter 25

From what has been said before, I infer that with two mobiles located in different parts of the circle, it is necessary that they be conjuncted and that this conjunction repeat an infinite number of times—assuming that nothing obstructs them and that they are moved unequally; if, however, [they are moved] equally without a conjunction, they will be perpetually equidistant.

Nevertheless, when three or more mobiles have been arranged in different parts of the circle, it is possible that they will never be conjuncted at the same time. This is

---

mg. hab. W numquam
52 in² *om. W* / ?sciente *W*

Capitulum 25
  1 Capitulum 25: 25ᵃ *Pa* 25 *V*

2 *post* partibus *scr. et del. V* s *et mg. hab.*
  W quatuor
3 necessarium: necessim *W*
5 erunt: nec *W sed mg. hab. W* erunt
7 insimul *om. VW*

scitur prompte sic: Iunge primum secundo et secundum tertio ac primum tertio
per iam dicta et si tempora harum trium coniunctionum sint equalia necessario
in eodem puncto circuli ea posse iungi. Non dubites et eam coniunctionem trium
reverti quamdiulibet necesse est per additionem communis temporis coniunctionis ab eorum coniunctionibus singulis numerati sicut docuimus ante nuper.
Quod si tempora trium coniunctionum inequalia reperta sint sicut est in tribus
mobilibus antedictis scias ea numquam in preteritum nec in futuro tempore
posse iungi, et si sine temporis termino moverentur. Cuius causa est quoniam
postquam eorum coniunctiones singule dico autem bine una alteri non equatur
et super eas additur tempus commune coniunctionis eorum quod est equale. Si
iungi deberent necessarium est tempus coniunctionis inequale redire, nam si
super inequalia equalia coniungantur semper inequalia producuntur.

Nunc autem sic est quod tempus commune coniunctionis trium mobilium predictorum si umquam iuncta fuerunt in preteritum aut iungi debeant in futurum
est 30 dies ut visum est. Ergo si tempus hoc super eorum coniunctiones singu-
lares addatur, que sunt inequales quoniam coniunctio primi et secundi est dies
$2\frac{1}{2}$; coniunctio quoque primi et tertii dies $2\frac{1}{4}$; coniunctio vero secundi et tertii
dies 2, noli mirari si a modo non iungantur. Sed dic etiam quod in preterito si
numquam moveri cessaverunt stante tempore sine initio numquam in eodem
puncto circuli coniuncta fuerunt, quapropter ypothesis data prius volens hec
tria mobilia quandoque fore iuncta impossibile supponebat. Si autem satis
prope aut quod insensibiliter appropinquent sic quod sensus distantiam discer-
nere nesciat nil ad respondum.

Capitulum 26
[De iterata coniunctione]

Sit adhuc circulus 60 in cuius initio signato 2 mobilia situentur moveanturque
ad eandem partem unum ad reliquum inequaliter omni die; necessario ex predictis in aliquo puncto circuli coniungentur. Nunc queritur quociens unientur
antequam totius diversitatis revolutio ad idem punctum circuli reducatur ut
habemus currere penitus sicut ante.

9 coniunctionum *om. Pa*
10 trium: tertiam *Pa et supra* tertiam *hab. Pa* ?iterum
11 quamdiulicet *V*/necessarie *Pa*
11–12 communis temporis coniunctionis: coniunctionis communis temporis *V* communis temporis *W*
13 trium coniunctionum *tr. W*/*post* inequalia *hab. Pa* et in ?coniunctum ?cia
14 preterito *W*/nec: vel *V*
15 iungi: coniungi *VW*/moveretur *W*

16 autem: aut *V*
17 commune *om. Pa*
18 necessim *W*
19 *ante* super *scr. et del. V* ?propter/semper: ?sivit *V*/*ante* producuntur *hab. V* que
21 umquam: numquam *VW* / iuncta: coniuncta *W*
22 tempus hoc *tr. VW*
24 $2\frac{1}{2}$: $1\frac{1}{4}$ *Pa*
25 mirari: minui *W sed mg. hab. W* ?mirari/a modo: ?ammodo *V* simili modo *W*

readily known as follows: By what has already been said, conjoin the first [mobile] with the second, the second with the third, and the first with the third, and if the times of these three conjunctions were equal, they are necessarily able to be conjuncted in the same point of the circle. And you should not doubt that the conjunction of the three is repeated as long as necessary by the addition of the common time of conjunction numbered [or measured] by the particular conjunctions of these [mobiles taken two by two], just as we taught before. But if the times of these conjunctions were found to be unequal, as in the aforementioned three mobiles, you should know that they could never be capable of conjunction in past or future time—[even] if they were moved without any time limit. The reason for this [lack of conjunction] is that after their particular conjunctions, which I say are not equal taken two at a time one with another, the common time of conjunction, which is equal, is added on to these [particular conjunctions]. If they were destined to be conjuncted, it is necessary that the time of conjunction repeat unequally, for if to unequal times equal times be added, unequal times are always produced.

Now, furthermore, if they were always conjuncted in the past, or must always be conjuncted in the future, it is necessary that the common time of conjunction of the three aforementioned mobiles be 30 days, as was seen. Therefore, if this time were added to [the times of] their particular conjunctions, which are unequal, do not be surprised if henceforth they were not conjuncted, since the conjunction of the first and second is $2\frac{1}{2}$ days, the conjunction of the first and third is $2\frac{1}{4}$ days, and the second and third is 2 days. But also say that if they never ceased to be moved in the past, and time had no beginning, they were never in conjunction in the same point of the circle, for which reason the hypothesis given before that these three mobiles would be conjuncted sometime was assuming the impossible. Moreover, if they came sufficiently close or imperceptibly near, so that the senses could not know how to discern the distance, there would be nothing to say.

Chapter 26
[On Repeated Conjunction]

Let there be a circle of 60 [parts] in the beginning of which 2 mobiles are situated, and every day let them be moved, one and the other, unequally in the same direction. From what has been said previously, they will be conjuncted necessarily in some point of the circle. Now let it be found how often they are joined before a revolution of complete diversity is returned to the same point of the circle so that we have them course exactly as before.

---

sed mg. hab. W a modo
26 movere *W*
28 fore *om. VW*
29 quod[1]: que *W*/insensebiliter *V*
30 respondum *?V ?*rhombum *W*

Capitulum 26
1 Capitulum 26: 26ª *Pa* 26 *V*
4 unum: unde *W sed mg. hab. W* unum
6 ut: unde *VW*

Responde: Sit exemplum; moveatur primum mobile per 7, secundum per 2
singulis diebus ex prehibitis in 24ª parte circuli coniungentur. Sunt autem 24$\frac{2}{5}$
10   circuli supradicti, dic ergo quod in omnibus quintis circuli, scilicet quinquies,
erunt in coniunctione et tunc redibunt coniunctiones pristine ordine iam transacto. Si autem in 27ª parte circuli iuncti fuerint cum sint $\frac{9}{20}$ circuli prenotati in
omnibus vicesimis, scilicet vigesies, coniungentur. Eodem modo de tribus aut
quot vis mobilibus si ea in eadem parte circuli contingerit adiuvari.
15   Ex hoc scire potes quociens et ubi planete superiores et ceteri antequam revolvatur diversitas coniunguntur. Sic ergo reductio fractionum circuli ad contra se
primas est huius regule fundamentum.

Capitulum 28
[De inquisitione coniunctionis]

Cum 2 mobilia noti motus girantur in circulo note quantitatis continue pro-
cedendo quando corporaliter coniunguntur. Modus 24° capitulo signatus est.
5   Sed quando primum mobile, quod est levius, sit 8, sed secundum vero, scilicet
tardius, sit 18 quod ad primum mobile reversatur, quando fiet eorum coniunctio?
Responsio: Iunge motus eorum diurnos et sunt 16 per quos divide differentiam inter eos, que est 10, exit $\frac{5}{8}$ diei in quibus pariter coniungentur. Diurnum
10   itaque motum levioris, qui est 9, duc in $\frac{5}{8}$ exit 5 et $\frac{5}{8}$ addenda super locum eius,
qui est 8, exit 13$\frac{5}{8}$, locus mobilis levioris. Similiter duc diurnum tardioris, qui
est 7, in $\frac{5}{8}$ exit 4$\frac{3}{8}$ demenda de loco eius, qui est 18, exit 13$\frac{5}{8}$, sicut nuper, locus
mobilis tardioris. Ergo sunt coniuncti loco et tempore supradictis.
Si autem mobile levius sit 18 et retrogradum tardius vero 8 et processivus
15   eodem tempore iuncta erunt sed non ubi supra, sed deme 5$\frac{5}{8}$ a 18 remanet 12$\frac{3}{8}$,
locus mobilis levioris; et adde 4 et $\frac{3}{8}$ super 8 exit 12$\frac{3}{8}$, locus tardioris.
Si tamen ambo retrograde moveantur idem capio ac si suos motus agerent
processive. Si levius sit 8 tardius sit 18, coniunctio preteriit; si econverso futura

---

10 quintis: quartis *W sed mg. hab. W* quintus
12 autem: aut *Pa*/iuncta *V* coniuncta *W*
14 quot: quod *V*/ante si *hab. W* sed / contingit *V* contingat *W*
15–16 antequam revolvatur: cum quam
nolitatur *W sed mg. hab. W* antequam
revolvatur
16 reductio fractionum *tr. VW*

Capitulum 28
 1 Capitulum 28: 28ª *Pa*
 2 [De inquisitione coniunctionis] *om. VW*
 4 coniungantur *Pa*

5 sed² *om. VW*/scilicet *om. VW*
 6 primum: secundum *Pa*/reversatur: quando revertatur *VW*
 8 respondio *Pa*/supra 16 *hab. Pa* 26
10 motum *om. VW*/exit: exiet *VW*
11 est 8: 28 *W*/exit: exiet *VW*
12 exit¹: exeunt *W*/exit²: exiet *VW*
14 *ante* levius *scr. et del. V* ?levis/processi-
vum *VW*
15 remanent *V*
16 locus¹ . . . exit 12$\frac{3}{8}$ *om. VW*
17 tamen: autem *VW*
18 processivo *Pa*/sit² *om. VW*

Answer [in this way]: Let there be an example; every day, the first mobile is moved through 7 [parts], the second through 2, [and] from what has been set forth before, they will be conjuncted in the 24th part of the circle. Moreover, [if] they are at $24\frac{2}{5}$ [parts] of the aforementioned circle, say, therefore, that they will be in conjunction in all fifths of the circle, i.e. [they will conjunct] five times, and the conjunctions will return to the original sequence through which they have already passed. If, however, they had been conjuncted in the 27th part of the circle plus $\frac{9}{20}$ [of a part] of the circle mentioned before, they will be conjuncted in all the twentieths, that is, 20 times. The same thing will happen with three mobiles, or any number you please, if it happens that these are located in the same part of the circle.

From this you can know how often, and where, the superior planets and others will be conjuncted before the diversity [of conjunctions] is completed through one revolution. Therefore, the reduction of fractions of the circle to prime [numbers] is the basis of this rule.

Chapter 28
[On the Finding of a Conjunction]

When two mobiles of known motion [or speed] are turned around continually on a circle of known quantity [or dimension], when will they be physically conjuncted? The method [for determining this] was shown in the 24th chapter. But when the first mobile, which is quicker [levius], is [in the] 8[th part of the circle] and the second, namely, the slower [tardius], is [in the] 18[th part] and is moved in an opposite direction with respect to the first [mobile], when will a conjunction of them occur?

The answer: Add their daily motions and get 16 [i.e., 9 parts + 7 parts], by which you divide the difference between them, which is 10, and get $\frac{5}{8}$ [i.e., $\frac{10}{16} = \frac{5}{8}$] of a day, [the time] in which they will be conjuncted together. Then multiply the daily motion of the quicker, which is 9 [parts], by $\frac{5}{8}$ and get $5\frac{5}{8}$, which is to be added to the place of the quicker, which is 8, and this gives $13\frac{5}{8}$, the place of the quicker mobile. Similarly, multiply the daily motion of the slower [mobile], which is 7 [parts], by $\frac{5}{8}$ and get $4\frac{3}{8}$, which must be substracted from its place, which is 18, and this gives, as before, $13\frac{5}{8}$, the place of the slower mobile. Therefore, they have been conjuncted in the place and time stated above.

If, however, the quicker mobile is [in the] 18[th part] and retrograde [in its motion], while the slower is [in the] 8[th part] and progressive, they will be conjuncted in the same time but not where they were [as shown] above. But subtract $5\frac{5}{8}$ from 18 and there remains $12\frac{3}{8}$, which is the place of the quicker mobile; and then add $4\frac{3}{8}$ to 8 and get $12\frac{3}{8}$, the place of the slower mobile.

If, however, both were moved with a retrograde motion, I do the same as if their motions were progressive. If the quicker is [in the] 8[th part and] the slower [in the] 18th, the conjunction has passed; if conversely [i.e., if the quicker is in the 18th and the slower in the 8th], the conjunction is yet to come. If the first case obtains, the

est. Si primo per 5 dies coniunctio fuit ante, si secundo per totidem fuit retro,
20 deinde duc 5 in diurnum levioris exit 45 quibus additis super locum suum exit 53;
aut duc 5 in diurnum tardioris exit 35 quibus iunctis super locum suum exit 53
locus amborum tempore preterite coniunctionis. Aut sic: Deme locum levioris
a circulo remanet 52 a quibus deme 45 remanet 7; quibus demptis ex circulo
remanet 53. Item deme similiter locum tardioris a circulo remanet 42 a quibus
25 deme 35 remanet 7 quibus ex circulo demptis remanet 53 ordine computationis
locus amborum qui prius.

   Item locum levioris deme ex circulo in secundo modo remanet 42 quibus
iunge 45 exit 87 dempto circulo remanet 27 quibus demptis de circulo remanet
33. Similiter motum tardioris deme ex circulo remanet 52 quibus additis 35
30 exit 87 quibus ex circulo demptis remanet 27 quibus ex circulo demptis remanet
33 sicut ante locus amborum tempore future coniunctionis.

20 *ante* super *scr. et del. Pa* 8
22 sic *om. W*
23 remanet[1]: remanent *V*
24 remanet[1]: remanent *V*
25 remanet[2]: remanent *V*
27 remanent *V*/42: 72 *Pa*

28 exeunt *W*/remanet[1]: remanent *V*/27 . . .
   remanet[2] *om. Pa*
29 remanet *VW*
30 remanet[1]: remanent *V* / quibus[2]: que *Pa*/
   demptis[2] *om. Pa*/remanet[2]: remanent *V*

conjunction was before by 5 days [and] if the second case obtains, it was after by just as many days. Then multiply 5 by the daily motion of the quicker body and get 45, which when added to its place gives 53; or multiply 5 by the daily motion of the slower mobile and get 35, which when added to its place gives 53, the place of conjunction of both in a past time. Or you can proceed as follows: Subtract the place of the quicker from the circle and 52 remains, from which you subtract 45, leaving 7; when 7 has been subtracted from the circle, 53 remains. Similarly, again subtract the place of the slower from the circle and 42 remains from which you subtract 35 and 7 remains, which when subtracted from the circle leaves 53. In this sequence of calculations, the place of both is as before.

Again, subtract the place of the quicker from the circle in the second way and 42 remains, to which you add 45 and get 87; and after the circle [i.e., 60 parts] has been subtracted, 27 remains, which when subtracted from the circle leaves 33. Similarly, subtract the motion of the slower [mobile] from the circle [and] 52 remains, to which 35 is added and gives 87, from which the circle is subtracted, leaving 27, which when subtracted from the circle leaves 33, [giving] just as before, the place of conjunction of both mobiles in the future.

Appendix 2

Pierre Duhem on Oresme and
Celestial Incommensurability

As with so many other ideas and concepts concerning the history of medieval science, Pierre Duhem was the first to describe and evaluate medieval discussions on celestial incommensurability. Because his influence has been, and is yet, very great, it is worthwhile to present a brief summary and critique of his judgments and interpretations of Oresme's contributions on this subject.

In *Le Système du monde*, vol. 8, pp. 448–52, in a section titled "Nicole Oresme et les durées incommensurables des circulations célestes," Duhem translated and evaluated a small part of Oresme's *Ad pauca respicientes*,[1] which he called *Traité de la proportionnalité des mouvements célestes*, thus translating *Tractatus de proportionalitate motuum celestium*, the title which appeared in the colophon of the only manuscript he used.[2] Since the colophon cites Oresme as the author,[3] Duhem saw no reason to question this attribution, and, as it happens, was proved right. But when he discovered an anonymous treatise in Bibliothèque Nationale, fonds latin, 7281 (fols. 259r–273r), titled *Tractatus de commensurabilitate motuum celi*,[4] he tentatively suggested Pierre d'Ailly as its author,[5] rather than Oresme, its true author.

Duhem's error can be traced to a statement by Jean Gerson in which Oresme and Pierre d'Ailly are said to have shown that rhetorical probability, not certitude, was the only answer to the question of whether or not the celestial motions are commensurable.[6] Since he was already aware of Oresme's authorship of the *Tractatus de proportionalitate motuum celestium* (*Ad pauca respicientes*), he thought it reasonable to suggest d'Ailly as the author of the anonymous *De commensurabilitate*, thereby neatly assigning one work to each of the two men mentioned by Gerson as having concerned themselves with the problem of celestial commensurability and incom-

1. This is edited and translated in Grant, *Oresme PPAP*, pp. 382–429.

2. This manuscript is Bibliothèque Nationale, fonds latin, 7378A fols., 14v–17v, which is briefly described in Grant, *Oresme PPAP*, pp. 379–80. Although Duhem seems to have known only this manuscript of the treatise, there are at least six complete versions of it, some bearing different titles, some being anonymous.

3. Duhem, *Le Système du monde*, p. 448.

4. Ibid., p. 455, n. 1. In the present edition of the *De commensurabilitate*, this manuscript bears the siglum *B* (see above, p. 162).

5. Ibid., p. 455.

6. For the translation and Latin text of this statement, as well as the full title of the work in which it appears, see above, chap. 3, n. 122. A French translation of this passage was provided by Duhem (ibid., pp. 454–55).

mensurability. Since Gerson saw fit to mention only the names but not the relevant works of Oresme and d'Ailly, Duhem did not realize that Gerson had in mind d'Ailly's *Tractatus contra astronomos*, in which, as already noted, long sections from Part III of Oresme's *De commensurabilitate* were shamelessly plagiarized.[7] Ironically, had Duhem been familiar with d'Ailly's *Tractatus contra astronomos*, any doubts that he may have had about d'Ailly's authorship of the anonymous *De commensurabilitate* would surely have been dispelled upon perceiving the near identity of passages in Part III of the *De commensurabilitate* with those in *Tractatus contra astronomos*. With Gerson's remarks in mind, it seems a virtual certainty that Duhem would have concluded that d'Ailly had composed both treatises, finding it convenient to use nearly identical, but relevant, material from the one in the other.

By assigning the two treatises in the manner just described, Duhem was committed to a rigid position. When he quotes a section from *Le Livre du ciel et du monde* in which Oresme twice mentions by full and complete title, and claims as his own, the *De commensurabilitate vel incommensurabilitate motuum celi*,[8] Duhem suppressed completely whatever curiosity may have been aroused by the near identity of this title with that of the anonymous *Tractatus de commensurabilitate motuum celi* in BN 7281. Indeed, he implies that Oresme was citing the *Tractatus de proportionalitate motuum celestium* (*Ad pauca respicientes*). Thus Duhem did not deem it necessary or desirable to explain this curious coincidence of titles, preferring instead to interpret Gerson's vague statement as indicating one work on celestial commensurability by Oresme and one by d'Ailly. Oresme's work could only be the *Tractatus de proportionalitate motuum celestium*, which was assigned expressly and correctly to Oresme in the colophon of the single manuscript known to Duhem; d'Ailly's most plausible title was the anonymous *Tractatus de commensurabilitate motuum*, the only other treatise of this kind of which Duhem was aware.

In his critique of these two treatises, Duhem rightly believed that the *Ad pauca respicientes* (this title is used hereafter, instead of *Tractatus de proportionalitate motuum celestium*, as Duhem knew it) was inferior to the *De commensurabilitate*,[9] but his reasons for so believing reveal a lack of understanding of the essential objectives of both treatises.

Duhem was highly critical of Oresme's assumption in the *Ad pauca respicientes* that the times of celestial motions are probably incommensurable and equally critical of the consequence which Oresme derived from this assumption, namely, that the foundations of astrology were thereby destroyed, since exact predictions of celestial positions and configurations would be impossible. At this point, I shall quote my previous summary of Duhem's criticism, a criticism that was based on the final two propositions of the *Ad pauca respicientes*:

---

7. See above, p. 131 and especially n. 121.

8. Duhem, *Le Système du monde*, vol. 8, pp. 453–54. The quoted passages are from Book 1, Chapter 29, 44b–44d and 45c of *Le Livre du ciel et du monde* (see Menut,

Oresme *du ciel*, pp. 194–96 for the first quotation, and p. 200 for the second).

9. In Grant, *Oresme PPAP*, pp. 74–80, I have argued that it was an earlier and inferior version of the *De commensurabilitate*.

In order to argue against the strongest possible position of the astrologers, Oresme, in the final two propositions, asserts that even if the universe were determined, and the celestial motions uniform and eternal, astrological prediction would be futile because of the improbability that any future celestial disposition would be similar to any past celestial configuration (AP2.228–35).

Duhem insists that any astrologer would have had a ready reply. Oresme, says Duhem, assumed that the times of any two celestial revolutions are probably incommensurable. But the determination of times, Duhem tells us, depends on observations that Oresme realized could only be approximate (Duhem may have had in mind AP2.263–64). Because of this, Duhem implies that Oresme maintained that any two numerical evaluations of some observable quantity, or quantities, should not be considered as separate and distinct when their difference falls below a certain value. But how was it possible, Duhem queries, for Oresme to determine whether the times of any two celestial revolutions were commensurable or incommensurable? In reply to his own query, Duhem notes first that if we are given an irrational number, it is always possible to find an infinite number of rational numbers that approach and differ from it by as little as you please. Oresme, Duhem seems to argue, would capitalize on this and choose to represent the time of any celestial motion by an irrational, rather than rational, number, on the grounds that the difference between some initially assigned rational value and the successively approaching irrational values could be made sufficiently small as to render indistinct the difference between the rational and irrational values. But the astrologers, according to Duhem, could be just as arbitrary, and, by the same reasoning, choose to represent that very same time by a rational number. Thus, Duhem reduces the issue to one of preference and prejudice. The difference between the irrational and rational values could be made so minimal that, with equal reason, one could choose to label the time rational or irrational.[10]

But even if the celestial motions were incommensurable, argues Duhem, and no configuration could repeat identically, nevertheless a configuration would eventually appear that would differ from a previous one by as little as you please; for an astrologer this would be as accommodating as an exact repetition, since the effects of two celestial configurations so nearly identical would be indistinguishable.[11] At best, then, Oresme's belief that the celestial motions are probably incommensurable would be ineffectual.

By contrast with Oresme's attitude in the *Ad pauca respicientes*, Duhem is filled with admiration for the author of the *De commensurabilitate*, who was sufficiently perspicacious to realize that no decision could be made about the commensurability

10. Grant, *Oresme PPAP*, pp. 438–40. This quotation summarizes Duhem's argument in *Le Système du monde*, vol. 8, p. 452.

11. "Que les durées des révolutions célestes soient commensurables ou incommensurables, qu'importe, d'ailleurs, à l'astrologue? Si elles sont incommensurables entre elles, jamais, c'est entendu, les astres ne reprendront exactement la configuration qu'ils ont prise une première fois; mais au bout d'un temps suffisant, ils dessineront une constellation qui différera aussi peu qu'on voudra de la constellation autrefois formée; sans être, dans la seconde circonstance, rigoureusement identiques à ce qu'ils étaient dans la première, les effets que ces astres produisent ici-bas se ressembleront d'aussi près qu'on le désirera, de si près qu'aucun observateur ne les pourra distinguer; n'est-ce pas, pour l'astrologue, tout comme s'ils se reproduisaient exactement?"—*Le Système du monde*, vol. 8, p. 452.

or incommensurability of any two physical magnitudes, even if this was possible, as indeed it is, in mathematics. Given any two physical magnitudes, one could, indifferently and with equal reason, assert their commensurability or incommensurability.[12] Moreover, the author of the *De commensurabilitate* surpasses Oresme in critical acumen. Whereas Oresme believed that if the celestial motions were actually incommensurable this would ipso facto destroy the foundations of astrology, the author of the *De commensurabilitate* recognized that this would have no effect on an observational science such as astrology. Indeed this author might well have believed in judicial astrology, and if it was Pierre d'Ailly, he did so believe.[13]

Duhem's critique is simply untenable, for it represents a serious misinterpretation of both treatises. Not only is Oresme the author of both, but in both his objectives are identical. In the *Ad pauca respicientes*, Oresme does not base his arguments on observations or approximations. His assumption that the celestial motions are probably incommensurable depends in no way upon observations or approximations, as Duhem would have it, "but the basis for the assumption is a mathematical demonstra-

12. "Ce que nous venons de lire marque avec une grande finesse le disparate que les bornes de notre sensibilité mettent entre une proposition de Mathématiques et une proposition formulée par une science expérimentale; une proposition de Mathématiques énonce que deux grandeurs sont égales entre elles; une proposition tirée de l'observation énonce que deux grandeurs sont égales *aux erreurs* d'expérience près.

"De là cette conséquence: Pour un mathématicien, tout rapport est commensurable ou incommensurable; pour le physicien, du rapport de deux grandeurs observées, on peut indifférement affirmer qu'il est commensurable ou incommensurable."—Ibid., pp. 460–61.

Immediately following this, Duhem is again full of praise for the author of the *De commensurabilitate*. After citing the law of definite proportions in chemistry and the law of rational indices in crystallography as instances of laws involving commensurable relationships, Duhem comments: "On rencontre, de nos jours, dans les laboratoires et les Sociétés savantes, des gens qui prétendent et soutiennent que la loi des proportions définies, que la loi des indices rationnels sont des vérités *directement* établies par l'expérience. Pierre d'Ailly où l'auteur, quel qu'il soit, du traité que nous venons d'ana-

lyser, n'eût point commis pareille erreur de logique."—Ibid., p. 461.

13. "La justesse d'esprit de cet auteur est donc très grande; elle enchérit sur celle même d'Oresme, et l'on serait presque tenté de dire que c'est tant pis; Oresme avait manqué de sens critique lorsqu'il avait pensé qu'en privant de commune mesure les durées des révolutions célestes, on renversait les fondements de l'Astrologie; mais cette heureuse erreur nous avait valu la vigoureuse condamnation de l'Astrologie judiciaire que, seul en son temps, le Maître normand a l'audace de formuler; le maître dont le livre vient de retenir notre attention a usé d'un sens critique plus pénétrant; il a fort bien reconnu que l'incommensurabilité des mouvements célestes serait de nulle conséquence pour les sciences d'observation comme pour celles qui s'affublent de ce titre; mais, sous sa plume plus exactement informée, qu'est devenue la condamnation de l'Astrologie? Elle s'est evanouie; tout au plus en retrouve-t-on un souvenir vague et presque effacé dans l'interdiction timide, faite à l'Astrologie, de prévoir certains futurs contingents. Même après sa très fine discussion, l'auteur pouvait encore croire à l'Art judiciaire; et si cet auteur est Pierre d'Ailly, nous pouvons assurer qi'il y croyait."—Ibid., pp. 461–62.

tion in Ch. III, Prop. X [of the *De proportionibus proportionum*], where it is shown that there are more irrational than rational ratios of ratios. Then in Ch. IV [of the *De proportionibus*] it is shown that ratios of quantities such as time, distance, and velocity vary as ratios of ratios. The ultimate purpose of all this is to show convincingly that even if we cannot determine whether the celestial motions are commensurable or incommensurable, *it is mathematically probable* that they are incommensurable. It is abundantly clear that whatever else may be said about Oresme's arguments, they are independent of observation and unconnected with approximations."[14] Duhem's talk of successive approximations and the arbitrariness of any choice between celestial commensurability or incommensurability is wholly irrelevant and unrelated to the substance and spirit of Oresme's *Ad pauca respicientes*.

Nor indeed would it apply to the *De commensurabilitate*, whose purpose differs in no way from the *Ad pauca respicientes*. Only the literary form and the mode of presentation differ in the former treatise, but these characteristics seriously misled Duhem, who appears to have had only a superficial acquaintance with the *De commensurabilitate*. In this treatise, as in the *Ad pauca respicientes*, only exact punctual and precise mathematical relationships are involved (see I.45–49)—not observations or approximations. Oresme's refusal in the *De commensurabilitate* to choose between commensurability or incommensurability for the celestial motions—and it is this which evoked Duhem's admiration—is explained simply by the fact that Oresme was fully aware that a categorical decision was incapable of conclusive demonstration, an awareness that was almost certainly a part of his outlook in the earlier *Ad pauca respicientes*, for otherwise he would not have asserted that the celestial motions are "probably" incommensurable, but would have claimed this categorically. But I have argued above (see pp. 69–77) that omission of a categorical decision must not be construed as indicating that Oresme believed either position to be equally valid and that any inclination to one position or the other was unjustifiable. On the contrary, the manner in which Oresme marshaled his arguments and the order in which he presented them leave little doubt to a discerning reader that he believed the celestial motions were probably incommensurable—a belief that he never abandoned, not even in his final scientific treatise, the *Le Livre du ciel et du monde*, where the *De commensurabilitate* is cited in support of this probability. The sharp contrast in outlook and attitude which Duhem distinguished between the *Ad pauca respicientes* and *De commensurabilitate* is non-existent. Duhem was victimized and misled by his conviction that a different author was responsible for each treatise, and by his ignorance of the mathematical—not observational—origin of Oresme's belief in the probable incommensurability of the celestial motions, as well as by the dialogue form of Part III of the *De commensurabilitate*, which gave the misleading impression that not only did its author find no reasonable grounds for a categorical decision, but no grounds existed for inclining to one side or the other.

14. Grant, *Oresme PPAP*, p. 441. I have added the titles in square brackets. For a summary of Chapter 3, Proposition X of the *De proportionibus proportionum* and the steps by means of which it was subsequently applied to celestial motions, see above, chap. 2, n. 113.

Bibliography

Abraham ibn Ezra. *Abrahe Avenaris Judei Astrologi peritissimi in re iudiciali opera ab excellentissimo Philosopho Petro de Abano post accuratam castigationem in latinum traducta: Introductorium quod dicitur principium sapientiae; Liber rationum;...; Liber coniunctionum planetarum et revolutionum annorum mundi qui dicitur de mundo vel seculo;....* Venice, 1507.

————. See also Levy, Raphael.

al-Battani. See Battani, al-.

al-Bitruji. See Bitruji, al-.

Apuleius Madaurensis. *De Platone et eius dogmate,* in *Apulei Madaurensis Opera quae supersunt.* Vol. 3: *De philosophia libri,* edited by Paul Thomas. Leipzig: B. G. Teubner, 1907; reprinted 1938.

Aquinas, Thomas. *Commentum in quatuor libros Sententiarum Magistri Petri Lombardi,* in *Sancti Thomae Aquinatis Doctoris Angelici Ordinis Praedicatorum opera omnia secundum impressionem Petri Fiaccadori,* Parma, 1852–73. Vol. 7, pt. 2. New York: Musurgia Publishers, 1948.

————. *In Aristotelis libros De caelo et mundo, De generatione et corruptione, Meteorologicorum expositio.* Cum textu ex recensione leonina. Edited by Raymond M. Spiazzi, O.P. Turin and Rome, 1952.

————. *In Aristotelis libros De sensu et sensato, De memoria et reminiscentia commentarium.* 3d ed. Edited by Raymond M. Spiazzi, O.P. Turin and Rome, 1949.

————. *In octo libros De physico auditu sive physicorum Aristotelis commentaria.* Edited by A. M. Pirotta. Naples, 1953.

Aristotle. *The Works of Aristotle.* Translated under the editorship of J. A. Smith and W. D. Ross. Vol. 2: *Physica* (translated by R. P. Hardie and R. K. Gaye), *De caelo* (translated by J. L. Stocks), *De generatione et corruptione* (translated by H. H. Joachim). Oxford, 1930; vol. 3: *Meteorologica* (translated by E. W. Webster), *De mundo* (translated by E. S. Forster). Oxford, 1931; vol. 7: *Problemata* (translated by E. S. Forster). Oxford, 1927; vol. 8: *Metaphysica* (translated by W. D. Ross). Oxford, 1908.

————. See also Aquinas, Thomas; Averroes; Simplicius.

Aristotle (pseudo-). *De mundo,* in *Aristoteles latinus,* vol. 11. Edited by W. L. Lorimer. Rome, 1951.

Arnold, E. Vernon. *Roman Stoicism.* Being lectures on the history of the Stoic philosophy with special reference to its development within the Roman Empire. Cambridge University Press, 1911. Reissue. London: Routledge and Kegan Paul, 1958.

Augustine. *Sancti Aurelii Augustini episcopi De civitate Dei libri XXII.* Edited by Bernhard Dombart. 2 vols. Leipzig: Teubner, 1877.

————. *Sancti Aurelii Augustini episcopi De civitate Dei libri XXII,* in *Corpus scrip-*

torum ecclesiasticorum latinorum, vol. 40, 2 pts. Edited with critical commentary by Emanuel Hoffmann. Vienna, 1899, 1900.

————. *The City of God by Saint Augustine.* Translated and edited by Marcus Dods. 2 vols. Edinburgh, 1872. Reprint. New York, 1948.

Averroes. *Aristotelis opera cum Averrois commentariis.* 9 vols. in 11 pts. Venice, Junctas ed., 1562–74. Reprint. Frankfurt-am-Main: Minerva Gmbh., 1962.

————. *Averroes on Aristotle's "De generatione et corruptione," Middle Commentary and Epitome.* Translated from the original Arabic and the Hebrew and Latin versions, with notes and introduction by Samuel Kurland. Mediaeval Academy of America. Cambridge, Mass., 1958.

Balić, Carl. *Les Commentaires de Jean Duns Scot sur les quatre livres des sentences.* Étude historique et critique. *Bibliothèque de la Revue d'Histoire Ecclésiastique,* fasc. 1. Louvain, 1927.

Bate, Henry. *Henricus Bate Speculum divinorum et quorundam naturalium.* Édition critique. Vol. 1: Introduction; Lettera dedicatoria; Tabula capitulorum; Prooemium; Pars 1. Edited by E. Van de Vyver, in *Philosophes médiévaux,* vol. 4. Louvain and Paris, 1960.

————. *Henri Bate de Malines Speculum divinorum et quorundam naturalium.* Étude critique et texte inédite, fasc. 1, pts. 1–2. Edited by G. Wallerand, in *Les Philosophes belges,* vol. 11. Louvain, 1931.

Battani, al-. *Rudimenta astronomica Alfragrani* [sic]; *item Albategnius ... De motu stellarum....* Nuremberg, 1537.

Bitruji, al-. *De motibus celorum.* Critical edition of the Latin translation of Michael Scot by Francis J. Carmody. Berkeley and Los Angeles, 1952.

Blasius of Parma. *Questiones super tractatu de proportionibus magistri Thome Berduerdini.* MS Vat. lat. 3012, fols. 137r–164r.

Boethius. *Boetii De institutione arithmetica libri duo; De institutione musica libri quinque; accedit geometria quae fertur Boetii.* Edited by Godofredus Friedlein. Leipzig, 1867.

————. *The Consolation of Philosophy.* With the English translation of "I. T." (1609) revised by H. F. Stewart, in *The Theological Tractates,* with an English translation by H. F. Stewart and E. K. Rand. Loeb Classical Library. New York and London, 1918. Reprint. Cambridge, Mass., 1962.

Boncompagni, Baldassare. "Catalogo de' lavori di Andalo di Negro," *Bullettino di Bibliografia e di storia delle scienze matematiche e fisiche,* vol. 7. Rome, 1874.

Burkhard, Karl. See Nemesius.

Buron, Edmond. *Ymago mundi de Pierre d'Ailly.* Texte latin et traduction française des quatre traités cosmographiques de d'Ailly et des notes marginales de Christophe Colomb. Étude sur les sources de l'auteur. 3 vols. Paris, 1930.

Busard, Hubertus L. L. See Oresme, Nicole.

Campanus of Novara. See Euclid.

Cantera, Francisco. See Levy, Raphael.

Cardano, Jerome. *Hieronymi Cardani ... Opus novum de proportionibus numerorum, motuum, ponderum, sonorum, aliarumque rerum mensurandarum, non solum geometrico stabilitum, sed etiam variis experimentis et observationibus rerum in natura solerti demonstratione illustratum ad multiplices usus accommodatum et in V libros digestum. Praeterea Artis magnae, sive de regulis algebraicis liber unus....* Item *De Aliza regula liber, hoc est algebraicae....* Basel, [1570].

Carmody, Francis J. "Autolycus," *Catalogus translationum et commentariorum: Mediaeval and Renaissance Latin Translations and Commentaries.* Edited by P. O. Kristeller. Washington, D.C., 1960. Vol. 1: 167–72.

———. See also Bitruji, al-.

Cassiodorus. *Cassiodori Senatoris Variae,* in *Monumenta Germaniae historica,* vol. 12. Edited by Theodore Mommsen. Berlin, 1894. Reprint. Berlin, 1961.

Catalogus codicum manuscriptorum Bibliothecae Regiae. 4 vols. Paris, 1744.

Censorinus. *Censorini De die natali liber.* Edited by Friedrich Hultsch. Leipzig, 1867.

———. *De die natale (The Natal Day)* by Censorinus (A.D. 238); *Life of the Emperor Hadrian* by Aelinus Spartianus (circ. A.D. 300). Translated by William Maude. New York, 1900.

Chadwick, Henry. See Origen.

Chalcidius. See Plato.

Cicero. *De natura deorum; Academica.* With an English translation by Harris Rackham. Loeb Classical Library. London and New York, 1933.

———. *De re publica, De legibus.* With an English translation by Clinton Walker Keyes. Loeb Classical Library. London and New York, 1928.

Clagett, Marshall. "The Medieval Latin Translations from the Arabic of the *Elements* of Euclid, with Special Emphasis on the Versions of Adelard of Bath," *Isis,* vol. 44 (1953), 16–42.

———. *The Science of Mechanics in the Middle Ages.* Madison, Wis., 1959.

———. See also Moody, Ernest A.; Oresme, Nicole.

Claudian. *Claudian with an English Translation* by Maurice Platnauer. 2 vols. Loeb Classical Library. London and New York, 1922.

Clavius, Christopher. *Christopher Clavii Bambergensis ex societate Iesu in Sphaeram Joannis de Sacro Bosco commentarium.* Rome, 1570.

Coleman-Norton, P. R. "Cicero's Doctrine of the Great Year," *Laval théologique et philosophique,* vol. 3 (1947), 293–302.

Coopland, G. W. See Oresme, Nicole.

Cornford, Francis M. See Plato.

Cranz, F. Edward. "Alexander Aphrodisiensis," *Catalogus translationum et commentariorum: Mediaeval and Renaissance Latin Translations and Commentaries.* Edited by P. O. Kristeller. Washington, D.C., 1960. Vol. 1: 77–135.

Curtze, Maximilian. See Oresme, Nicole.

d'Ailly, Pierre. *Petri de Alliaco Cardinalis Camaracensis Tractatus contra astronomos,* in *Joannis Gersonii... Opera omnia.* Antwerp, 1706. Vol. 1, cols. 778–804. For full title of edition, see Gerson, Jean.

Delisle, Léopold V. "Notice sur des manuscrits du fonds Libri conservés à la Laurentienne à Florence," *Notices et extraits des manuscrits de la Bibliothèque Nationale et autres bibliothèques,* vol. 32, pt. 1. Paris, 1886.

Denifle, Heinrich, and Chatelain, Emil. *Chartularium Universitatis Parisiensis.* 4 vols. Paris, 1889–97.

Denomy, Alexander J. See Oresme, Nicole.

Dods, Marcus. See Augustine.

Dombart, Bernhard. See Augustine.

Doucet, Victorin. *Commentaires sur les sentences, supplément au répertoire de M. Frédéric Stegmueller*. Quaracchi, Florence, 1954.

Dreyer, J. L. E. *A History of Astronomy from Thales to Kepler*. 2d ed. New York, 1953.

Duhem, Pierre. *Le Système du monde*. 10 vols. Paris, 1913–59.

Duns Scotus, John. *Quaestiones in quartum librum Sententiarum* in Vols. 20 and 24 of *Joannis Duns Scoti... Opera omnia editio nova juxta editionem Waddingi XII tomos continentem a patribus Franciscanis de observantia accurate recognita*. 26 vols. Paris: Luis Vivès, 1891–95.

———. *Duns Scotus Philosophical Writings*. A selection edited and translated by Allan Wolter, O.F.M. London: Nelson, 1962.

Euclid. *Euclidis Megarensis mathematici clarissimi Elementorum geometricorum libri XV. Cum expositione Theonis in priores XIII a Bartholomaeo Veneto Latinitate donata; Campani in omnes, et Hypsicles Alexandrini in duos postremos*. Basel, 1546.

———. *The Thirteen Books of Euclid's Elements*. Translated from the text of I. L. Heiberg, with introduction and commentary, by Sir Thomas L. Heath. 3 vols. 2d edition revised with additions. Cambridge, 1926. Reissue. Dover Publications, New York, 1956.

Eyssenhardt, Franciscus. See Macrobius.

Fairclough, H. Rushton. See Virgil.

Fecht, Rudolf. See Theodosius of Bithynia.

Festugière, A. J. See Hermes Trismegistus.

Forster, E. S. See Aristotle.

Frazer, Sir James George. See Ovid.

Friedlein, Godofredus. See Boethius.

Fundis, John de. *Tractatus reprobationis eorum que scripsit Nicolaus Orrem in suo libello intitulato de proportionalitate motuum celestium contra astrologos et sacram astrorum scientiam*. MS Bibliothèque Nationale, fonds latin, 10271, fols. 63r–153v. Date of composition given as 1451.

Gaye, R. K. See Aristotle.

Gerard of Cremona. *Theorica planetarum*, in *Spherae Tractatus Ioannis de Sacrobusto Anglici viri clariss. Gerardi Cremonensis Theoricae planetarum veteres. Georgii Purbachii Theoricae planetarum novae. Prosdocimi de beldomandi.... N.p., 1531.

Gerson, Jean. *Trilogium Astrologiae theologizatae*, in *Joannis Gersonii doctoris theologi et cancellarii Parisiensis opera omnia novo ordine digesta et in V tomos distributa*. Antwerp, 1706. Vol. I, p. 2.

Gilson, Etienne. *History of Christian Philosophy in the Middle Ages*. New York, 1955.

Grant, Edward. "The Mathematical Theory of Proportionality of Nicole Oresme (*ca.* 1320–1382)." Ph.D. dissertation, University of Wisconsin, 1957.

———. "Nicole Oresme and the Commensurability or Incommensurability of the Celestial Motions," *Archive for History of Exact Sciences*, vol. 1 (1961), 420–58.

———. See also Oresme, Nicole.

Hardie, R. P. See Aristotle.

Heath, Thomas L. *Aristarchus of Samos.* Oxford, 1913.

——. *A History of Greek Mathematics.* 2 vols. Oxford, 1921.

——. See also Euclid.

Henry of Hesse. See Langenstein, Heinrich von.

Hermes Trismegistus. *Corpus hermeticum.* Vol. 2: *Asclepius.* Texte établi par A. D. Nock et traduit par A. J. Festugière. Paris, 1945.

——. *Hermetica, the Ancient Greek and Latin Writings Which Contain Religious or Philosophic Teachings Ascribed to Hermes Trismegistus.* Edited with English translation and notes by Walter Scott. 4 vols. Oxford, 1924–36.

Hermes, Emil. See Seneca.

Hoffmann, Emanuel. See Augustine.

ibn Ezra, Abraham. See Abraham ibn Ezra.

Isidore of Seville. *Isidori Hispalensis Episcopi Etymologiarum sive originum libri XX.* Edited by W. M. Lindsay. 2 vols. Oxford, 1911.

James, Montague R. *A Descriptive Catalogue of the Manuscripts in the Library of Peterhouse.* Cambridge, 1899.

——. *Lists of Manuscripts Formerly Owned by Dr. John Dee with Preface and Identifications.* Transactions of the Bibliographical Society, supplement no. 1. Oxford, 1921.

Joachim, H. H. See Aristotle.

Johannes de Muris. *Quadripartitum numerorum.* For manuscripts of, see p. 359.

John Major. *Quartus sententiarum Johannis Maioris.* Paris: Ponset le Preux, [1509].

John of Salisbury. *Ioannis Saresberiensis Episcopi Carnotensis Policratici siue De nugis curialium et vestigiis philosophorum libri VIII.* Edited by Clemens C. I. Webb. 2 vols. Oxford, 1909.

——. *Frivolities of Courtiers and Footprints of Philosophers.* Being a translation by Joseph B. Pike of the first, second, and third books and selections from the seventh and eighth books of the *Policraticus* of John of Salisbury. Minneapolis, 1938.

Jones, W. H. S. See Pliny.

Kaltenbrunner, Ferdinand. "Die Vorgeschichte der Gregorianischen Kalenderreform," in *Sitzungsberichte der kaiserlichen Akademie der Wissenschaften,* philosophisch-historische Klasse, vol. 82, pp. 289–414. Vienna, 1876.

Keyes, Clinton W. See Cicero.

Knight, W. F. Jackson. *St. Augustine's De musica, A Synopsis.* The Orthological Institute. London, 1949.

Kren, Claudia. See Oresme, Nicole.

Kurland, Samuel. See Averroes.

Langenstein, Heinrich von (Henry of Hesse). *Studien zu den astrologischen Schriften des Heinrich von Langenstein.* Edited by Hubert Pruckner. Leipzig, 1933.

Levy, Raphael. *The Astrological Works of Abraham ibn Ezra.* Baltimore, 1927.

Levy, Raphael, and Cantera, Francisco, eds. *The Beginning of Wisdom; An Astrological Treatise by Abraham ibn Ezra.* Baltimore, 1939.

Lewis, C. S. *The Discarded Image; An Introduction to Medieval and Renaissance Literature.* Cambridge, 1964.

Lindsay, W. M. See Isidore of Seville.

Lorimer, W. L. See Aristotle.

McCarthy, Lillian M. See Oresme, Nicole.

McCue, James F. See Oresme, Nicole.

Macrobius. *Ambrosii Theodosii Macrobii Commentarii in somnium Scipionis.* Edited by Jacob Willis. Leipzig; Teubner, 1963.

————. *Commentary on the Dream of Scipio.* Translated with an introduction and notes by William Harris Stahl. New York, 1952.

————. *Macrobius.* Edited by Franciscus Eyssenhardt. Leipzig, 1868.

Maier, Anneliese. *Studien zur Naturphilosophie der Spätscholastik.* Vol. 4: *Metaphysische Hintergründe der spätscholastischen Naturphilosophie.* Rome, 1955.

Marsilius of Inghen. *Questiones Marsilij* [de Inghen] *super quattuor libros sententiarum.* 2 vols. Strasbourg, 1501.

————. *Questiones subtilissime Johannis Marcilii Inguen super octo libros physicorum secundum nominalium viam. Cum tabula in fine posita....* Lyon, 1518. Reprint. Frankfurt-am-Main, 1964.

Martin, Henry. *Catalogue des manuscrits de la Bibliothèque de l'Arsenal,* vol. 1. Paris, 1885.

Maude, William. See Censorinus.

Menut, Albert D. See Oresme, Nicole.

Migne, Jacques Paul, ed. *Patrologiae cursus completus. Series latina.* 221 vols. Paris, 1844–64.

Miller, Frank J. See Ovid.

Mommsen, Theodore. See Cassiodorus.

Moody, Ernest A., and Clagett, Marshall, eds. and trans. *The Medieval Science of Weights.* Madison, Wis., 1952.

Mugler, Charles. *La Physique de Platon.* Vol. 35: *Études et Commentaires.* Paris, 1960.

Murdoch, John E. Review of *Nicole Oresme: "Quaestiones super geometriam Euclidis,"* edited by Hubertus L. L. Busard (2 fasc., Leiden, 1961), in *Scripta Mathematica,* vol. 27 (1964), 67–91.

Muris, Johannes de. See Johannes de Muris.

Nemesius, Bishop of Emesa. *Gregorii Nysseni (Nemesii Emeseni) Peri Phuseos Anthropou liber a Burgundione in latinum translatus.* Edited by Karl Burkhard, in *Programme des Carl-Ludwig-Gymnasiums.* Vienna, 1891, 1892, 1896, 1901, and 1902.

————. *Nemesii Episcopi Premnon Physicon... a N. Alfano Archiepiscopo Salerni in latinum translatus.* Edited by Karl Burkhard. Leipzig, 1917.

Neugebauer, Otto. See Thabit ibn Qurra.

Nicholas of Cusa. *Nicolai Cusae Cardinalis Opera.* 3 vols. Paris, 1514. Reprint. Minerva Gmbh.: Frankfurt-am-Main, 1962.

Nock, Arthur D. See Hermes Trismegistus.

Oresme, Nicole. *Der "Algorismus proportionum" des Nicolaus Oresme. Zum ersten Male nach der Lesart der Handschrift R. 4° 2 der Königlichen Gymnasialbibliothek zu Thorn.* Edited by Maximilian Curtze. Berlin, 1868.

————. *"Maistre Nicole Oresme: Traitié de l'espere."* Critically edited by Lillian Margaret McCarthy. Ph.D. dissertation, University of Toronto, 1943.

————. *Nicole Oresme: "De proportionibus proportionum" and "Ad pauca respicientes."*

Edited with introductions, English translations, and critical notes by Edward Grant. Madison, Wis., 1966.

————. *Nicole Oresme: "Le Livre du ciel et du monde."* Edited by Albert D. Menut and Alexander J. Denomy, C.S.B.; Translated with an introduction by Albert D. Menut. Madison, Wis., 1968.

————. *Nicole Oresme and the Astrologers; A Study of His "Le Livre de Divinacions."* Edited and translated by G. W. Coopland. Cambridge, Mass., 1952.

————. *Nicole Oresme and the Medieval Geometry of Qualities and Motions; A Treatise on the Uniformity and Difformity of Intensities Known as "Tractatus de configurationibus qualitatum et motuum."* Edited with an introduction, English translation, and commentary by Marshall Clagett. Madison, Wis., 1968.

————. *Nikolaĭ Orem i ego matematiko-astronomicheskiĭ traktat "O soizmerimosti ili nesoizmerimostĭ dvizheniĭ neba."* A complete translation of Oresme's *De commensurabilitate* into Russian by Vassily P. Zoubov, in *Istoriko-astronomicheskie issledovaniiā*, no. 6, pp. 301–400. Moscow, 1960.

————. "Part I of Nicole Oresme's *Algorismus proportionum.*" Translated and annotated by Edward Grant. *Isis*, vol. 56 (1965), 327–41.

————. *Quaestiones super geometriam Euclidis.* Edited by Hubertus L. L. Busard. 2 fasc. Leiden, 1961.

————. *Questiones de sphera.* MS Florence, Bibl. Riccardiana 117, fols. 125r–135r.

————. "The *Questiones super de celo* of Nicole Oresme." Edited and translated by Claudia Kren. Ph.D. dissertation, University of Wisconsin, 1965.

————. *Quodlibeta.* MSS Paris, Bibliothèque Nationale, fonds latin, 15126, 39r–158r; Florence, Laurentian Library, Ashburnham 210, 21r–70v.

————. "The Treatise *De Proporcionibus velocitatum in motibus* Attributed to Nicholas Oresme." Edited by James F. McCue. Ph.D. dissertation, University of Wisconsin, 1961.

Origen. *Origen: Contra Celsum.* Translated by Henry Chadwick. Cambridge, 1953.

Ovid. *Fasti.* With an English translation by Sir James George Frazer. Loeb Classical Library. London and New York, 1931.

————. *Metamorphoses.* With an English translation by Frank Justus Miller. 2 vols. Loeb Classical Library. Cambridge, Mass., and London, 1946.

Paoli, Cesare. *I Codici Ashburnhamiani della R. Biblioteca Mediceo-Laurenziana di Firenze*, vol. 1, fasc. 3. Ministero della Pubblica Istruzione, *Indici e cataloghi*, vol. 8. Rome, 1891.

Paul of Venice. *Expositio super libros de generatione et corruptione Aristotelis.* Bologna, 1498.

Pedersen, Olaf. "The Theorica Planetarum-Literature in the Middle Ages," *Classica et Mediaevalia*, vol. 23 (1962), 225–32.

Petrus Aureoli. *Petri Aureoli Verberii ordinis minorum Archepiscopi Aquensis S.R.E. Cardinalis Commentariorum in quartum librum sententiarum.* Rome, 1605.

Petrus de Tarantasia. *Innocentii Quinti Pontificis Maximi ex ordine Praedicatorum assumpti qui antea Petrus de Tarantasia dicebatur in IV libros Sententiarum commentaria.* 2 vols. Toulouse, 1652.

Petrus Tataretus. *D. Petri Tatareti artium et sacrae theologiae doctoris praeclarissimi lucidissima commentaria sive (ut vocant) reportata, in quartum librum sententiarum Ioannis Duns Scoti subtilium principis....* Venice, 1583.

Pico della Mirandola, G. *Disputationes adversus astrologiam divinatricem.* Edited by Eugenio Garin. 2 vols. Florence, 1946 and 1952.

Pike, Joseph B. See John of Salisbury.

Pines, Shlomo. "The Semantic Distinction Between the Terms *Astronomy* and *Astrology* According to Al-Bīrūnī," *Isis,* vol. 55 (1964), 343–49.

Platnauer, Maurice. See Claudian.

Plato. *Plato's Cosmology; The Timaeus of Plato.* Translated with a running commentary by Francis M. Cornford. London and New York, 1937. Reprint. Liberal Arts Press: New York, 1957.

––––. *Timaeus a Calcidio translatus commentarioque instructus.* Edited by J. H. Waszink. *Corpus Platonicum medii aevi; Plato Latinus* (under the general editorship of Raymond Klibansky). Vol. 4. London, 1962.

Pliny. *Natural History.* Translated into English in 10 volumes by Harris Rackham (1–5, 9), W. H. S. Jones (6–8), and D. E. Eichholz (10). Loeb Classical Library. London and Cambridge, Mass., 1938–63.

Price, Derek J., ed. and trans. *The Equatorie of the Planetis.* Edited from Peterhouse MS 75.I with a linguistic analysis by R. M. Wilson. Cambridge, 1955.

Pruckner, Hubert. See Langenstein, Heinrich von.

Ptolemy, Claudius. *Almagest.* Translated by R. Catesby Taliaferro, in Great Books of the Western World, vol. 16. Chicago: Encyclopaedia Brittanica, 1952.

––––. *Liber Ptholomei quattuor tractatuum cum Centiloquio eiusdem Ptholomei et commento Haly....* Venice: Erhard Ratdolt, 1484.

––––. *Tetrabiblos.* Edited and translated into English by F. E. Robbins. Loeb Classical Library. London and Cambridge, Mass., 1940.

Rackham, Harris. See Cicero; Pliny.

[Richard de Fournival?]. *Libri tres De vetula Ovidii, falso sic dicti* published with separate pagination in *Brunellus Vigelli et Vetula Ovidii seu opuscula duo auctorum incertorum: prius quidem Vigelli, qui fertur Speculum stultorum; posterius vero Libri tres De vetula Ovidii, falso sic dicti.* [Edited by S. Closius?] Wolfenbüttel, 1662.

Richard of Middleton. *Ricardus mediavilla super quarto sententiarum.* N.p., 1499.

Robathan, D. M. "Introduction to the Pseudo-Ovidian *De Vetula.*" *Transactions of the American Philological Association,* vol. 88 (1957), 197–207.

Robbins, F. E. See Ptolemy, Claudius.

Sacrobosco. See Thorndike, Lynn.

Sarton, George. *Introduction to the History of Science.* 3 vols. in 5 parts. Baltimore, 1927–48.

Scott, Walter. See Hermes Trismegistus.

Scotus. See Duns Scotus, John.

Seneca, Annaeus. *L. Annaei Senecae Opera quae supersunt.* 10 vols. *Dialogorum libros XII,* vol. 1, fasc. 1. Edited by Emil Hermes. Leipzig, 1905.

Simplicius. *Simplicii philosophi acutissimi commentaria in quatuor libros De celo Aristotelis.* Translated by Wm. Moerbeke. Venice, 1540.

Stahl, William H. See Macrobius.

Stegmüller, Friedrich. *Reportorium commentariorum in sententias Petri Lombardi.* 2 vols. Würzburg, 1947.

Steinschneider, Moritz. *Die arabischen Übersetzungen aus dem Griechischen.* Reproduced in Graz, 1960; originally published as journal articles between 1889 and 1896.

——. *Die europäischen Übersetzungen aus dem Arabischen bis mitte des 17. Jahrhunderts.* Reproduced in Graz, 1956.

Stewart, H. F. See Boethius.

Suter, Heinrich. "Die Quaestio *De proportione dyametri quadrati ad costam ejusdem* des Albertus de Saxonia," *Zeitschrift für Mathematik und Physik*, historisch-literarische Abteilung, vol. 32 (1887), 41–56.

Taliaferro, R. Catesby. See Ptolemy, Claudius.

Thabit ibn Qurra. "Thâbit ben Qurra *On the Solar Year* and *On the Motion of the Eighth Sphere*," translation and commentary by Otto Neugebauer, *Proceedings of the American Philosophical Society*, vol. 106, no. 3 (June, 1962), 264–99.

Theodosius of Bithynia (also called Theodosius of Tripoli[s]). *Theodosii De habitationibus liber; De diebus et noctibus libri duo.* Edited and translated by Rudolf Fecht, in *Abhandlungen der Gesellschaft der Wissenschaften zu Göttingen*, philologisch-historische Klasse, N.S., vol. 19, 4. Berlin, 1927.

Thomas, Paul. See Apuleius Madaurensis.

Thorndike, Lynn. *A History of Magic and Experimental Science.* 8 vols. New York, 1923–58.

——, ed. and trans. *The Sphere of Sacrobosco and Its Commentators.* Chicago, 1949.

——. "The Latin Translations of the Astrological Tracts of Abraham Avenezra," *Isis*, vol. 35 (1944), 293–302.

——. "Notes upon Some Medieval Latin Astronomical, Astrological and Mathematical Manuscripts at the Vatican," *Isis*, vol. 47 (1956), 391–404.

——. "A Summary Catalogue of Reproductions 296–383 of Medieval Manuscripts Collected by Lynn Thorndike," *Medievalia et Humanistica*, fasc. 13 (1960), 81–100.

Thorndike, Lynn, and Kibre, Pearl. *A Catalogue of Incipits of Mediaeval Scientific Writings in Latin.* Revised and augmented edition. Mediaeval Academy of America, Publication no. 29. Cambridge, Mass., 1963.

Tiele, P. A. *Catalogus codicum manu scriptorum Bibliothecae Universitatis Rheno-Trajectinae*, vol. 1. Utrecht, 1887.

Van de Vyver, E. See Bate, Henry.

Vescovini, Graziella F. *Studi sulla prospettiva medievale.* Turin, 1965.

De vetula. See Richard de Fournival.

Virgil. *Virgil with an English Translation.* By H. Rushton Fairclough. 2 vols. Loeb Classical Library. London and New York, 1916.

Vitello (or Witelo). *Opticae libri decem*, in *Opticae thesaurus....* Basel, 1572.

Wallerand, G. See Bate, Henry.

Waszink, J. H. See Plato.

Webb, Clemens C. I. See John of Salisbury.

Weinberg, Julius R. *Nicolaus of Autrecourt*. Princeton, 1948.

Willis, Jacob. See Macrobius.

Witelo. See Vitello.

Wolter, Allan, O.F.M. See Duns Scotus, John.

Zinner, Ernst. *Leben und Wirken des Johannes Müller von Königsberg genannt Regio-montanus*. Schriftenreihe zur bayerischen Landesgeschichte, vol. 31. Munich, 1938.

Zoubov, V. P. "Nicole Oresme et la musique," *Mediaeval and Renaissance Studies*, vol. 5 (1961), 96–107.

————. See also Oresme, Nicole.

Index of Manuscripts Cited

Selective Index of Latin Mathematical, Astronomical, and Astrological Terms

My objective in this index has been to include all Latin terms occurring in the texts, appendixes, and quotations which seemed relevant—taking relevant in the broadest sense—to mathematics, astronomy, astrology, and cosmology in general. I have not endeavored to cite all instances of a given term, but rather have attempted to represent the many shades of meaning that came to my attention. To convey something of the variety of contexts in which many terms occur, I have often included entire expressions and phrases to supplement or replace single word entries. Because of the nature of the Latin language, some of these could be formed only by the use of ellipses. Verb forms are usually indexed under the infinitive, while the different noun forms are cited in the nominative singular or plural. Comparatives and superlatives are given in the positive form. Numbers in roman type refer to page numbers; numbers in italic type represent line numbers.

General Index

All Latin terms appear in the Selective Index of Latin Mathematical, Astronomical, and Astrological Terms. Medieval scholastic authors are usually indexed under their first names except for those who are better known by their last names (as, for example, Roger Bacon and Nicole Oresme). The Bibliography has not been indexed since it is arranged alphabetically by author.

THE UNIVERSITY OF WISCONSIN
PUBLICATIONS IN MEDIEVAL SCIENCE
Marshall Clagett, *General Editor*

1

The Medieval Science of Weights (Scientia de ponderibus):
Treatises Ascribed to Euclid, Archimedes, Thābit ibn Qurra,
Jordanus de Nemore, and Blasius of Parma
Edited by Ernest A. Moody and Marshall Clagett
448 pages

2

Thomas of Bradwardine: His "Tractatus de proportionibus."
Its Significance for the Development of Mathematical Physics
Edited and translated by H. Lamar Crosby, Jr.
216 pages

3

William Heytesbury: Medieval Logic and the Rise of Mathematical Physics
By Curtis Wilson
232 pages

4

The Science of Mechanics in the Middle Ages
By Marshall Clagett
742 pages

5

Galileo Galilei: "On Motion" and "On Mechanics"
Edited and translated by I. E. Drabkin and Stillman Drake
204 pages

6

Archimedes in the Middle Ages. Volume I: The Arabo-Latin Tradition
By Marshall Clagett
752 pages

7

The "Medical Formulary" or "Aqrābādhīn" of al-Kindī
Translated with a study of its materia medica by Martin Levey
424 pages

8

Kūshyār ibn Labbān: "Principles of Hindu Reckoning"
A translation with introduction and notes by Martin Levey and
Marvin Petruck of the *Kitāb fī uṣūl ḥisāb al-hind*
128 pages

9

Nicole Oresme: "De proportionibus proportionum" and "Ad pauca respicientes"
Edited with introductions, English translations, and critical notes
by Edward Grant
488 pages

10

The "Algebra" of Abū Kāmil, "Kitāb fī al-jābr wa'l-muqābala,"
in a Commentary by Mordecai Finzi
Hebrew text, translation, and commentary, with special reference
to the Arabic text, by Martin Levey
240 pages

11

Nicole Oresme: "Le Livre du ciel et du monde"
Edited by Albert D. Menut and Alexander J. Denomy, C.S.B.
Translated with an introduction by Albert D. Menut
792 pages

12

Nicole Oresme and the Medieval Geometry of Qualities and Motions:
A Treatise on the Uniformity and Difformity of Intensities
Known as "Tractatus de configurationibus qualitatum et motuum"
Edited with an introduction, English translation, and commentary
by Marshall Clagett
728 pages

13

Mechanics in Sixteenth-Century Italy:
Selections from Tartaglia, Benedetti, Guido Ubaldo, and Galileo
Edited and translated by Stillman Drake and I. E. Drabkin
440 pages

14

John Pecham and the Science of Optics: "Perspectiva communis"
Edited with an introduction, English translation, and critical notes
by David C. Lindberg
320 pages

15

Nicole Oresme and the Kinematics of Circular Motion:
"Tractatus de commensurabilitate vel incommensurabilitate motuum celi"
Edited with an introduction, English translation, and commentary
by Edward Grant
438 pages